W9-CNJ-470

HORTICULTURAL REVIEWS
VOLUME 7

Horticultural Reviews is co-sponsored by the American Society for Horticultural Science and The AVI Publishing Company

HORTICULTURAL REVIEWS

VOLUME 7

edited by
Jules Janick
Purdue University

AVI PUBLISHING COMPANY, INC.
Westport, Connecticut

ISSN-0163-7851
ISBN-0-87055-492-1

Printed in the United States of America

A B C D E 4 3 2 1 0 9 8 7 6 5

Contents

Contributors

G.D. BLANPIED. Department of Pomology, Cornell University, Ithaca, New York 14853

W.L. COLLINS. Environmental Research Laboratory, University of Arizona, Tucson, Arizona 85706

JALEH DAIE.[1] Department of Biology, Utah State University, Logan, Utah 84322

BERNARD GRODZINSKI. Department of Horticultural Science, Ontario Agricultural College, University of Guelph, Guelph, Ontario, Canada N1G 2W1

WESLEY P. HACKETT. Department of Horticultural Science and Landscape Architecture, University of Minnesota, St. Paul, Minnesota 55108

MERLE H. JENSEN. Environmental Research Laboratory, University of Arizona, Tucson, Arizona 85706

HAMLYN G. JONES. East Malling Research Station, Maidstone, Kent ME19 6BJ, United Kingdom

J. BENTON JONES, JR. Department of Horticulture, University of Georgia, Athens, Georgia 30602

ALAN N. LAKSO. Department of Horticultural Science, New York State Agricultural Experiment Station, Geneva, New York 11456

ROY A. LARSON. Department of Horticultural Science, North Carolina State University, Raleigh, North Carolina 27650

RICHARD E. LITZ. Tropical Research and Education Center, Institute of Food and Agricultural Sciences, University of Florida, Homestead, Florida 33031

[1]Present address: Department of Soils and Crops, Cook College, Rutgers University, New Brunswick, NJ 08903.

GLORIA A. MOORE. Fruit Crops Department, Institute of Food and Agricultural Sciences, University of Florida, Gainesville, Florida 32611

MICHAEL A. PORTER.[2] Department of Horticultural Science, Ontario Agricultural College, University of Guelph, Guelph, Ontario, Canada N1G 2W1

MAX C. SAURE. Diplom-Agraringenieur, Dorfstrasse 17, 2151 Moisburg, Federal Republic of Germany.

CHINNATHAMBI SRINIVASAN. Tropical Research and Education Center, Institute of Food and Agricultural Sciences, University of Florida, Homestead, Florida 33031

J.P. SYVERTSEN. Citrus Research and Education Center, University of Florida, Lake Alfred, Florida 33850

GEORGE YELENOSKY. U.S. Department of Agriculture, Agricultural Research Service, Horticultural Research Laboratory, Orlando, Florida 32803

[2]Present address: Oak Ridge National Laboratory, Biology Division, Oak Ridge, TN 37830.

Robert Smock

Dedication

The name has remained the same on the office door for nearly 50 years. The bearer of that name, whose professional life has changed emphasis through the years and is punctuated with a variety of rewarding personal adventures, is Robert Smock. He has enjoyed a bountiful career of personal interaction and postharvest pomology.

Although Bob has an interest in laboratory physiology, exemplified by his published observations of orchard, handling, and storage factors that influence the respiration rate of apples, applied research has been the hallmark of his pomology career. Most practical pomologists would consider their career to be a success if they made one significant contribution to the fruit industry. Bob has made two.

The first was his research on and development of recommendations for the controlled-atmosphere (CA) storage of apples. From the late 1930s to the early 1950s his CA research was carried out under the physically adverse conditions imposed by the location of his storage laboratory in the wet basement of an old barn. He developed recommendations for the temperature and concentrations of oxygen and carbon dioxide to be used during CA storage of most apple cultivars grown in the United States. These recommendations are currently followed by most CA operators in North America and by many CA operators in other apple-growing regions. In addition to developing recommendations for the CA temperature and atmosphere, he personally worked with fruit growers for several years to develop recommendations for the construction, gas sealing, testing for air tightness, and operation of commercial CA rooms. The growth of the commercial CA industry in New York and New England, which preceded and set the example for the establishment of CA operations elsewhere in North America, can be attributed almost exclusively to Bob Smock's adherence to the philosophy that changes are brought about by the actions of dedicated people.

The physiological disorder storage scald caused multimillion dollar losses to the world's apple industry every year until Bob discovered that diphenylamine and ethoxyquin, used as postharvest treatments, controlled the disorder. He spent several years screening scores of antioxidants before he found these two compounds, which consistently gave complete control of storage scald. He then acted through the U.S. Department of Agriculture to obtain the toxicological data to clear the compounds with the Food and Drug Administration, and he cooperated with commercial chemists to develop suitable formulations. Finally, since these were the first postharvest-prestorage materials to be applied to apples, he worked with growers to develop suitable application equipment.

Other noteworthy research contributions include his early work with waxing apples and bitter pit, and his later work with enhancement of red color development and with the influence of mineral nutrition and plant growth regulators on apple quality and condition. Bob was a member for several years of the small school of researchers who thought that nonethylenic apple volatiles may influence the development of storage scald. The lack of consistency in his research data led to his withdrawal from that school, which soon thereafter became defunct. Current research suggests that this theory was correct—that a volatile from apples may induce other apples to develop storage scald.

Although his formal credentials classify him as a professor of pomology, a semester in his classroom or a few intimate visits to his office leave most students and professionals with the impression that he is also a distinguished professor of human interactions. His philosophy that education is supposed to engender a little curiosity and that research should be fun, not work, has inspired several generations of undergraduate and graduate students. It is not at all surprising that he is the first person to have received the L.M. Ware Award for Distinguished Teaching (1964).

Bob was mistaken in thinking "the only reward [*Professor perfectus*] can look forward to is to be flattened into a herbarium specimen and put away on a shelf and never looked at again."[1] Although he will not admit it, he is a *Professor perfectus* (cv. Emeritus). Since his retirement 10 years ago, there has been a meshing of his vocation

[1] R.M. SMOCK. 1969. A taxonomy of college professors. Improving College and University Teaching 17:226–227.

(pomology) and avocation (human interactions). He still has an active research program, but most of his time, I observe, is spent following the calling of his heart, i.e., counseling undergraduate students and teaching English to foreign students.

G.D. Blanpied
Department of Pomology
Cornell University
Ithaca, New York

Related AVI Books

Breeding Field Crops
Poehlman
Breeding Vegetable Crops
Bassett
Introduction to Plant Diseases: Identification and Management
Lucas et al.
Landscape Plants in Design: A Photographic Guide
Martin
Natural Toxicants in Feeds and Poisonous Plants
Cheeke and Schull
Plant Breeding Reviews, Vols. 1-3
Janick
Plant Physiology in Relation to Horticulture, 2nd Edition
Bleasdale
Protecting Farmlands
Steiner
Tomato Production, Processing and Quality Evaluation, 2nd Edition
Gould
Vegetable Growing Handbook, 2nd Edition
Splittstoesser
World Vegetables
Yamaguchi

HORTICULTURAL REVIEWS
VOLUME 7

1

Soil Testing and Plant Analysis: Guides to the Fertilization of Horticultural Crops

J. Benton Jones, Jr.
Department of Horticulture, University of Georgia, Athens, Georgia 30602

Horticultural Reviews, Volume 7

ISBN 0-87055-492-1

I. INTRODUCTION

The results from soil tests and plant analyses should play a major role in the crop production decision making of growers. These procedures provide the only means for assessing the nutrient element status of the soil–crop environment and for establishing the basis of lime and fertilizer recommendations as well as the need for supplemental fertilizer applications. In addition, soil tests and plant analyses are used for diagnosing suspected nutrient element insufficiencies (i.e., deficiencies, toxicities, or imbalances). A little-understood and used application of these procedures is in "monitoring," an application that may be far more significant than the conventional uses of soil tests and plant analyses.

Soil-testing and plant analysis services are readily available in the United States and in many other countries. There is considerable scientific literature about the various techniques involved from sampling, sample preparation, and laboratory analysis, through methods of interpretation. Semitechnical and commercial publications are equally numerous on the subjects of soil testing and plant analysis. Soil testing as practiced today has been studied and used for about 50 years, whereas plant analysis has a considerably longer history. Soil testing probably has a greater agronomic application than plant analysis for annual row crops, but for perennial horticultural crops plant analysis is the more significant testing procedure.

Surveys reported by Owens (1980) and Jones (1984) have found that since 1968 around 3 million soil and 350,000 plant tissue samples are analyzed annually for farmers and growers by the 47 state and about 300 commercial soil-testing and plant analysis laboratories in the United States. These sample numbers represent a significant underutilization of both techniques. Lack of confidence may be a primary cause for the low usage of soil tests, as was recently discussed in a series of articles appearing in *The New Farm* magazine (Liebhardt 1982–1983). This series emphasized the lack of agreement in the test results from a selected number of soil-testing laboratories, as well as their differences in lime and fertilizer recommendations. Such concerns are not new and have been voiced before (Jones 1970a). Even though there has been considerable improvement in the accuracy, precision, and standardization of laboratory methodology, much remains to be done to raise the standards of soil testing even higher. In spite of deficiencies in the procedures themselves, growers who do not regularly use soil tests and/or plant analyses appear more likely to face unexpected nutrient element problems than those who rely on these tests as a regular component of their crop production management program.

This review deals with the four basic aspects of soil testing and plant analysis—sampling, sample preparation, laboratory analysis, and interpretation and use of the results. The primary emphasis is on the application of these techniques to horticultural crops. However, pertinent literature for agronomic crops is also included when it is the only information available on the topic being discussed, or the example cited has universal application for all crops.

This review is not intended to completely cover each subject discussed, although the reader should find sufficient information and references to the current literature to obtain an understanding on how these testing techniques are conducted and used. Material selected for presentation in the tables and figures is primarily for illustrative purposes. Although plant analysis is the all-inclusive term that refers to both leaf analysis and tissue testing, in this review, plant analysis and leaf analysis are used as synonymous terms, while tissue testing is treated as an entirely separate subject. Fertilizer ratios are expressed as elemental P and K.

II. SOIL TESTING

A. Principle

"There is good evidence that the competent use of soil tests can make a valuable contribution to the more intelligent management of the soil." This statement made by the National Soil Test Workgroup in its 1951 report (Nelson *et al.* 1951) is still applicable today. The objectives of soil testing have changed little since they were summarized some years ago by Fitts and Nelson (1956):

1. To group soils into classes for the purpose of suggesting fertilizer and lime practices
2. To predict the probability of getting a profitable response to the application of plant nutrients
3. To help evaluate soil productivity
4. To determine specific soil conditions that may be improved by addition of soil amendments or cultural practices

To some, soil testing is the only means of specifying lime and fertilizer needs and of describing the soil's nutrient element status correctly. The use of lime and fertilizer, in the absence of a soil test and the recommendations based on that test, is considered indiscriminate and hazardous to a successful crop free from nutrient element stress. Such stresses are commonplace on most cropped soils. About one-quarter of the world's land surface is affected by some type of naturally occurring mineral stress (Dudal 1976). With intensive cropping even the best natively fertile soils eventually deteriorate if the proper procedures are not followed to replace nutrient elements removed by crops, to counter acidification, and to maintain a balance of nutrient elements that is optimal for plant growth.

The primary goal of most soil testing done today is to determine lime and fertilizer needs (Objective 1). Although acceptance of Objective 2 is also nearly unanimous, there is still considerable dispute as to the feasibility of predicting responses to the recommended fertilizer application rates. Even with soils having similar soil test results, recommendations may have to be adjusted on the basis of crop requirement, anticipated yield, management skill of the grower, and economic goals.

Objectives 3 and 4 are diagnostic in nature; that is, soil tests are used to describe the general condition of the tested soil. From the long-term standpoint, these objectives may have far more importance than generally recognized. For example, many cropland soils are too acid, the result of inadequate liming due in part to the failure of growers to properly monitor the soil's changing pH. An acid soil, in turn, creates an Al and/or Mn toxicity coupled with a possible Ca and/or Mg deficiency in the crop; these are frequently occurring stress conditions.

Monitoring the soil for other nutrient elements has uncovered a variety of stress conditions. For example, in the United States, many cropland soils have soil P test levels in the "high" range, or higher, due to several decades of fertilization with P. Some consider this a major cause of the increasing incidence of Zn deficiency in many areas (Olsen 1972). In some soils and crops Fe and Zn deficiencies are prevalent, while other selected soil areas and crops are also deficient in either B, Mn, or Mo. The increasing incidence of S deficiencies now being reported results from heavy N fertilization, the widespread use of S-free fertilizers (such as the shift away from ordinary superphosphate, O-8.8-O NPK, which contains 12-14% S), and decreased S additions from atmospheric deposition from the burning of fossil fuels. Most of these deficiencies and imbalances can be detected and corrective recommendations made on the basis of soil tests and/or plant analyses.

Soil tests for N and S, as well as for the micronutrients, are either limited in their application to a narrow range of soil–crop conditions or are nonexistent. Therefore, the combination of a soil test and plant analysis often is necessary to assess the levels of all the essential elements. The use of a soil test as a diagnostic device in conjunction with a plant analysis best relates to Objectives 3 and 4 described previously. Soil tests combined with plant analyses permit a more specific evaluation of the soil–crop nutrient element status than do soil tests alone, and the combined results can be used to develop specific corrective treatments.

Few growers have adopted a system of regular testing, using both soil and plant tests, to follow and monitor changes in the nutrient element status of their soils and crops. The most common objective of soil testing (i.e., to develop lime and fertilizer recommendations) fails to fully utilize the real advantage of the soil test—the ability to monitor changes in soil fertility status in order to apply corrective treatments before nutrient element stress occurs. When used in a regular, systematic way, soil testing can serve as both a diagnostic and a planning device in crop production systems.

B. Soil Sampling

Because most field soils, and even soilless plant-growing mixes, are not homogeneous, it is necessary to obtain a sample(s) that is *representative* of the field, orchard, or growing media to be tested. The common procedure is to take a number of individual cores to form a composite; the number of cores required to make one composite sample ranges from 4 to 16.

In the field or orchard, cores should be taken at random locations; areas that are markedly different in elevation and soil type should be avoided. Coring should not be done near roads, fence rows, buildings, or adjacent forest trees. As a general rule, the total area represented by one composite sample should not exceed 8 ha (Peck and Melsted 1973), although this limit may be exceeded if the field under test is reasonably homogeneous in its soil characteristics, and if the field is being treated as a single unit in terms of lime and fertilizer applications and cropping sequence. In fields being treated as a single unit but with variations in soil type, cores from the different soil types should not be mixed; that is, composites made from each major soil type should be kept separate for analysis.

The recommended procedure for field sampling is to core to the plow depth or to that soil depth occupied by the majority of plant roots in unplowed soils. Surface soil and subsoil should be kept separate for individual analysis and interpretation. Specific soil-sampling instructions are given in Table 1.1.

Normally, sampling instructions do not specify a particular time to collect soil samples, although there are seasonal cycles in soil test parameters (Peck and Melsted 1973). Generally, the best time to take samples (i.e., when seasonal effects are minimal) is in midsummer to early fall. Some workers recommend taking soil samples at the same time plant tissue samples are being collected for analysis, normally during the mid- or late-summer period. However, the actual time of soil sampling may be dictated more by the cropping sequence than by the season of the year, particularly in multicropping systems.

Various devices can be used to collect soil cores. The Hoffer Soil Sampling Tube and soil augers are the easiest to use. However, a shovel, such as a tile spade, can be used: first, cut a v-shaped slice to the proper soil depth, and then with a knife, cut and lift out a center section to make a core of soil similar to what would be obtained using a coring-type device. The collected cores are put into a soil sample bag for transport to the laboratory. If only a portion of the collected cores are being saved for analysis, the sample cores are put into a clean plastic bucket, thoroughly mixed, and then transferred to the soil sample bag.

When sampling a bench or pot of soil or soilless growing media, core to the bottom and take sufficient cores to obtain a representative sample. A small coring device, 0.5–1.0 cm in diameter, is the best.

C. Laboratory Sample Preparation

All mineral soils must be air-dried before analysis. To speed the drying process, samples may be placed in moving air, but *not* heated air. Once a soil sample is dry, it is gently crushed and passed through a 10-mesh (2-mm) screen. If Cu and Zn levels are to be determined, the screen should be made of stainless steel, not brass, in order to avoid possible contamination.

TABLE 1.1 Field-Sampling Procedures for Obtaining Soil Test Samples

Location	Procedure
Plowed fields	Core to the plow depth. In fields planted or to be planted in row crops but not plowed, core to the depth where at least 75% of the plant roots will be found.
Turf	Core to 10 cm (4 in.) into the soil (the surface of the soil would begin just below the root mat).
Orchards and vineyards	Core to 46–51 cm (18–20 in.).

Organic soils or soilless mixes are analyzed as received; they should not be dried.

D. Laboratory Analyses

In most soil-testing laboratories providing services to growers, sample size is determined by volume, with a standard scoop, because of the time required to weigh large numbers of samples quickly. Scoop size is determined by the average bulk density of the soils being tested so as to give an estimated sample weight. Peck (1980), in a study of bulk density estimates for soils typical of the north central region of the United States, defined the typical soil of this region as a medial silt loam in texture with 2.5% organic matter content. He found that the bulk density of this typical soil, crushed to pass a 10-mesh screen, was approximately 1.18, compared with 1.32 for "undisturbed" soil. In comparison, the approximate bulk density of the sandy soils of the southeastern coastal plain of the United States is 1.25.

The design of the scoop itself is an important factor that can affect the ability of the scoop to deliver the same "estimated" weight of sample each time. In general, a scoop whose diameter is double its height is more consistent in its delivery. Peck (1980) has described the best scoop design for use with prepared soils (air dried and passed through a 10-mesh screen) that have an approximate bulk density of 1.18, as well as the proper procedures for using soil scoops. Tucker (1984) has also given specific design characteristics for soil scoops based on the volume concept of sample size (Mehlich 1972).

If soil samples are measured out with a scoop, rather than weighed, care must be taken in selecting the scoop design and size that will consistently give the desired weight of sample.

1. Water pH and Lime Requirement. Soil water pH is determined by placing a soil sample into a specificed volume of water and measuring the pH of the resultant slurry. Normally, the ratio of soil to water is 1:1, although a 1:2 ratio has also been suggested. Although the entire procedure sounds simple, it is as difficult as any other soil-testing procedure. Inaccurate calibration of the pH meter, failure to calibrate on the basis of known soil standards, and electrode problems are frequent sources of difficulty. Obtaining the proper flow rate of the filling solution from the calomel, or reference, electrode can be a major difficulty: a low flow rate results in higher pH readings; a high flow rate in lower pH readings. Combination pH electrodes are not well suited for soil pH determinations.

In order to mask the variability in salt content of soils, to maintain the soil in a flocculated condition, and to decrease the junction potential effect, Schofield and Taylor (1955) have recommended measuring the soil pH in 0.01 *M* $CaCl_2$. Similarly, the determination of soil water pH in a 1 *N* KCl solution will also effectively mask the effect of differences in salt concentration on a pH reading. However, neither of these solutions is commonly used when measuring soil water pH.

Soil water pH can also be estimated using a 0.04% brom cresol purple indicator covering the pH range from 5.2 (yellow) to 6.5 (purple) (Jackson 1958).

If the soil water pH is less than desired for best plant growth, the amount of lime required to adjust the pH can be determined by adding a buffer to the same soil-water slurry and determining the pH again. Over the years, a number of buffer procedures for determining the lime requirement have been proposed and used. Currently, the two most common buffer procedures are the Adams–Evans (Adams and Evans 1962) for acid sandy soils with a low content of organic matter and the SMP (Shoemaker *et al.* 1962) for acid soils with less than 10% organic matter content and high in soluble Al. Descriptions of the procedures for determining water and buffer pHs using these two buffers are given in Table 1.2.

There are other procedures for water and buffer pH determinations suited for specific situations. An excellent review of this entire subject has recently been published by McLean (1982a).

2. Extractable Major Elements. The normal soil-testing procedure for determining the level of a particular element is to extract a portion of it from the soil and then to assay the amount extracted. The details of various extraction procedures are beyond the scope of this review; however, references given in the text provide the background for the extraction procedures discussed. The reader should be aware that the quantity of an element extracted by a particular procedure is an indicator value whose significance is based on its correlation with plant growth, yield, or quality.

The selection of an extraction procedure is based to a large degree on the soil's characteristics, especially its water pH and cation exchange capacity (the latter is partially a measure of the soil's texture and organic matter content). For soils typical of those found in the north central United States, P is extracted by the Bray P1 procedure (Bray and Kurtz 1945), and the exchangeable cations, Ca, K, Mg, and Na, are extracted with neutral normal ammonium acetate (Schollenberger and Simon 1945). For the sandy soils of the eastern coastal plain, P and the exchangeable cations are extracted by the Mehlich No. 1 procedure (Mehlich 1953). For alkaline soils, P is extracted by the

TABLE 1.2 Procedures for Determining Soil Water and Lime Buffer pH by the SMP and Adams-Evans Buffer Methods

Soil test	Sample size	Extraction reagent	Volume used (ml)	Shaking time	Method of determination
Soil water pH (for SMP buffer)	5.0 g (4.25 cm^3)	Water	5	10 min with intermittent stirring	Read pH to nearest 0.1 with glass electrode pH meter calibrated with pH 4.0 and 7.0 buffers
Soil water pH (for Adams-Evans buffer)	10 cm^3	Water	10	10 min with intermittent stirring	Read pH to nearest 0.1 with glass electrode pH meter calibrated with pH 4.0 and 7.0 buffers
Soil pH in 0.01 *M* $CaCl_2$	5.0 g (4.25 cm^3)	0.01 *M* $CaCl_2$	5	30 min with intermittent stirring	Read pH to nearest 0.1 with glass electrode pH meter calibrated with ph 4.0 and 7.0 buffers
SMP buffer pH	5.0 g (4.25 cm^3) soil in 5 ml water	SMP buffer	10	10 min with intermittent stirring for 20 min	Read pH to nearest 0.1 with glass electrode pH meter calibrated with pH 4.0 and 7.0 buffers
Adams-Evans buffer pH	10 cm^3 soil in 10 ml water	Adams-Evans buffer	10	10 min with intermittent stirring for 20 min	Read pH to nearest 0.5 with glass electrode pH meter calibrated to read pH 8.00 in mixture of 10 ml buffer reagent and 10 ml water

Source: (Jones 1980).

Olsen sodium bicarbonate procedure (Olsen *et al.* 1954). A description of these extraction procedures and others is given in Table 1.3 for P and in Table 1.4 for Ca, K, Mg, and Na. There are recent review articles on the determination of extractable P (Olsen and Sommers 1982) and of exchangeable cations (Thomas 1982).

3. Extractable Micronutrients. Many of the procedures for determining the micronutrient status of a soil are quite poor and frequently limited to special soil–crop situations (Table 1.5). Cox and Kamprath (1972) have written a review on the more common soil tests for the micronutrients, and their summary of these methods is given in Table 1.6. It is important to note the possible interacting factors that can influence the interpretation of test results.

The most useful soil test for B involves hot water extraction, first proposed by Berger and Troug (1940). For the heavy metals Cu, Fe, Mn, and Zn, the DTPA (diethylenetriaminepentaacetic acid) extraction method proposed by Lindsay and Norvell (1969) seems to be the best method. Molybdenum is best determined by extraction with ammonium oxalate (Grigg 1960), although the method is not entirely reliable since Mo availability seems to be more highly correlated with soil pH than that measured by ammonium oxalate extraction (Anderson and Mortvedt 1982).

Although soil-testing procedures for the micronutrients exist, some workers suggest that these elements and their soil–crop status are best determined by means of plant analysis or a combination of soil and plant tests.

4. Universal Extraction Reagents. Several workers have tried to develop a single extraction procedure that could be used in determining all the major elements and most of the micronutrients and that would be effective with most soils. Such a universal extraction reagent would greatly simplify soil test laboratory procedures and facilitate the use of multielement analyzers (e.g., the plasma emission spectrometer). Adolph Mehlich, one of the most active researchers in this quest, has recently proposed his No. 3 extractant as a universal reagent (Mehlich 1982). Wolf (1982a) has modified the Morgan extractant so that both the major elements and the micronutrients can be extracted by one procedure. The AB-DTPA extraction reagent developed by Soltanpour and Schwab (1977) combines the Olsen bicarbonate procedure for P (Olsen *et al.* 1954) with the DTPA micronutrient soil test (Lindsay and Norvell 1969). No doubt other extraction reagents will be proposed to accommodate specific soil–crop needs and suited for extraction of most of the essential elements from the soil regardless

TABLE 1.3 Commonly Used Methods for Determining Extractable Phosphorus (P) in Soils

Parameter	Method				
	Mehlich No. 1 (double acid)	Bray P1	Mehlich No. 2	Olsen	AB-DTPA
Soil type	Sandy soils, acid, low in CEC	Acid soils with moderate CEC	All soils	Alkaline soils	Alkaline soils
Sample size	5 g (4 cm³)	2 g (1.70 cm³	2.5 cm³)	2.5 g (2 cm³)	10 g (8.5 cm³)
Volume of extractant (ml)	25	20	25	50	20
Extraction reagent	0.05 *N* HCl in 0.025 *N* H_2SO_4	0.03 *N* NH_4F in 0.025 *N* HCl	0.2 *N* $HC_2H_3O_2$ in 0.015 *N* NH_4F in 0.2 *N* NH_4Cl in 0.012 *N* HCl	0.5 *N* $NaHCO_3$ at pH 8.5	1 *M* NH_4 HCO_3 0.005 *M* DTPA at pH 7.6
Shaking time (min)	5	5	5	30	15
Shaking action and speed	Reciprocating 180+ oscillations/min	Reciprocating 180+ oscillations/min	Reciprocating 180+ oscillations/min	Reciprocating 180+ oscillations/min	Reciprocating 180+ oscillations/min
Method of P determination in extract	Molybdenum blue	Molybdenum blue	Molybdenum blue	Molybdenum blue	Molybdenum blue
Range in soil P concn without dilution (kg/ha)	2–100	2–250	2–200	2–200	2–100
Sensitivity (kg/ha)	1	1	1	1	2
Primary reference	Mehlich 1953	Bray and Kurtz 1945	Mehlich 1978	Olsen *et al.* 1954	Soltanpour and Schwab 1977

Source: (Jones 1980).

TABLE 1.4 Commonly Used Methods for Determining Exchangeable Cations (Ca, Mg, K, and Na) in Soils

	Method				
Parameter	Mehlich No. 1 (double acid)	1 *N* NH_4OAc pH 7.0	Mehlich No. 2	Water	AB-DTPA[1]
Soil type	Sandy soils, acid, low in CEC	Wide range of soils	Wide range of soils	Wide range of soils	Alkaline soils
Sample size	5 g (4 cm^3)	5 g (4.25 cm^3)	2.5 cm^3	5 g (4.25 cm^3)	10 g (8.5 cm^3)
Volume of extractant (ml)	25	25	25	25	20
Extraction reagent	0.05 *N* HCl in 0.025 *N* H_2SO_4	1 *N* $NH_4C_2H_3O_2$ at pH 7.0	0.2 *N* $HC_2H_3O_2$ in 0.015 *N* NH_4F in 0.2 *N* NH_4Cl in 0.012 *N* HCl	Pure water	1 *M* NH_4HCO_3 0.005 *M* DTPA at pH 7.6
Shaking time (min)	5	5	5	30	15
Shaking action and speed	Reciprocating 180+ oscillations/min	Reciprocating 180+ oscillations/min	Reciprocating 180+ oscillations/min	Reciprocating 180+ oscillations/min	Reciprocating 180+ oscillations/min
Method of K determination in extract	Flame emission spectroscopy	Flame emission spectroscopy	Flame emission spectroscopy	Flame emission spectroscopy	Atomic adsorption
Range in soil K conc without dilution (kg/ha)	50–400	50–1000	50–1000	50–500	50–750
Sensitivity (kg/ha)	5	5	5	5	1
Primary reference	Mehlich 1953	Schollenberger and Simon 1945	Mehlich 1978	Bower and Wilcox 1965	Soltanpour and Schwab 1977

Source: Jones (1980).

[1] The AB-DTPA method is not suited for the determination of Ca or Mg.

TABLE 1.5 Soil Conditions and Crops Where Micronutrient Deficiencies Most Often Occur

Micronutrient	Sensitive crops	Soil conditions for deficiency
B	Alfalfa, clover, cotton, peanuts, sugar beet, cabbage	Acid sandy soils low in organic matter; overlimed soils; organic soils
Cu	Corn, onions, small grains, watermelon	Organic soils; mineral soils high in pH and organic matter
Fe	Citrus, clover, pecan, sorghum, soybean	Leached sandy soils low in organic matter; alkaline soils; soils high in phosphate
Mn	Alfalfa, small grains, soybean, sugar beet	Leached acid soils; neutral to alkaline soils high in organic matter
Mo	Alfalfa, cauliflower, soybean	Highly weathered acid soils
Zn	Corn, field beans, pecan, sorghum	Leached acid sandy soils low in organic matter; neutral to alkaline soils and/or high in phosphate

of its physical and chemical characteristics. However, neither the Mehlich No. 3 extractant nor the modified Morgan reagent is widely accepted or used at the present time.

Baker (1971, 1973) has proposed an entirely different soil-testing procedure which employs an equilibrium solution that is interacted with the soil under test. The method has considerable promise and has been developed into a diagnostic approach to soil testing (Baker and Amacher 1981). However, this soil-testing technique has yet to be put into general use as a replacement for current testing procedures for the major elements and micronutrients.

5. Methods of Elemental Analysis. The analytical techniques for determining the concentrations of elements in the extract from a soil sample have changed significantly in the last decade. Wet chemistry procedures were initially used and are still in wide use today (Piper 1942; Jackson 1958). The first major change came with the introduction of flame emission spectrophotometry for determining K and Na, and to a certain degree Ca and Mg, in soil extracts, a technique that was both easy and fast (Isaac and Kerber 1971). Technicon AutoAnalyzers were adopted by many soil-testing laboratories for the determination of Ca, Mg, P, and K (Isaac and Jones 1970; Flannery and Markus 1971), but they have proven too slow during peak demand periods.

The next major change came when atomic absorption spectrometry offered the analyst a relatively easy and fast method for determining Ca and Mg and the micronutrients Cu, Fe, Mn, and Zn in soil extracts (Isaac and Kerber 1971). However, the determination of B and P is not possible by atomic absorption, and the analysis of K and Na by this

TABLE 1.6 Soil Test Methods, Soil Factors Influencing Their Interpretation, and Typical Ranges in Critical Level for Micronutrients

Element	Interacting factors[1] Essential	Probable	Method	Range in critical level[2] (ppm)
B	Texture, pH	Lime	Hot H_2O	0.1–0.7
Cu		O.M., Fe	$NH_4C_2H_3O_2$(pH 4.8)	0.2
			0.5 *M* EDTA	0.75
			0.43 *N* HNO_3	3–4
			Biological assay	2–3
Fe		pH, lime	$NH_4C_2H_3O_2$ (pH 4.8)	2
			DTPA + $CaCl_2$ (pH 7.3)	2.5–4.5
Mn	pH	O.M.	0.05 *N* HCl + 0.025 *N* H_2SO_4	5–9
			0.1 *N* H_3PO_4 and 3 *N* $NH_4H_2PO_4$	15–20
			Hydroquinone + $NH_4C_2H_3O_2$	25–65
			H_2O	2
Mo	pH	Fe, P, S	$(NH_4)_2C_2O_4$(pH 3.3)	0.04–0.2
Zn	pH, lime	P	0.1 *N* HCl	1.0–7.5
			Dithizone + $NH_4C_2H_3O_2$	0.3–2.3
			EDTA + $(NH_4)_2CO_3$	1.4–3.0
			DTPA + $CaCl_2$ (pH 7.3)	0.5–1.0

Source: Cox and Kamprath (1972).
[1] Climatic crop factors, although highly important, are not considered here. O.M. = organic matter.
[2] Test value(s) below which deficiency occurs and above which sufficiency occurs (see Section II.E.3).

technique is unsatisfactory. Although atomic absorption spectrometry enjoys wide acceptance and use, its inability to adequately assay several important elements has turned analysts back to emission spectrometry in the form of plasma emission spectrometry (Soltanpour *et al.* 1982). This technique, referred to as ICAP (inductively coupled agron plasma), can be used to determine most of the elements of interest in soil extracts. Besides having multielement capability, this technique has the advantage of excellent sensitivity and high speed. The ICAP analysis technique has stimulated interest in the development and use of universal extraction reagents.

6. Other Soil Test Determinations. In addition to determinations for soil water pH and the extractable elements, tests for organic matter and soluble salt content and for extractable nitrate and sulfate levels are commonly done as part of a soil test.

The organic matter content of the soil can be determined by a number of procedures (Nelson and Sommers 1982), although the Walkley and Black method (1934) is the most widely used. The soil organic matter content forms the basis for determining the potential release of plant-available N during the growing season (Dahnke and Vasey 1973) and is a factor in adjusting herbicide recommendations for those chemicals inactivated by soil adsorption.

A high concentration of soluble salts, due to the accumulation of fertilizer elements, is a common problem when plants are grown in confined containers. Also, in those areas where irrigation water contains sizable quantities of "salts," the ever-increasing salinity condition in the soil can present problems. In addition, many of the soils in the semiarid and arid regions of the world are usually saline.

The soluble salt level in the soil can be determined with several water extraction procedures (Bower and Wilcox 1965; Rhoades 1982). In the most common procedure, a 1:2 volume ratio of soil to water is used for extraction of soluble salts, and then the specific conductance of the extract is measured. The interpretation of conductance values in terms of salinity effects is shown in Table 1.7.

The soil test for nitrate content has limited but useful applications when attempting to regulate N fertilizer rates for those crops sensitive to excess N. The test can also be used to determine the need for supplemental N fertilizer during the growing season and for modifying a N fertilizer recommendation based on crop requirement. Dahnke and Vasey (1973) discuss some of these applications for a nitrate soil test.

Concern about adequate soil levels of S, a major essential element, is increasing, due in part to the decreasing use of S-containing fertilizers, such as ordinary superphosphate (0–8.8–0). Increased usage of air

TABLE 1.7 Interpretation of Specific Conductance Values of Soil Extracts (1:2 Soil to Water Ratio)

Specific conductance (mmho/cm at 25° C)	Interpretation
<0.40	Salinity effects mostly negligible, except possibly in beans and carrots
0.40–0.80	Very slightly saline. Yields of very salt-sensitive crops such as flax, clovers (alsike, red), carrots, onions, bell pepper, lettuce, and sweet potato may be reduced by 25–50%.
0.81–1.20	Moderately saline. Yield of salt-sensitive crops restricted. Seedlings may be injured. Satisfactory for well-drained greenhouse soils. Crop yields reduced by 25–50% may include broccoli, potato, and the other plants listed above.
1.21–1.60	Saline soils. Tolerant crops include cotton, alfalfa, cereals, grain sorghums, sugar beets, bermuda grass, tall wheat grass, and Harding grass. Salinity higher than desirable for greenhouse soils.
1.61–3.20	Strongly saline. Only salt-tolerant crops yield satisfactorily. For greenhouse crops, leach soil with enough water so that 2–4 qt pass through each square foot of bench area or 1 pint of water/6-in. pot; repeat after about 1 hr. Repeat again if readings are still in the high range.
>3.2	Very strongly saline. Only salt-tolerant grasses, herbaceous plants, certain shrubs, and trees will grow.

Source: Jones (1980).

pollution control technology has also reduced the amount of S deposited on soils downwind of major industrial areas. There seems to be a relationship between N and S in the plant such that high application rates of N fertilizer may demand a greater S supply. However, proposed techniques for determining S availability in soils are quite poor based on their correlation to plant response. Hoeft (1982) has listed the more common procedures for extracting S from the soil and the critical levels of S (Table 1.8).

7. Test Procedures for Organic Soils and Soilless Media. Organic soils and soilless media are not suited for analysis by those procedures prescribed for mineral soils. These soils should not be air-dried or sieved prior to analysis. The procedure described by Warncke (1980), which involves analysis of a saturated water extract, is widely used for testing organic soils and soilless media. An interpretation of the test results is given in Table 1.9.

Sonneveld *et al.* (1974) have suggested an extraction ratio of 1:1.5 medium to water volume for evaluating the nutrient element status of organic potting media. The results obtained compared favorably with other previously used procedures that are difficult to perform in the laboratory. Prasad *et al.* (1983) have used this same procedure on peat, pine bark and peat, and peat and soil mixes with good results, the concentrations of extracted elements correlating well with plant response.

8. Methods of Expressing Test Results. Soil test results based on element concentrations determined on soil extracts have in the past been expressed primarily in pounds per acre (lb/acre), and more re-

TABLE 1.8 Extractant and Critical Level for Various Sulfur Soil-Testing Procedures

Location	Extractant	Critical level of S (ppm)	Additional comments
Nebraska	$Ca(H_2PO_4)_2$	8	<1% organic matter
		5	>1% organic matter
Minnesota	$Ca(H_2PO_4)_2$	7	Sandy soils
		7–12	Alfalfa—trial basis only
Wisconsin	$Ca(H_2PO_4)_2$–HOAc	6	
		6–10	Use equation including pH (see Hoeft *et al.* 1973)
Virginia	$Ca(H_2PO_4)_2$–HOAc	2.5	
New Zealand	NaH_2PO_4–HOAc	10	
Scotland	KH_2PO_4	8–10	

Source: Hoeft (1982).

TABLE 1.9 Guidelines for Interpreting Analyses of Soilless Growth Media by the Saturated Media Extract Method

Parameter	Category				
	Low	Acceptable	Optimum	High	Very high
Soluble salt (mmho/cm)	0–0.75	0.75–2.0	2.0–3.5	3.5–5.0	5.0+
Nitrate-N (ppm)	0–39	40–99	100–199	200–299	300+
Phosphorus (ppm)	0–2	3–5	6–9	11–18	19+
Potassium (ppm)	0–59	60–149	150–249	250–349	350+
Calcium (ppm)	0–79	80–199	200+	—	—
Magnesium (ppm)	0–29	30–69	70+	—	—

Source: Warncke (1980).

cently as kilograms per hectare (kg/ha), the equivalent in the metric system. The soil concentration is arrived at by multiplying the element concentration found in the extract times the soil to solution ratio times 2, a value based on the assumption that an acre (hectare) of soil taken to a depth of 6.7 in. (16.9 cm) weighs 2 million lb (or 2 million kg). This assumes an average bulk density of 1.32 g/cm^3. This value is the same if expressed as parts in 2 million parts (pp2m). Some analysts express element concentrations as ppm in the soil and do not convert them to the weight basis.

Mehlich (1972) has questioned these weight methods of expression and has suggested a volume system based on a standard volume of soil of 2 million cubic decimeters per hectare (dm^3/ha). Therefore, a soil test result for an extractable element would be expressed in milligrams per cubic decimeters (mg/dm^3). Although Mehlich (1973) has eloquently put forth his suggested method, tradition seems for the moment to prevail.

A unitless value for element concentration is the trend today, either assigning a given category or percentage sufficiency based on the relationship between concentration and crop response. This topic will be discussed in greater detail later (Section II. E. 2).

For the major cations (Ca, K, Mg, and Na), the extractable concentration may be expressed as their equivalents in milliequivalents per 100 grams of soil (meq/100 g). Knowing the milliequivalents of exchangeable hydrogen (determined from the buffer pH measurement), one can estimate the soil's cation exchange capacity (CEC) and calculate the percentage base saturation and percentage occupation of the CEC by each cation. Based on International System of Units, these same cations can be expressed as mol (P^+)/kg, a requirement in much of the current scientific literature.

Therefore, there are various ways of expressing the extractable element content of a soil with no one method of expression universally accepted and used.

E. Interpretation

The interpretation of a soil test result has two aspects: (1) an evaluation of the result itself in terms of its relationship to plant growth, and (2) determination of the amount of lime and/or fertilizer needed to correct a deficiency revealed by the test result and to meet the crop requirement. There is considerable literature on the evaluation of results and reasonable agreement among soil and plant scientists about what constitutes adequacy and inadequacy. It is the determination of what is needed to correct soil deficiencies and to supply the crop requirement that varies considerably among these same scientists. Therefore, those who only measure the value of a soil test result in terms of the given lime and fertilizer recommendations may misjudge the full value of the soil test.

1. Soil pH and Liming. The soil water pH can be categorized as shown in Fig. 1.1. With increasing soil acidity, it is not the hydrogen ion concentration itself that affects plant growth, but the increasing concentration, to toxic levels, of Al and Mn in the soil solution, and to a limited degree the decreasing availability of P and Mg. As the soil becomes more alkaline, the solubility and therefore the concentration in the soil solution of several micronutrients (Cu, Fe, Mn, and Zn) decreases to deficiency levels. Therefore, in a soil management program, the extremes of acidity or alkalinity should be avoided. Changes in soil water pH also affect the major element composition of the soil solution; this in turn affects plant growth. With increasing acidity, for example, the concentration of Mg and P in the soil solution decreases.

The effects of soil pH on the availability of the essential elements are summarized in Fig. 1.2. The maximum availability for most of the essential elements occurs when the pH is between 6.0 and 7.0. For sandy soils, low in cation exchange capacity (<10 meq/100 g) and organic matter (<2%), it is best to keep the pH near 6.0; for silt and clay loam soils, higher in cation exchange capacity (10–30 meq/100 g) and organic matter (2–5%), a pH between 6.5 and 7.0 is recommended.

Soil acidity can be corrected by application of agricultural limestone in the amount specified by a buffer pH determination (see Section II. D.1; Table 1.2). Lime requirements based on the Adams–Evans and SMP buffer methods are found in the *Handbook on Reference Methods for Soil Testing* (Jones 1980). Adjustments to the lime requirement given in this handbook may be necessary to account for differences in the liming material (i.e., fineness and chemical composition) and in the depth of soil incorporation. Although much has been written about

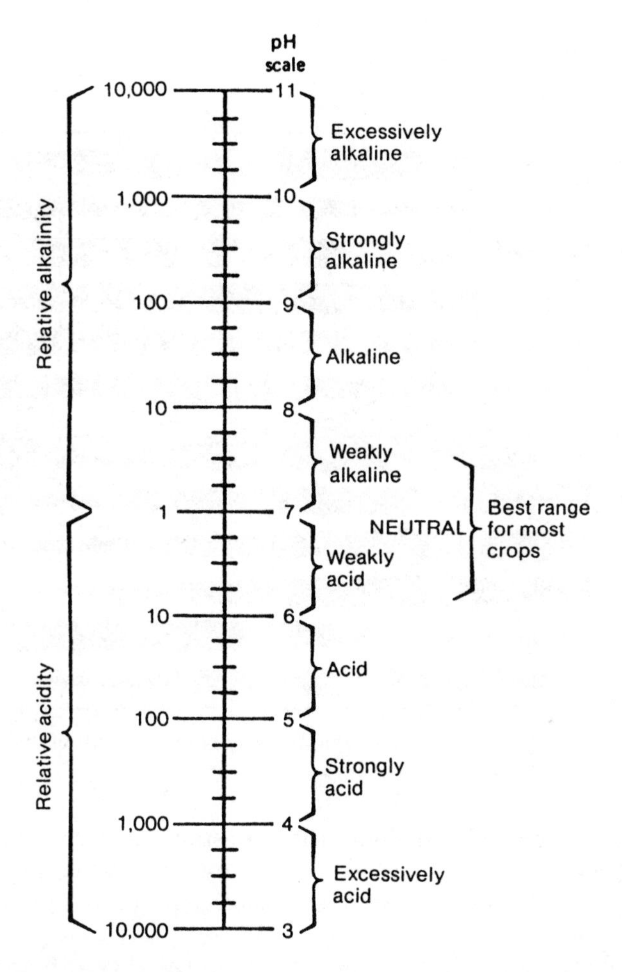

FIG. 1.1. Relationship among soil pH, alkalinity, acidity, and plant growth. *From Lorenz and Maynard (1980).*

soil pH and liming (Coleman *et al.* 1958; Pearson and Adams 1967; McLean 1973), soil acidity still remains a frequently occurring soil fertility problem for many cropland soils (Follett *et al.* 1981), in orchards (Korack 1980), and in the growing of ornamental plants (Nesmith and McElwee 1974). Growers are being advised to lime more frequently (annually if necessary) and in smaller doses to more effectively counter the acidifying effects of applied N fertilizer and the losses of Ca and Mg due to leaching and crop removal.

Liming not only increases the pH of a soil, it also adds to the soil an

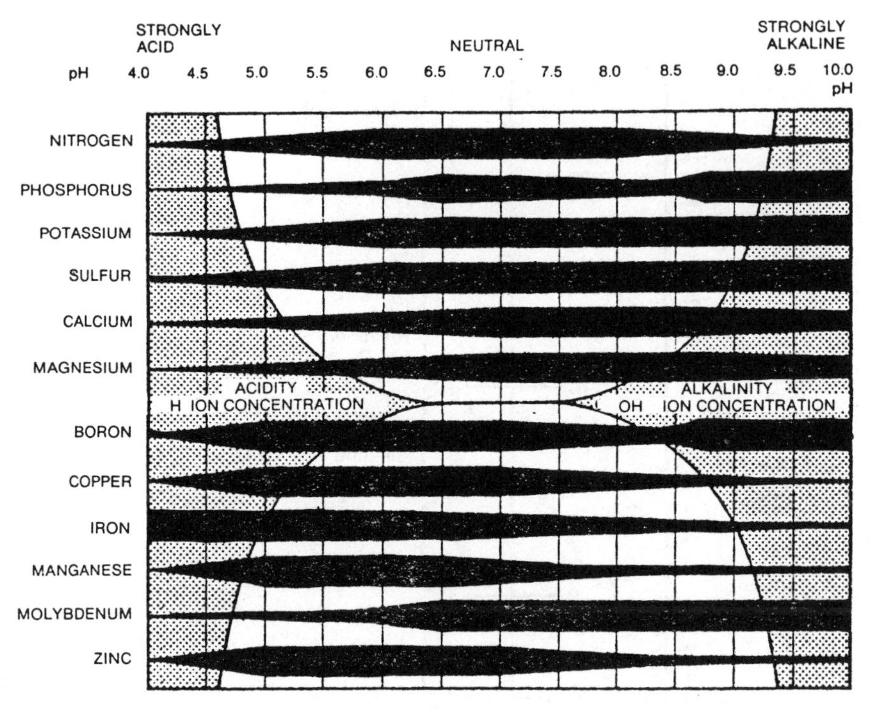

FIG 1.2. Effect of soil pH on the availability of plant nutrient elements. *From Hanna and Hutcheson (1968).*

essential element, Ca; if dolomitic limestone is used, Mg, another essential element, is also added. Therefore, the liming can have two beneficial effects—correction of soil acidity and resupply of Ca and Mg.

Some plant species have a tolerance to soil acidity; others have, to a limited degree, tolerance to soil alkalinity. Plant species can be grouped based on their tolerance to pH as shown in Table 1.10. Such groupings essentially reflect the species tolerance and/or intolerance to those elements (mainly Al and Mn) whose concentration in the soil solution is significantly affected by the soil's pH.

The pH of organic soilless media also can affect the growth of plants, but the effects are different from those observed for mineral soils. The danger here is from too high a pH, i.e., greater than 6.0. A pH of less than 5.0 is of little consequence since there is normally too little Al and Mn in the organic media to reach toxic proportions. The relationship between the pH of an organic soilless medium and the availability of the essential elements is shown in Fig. 1.3. Note the marked differences in element availability (especially for P, Mn, and

TABLE 1.10 Approximate Soil pH Ranges Suitable for Various Plant Species

pH 4.8–5.2 (acid)	pH 5.8–6.2 (moderately acid)		pH 6.3–6.7[1] (nearly neutral)
Azaleas	Apples	Peaches	Alfalfa
Blackberries	Beans, lima	Peanuts	Asparagus
Blueberries	snap	Peppers, pimiento	Cabbage
Grass, carpet	velvet	Pine, yellow	Carrots
centipede	Cantaloupes	Radishes	Ladino—grass
Hydrangia, blue	Corn	Rice	Lettuce
Irish potatoes	Cotton	Small grain	Onions
Juniper, Irish	Cowpeas	Sorghum	Peas
Pine, longleaf	Crimson clover	Soybeans	Red clover
Watermelon (<5.8)	Cucumber	Squash	Spinach
	Grass, most kinds	Strawberries	Sweet clover
	Greens, mustard, etc.	Sudan grass	Timothy
	Iris, blueflag	Sweet potatoes	White clover
	Kale	Tobacco	
	Lespedeza	Tomatoes	
	Parsnips	Trefoil, birdsfoot	
	Redtop	Vetch	

Source: Jones (1979).

[1] On highly organic soils, it is not advisable to lime to a high pH. Good plant growth on these soils may be obtained where the pH is around 5.5. Approximate pH ranges refer to those suitable for mineral soils in humid regions where lime is often added. Many of these species, (for example, alfalfa, apples, iris, cotton, rice, sorghum, sweet clover, and timothy) grow well on calcareous soils with pH values up to 7.5–8.0.

B) compared with those for mineral soils shown in Fig. 1.2. Therefore, liming an organic soilless mix is probably not wise, and other sources of Ca and Mg, such as their sulfates (gypsum and epsom salts, respectively) should be used. Techniques for establishing and maintaining sufficient concentrations of nutrient elements in organic soilless growing mixes is discussed in considerable detail by Jones (1983).

2. Major Elements. The interpretation of a soil test result is based on the test value's relationship (correlation) to plant growth or response to application of the test element as fertilizer. Much has been written about this subject; a good summary is given by Peck *et al.* (1977).

There is considerable diversity of opinion about how to express soil test results for the major elements as well as the micronutrients (see Section II. D. 8). Although a laboratory-obtained test value may be given in weight or volume terms, this is not the actual amount of element available for plant use. Therefore, the common procedure is to assign a descriptive word or numerical category to a range of test values, a procedure called *indexing.* Cope (1972) has developed an indexing system for categorizing soil test results for the Alabama Soil Testing Program (Table 1.11), which is a good illustration of how both categorization systems work. With either system, the objective is to

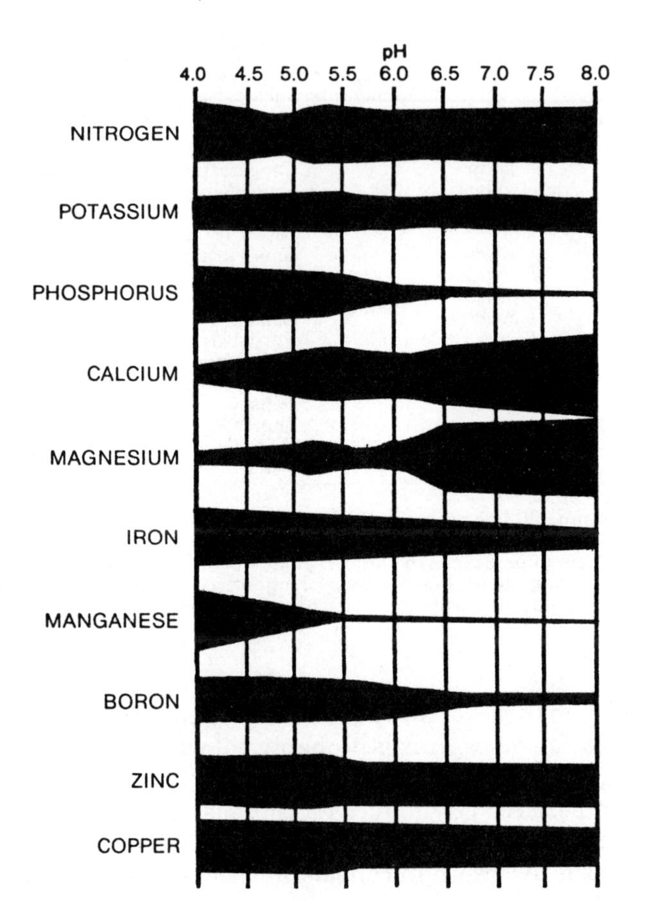

FIG 1.3. Availability of plant nutrient elements at different soil pH levels with W.R. Grace Metro Soilless Mix 300 and organic soils.
From Peterson (1982).

translate actual laboratory-obtained soil test values into terms that have biological significance. The basis for this association is the Mitscherlich-Bray growth function (Melsted and Peck 1977).

A simpler way to interpret soil test results is the Cate–Nelson method (Cate and Nelson 1965), which provides a definite decision point. This graphical method involves dividing a scatter diagram of percentage yield vs soil test value into four quadrants so as to maximize the number of points in the positive quadrants and minimize the number of points in the negative quadrants, as illustrated in Fig. 1.4. This method was designed for use in those situations where less sophistication in the interpretation of test results is required.

TABLE 1.11 Soil Test Rating and Fertility Index System Used in the Alabama Soil-Testing Program

Soil test rating	Fertility index (%)	Crop response
Very low	0–50	Yield may be less than 50% of potential
Low	60–70	Yield may be 50–75% of potential
Medium	80–100	Yield may be 75–100% of potential
High	110–200	Desirable level that is the objective of most soil-building programs
Very high	210–400	Soil supply double that considered adequate
Extremely high	410+	Level is excessive and further additions may be detrimental

Source: Cope (1972).

Thomas and Peaslee (1973) have summarized the interpretation for three commonly used extraction procedures for P and Mehlich (1978) has provided interpretive categories for Mehlich Extractant No. 2 (Table 1.12). Lorenz and Maynard (1980) have also given soil test interpretative data for three extraction procedures and various combinations of the P, K, Mg, and Zn (Tables 1.13 and 1.14).

Two quite different concepts can be employed in interpreting soil

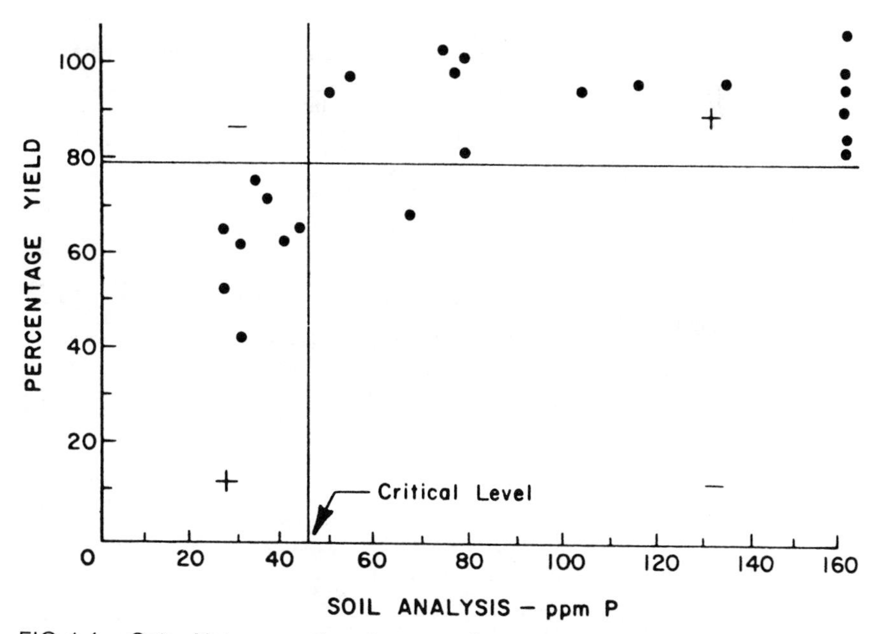

FIG 1.4. Cate–Nelson scatter diagram of percentage yield vs soil test P for maize.
From Cate and Nelson (1965).

TABLE 1.12 Interpretation of P Soil Test Values Obtained with Four Common Soil-Testing Extractants

Relative soil level	Extractable P in soil (ppm)			
	Mehlich No. 1	Mehlich No. 2	Bray P1	Olsen
Low	0–16	0–16	0–15	0–5
Medium	17–37	17–44	16–30	6–10
High	>37	>44	>30	>10

Source: Thomas and Peaslee (1973); Jones (1980).

TABLE 1.13 Interpretation of Soil Test Values Obtained with Mehlich No. 1 and Bray P1 Extraction Procedures

	Extractable elements in soil (ppm)				
	Melich No. 1			Bray P1	
Relative soil level	P	K	Mg	P	K
Very low	0-6	0-14	0-17	0-12	0-60
Low	7-13	15-35	18-25	13-25	61-90
Medium	14-23	36-37	26-59	26-37	91-150
High	23-44	68-133	60-126	38-75	151-250
Very high	45+	134+	127+	76+	251+

Source: Lorenz and Maynard (1980).

TABLE 1.14 Interpretation of Soil Test Values for Phosphorus, Potassium, Magnesium, and Zinc

Nutrient need	Amount in soil (ppm)			
	PO_4–P[1]	K[2]	Mg[2]	Zn[3]
Deficient levels for most vegetables	0–10	0–60	0–25	0–0.3
Deficient for susceptible vegetables	10–20	60–120	25–50	0.3–0.6
A few susceptible crops may respond	20–40	120–200	50–100	0.6–1.0
No crop response	Above 40	Above 200	Above 100	Above 1.0
Levels are excessive and could cause problems	Above 150	Above 2000	Above 1000	Above 3.0

Source: Adapted from Reisenauer (1978), cited in Lorenz and Maynard (1980).
[1] Olson (5 *M*, pH 8.5) sodium-bicarbonate extractant.
[2] Exchangeable with (1 *N*, pH 7.0) ammonium acetate extractant.
[3] DPTA extractable Zn.

TABLE 1.15. Cation Base Saturation Percentages Suggested as Being "Ideal" for Plant Growth

Cation	Cation base saturation (%)	
	Bear *et al.* (1945)	Graham (1959)
Ca	65	65–85
Mg	10	6–12
K	5	2–5

tests for the three major cations, Ca, K, and Mg. The Sufficiency Levels of Available Nutrients (SLAN) method of interpretation is based on a regression of soil test level with plant growth, which is best described by the Mitscherlich-Bray growth function (Melsted and Peck 1977). The other concept, Basic Cation Saturation Ratios (BCSR), suggests that there is a "best" cation ratio on the soil cation exchange complex. The possibility that an "ideal" cation ratio does exist was first suggested by Bear *et al.* (1945), and then later modified to cover a range of ratios by Graham (1959). These ratios are given in Table 1.15. Both methods of interpretation have their strengths and weaknesses, as discussed by McLean (1977a). It is probably fair to say that the BCSR system of soil test interpretation for the major cations is most frequently misused as interpreters attempt to "force" a soil into a fixed cation ratio status.

The trend today is away from SLAN and BCSR interpretative procedures to one based on chemical equilibria concepts, i.e., looking at the soil and its measured soil fertility level, determined by laboratory tests, as being the result of interacting chemical and biological forces. Soil-testing procedures and interpretation methods based on ion activities and ratios suggested by Baker (1977), and to a certain extent by Geraldson (1977), are attempts to put into practice ion activity concepts. McLean (1982b) suggests looking at the soil as a buffered system in which the ionic composition of the soil solution is deter-

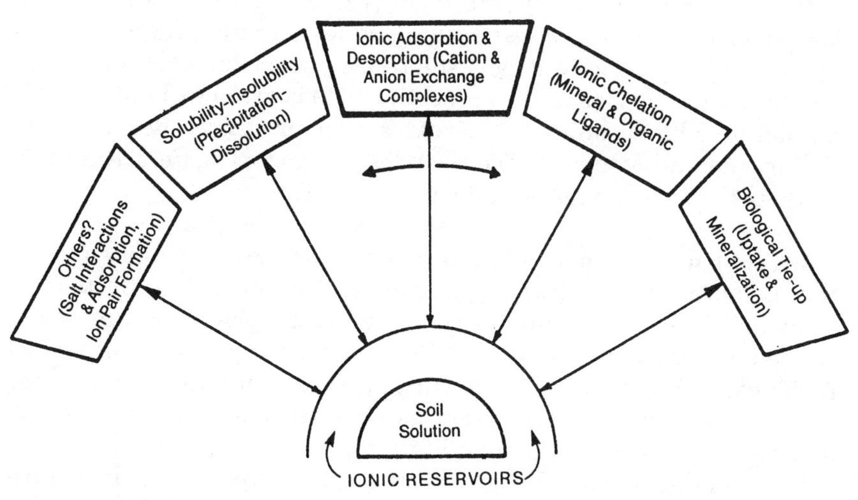

FIG 1.5. Schematic diagram of the ionic reservoirs removing ions from and returning ions to the soil solution.
From McLean (1977b).

mined by equilibria phenomena. Figure 1.5 presents a schematic illustration of ions moving into and out of the soil solution from various ionic reservoirs. Test procedures that can measure the capacity of these ionic reservoirs will become the soil-testing procedures of the future (McLean 1977 and McLean *et al.* 1979). These newer soil-testing procedures have been proposed by McLean *et al.* (1982) for K and by McLean and Mostaghimi (1983) for P.

Jones and Budzynski (1980) have suggested a sufficiency range concept of interpretation that uses the soil test result as a monitor to detect changes that would require adjustment in the lime and/or fertilizer recommendation in order to maintain test values within prescribed limits. This technique also relies heavily on the use of plant analyses and requires exact field observations and yield determinations. Jones (1983) has recently described how this technique can be applied to a crop production system.

3. Micronutrients. Soil test results for micronutrients are interpreted in a somewhat different manner from those for the major elements. Interpretation is based on a *critical value* below which deficiency occurs and above which there is sufficiency. Trierweiller and Lindsay (1969) illustrated this interpretive approach in their calibration of the EDTA-ammonium bicarbonate extraction procedure for Zn (Fig. 1.6). Therefore, for most of the micronutrients, interpretation is based on a single value that delineates the point between deficiency and sufficiency. Mortvedt (1977) reviewed the interpretation of micronutrient soil tests, giving tables of critical values for most of the common micronutrient soil-testing procedures. A somewhat similar listing by Cox and Kamprath (1972) is given in Table 1.6. The ammonium bicarbonate-DTPA extraction procedure developed by Soltanpour and Schwab (1977) includes assays for Cu, Fe, Mn, and Zn; their sufficiency ranges are given in Table 1.16.

4. Fertilizer Recommendations. The fertilizer recommendation that is based on a particular soil test result frequently reflects more than just a strict interpretation of the test value; this has led to confusion among growers about the real potential of soil testing (see Liebhardt 1981–1982). The fertilizer recommendation, which embodies a strategy for the economic use of fertilizers, takes into account a number of modifying factors (Fig. 1.7).

Two important modifying factors are soil type (i.e., pH, texture, CEC, organic matter content) and crop requirement. In most instances, the effect of soil type is incorporated into the soil test ratings and interpretive categories, although this is not always true. Crop

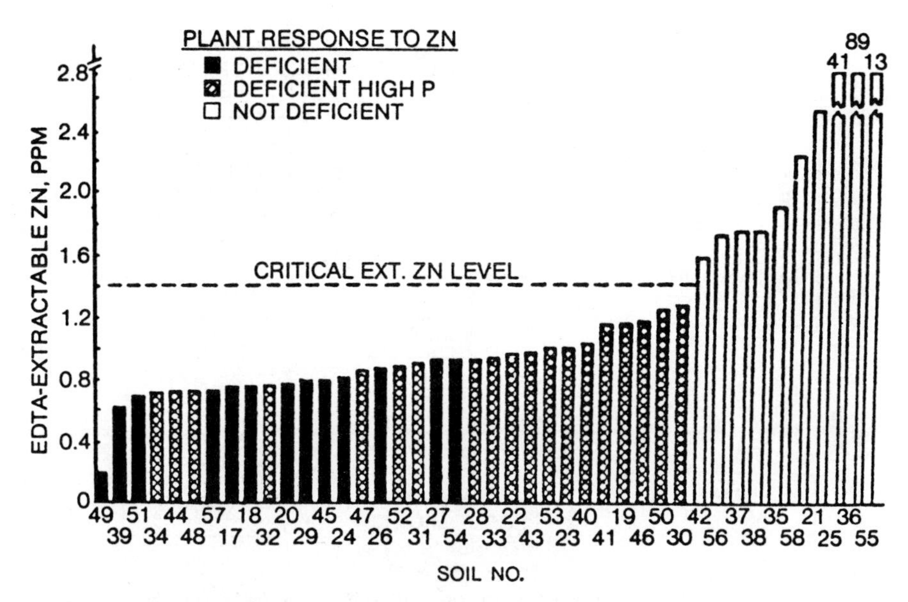

FIG 1.6. EDTA-ammonium bicarbonate extractable Zn for 42 Colorado soils in relation to Zn response of maize in the greenhouse.
From Trierweiler and Lindsay (1969).

requirement considerations are necessary to fit the fertilizer recommendation to the specific needs of the crop. Thus, fertilizer recommendations may be derived from the difference between soil test values and the total content of the essential elements in a healthy crop, or they may be based on results from practical experience.

Another major modifying factor is economics, that is, balancing the cost of applied fertilizer versus the benefits derived in terms of yield and quality. The best treatise on this subject is by Heady *et al.* (1961). Davidescu and Davidescu (1982) also discuss this subject in some detail, illustrating the relationship between yield and fertilizer application rate as shown in Fig 1.8, where $0x_3$ is the limit at which total production is maximum, and $0x_2$ is the limit at which average produc-

TABLE 1.16 Interpretation of Soil Test Values for Ammonium Bicarbonate-DTPA Extractable Copper, Iron, Manganese, and Zinc

Sufficiency category	Extractable Elements (ppm)			
	Cu	Fe	Mn	Zn
Low	0.5	0–2.0	1.8	0–0.9
Marginal	—	2.1–4.0	—	1.0–1.5
Adequate	>0.5	>4.0	>1.8	>1.5

Source: (Jones 1980).

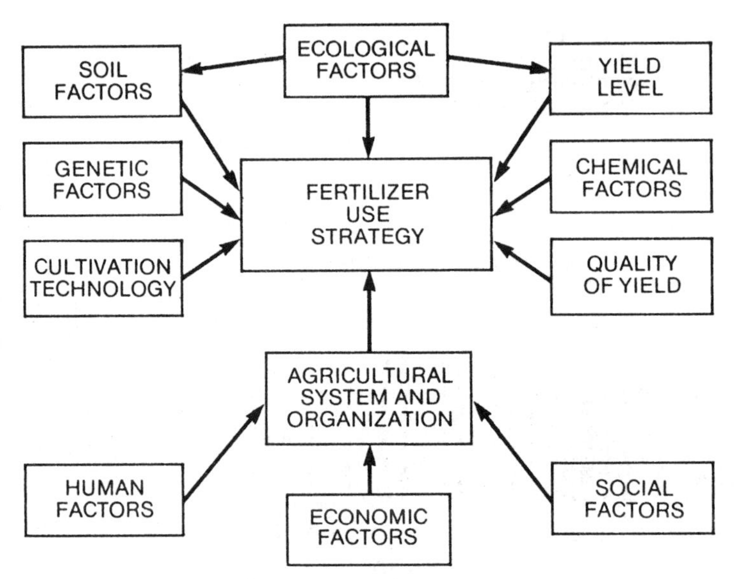

FIG 1.7. Factors that may be considered in developing a fertilizer use strategy. *From Davidescu and Davidescu (1982).*

tion is maximum. Therefore, the fertilizer rate for optimum return on cost of applied fertilizer lies between $0x_2$ and $0x_3$.

For most horticultural crops (fruits, vegetables, and ornamental plants), the economic concern is not so much fertilizer cost, but the market value of the crop yield and quality. Therefore, growers are likely to apply more fertilizer than called for by soil tests in order to ensure maximum yield and quality. This practice can lead to excesses and imbalances, which are more likely than deficiencies. In such cases, soil tests are most useful as a monitor of the soil fertility level, alerting growers to developing or occurring situations that can adversely affect crops. Growers are advised to soil-test yearly in order to develop a "track" of soil test values. By using established sufficiency ranges for each soil test parameter, adjustments in lime and fertilizer treatments can be made when necessary to keep soil test values within the sufficiency range, a technique recommended by Jones and Budzynski (1980).

Two philosophies have been used in developing fertilizer recommendations based on soil tests; one is based on the concept of fertilizing the soil, the other on fertilizing the crop. For soils other than very sandy ones with low cation exchange capacities, the trend in the past has been to recommend more fertilizer than called for by the soil test in

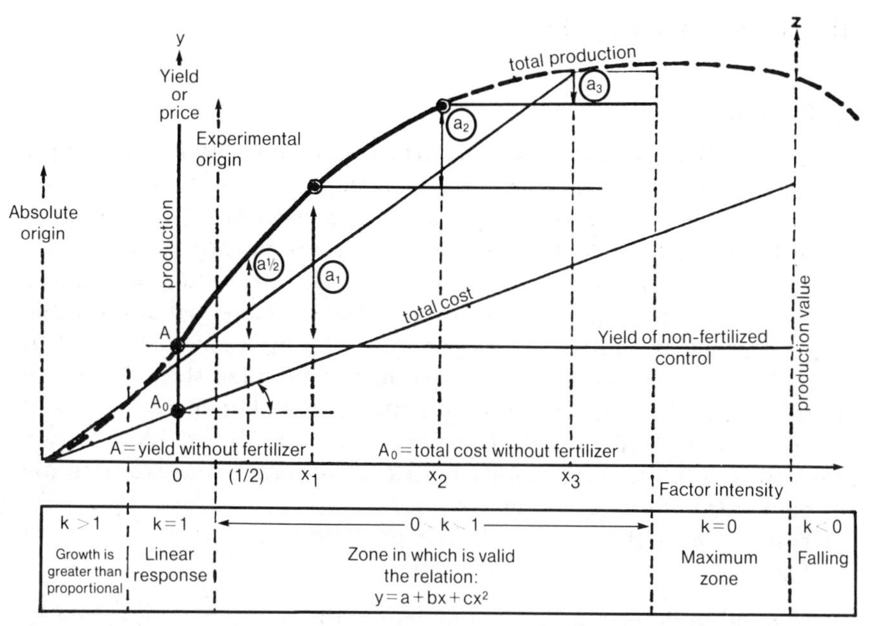

FIG 1.8. Production curve when a single production factor varies while the others are considered to be constants. x_1—Midpoint of response (a_1) to applied fertilizer. x_2—Limit at which average production a_2 is maximum. x_3—Limit at which total production (a_3) is maximum.
From Davidescu and Davidescu (1982).

order to increase the test level, the concept being to fertilize the soil. Much of the increase in P levels in cropland soils in the United States can be attributed to this philosophy of soil fertilization (Jones 1965). Applying just enough fertilizer to satisfy the crop requirement has no appreciable residual effect on the soil test level. The general rule of thumb is that each kilogram of applied P or K above the crop requirement will result in a possible increase of the soil test level of 5–10 units. A more thorough discussion of philosophies of fertilizer use is given by Nelson and Hansen (1968).

Since the subject of fertilization of horticultural crops is vast and complex, the reader may wish to refer to the review articles by Lorenz and Bartz (1968) for vegetables, Reitz and Stiles (1968) for orchard crops, and Joiner *et al.* (1983) for floricultural plants.

Sumner (1982) has suggested that the Diagnosis and Recommendation Integrated System (DRIS) for plant analysis interpretation can be applied to interpreting soil test results. A discussion of the DRIS technique is given in section III. F.

III. PLANT ANALYSIS[1]

A. Principle

Plant analysis, sometimes referred to as leaf analysis, is a technique for correlating the elemental content of the whole plant or one of its parts with either its physical appearance, growth rate, yield, or quality of harvested product. The technique requires precise sampling procedures in terms of the plant part selected and stage of growth (time of sampling). The elemental content of the collected plant tissue is determined by analytical procedures, which will be discussed in detail later. Interpretation is based on the premise that there is a significant biological relationship between the elemental content of the plant and plant growth (Lundegardh, 1951, Ulrich 1952; Ulrich and Hills 1973). Therefore, plant (leaf) analysis can be used to evaluate the elemental status of the plant.

Krantz *et al.* (1948) outlined their principal objectives for a plant analysis:

1. To aid in determining the nutrient-supplying power of the soil
2. To aid in determining the effect of treatment on the nutrient supply in the plant
3. To study relationships between the nutrient status of the plant and crop performance as an aid in predicting fertilizer requirements
4. To help lay the foundation for approaching new problems or for surveying unknown regions to determine where critical plant nutritional experimentation should be conducted

None of these early objectives reflects the primary use for plant analyses today—diagnosis of suspected nutrient element deficiencies. Objective 3 is related to the second most important use of plant analyses today, especially for tree fruits and nuts. Munson and Nelson (1973) describe the use of plant analyses for verification of deficiency symptoms, analysis of normal and abnormal plants, and crop logging, each one an important application for plant analysis.

The interpretation of the results of a plant analysis is based on the concept of *critical values*, which Ulrich (1948) defines as "that range of concentrations at which the growth of the plant is restricted in comparison to that of plants at a higher nutrient level." In a later review article, Ulrich (1952) discussed the "physiological bases for assessing

[1] Plant analysis and tissue testing are considered by some to be synonymous. However, in this review they are considered to be different and are discussed separately. See Section IV for a discussion of tissue testing.

the nutrient requirements for plants," the foundation on which the interpretation of a plant analysis is built.

The considerable literature on plant analysis dates back to the 1800s. Considerably more has been written about plant analysis than soil testing even in more recent times. As stated earlier, plant analysis has had greater application to those crops considered horticultural (fruits, nuts, vegetables, and ornamental plants) than to field crops (corn, soybeans, alfalfa, wheat, rice, cotton, etc.). Therefore, it would be quite difficult to cover all the pertinent literature on plant analysis in this review, and only a relatively brief overview is attempted.

B. Plant Sampling

Probably no other single aspect of the plant analysis technique can have as much effect on the final result as the procedure used to collect the sample for analysis. Plants are made up of complex structures that lack homogeneity in elemental content. Sayre (1952), using radioactive isotopes, found that some elements are not uniformily distributed in corn leaf blades, and Jones (1963) found that B, Mn, and Si accumulate in the leaf margins. It is well known that leaves, petioles, and stems—as well as the upper and lower portions of the plant—vary considerably in their elemental makeup (Jones 1970b). Therefore, the mixing of plant parts, or the taking of whole plants, is not a recommended sampling procedure. The plant leaf is the most commonly recommended plant part for sampling, but Jones (1963) suggests removing the leaf margin in order to minimize the effect of elemental accumulation in the margin. Taking samples from the leaf blade with a puncheon would minimize the effect of uneven elemental distribution found in leaves; however, such a sampling procedure would be tedious. The lack of elemental homogeneity, even in the leaves of plants, may partially explain why Steyn (1961) and Colonna (1970) found it necessary to take fairly large numbers of replicate leaves in order to obtain a high level of precision in their measurements of elemental concentration. Steyn (1961) found that heterogeneity was so great in some instances as to make it physically impossible to collect enough leaves to obtain a certain desired level of precision.

The decision as to what plant part to collect for elemental assay generally is based on two considerations: the best correlation between plant appearance or performance with elemental content, and the ease of identification and collection. Therefore, most sampling procedures call for the collection of mid-shoot leaves from woody perennial plants and trees, and for recently mature leaves from annual plants. These leaves reflect the plant's current nutrient element status and are easy to identify and collect. The time of sampling also is important,

TABLE 1.17 Plant Tissue Sampling Procedures for Horticultural Crops

Crop	Stage of Growth	Plant part to sample	No. plants to sample
VEGETABLES			
Potato	Prior to or during early bloom	3rd to 6th leaf from growing tip	20–30
Head crops (cabbage, etc.)	Prior to heading	1st mature leaves from center of whorl	10–20
Tomato (field)	Prior to or during early bloom	3rd or 4th leaf from growing tip	20–25
Tomato (greenhouse)	Prior to or during fruit set	(1) Young plants: leaves adjacent to 2nd and 3rd clusters (2) Older plants: leaves from 4th to 6th clusters	20–25
Beans	(1) Seedling stage (less than 12 in.)	All the above-ground portion	20–30
	(2) Prior to or during initial flowering	2 or 3 fully developed leaves at the top of the plant	20–30
Root crops (carrots, onions, beets, etc.)	Prior to root or bulb enlargement	Center mature leaves	20–30
Celery	Midgrowth (12–15 in. tall)	Petiole of youngest mature leaf	15–30
Leaf crops (lettuce, spinach, etc.)	Midgrowth	Youngest mature leaf	35–55
Peas	Prior to or during initial flowering	Leaves from the 3rd node down from the top of the plant	30–60
Sweet corn	(1) Prior to tasselling	The entire fully mature leaf below the whorl	
	(2) At tasselling	The entire leaf at the ear node	20–30
Melons (water, cucumber, muskmelon	Early stages of growth prior to fruit set	Mature leaves near the base portion of plant on main stem	20–30
FRUITS AND NUTS			
Apple, apricot, almond, prune, peach, pear, cherry	Midseason	Leaves near base of current year's growth or from spurs	50–100

and recommended sampling times vary with crop. For fruit and nut tree crops, mid-season is the best sampling period, while for other crops, sampling may be recommended at flowering or just prior to or at seed set. These times represent periods in the growth cycle when elemental status could make a difference in plant performance, as well as times when accumulation is at its maximum or in steady state.

Therefore, sampling procedures have evolved from both practical requirements and the association of plant performance with the elemental content of that plant part selected for assay. A listing of

TABLE 1.17 (Continued)

Crop	Stage of Growth	Plant part to sample	No. plants to sample
Strawberry	Midseason	Youngest fully expanded mature leaves	50–75
Pecan	6 to 8 weeks after bloom	Leaves from terminal shoots, taking the pairs from the middle of the leaf	30–45
Walnut	6 to 8 weeks after bloom	Middle leaflet pairs from mature shoots	30–35
Lemon, lime	Midseason	Mature leaves from last flush of growth on non-fruiting terminals	20–30
Orange	Midseason	Spring-cycle leaves, 4 to 7 months old from fruit-bearing terminals	20–30
Grapes	End of bloom period	Petioles from leaves adjacent to fruit clusters	60–100
Raspberry	Midseason	Youngest mature leaves on laterals or "primo" canes	20–40
ORNAMENTALS AND FLOWERS			
Ornamental trees	Current year's growth	Fully developed leaves	30–100
Ornamental shrubs	Current year's growth	Fully developed leaves	30–100
Turf	During normal growing season	Leaf blades. Clip by hand to avoid contamination with soil or other material	½ pt of material
Roses	During flower production	Upper leaves on the flowering stem	20–30
Chrysanthemums	Prior to or at flowering	Upper leaves on flowering stem	20–30
Carnations	(1) Unpinched plants	4th or 5th leaf pairs from base of plant	20–30
	(2) Pinched plants	5th and 6th leaf pairs from top of primary laterals	20–30
Poinsettias	Prior to or at flowering	Most recently mature, fully expanded leaves	15–20

Source: Jones *et al.* (1971).

sampling procedures has been prepared by Jones *et al.* (1971); those for a number of horticultural crops are given in Table 1.17. It is important that these instructions be carefully followed since interpretation of element concentrations are based on the plant part and time of collection specified in the sampling instructions.

Knowing what not to sample is as important as knowing what to sample for plant analyses. Plants that have been under a long period of stress, or disease infested or damaged by insects, and/or are mechanically injured should not be selected for sampling. Plants heavily coated with soil or dust particles, or that have been sprayed and possibly contaminated with Cu, Zn, or other elements, should be avoided unless steps are taken to remove these contaminates. As in soil sampling, areas that are not typical of the general conditions present should be avoided and care should be taken to ensure that the plant tissue selected for assay is truly representative of the site being evaluated.

When plant analyses are being done to diagnose a suspected nutrient deficiency, it is advisable to collect separate samples from both affected plants and plants that are normal in appearance so that a comparison of the analysis results can be made (Munson and Nelson 1973). However, there is some danger in this approach if the deficient plants have been in that condition for an extended period of time, which can affect dry weight and elemental accumulation.

C. Plant Tissue Preparation

Fresh plant tissue is quite perishable and, therefore, must be kept cool and in a drying atmosphere before delivery to the laboratory. It is best to transport plant tissue in clean paper or cloth bags, not in tight containers or plastic bags. If possible, the tissue should be air-dried before shipment, particularly if the time in transit is going to be greater than 24 hrs. Keeping the tissue at a reduced temperature will prevent decay. Any deterioration will result in reduced dry weight, which in turn will affect the analysis result (Lockman 1970).

Fresh plant tissue that is covered with dust, soil particles, or applied spray materials may require decontamination prior to drying. Normally, washing is not recommended unless it is absolutely necessary. Mechanical wiping or brushing may be sufficient to remove large soil particles. Washing the fresh plant tissue in a 0.1–0.3% detergent solution followed by a rinse in pure water can effectively remove most extraneous materials. The effect of different washing procedures on elemental concentrations is shown in Table 1.18. As can be seen, Fe and Mn concentrations tend to be most affected by washing.

Drying of fresh plant tissue is done in a dust-free, forced draft oven at a temperature of 80°C. Drying temperatures less than 80°C may not be sufficient to remove all the moisture, and temperatures above 80°C can result in thermal decomposition. Once dried (which may take 24 hrs. or more), the dried tissue should be stored in a moisture-free atmosphere until further processing is undertaken.

TABLE 1.18 Elemental Content of Unwashed and Washed Apple, Maize, and Orange Leaves

	Apple leaves		Maize leaves				Orange leaves		
Element	Unwashed	Washed with detergent	Unwashed	Wiped with detergent solution	Unwashed	Washed in distilled water	Unwashed	Detergent washed	Detergent plus acid wash
					%				
Ca	—	—	0.87	0.85	0.48	0.45	3.97	3.97	3.96
K	1.78	1.80	2.04	2.17	1.22	1.26	1.07	1.08	1.07
Mg	0.25	0.23	0.16	0.16	0.39	0.41	0.42	0.41	0.42
N	2.22	2.19	—	—	2.93	3.11	2.53	2.56	2.55
P	—	—	0.27	0.29	0.22	0.22	0.15	0.15	0.15
					ppm				
B	28.0	27.0	17.2	17.0	11.0	10.0	367.0	368.0	369.0
Cu	—	—	29.5	28.8	9.0	10.0	5.6	5.1	5.0
Fe	77.0	72.0	136.0	134.0	96.0	85.0	186.0	61.0	61.0
Mn	38.0	36.0	—	—	73.0	64.0	182.0	94.0	92.0
Mo	—	—	—	—	1.1	1.2	—	—	—
Zn	14.1	15.6	23.1	30.5	22.0	22.0	123.0	68.0	65.0

Source: Jones and Steyn (1973).

Plant tissues high in soluble sugars are not easily oven-dried. Therefore, moisture is removed from these tissues by either freeze drying or vacuum–oven drying (Horwitz 1980a).

In order to reduce the dried plant tissue to a particle size suitable for laboratory analysis and also to ensure a greater degree of uniformity in sample composition, the tissue is mechanically ground or crushed. This can be done by cutting the tissue with a Wiley or hammer mill, or by crushing it in a ball mill. Particles from the contact surfaces of most mills will contaminate the sample: for example, Cu and Zn from brass fittings, and even Fe, when fittings and the cutting and crushing surfaces are made of tool or stainless steel. Therefore, to avoid Fe contamination during sample size reduction, dried tissue should be cut by hand or crushed in an agate mortar and pestle or ball mill. If the accuracy of the Fe determination is not of consequence, then grinding in a mill with tool or stainless steel fittings is suitable.

A Wiley mill fitted with a 20-mesh screen is commonly used for grinding tissue samples. The finer the screen (40- or 60-mesh), the more homogeneous the sample will be; however, contamination with elements from the cutting surfaces is greater with finer screens due to the longer contact time between the tissue sample and the cutting mill itself (Hood *et al.* 1944). During the grinding of dry plant material, segregation of the finer particles must be controlled by eliminating static electricity buildup and particle segregation (Nelson and Boodley 1965; Smith *et al.* 1968).

Some types of samples are not easy to grind due to the presence of pubescence (such as on apple leaves) or because the tissues are very fibrous or highly deliquescent. Special care is required when reducing the particle size of these types of tissues.

D. Organic Matter Destruction

Probably no other aspect of plant tissue preparation prior to elemental analysis has stirred as much controversy as that which surrounds how best to destroy the organic matter portion of tissue. The best treatises on the subject are the books by Gorsuch (1970) and Block (1979). These two authors describe in considerable detail the advantages and difficulties associated with the more commonly used organic matter destruction procedures.

The two most common decomposition procedures are wet oxidation and dry ashing, the latter being frequently criticized for several reasons. When dry ashing, elemental loss by volatilization can occur, as well as elemental acclusion to silica particles in the sample. Wet oxidation, on the other hand, is the longer and more tedious of the two procedures with greater potential for sample contamination and danger to the analyst. However, B is loss due to volatilization in most wet-oxidation procedures, requiring the analyst to use the dry-ashing procedure. Gorsuch (1976) has given a comparison between wet oxidation and dry ashing (Table 1.19).

Numerous wet-oxidation procedures have been proposed, but they all make use of one of three basic digestion reagents: (1) nitric and sulfuric acids, (2) sulfuric acid and hydrogen peroxide, or (3) mixtures containing perchloric acid.

With nitric and sulfuric acid digestion, which is suitable for a wide range of sample types, most all elements except Se are recovered. Problems may occur with samples high in Ca due to the formation of calcium sulfate precipitates that cause losses due to coprecipitation.

Sulfuric acid and hydrogen peroxide digestion is a vigorous procedure with potential for losses of some elements in the presence of chlorides. The amount of sample handled can affect the procedure if the sample is high in involatile hydrocarbons. The user needs to experiment with the procedure to determine the amounts of each component required to complete the digestion. Wolf (1982b) has given instructions for this technique of organic matter destruction for plant tissue.

A mixture of nitric (sometimes sulfuric) and perchloric acids is a widely used digestion reagent, although its use requires extreme care and a specially designed hood. Hot perchloric acid in the presence of easily oxidizable organic matter can be an explosive mixture. A list of the necessary precautions to be observed when using perchloric acid may be found in the AOAC Manual (Horwitz 1980b). Tolg (1974) has listed the characteristics of the common wet-oxidation procedures (Table 1.20).

It is becoming common practice to carry out a wet oxidation using a block digestor, which can be obtained commercially or constructed by the analyst (Gallaher *et al.* 1975). With the proper control devices, the temperature during digestion can be carefully regulated, avoiding

TABLE 1.19 Comparison of Wet-Oxidation and Dry-Ashing Techniques

Wet oxidation	Dry ashing
More rapid	Rather slow
Temperature lower; less volatilization and retention	Temperature higher; more volatilization and retention
Generally less sensitive to nature of sample	Generally more sensitive to nature of sample
Relatively more supervision	Relatively less supervision
Reagent blank larger	Reagent blank smaller
Large samples inconvenient	Large samples easily handled

Source: Gorsuch (1976).

TABLE 1.20 Common Methods of Wet Oxidation

Digestion reagent	Applicability to organic matrix	Remarks
H_2SO_4/HNO_3	Vegetable origin	The most used; danger of volatilization of As, Se, Hg, etc.
H_2SO_4/H_2O_2	Vegetable origin	Pb loss on co-precipitation with $CaSO_4$; loss of Ge, As, Ru, Se, etc.
HNO_3	Biological origin	Easily purified reagent; digestion temp. 350° C; short digestion time; soluble metal-nitrates
$HClO_4$	Biological origin	Catalysts; $(NH_4)_2MoO_4$, etc.
$H_2SO_4/HClO_4$	Biological origin	Suitable only for small samples; danger of explosion
$HNO_3/HClO)_4$	Protein, carbohydrate (no fat)	Less explosive; no loss of Pb
$H_2SO_4/HNO_3/HClO_4$	Universal (also fat and carbon black	No danger with exact temp. control; As, Sb, Au, Fe, etc., are volatile ash under reflux under some conditions

Source: Tolg (1974).

losses due to rapid boiling and excessive temperatures. An oxidation procedure using a mixture of nitric and perchloric acids and employing a digestion block is given by Zasoski and Burau (1977). Halvin and Soltanpour (1980) have also described a wet-oxidation procedure using a digestion block but with nitric acid alone. The success of their technique may be due in part to the pretreatment step during which the plant tissue and nitric acid are allowed to interact overnight before the high temperature (125°C) digestion is begun.

Wet oxidation under pressure can be performed, placing sample and digestion acids into a Parr Bomb (Vigler *et al.* 1980), or placing sealed ampules into an autoclave at 125°C under pressure (Sung *et al.* 1984). Both techniques are useful for biological materials difficult to digest completely without pressure.

Organic matter destruction can also be accomplished by high-temperature dry ashing. The critical requirements are (1) the nature of the ashing vessel, (2) its placement in the muffle furnace, (3) ashing temperature, and (4) time.

Although the shape and size of the ashing vessel are not generally specified, I recommend the use of high-walled vessels sufficiently large to adequately accommodate the sample while keeping the sample depth in the vessel minimal. Covering of the sample vessel is not recommended unless ashing is done in a forced circulating air muffle furnace. Silica (quartz) is probably one of the best materials for the ashing vessel, although pyrex glass high-form beakers (not satisfactory if B and Na are to be assayed) and well glazed acid-washed porcelain high-form crucibles are equally suitable vessels for most uses. Munter *et al.* (1984) found small additions of B, Cu, Fe, and Mn, and a large addition of Na to corn tissue ashed in porcelain as compared with samples ashed in quartz crucibles, whereas the elements Al, Ca, Cr, K, Mg, Ni, and P were unaffected.

Ashing aids may be used to assist in the decomposition of the organic matter. Gorsuch (1970) recommends 10 ml of 10% sulfuric acid or 10 ml of 7% magnesium nitrate/5-g sample as ashing aids. The AOAC procedure for dry ashing calls for nitric acid as an ashing aid (Horwitz 1980c). However, with this procedure, there may be some loss of B after the ash is treated with nitric acid, the residue heated, and the ashing vessel placed back into the muffle furnace. The use of an ashing aid is more important for highly carbonaceous tissues in order to obtain complete organic matter destruction. The analyst should experiment with typical tissue samples to determine if an aid is needed to obtain an ash relatively free of carbon residue.

To prevent flaming of the tissue sample, the muffle furnace temperature must be increased slowly and the furnace door kept closed. Samples should never be placed in a hot muffle furnace.

According to Gorsuch (1970), as long as the ashing temperature does not exceed 500°C, volatilization losses should not occur for most elements. This conclusion is in close agreement with the results obtained by Isaac and Jones (1972) who studied ashing temperature and time effects on the determination of 13 elements in 5 different plant tissues. Too low an ashing temperature may result in incomplete destruction of organic matter, which blocks complete recovery of elements from the organic matrix. Therefore, the minimum ashing temperature is probably about 490°C.

Although the effect of the position of the ashing vessel in the muffle furnace has not been shown to be critical, I use a tray that suspends the vessel in the furnace and keeps it from contacting the muffle furnace floor, which is a heated surface. Also, analysts have frequently observed that samples near the furnace door may look "darker" due to less oxidation than those positioned in the center. Therefore, the analyst should determine if vessel position has any effect with the muffle furnace being used and make adjustments as required to ensure complete organic matter destruction.

The procedure used to solubilize the ash remaining after oxidation may also affect the elemental analysis. The common procedure is to take up the ash in moderately concentrated nitric or hydrochloric acid (nitric being preferred since it is less corrosive), at room temperature. However, acid dissolution at room temperature may not completely release Al, Cr, and Fe (Dahlquist and Knoll 1978). Munter and Grande (1981) and Munter *et al.* (1984) have reviewed some of the more important literature on this subject. They suggest that the ash be heated in an aliquot of 2 *N* hydrochloric acid till fumes of the acid evolve and then be allowed to cool, bringing the digest back to the original volume with pure water.

What effect the silica content in plant tissue has on the retention of heavy metals (e.g., Zn in rice tissue) also needs to be determined. It has been suggested that for those tissues high in silica, wet oxidation is the only satisfactory method for organic matter destruction unless silica is removed by treatment of the ash with hydrofluoric acid.

Since there is so much contradictory literature on dry ashing as a method of organic matter destruction, dry ashing is viewed by some as too risky for precise analytical work. It is evident that better defined techniques are needed before a significant degree of confidence can be placed in the dry-ashing method of organic matter destruction. The technique described by Munter and Grande (1981) should be followed as the best dry-ashing procedure.

Low-temperature dry ashing can be used to retain the more volatile elements in samples during the destruction of organic matter. Gleit and Holland (1962) introduced the basic procedure of ashing in the presence of electronically excited oxygen. The ashing temperature is 100°-150°C; at these low temperatures, ashing for as long as 5 days may be required to completely destroy the organic matter.

It is doubtful that the controversy over which procedure for destroying organic matter is best will ever be fully resolved. There is ample evidence that both wet oxidation and dry ashing give reasonably comparable results for most elements and organic matrices. In most instances, it is the skills and preference of the analyst that determine which method is employed. Therefore, the analyst should try both wet-oxidation and dry-ashing techniques on known samples and then select the technique that gives the expected results.

Extraction by 2% acetic acid (Ulrich 1948) to determine acid-soluble P in plant tissue may be a more suitable method for determining the status of other elements than is total content analysis following organic matter destruction. Elemental determination by extraction would simplify sample preparation, eliminating the need for either wet oxidation or dry ashing. Nickolas (1957) obtained good correlations between results obtained by extraction and by total content analysis for the elements Ca, K, Mn, and P. Elemental determination by extraction with water or dilute acid could be viewed as a form of tissue testing, a topic discussed more fully in Section IV. However, the evaluation of plant nutrient status based on the ratio of extractable to total nutrient concentration seems worthy of consideration and will be discussed in more detail later.

E. Laboratory Analysis

Advances in analytical chemistry in the last 20 years—particularly so-called "instrumental methods"—have had a greater impact on plant analysis than on soil testing. Traditional wet-chemistry procedures (Piper 1942; Jackson 1958) and even atomic absorption spectroscopy (Isaac and Kerber 1971) have not been universally accepted for elemental determination in plant tissue. As in soil testing, the Technicon AutoAnalyzer can be used to determine Ca, Mg, P, and K in plant tissue digests (Steckel and Flannery 1971), although this instrumentation has not been widely used. X-ray emission spectroscopy (Kubota and Lazar 1971) is another technique suitable for elemental assay of plant tissues. Although this technique is nondestructive, matrix effects have seriously hampered its acceptance and wide use.

By far the most common method for determining the elemental content of plant tissue has been emission spectroscopy. A number of excitation sources have been used, from AC and DC arcs (Mitchell 1956), to AC spark (Jones 1976), and most recently inductively coupled (Dahlquist and Knoll 1978; Jones 1977; Munter and Grande 1981; Soltanpour *et al.* 1982) or DC plasmas (DeBolt 1980), with either single, scanning, or multielement fixed-slit spectrometers. Most elements, in trace amounts or larger concentrations, can be easily and quickly determined in plant tissue digests by plasma emission spectroscopy. If the spectrometer is a vacuum model, S can also be determined. High speed, multielement capacity, and computer or microprocessor control are the analytical features sought by most analysts today, and a plasma emission spectrometer (Jones 1977; Munter and Grande 1981; Soltanpour *et al.* 1982) offers these features for doing plant analyses.

1. Nitrogen. The Kjeldahl digestion procedure dates back to the late 1800s, and the first paper describing the procedure was published in 1883 (Morries 1983). Since then, hundreds of papers have been written proposing numerous modifications to speed the analysis, to improve N recovery, and to increase the accuracy and precision of the method. A Kjeldahl digestion involves two steps: high-temperature (330°–350°C) digestion in concentrated sulfuric acid in the presence of a catalyst that converts organic N to inorganic ammonium-N; and then, determination of the formed ammonium. Depending on the weight of sample taken for analysis, a Kjeldahl digestion is classed as being either macro (1 g or greater), semimicro, (0.5–1.0 g), or micro (less than 0.5 g). Normally, precision declines with tissue sample weight due to sample heterogeneity, which can be influenced, to some degree, by the size of sample particles. The shift from macro- to micro-procedures has occurred in order to reduce the amount of laboratory space and reagents needed. Micro-Kjeldahls can be carried out with a digestion block (Gallaher *et al.* 1975) followed by ammonium-N determination with an AutoAnalyzer (Isaac and Johnson 1976), specific-ion electrode (Gallaher *et al.* 1976), or automated distillation (Munsinger and McKinney 1982).

An excellent review of the Kjeldahl digestion procedure is given by Nelson and Sommers (1980), who discuss the various modifications of the technique that have been proposed. The standard digestion/distillation procedure is given in the AOAC Manual (Horwitz 1980d); a semi-automated procedure using a digestion block and Technicon AutoAnalyzer also is described in the Manual (Horwitz 1980e).

The term "Kjeldahl N" can be misleading as the determination may or may not include nitrate-N; and therefore, without the inclusion of nitrate, the kjeldahl result cannot be called "total N." Unfortunately, the literature is confusing on this issue as most published nitrogen values are probably without nitrate inclusion.

Near infrared reflectance (NIR) spectroscopy looks promising as a nondestructive method for N determination in dried ground plant tissue (Dorsheimer and Isaac 1982). Isaac and Johnson (1983) have compared N concentrations in corn leaf tissue determined by Kjeldahl and by NIR using a Technicon InfraAlyzer 400 instrument. Nitrogen concentrations determined by the two methods were found to be within 0.1% of each other for 90% of the samples and within 0.2% for all samples.

Another recently proposed procedure for N determination in plant tissue is direct distillation.[2] The technique requires precise control of the distillation process in terms of alkali concentration, time, and temperature because the N content in plant tissue is determined indirectly based on the release of amide-N times a factor determined by correlation with Kjeldahl N.

2. Sulfur. Following wet oxidation, which converts organic S to sulfate-S, the sulfate can be determined by either spectrophotometric or turbidimetric procedures (Beaton *et al.* 1968). The turbidity method can be automated using a Technicon AutoAnalyzer as described by Wall *et al.* (1980). Sulfur also can be determined by combustion tech-

---

[2] Rapid determination of protein content in grains by using the Kjeltec Auto System DD. Tecator Application Note AN 33/81. Tecator, Inc., P.O. Box 405, Herndon, VA 22070.

niques (Bremner and Tabatabai 1971), first proposed by Jones and Isaac (1972) and more recently by Hern (1984), using a LECO Sulfur Analyzer equipped with either a titrator or infrared sulfur dioxide detector. The AOAC Manual (Horwitz 1980f) offers the analyst two alternative sample preparation procedures to form sulfate-S and then its determination by the barium sulfate gravimetric procedure.

3. Methods of Expressing Analytical Results. The elemental concentration in plant tissue is normally expressed as a percentage on a dry weight basis for the major elements—N, P, K, Ca, Mg, and S—and as parts per million (ppm) for all other elements. There seems to be a trend away from designating concentration in parts per million and substituting micrograms/gram (μg/g). In some publications, concentrations are expressed in milliequivalents. Since most of the past and current literature expresses elemental concentration as a percentage for the major elements and parts per million for the micronutrients, this convention will be followed in this review.

F. Interpretation

Interpretation of the results of plant (leaf) analyses is based on the relationship between elemental concentrations in the analyzed plant part and growth or yield. A graphic representation of this relationship

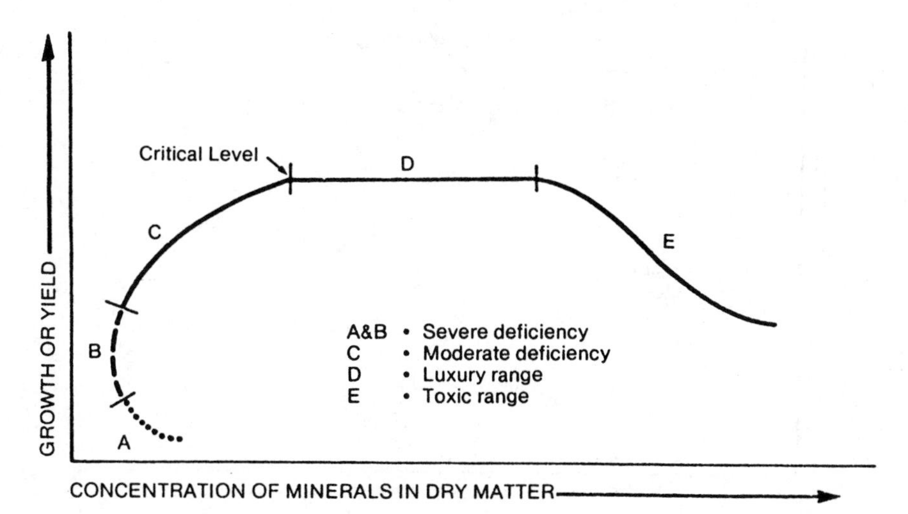

FIG 1.9. General relationship between plant growth or yield and elemental content of the plant.
From Smith (1962).

was first given by Prevot and Ollagnier (1956) and later by Smith (1962). Figures 1.9, 1.10, and 1.11 graphically display the relationship between nutrient concentration in the plant and yield or growth.

The major difference among these curves is in their slopes, i.e., the relative change in growth or yield as the elemental concentration increases from deficiency to sufficiency. In Fig. 1.10 (Ulrich 1961) and Fig. 1.11 (Ulrich and Hills 1973), there is a large, linear increase in plant growth with a very small increase in elemental concentration. In contrast, Fig. 1.9 (Smith 1962) shows a smaller change in growth over a larger range of elemental concentrations. When the plant is severely deficient, there is an increase in elemental concentration with a corresponding yield increase, referred to as the Steenbjerg effect because of its initial description by Steenbjerg (1954). I believe that Fig. 1.9 best represents the relationship between plant growth and the concentration of the major elements, whereas Fig. 1.10 best represents the relationship for the micronutrients.

The nature of these plant response vs elemental concentration curves presents several problems for the interpretation of a plant analysis result. Elemental concentrations in plants exhibiting symptoms of extreme deficiency may be greater than those found in plants free of deficiency, the Steenbjerg Effect (Fig. 1.9, part A). The range in

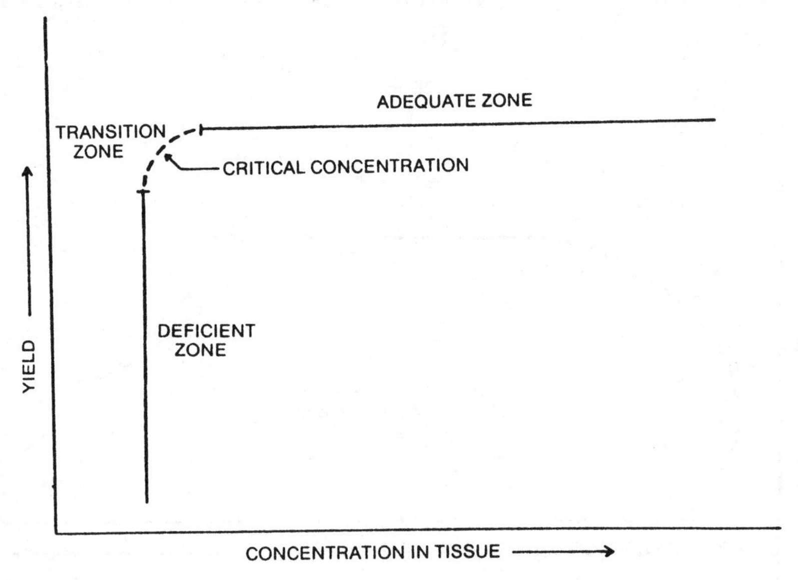

FIG 1.10. General relationship between plant growth or yield and elemental content of the plant.
From Ulrich (1961).

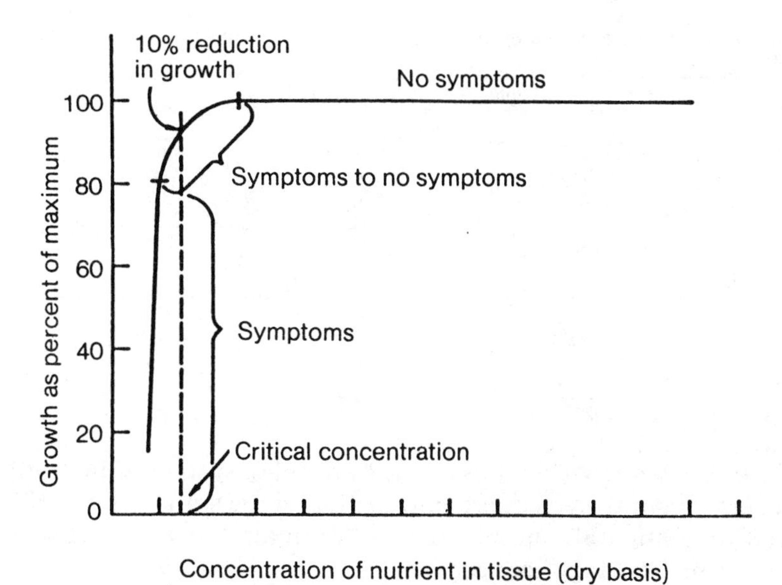

FIG 1.11. Relationship between plant growth or yield and elemental content of the plant.
From Ulrich and Hills (1973).

element concentration between deficiency (with visual symptoms) and sufficiency (no visual symptoms) can be small, as shown in Fig. 1.10. The difficulties in detecting these small differences with current sampling and analysis techniques may make interpretation of a plant analysis result unreliable. Since the elemental concentration is based on the dry weight of tissue, the effects of an uncoordinated relationship between element and dry weight accumulation either increases (concentrates) or decreases (dilutes) the elemental concentration. Elemental accumulation in excess of dry weight accumulation explains the Steenbjerg effect, while the coordinated increase in elements, particularly the micronutrients, and dry weight explains the steep left-hand character of the curves in Fig. 1.10 and 1.11.

Three techniques have been devised for interpreting a plant analysis result. These are based on the use of a *standard value, critical level,* or *sufficiency range.*

Kenworthy (1961) developed an interpretive system for fruit trees based on standard values obtained by the elemental analysis of large numbers of leaf samples taken from normal productive orchards. A standard value then serves as a guide for interpreting the results of other analyses. Standard values for apple, peach, and sour cherry in Michigan plantings are given in Table 1.21. Based on such standard

TABLE 1.21 Standard Values of Elemental Concentration for Fruit Trees in Michigan Plantings

Element	Apple	Peach	Sour cherry
		%	
N	2.33	3.87	2.95
P	0.23	0.26	0.25
K	1.53	1.68	1.67
Ca	1.40	2.12	2.09
Mg	0.41	0.67	0.68
		ppm	
Mn	98	151	150
Fe	220	166	203
Cu	23	18	57
B	42	48	50

Source: Kenworthy (1961).

values, Kenworthy (1961) developed an index system of interpretation, establishing percentage ranges around the standard value that define an elemental concentration as "normal," "above or below normal," or in "shortage" or "excess."

A critical level is the elemental concentration below which deficiency occurs and above which sufficiency exists; this level is graphically shown in Fig. 1.9 and 1.10. Critical levels have been widely published in the literature, although there is considerable disagreement about how to determine them and their values (Smith 1962). Beutel *et al.* (1978) have published critical levels for eight elements in a number of fruit and nut tree species (Table 1.22). It is important to note the specific time of sampling (July) and plant part analyzed, as these factors can influence critical levels.

Both standard values and critical levels suffer from the fact that they are single values representing an optimum concentration; thus, they do not describe the possible range in concentration that can sustain maximum plant growth. In addition, excessive element concentrations that are toxic may be as important to know as concentrations in the deficiency range. Therefore, expressing elemental concentrations as a range—from the concentration at the point just above deficiency to the concentration at that point beyond which excess or toxicity occurs—is the common practice. The limits of such a Sufficiency Range can be viewed as critical values, below which deficiency occurs and above which excess or toxicity occurs. It would be safe to say that we know far more about the dividing point in element concentration between deficiency and sufficiency than between sufficiency and excess (toxicity).

Sufficiency ranges have been published for many different crops. A major source of such elemental concentration ranges may be found in

TABLE 1.22 Critical Levels of Elements in the Leaves of Fruit and Nut Trees (July Samples)[1]

	%									ppm			
	N[2]		K[3]		Ca	Mg	Na[4]	Cl[4]	B				Zn
	Defic.	Adequate	Defic.	Adequate	Adequate		Excess		Defic.		Excess		Adequate
Crops	below	over	below	Over	Over	Over	Over	Over	below	Adequate	over		over
Almonds	2.0	2.2–2.7	1.0	1.4	2.0	0.25	0.25	0.3	25	30–65	85		18
Apples	1.9	2.0–2.4	1.0	1.2	1.0	0.25	—	0.3	20	25–70	100		18
Apricots (ship)	1.8	2.0–2.5	2.0	2.5	2.0	—	0.1	0.2	15	20–70	90		16
(can)	2.0	2.5–3.0	2.0	2.5	2.0	—	0.1	0.2	15	20–70	90		16
Cherries (sweet)	—	2.0–3.0	0.9	—	—	—	—	—	20	—	—		14
Figs	1.7	2.0–2.5	0.7	1.0	3.0	—	—	—	—	—	300		—
Olives	1.4	1.5–2.0	0.4	0.8	1.0	0.10	0.2	0.5	14	19–150	185		—
Nectarines and peaches (freestone)	2.3	2.4–3.3	1.0	1.2	1.0	0.25	0.2	0.3	18	20–80	100		20
Peaches (cling)	2.4	2.6–3.5	1.0	1.2	1.0	0.25	0.2	0.3	18	20–80	100		20
Pears	2.2	2.3–2.8	0.7	1.0	1.0	0.25	0.25	0.3	15	21–70	80		18
Plums (Japanese)	—	2.3–2.8	1.0	1.1	1.0	0.25	0.2	0.3	25	30–60	80		18
Prunes	2.2	2.3–2.8	1.0	1.3	1.0	0.25	0.2	0.3	25	30–80	100		18
Walnuts	2.1	2.2–3.2	0.9	1.2	1.0	0.3	0.1	0.3	20	36–200	300		18

Source: Beutel *et al.* (1978).

[1] Leaves are from nonfruiting spurs on spur-bearing trees, fully expanded basal to midshoot leaves on peaches and olives, and terminal leaflet on walnut. Adequate levels for all fruit and nut crops: P, 0.1–0.3%; Cu, over 4 ppm; Mc, over 20 ppm.

[2] Percentage N in August and September samples can be 0.2–0.3% lower than July samples and still be equivalent. Nitrogen levels higher than *underlined* values will adversely affect fruit quality and tree growth. Maximum N for Blenheims should be 3.0% and for Tiltons 3.5%.

[3] Potassium levels between deficient and adequate are considered "low" and may cause reduced fruit sizes in some years. Potassium fertilizer applications are recommended for deficient orchards but test applications only for "low" K orchards.

[4] Excess Na or Cl causes reduced growth at the levels shown. Leaf burn may or may not occur when levels are higher. Confirm salinity problems with soil or root samples.

Chapman's book (1966). Other compilations for various crops have been published by Goodall and Gregory (1947), Wallace (1961), Childers (1966), Chapman (1967), Neubert *et al.* (1969), Jones (1974), and Reisenauer (1978). Additional plant analysis interpretive data for pecans (Worley 1969; Jones 1972b), vegetable crops (Geraldson *et al.* 1973), orchard crops (Kenworthy 1973; Atkinson *et al.* 1980), deciduous tree fruits and nuts (Shear and Faust 1980), citrus (Embleton *et al.* 1978), ornamental and flowering crops (Criley and Carlson 1970), and ornamental greenhouse crops (Joiner *et al.* 1983) have been published. Interpretive data for specific elements, such as K for tree crops (Koo 1968) and vegetable crops (Lucas 1968), and for the micronutrients (Jones 1972a) have been compiled. These are only a few of the many useful references describing techniques for interpreting a plant analysis result.

When using any of these elemental concentration values, whether they be standard values, critical levels, or sufficiency ranges, the plant species, plant part collected, and time of sampling must match that specified in the procedure. Any deviation in sampling procedure invalidates the use of these concentration values for interpretation of plant analyses.

As with soil testing, the interpretation of plant analysis results depends considerably on the skill of the analyst and, to some degree, on how results are to be utilized. When plant analysis is done to verify a suspected elemental insufficiency, tables of interpretive concentration levels are essential, even if only a comparison is to be made between two sets of results from normal and abnormal appearing plants found in the same field. The goal of most plant analyses done today fall into this category.

Monitoring is the least understood and used, yet the most valuable, application of plant analysis. Jones (1983) has described a system for achieving nutrient element sufficiency by the utilization of soil tests and plant analyses. By a system of periodic sampling and tracking of the analysis results, the soil-plant chemical environment can be monitored, with lime and fertilizer requirements developed based on the goal of establishing and maintaining a prescribed nutrient element status for both the soil and plant. A more thorough discussion of the technique is given by Jones and Budzynski (1980). A similar system of tracking (data logging) has been proposed by Clements (1960) for sugarcane, a perennial crop, and by Jones *et al.* (1980) for peanut, an annual crop. Tracking probably is of greater value for perennial crops than annuals, unless a monoculture system is generally followed. This system of analysis and tracking requires consistent sampling and standardized analytical procedures for the soils and plant tissues, and specific designation of the limits for the sufficiency range of each analytical parameter. The goals are nutrient element sufficiency and the achievement of maximum yield and quality potentials.

A relatively new technique for the interpretation of plant analysis results is the Diagnosis and Recommendation Integrated System (DRIS), first proposed by Beaufils (1973), who had been developing the concept for some 20 years previously. Sumner (1982) has prepared a review of the technique as it is applied to the interpretation of both soil and plant analysis results; he states that "this system (DRIS) represents an holistic approach to the mineral nutrition of crops and is, in fact, an integrated set of norms representing calibrations of plant tissue composition, environmental calibrations of plant tissue composition, soil composition, environmental parameters and farming practices as functions of yield of the particular crop." He also states that the most important advantages of the DRIS approach are its ability to (1) make a diagnosis at any stage of the crop's development, and (2) list the nutrient elements in their order of limiting importance on yield.

The DRIS approach employs a survey technique to establish norms for the crop, the objective being to establish a data bank that can be applied to the interpretation of soil and plant analysis unknowns. DRIS norms have been established for a variety of crops, including sugarcane (Beaufils and Sumner 1976), potatoes (Meldal-Johnsen 1975), soybeans (Sumner 1977a), wheat (Sumner 1977b), pineapples (Langenegger and Smith 1978), sorghum (Sumner *et al.* 1983), alfalfa (Erickson *et al.* 1982), and sunflower (Grove and Sumner 1982). Letzch and Sumner (1983) have just recently prepared a computer program for calculating DRIS indices on a microcomputer for most of the crops listed above. DRIS norms need to be developed for the common horticultural crops.

Sumner (1977c) discusses in some detail the DRIS approach with crops at high yield levels, using corn as the crop for illustration, and the elements N, P, and K. Corn was chosen since there is a very large quantity of analytical information in the literature for this crop. From the large number of values available in the literature, indices were determined for each of the elements and their mean ratios, which form the apex for the DRIS chart shown in Fig. 1.12. The chart then becomes the basis for comparison with unknowns for diagnostic evaluation. The sign (positive or negative) of the index value defines its level of sufficiency and the size of the value, the degree of significance from the norm.

The advantages of the DRIS approach over the other conventional techniques for plant analysis interpretation are its ability to make

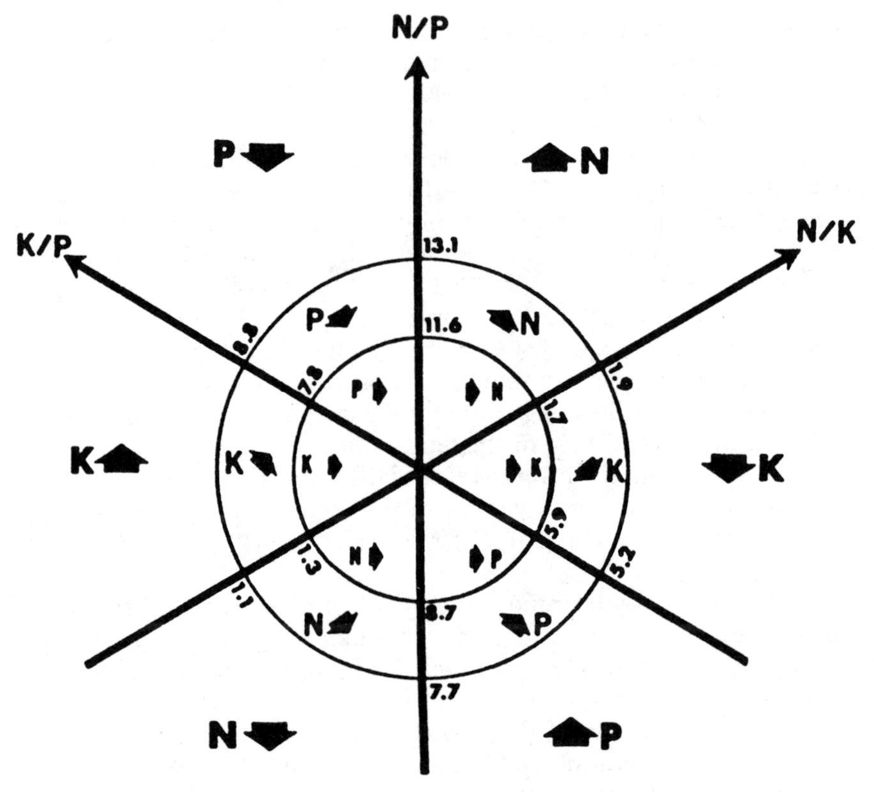

FIG 1.12. DRIS chart for obtaining the qualitative order of requirement for NPK in corn. Means of significant expressions (values at origin) are: N/P = 10.04, N/K = 1.49, K/P = 6.74.
From Sumner (1977a).

diagnoses over a range of sampling times and cultivars and to rank elements in order of limiting importance. Currently, DRIS systems are limited to a few crops since data bases for most are inadequate to establish the elemental concentration norms required to establish interpretive indices.

Although yield is of significance with all crops, quality is of particular importance with many horticultural crops and is related to the crop's physical appearance, storage life, and marketability. Elemental insufficiencies can cause various metabolic disorders that reduce product quality. Calcium has been widely studied because of its marked influence on the quality of many fruits and vegetables (Shear 1975; Millaway and Wiersholm 1979). Bitter pit in apple and blossom-end rot (BER) in tomato and other fruits have been intensively studied for Ca-related nutritional disorders. Similarly, the N and K status of a plant can affect its shelf-life as an ornamental plant or the quality of fruits produced. Even the micronutrients B and Mn have been found to produce undesirable characteristics in apple fruits (Boynton and Oberly 1966; Shear and Faust 1980).

These are only few examples of the consequences that nutrient element stress can have on characteristics other than yield. The question is whether soil tests or plant (leaf) analyses can be used to predict the possible occurrence of these metabolic disorders that affect plant and fruit quality. It is a topic not well addressed. In many instances, metabolic disorders result from a combination of events relating to the environment and cultural practices (Wiersum 1979) and, as such, are not easily evaluated by a single factor, such as a soil test or plant analysis (Kirkby 1979; Shear and Faust 1980).

IV. TISSUE TESTING

A distinction is being made in this review between plant (leaf) analysis and tissue testing: the former involves an analysis for the total elemental content; the latter involves an assessment of the elemental content of sap from fresh tissue or the elemental content determined by an extraction of either fresh or dried plant tissue. Ulrich (1948) describes a tissue-testing procedure by extraction of dried plant tissue with water for the determination of nitrate and chloride, and extraction with 2% acetic acid for P as the phosphate anion and K. Gallaher and Jones (1976) used a similar extraction procedure to evaluate the elemental status of pecan leaves, and Spenser *et al.* (1978) found that the ratio between total S and extractable sulfate-S was a better indicator of S status than either determination alone. An evaluation system of plant analysis based on extraction of elements from dried tissue may be promising, eliminating the need for organic matter destruction, and may provide a better means of evaluating nutrient element status.

Krantz *et al.* (1948) give instructions for the field testing of maize, cotton, and soybean plants using sap pressed from fresh tissue for the semiquantitative determination of nitrate, phosphate, and K employing test papers, vials, and color charts. Wickstrom (1967) has discussed the use of such tissue tests for field diagnosis. A test kit is commercially available for conducting such tests.[3] Syltie *et al.* (1972) have given procedural details for conducting tissue tests in the field

[3] PLANT CHECK, 4200 Woodville Pike, Urbana, OH 43078.

for maize and soybeans for N, P, K, Mg, and Mn. They give instructions for the preparation of required reagents and for conducting the tests. They correlated the results obtained in such tissue tests with results from plant analyses of the same plants, as well as with other site parameters. Correlations between the tissue tests for nitrate, phosphate, and K and plant analysis determinations were significant for maize, but only for K in soybeans.

Lorenz and Tyler (1978) have given deficiency and sufficiency levels for nitrate, phosphate, and K in petioles for 17 vegetable crops; a similar listing by Lorenz and Bartz (1968) for seven vegetables is given in Table 1.23. The effect of time of sampling on petiole nitrate, phosphate, and K concentrations for potato is shown in Fig. 1.13. All of these data on tissue levels for nitrate, phosphate, and K can be used to evaluate the nutritional status of the crop as well as form the basis for prescribing supplemental fertilizer applications to correct insufficiencies.

Since N does have a dramatic effect on plant growth, tissue testing for nitrate may be the more useful in the management of crops than is testing for phosphate and K. The tissue nitrate test serves to guide supplemental applications of N fertilizer in order to avoid N insufficiencies. Some have suggested that extractable nitrate content in the plant may be a better indicator of N status than total N determined by Kjeldahl digestion. Maynard *et al.* (1976) have reviewed the literature on nitrate accumulation in vegetables, listing the critical nitrate levels for eight different vegetables (Table 1.24). El-Sheikl *et al.* (1970) determined the critical nitrate level in petioles of squash, cucumber, and

TABLE 1.23 Tissue Test Guide for Sampling Time, Plant Part, and Deficiency Levels for Seven Vegetable Crops

Crop	Time of sampling	Plant part	Deficiency levels		
			NO_3–N (ppm)	PO_4–P (ppm)	K (%)
Cabbage	At heading	Leaf midrib of wrapper leaf	5000	2500	2
Celery	At midgrowth when 12–15 in. tall	Petiole of youngest fully elongated leaf	5000	2000	4
Lettuce	At heading	Leaf midrib of wrapper leaf	4000	2000	2
Melons	Early fruit set	Petiole of 6th leaf from growing tip	5000	1500	3
Potatoes	Midseason 35–45 days from emergence	Petiole of 4th leaf from growing tip	6000	800	7
Tomatoes	Early bloom	Petiole of 4th leaf from growing tip	2000	1500	1.5
Sweet corn	At tasseling	Leaf midrib of 1st leaf above primary ear	500	500	2

Source: Lorenz and Bartz (1968).

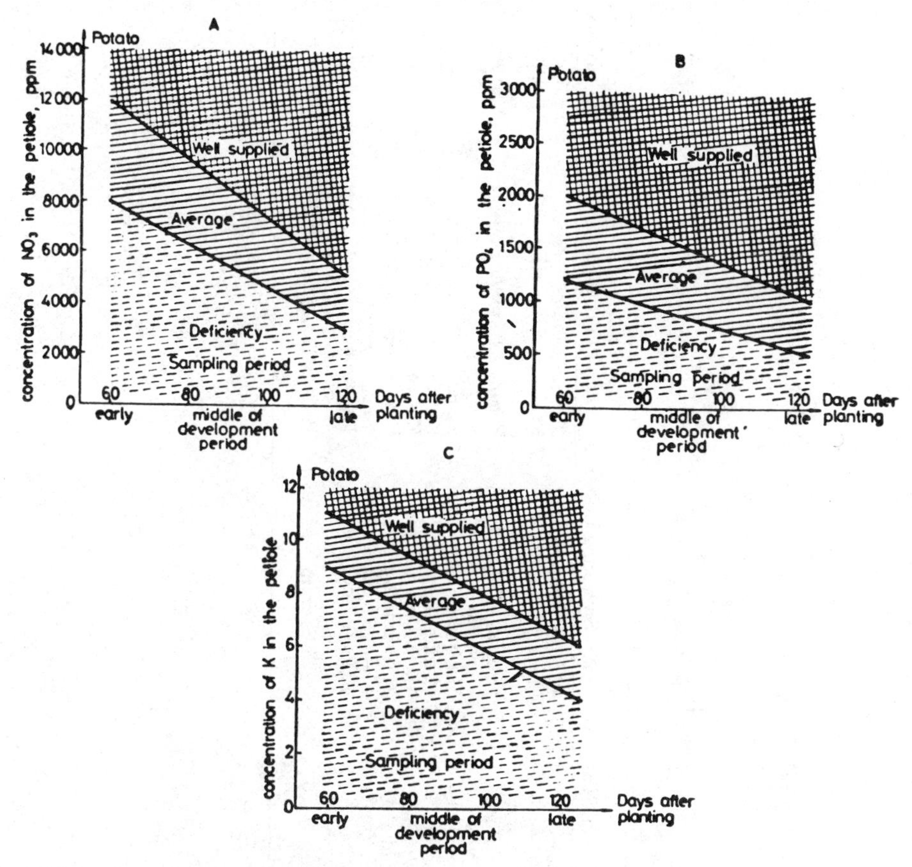

FIG 1.13. Effect of sampling time on petiole nitrate (A), phosphate (B), and potassium (C) concentration for potato.
From Davidescu and Davidescu (1982).

melon to be 1000, 2000, and 3000 ppm, respectively, on a dry weight basis. Their response curve for squash is shown in Fig. 1.14. Woodson and Boodley (1983) conducted a similar study with geranium, obtaining the response curve shown in Fig. 1.15. The critical nitrate concentration in geranium petioles was five times that in squash, although the concentration (about 14,000 ppm) at which nitrate became excessive was about the same for both species. Prasad and Spiers (1982) found a nitrate sap test to be a good indicator of the correlation between current growth rate and N nutrition for seven different ornamental plants.

The test for nitrate in plant tissue can be conducted in the field on sap extracted from fresh plant material, or tissue may be taken into

TABLE 1.24 Critical Nitrate-N Concentrations at a 10% Growth Restriction for Various Vegetable Crops

Crop	Species	Cultivar	Sampling time	Plant tissue	Critical Conc. % NO_3–N dry wt
Cucumber	*Cucumis sativus* L.	Cubit	42 days from seeding	Mature petioles	0.200
Lettuce	*Lactuca sativa* L.	Black-seeded Simpson	Market maturity	Entire aerial portion	0.200
Potato	*Solanum tuberosum* L.	White Rose	18 days vegetative growth	Petiole 2 from terminal	0.200
Radish	*Raphanus sativus* L.	Cherry Belle	Market maturity	Root	0.500
Spinach	*Spinacia oleracea* L.	America	Market maturity	Entire aerial portion	0.170
		Heavy Pack	Market maturity	Entire aerial portion	0.150
		Hybrid 424	Market maturity	Entire aerial portion	0.045
Squash	*Cucurbita pepo* L.	Black Zucchini	42 days from seeding	Mature petiole	0.100
Sweet melon	*Cucumis melo* var. Naud.	Ginza 4	42 days from seeding	Mature petiole	0.300
Tomato	*Lycopersicon esculentum* Mill.	VF145–21–4	46 days from seeding	Petiole 2 from terminal	0.050

Source: Maynard *et al.* (1976).

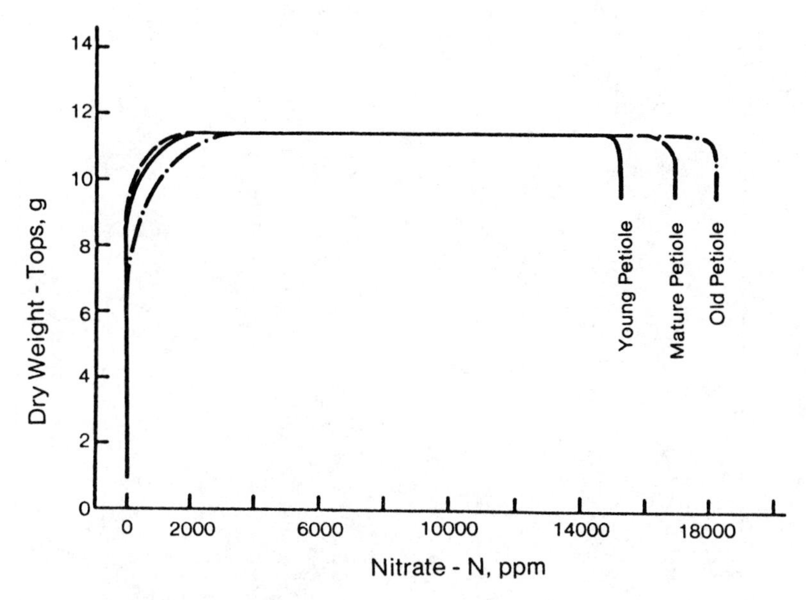

FIG 1.14. Relationship between dry weight of tops of squash and nitrate-N in young, mature, and old petioles.
From El-Sheikh et al. (1970).

the laboratory and the determination done on extracts from either fresh (Ulrich 1948) or dried (80°C) and ground plant material (Baker and Smith 1969). In the field, nitrate in plant sap can be determined using Bray's Nitrate Powder (Syltie *et al.* 1972), or with Mercko-quant test strips, a procedure recommended by Scaife and Stevens (1983). Keeney and Nelson (1982) have an excellent review article on the methods of determining nitrates.

Iron also can be determined by a tissue test in a procedure developed by Bar-Akiva *et al.* (1978) for field use. Peroxidase activity (Bar-Akiva 1984) is measured on leaf disks, with the development of a blue color indicating adequate Fe in the plant tissue.

Most of the test procedures mentioned in this section can be performed in the field, an advantage considered significant by some. Many of the tests themselves are not entirely quantitative, but provide the user with a qualitative "yes or no" evaluation of a crop; that is, the element being assayed by the tissue test is either present or not at the desired sufficiency level. An additional advantage may be the savings in time and cost compared with conventional laboratory plant analyses. For some, the disadvantages of tissue tests are their lack of quantification, insufficient interpretive data available, and the limited number of elements that can be determined.

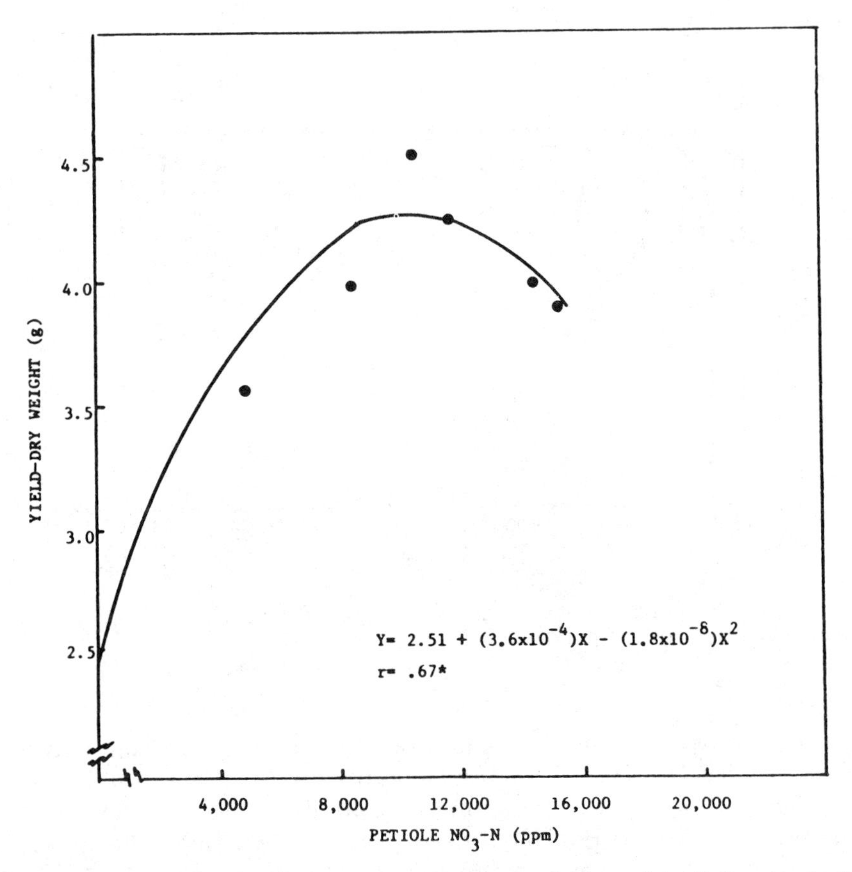

FIG 1.15. Relationship between geranium yield and petiole nitrate-N concentration.
From Woodson and Boodley (1983).

V. SUMMARY

The use of soil tests, plant (leaf) analyses, and tissue tests in a crop production system is essential if growers are going to be able to ensure against nutrient element insufficiencies, which can adversely affect crop yield and quality (Armstrong *et al.* 1984). Current estimates indicate that soil tests and plant analyses are being underutilized in the United States; similar underutilization is probably prevalent in most other countries. This underutilization may be greater for horticultural crops (fruits, vegetables, and ornamental) than for grain crops (maize, sorghum, soybeans, wheat, rice), and pasture crops (alfalfa and the grasses).

Although there has been considerable research on the techniques of soil testing and plant analysis, each has its problems in terms of sampling, sample preparation, laboratory analysis, and/or methods of interpretation. For example, soil-testing procedures have not kept pace with changes in soil fertility conditions due to continuous cropping and the repeated application of lime and fertilizer. How best to use and interpret a soil test result is still a topic of considerable controversy.

Probably one of the most significant events occurring today is the rapid adoption of the newest analytical technology in soil-testing and plant analysis laboratories. Mass analysis systems and computer applications are being quickly adopted, the goals being multielement determinations and rapid analytical service. There may come a shift away from the large central laboratory serving a large regional marketing area to smaller facilities, offering the same services but covering a smaller and more specialized market. Newer analytical instrumentation, such as the plasma emission spectrometer, and the increasing power of the microcomputer will help to make this shift possible.

There seems to be a renewed interest in soil fertility and plant nutrition research after several decades of relatively little interest and activity. This is evident, for example, from the fact that many of the soil-testing procedures currently in use originated during the 1940s and 1950s. Efforts are now under way to adopt newer procedures that provide a better evaluation of the soil-plant nutrient environment. With improved techniques for soil testing, plant analysis, and tissue testing, growers will find these test procedures more effective guides in the proper and economic use of lime and fertilizer. Better testing procedures, more effective and reliable interpretations, and laboratories physically closer to the user will all contribute to the increasing value and use of these testing procedures—the end result being more efficient use of lime and fertilizer, and higher yields of better quality crops.

LITERATURE CITED

ADAMS, F. and D. E. EVANS. 1962. A rapid method for measuring lime requirement of red-yellow podzolic soils. Soil Sci. Soc. Amer. Proc. 26:355–357.

ANDERSON, O. E. and J. J. MORTVEDT (eds.). 1982. Soybeans: diagnosis and correction of manganese and molybdenum problems. Southern Coop. Serv. Bull. 581.

ARMSTRONG, D., B. AGERTON and S. MARTIN (eds.). 1984. The diagnostic approach. Better Crops, Spring 1984.

ATKINSON, D., J. E. JACKSON, R. O. SHARPLES, and W. H. WALLER (eds.). 1980. Mineral nutrition of fruit trees. Butterworths, London.

BAKER, D. E. 1971. A new approach to soil testing. Soil Sci. 112:381-391.

BAKER, D. E. 1973. A new approach to soil testing: II. Ionic equilibria involving H, K, Ca, Mg, Fe, Cu, Zn, Na, P, and S. Soil Sci. Soc. Amer. Proc. 27:537-541.

BAKER, D. E. 1977. Ion activities and ratios in relation to corrective treatments of soils. p. 55-74. In: T. R. Peck, J. T. Cope, Jr. and D. A. Whitney (eds.), Soil testing: correlating and interpreting the analytical results. Amer. Soc. Agron., Madison, Wisconsin.

BAKER, D. E. and M. C. AMACHER. 1981. The development and interpretation of a diagnostic soil-testing program. Pennsylvania Agr. Expt. Sta. Bull. 826.

BAKER, A. S. and R. SMITH. 1969. Extracting solution for potentiometric determination of nitrate in plant tissue. J. Agr. Food Chem. 17:1284-1287.

BAR-KIVA, A., D. N. MAYNARD, and J. E. ENGLISH. 1978. A rapid tissue test for diagnosing iron deficiencies in vegetable crops. HortScience 13:284-285.

BAR-AKIVA, A. 1984. Substitutes for benzidine as H-donors in the peroxidase assay for rapid diagnosis of iron in plants. Comm. Soil Sci. Plant Anal. 15:929-934.

BEAR, F. E., A. L. PRINCE, and J. L. MALCOM. 1945. Potassium needs of New Jersey soils. New Jersey Agr. Expt. Sta. Bull. 721.

BEATON, J. D., G. R. BURNS, and J. PLATOU. 1968. Determination of sulphur in soils and plant material. Tech. Bull. 14. The Sulphur Institute, Washington, DC.

BEAUFILS, E. R. 1973. Diagnosis and recommendation intergrated system (DRIS). Soil Sci. Bull. 1. Univ. of Natal, South Africa.

BEAUFILS, E. R. and M. E. SUMNER. 1976. Application of the DRIS approach in calibrating soil and plant parameters for sugarcane. Proc. South Africa Sugar Technical Assoc. 50:118-124.

BERGER, K. C. and E. TROUG. 1940. Boron deficiencies as revealed by plant and soil tests. J. Amer. Soc. Agron. 32:297-301.

BEUTEL, J., K. URIU, and O. LILLELAND. 1978. Leaf analysis for California deciduous fruits. p. 11-14. In: H. M. Reisenaur (ed.), Soil and plant-tissue testing in California. Bull. 1879 (rev. June 1978). California Coop. Ext. Serv., Davis, Calif.

BLOCK, R. A. 1979. Handbook of decomposition methods in analytical chemistry. Wiley, New York.

BOWER, C. A. and L. V. WILCOX. 1965. Soluble salts. p. 935-945. In: C. A. Black (ed.), Methods of soil analysis, Part 2. Agronomy Publ. 9. Amer. Soc. Agron., Madison, Wisconsin.

BOYNTON, D. and G. H. OBERLY. 1966. Apple nutrition. pp. 1-50. In: N. F. Childers (ed.), Nutrition of fruit crops. Horticulture Publications. Rutgers—The State University, New Brunswick, New Jersey.

BRAY, R. H. and L. T. KURTZ. 1945. Determination of total, organic and available forms of phosphorus in soils. Soil Sci. 59:39-45.

BREMNER, J. M. and M. A. TABATABAI. 1971. Use of automated combustion techniques for total carbon, total nitrogen, and total sulfur analysis of soils. p. 1-16. In: L. M. Walsh (ed.), Instrumental methods for analysis of soils and plant tissue. Soil Sci. Soc. Amer., Madison, Wisconsin.

CATE, JR., R. B. and L. A. NELSON. 1965. A rapid method for correlation of soil test analyses with plant response data. North Carolina Agr. Expt. Sta., Intern. Soil Testing Series Technical Bull. 1.

CHAPMAN, H. D. (ed.). 1966. Diagnostic criteria for plants and soils. University of California, Division of Agricultural Sciences, Riverside.

CHAPMAN, H. D. 1967. Plant analysis values suggestive of nutrient status of selective crops. p. 77-92. In: G. W. HARDY (ed.), Soil testing and plant analysis. Part II. Plant analysis. SSSA Special Publications Series 2. Soil Sci. Soc. Amer., Madison, Wisconsin.

CHILDERS, N. F. (ed.). 1966. Temperate to tropical fruit nutrition. Rutgers—The State University, New Brunswick, New Jersey.

CLEMENTS, H. F. 1960. Crop logging of sugar cane in Hawaii. p. 131-147. In: W. REUTHER, (ed.), Plant analysis and fertilizer problems. Amer. Institute of Biological Sciences, Washington, DC.

COLEMAN, N. T., E. J. KAMPRATH, and S. B. WEBB. 1958. Liming. Adv. Agron. 10:475-522.

COLONNA, J. P. 1970. The mineral diet of excellsor coffee plants: natural variability of the mineral foliar composition on a homogenous plantation. Cashiers Office Research Science Technical Outre-Mer Serial Biology 13:67-80.

COPE, JR., J. T. 1972. Fertilizer recommendations and computer program key used by the soil testing laboratory. Auburn Agr. Expt. Sta. Cir. 176.

COX, F. R. and E. J. KAMPRATH. 1972. Micronutrient soil tests. p. 289-318. In: J. J. Mortvedt, P. M. Giordano and W. L. Lindsay (eds.), Micronutrients in agriculture. Soil Sci. Soc. Amer., Madison, Wisconsin.

CRILEY, R. A. and W. H. CARLSON. 1970. Tissue analysis standards for various floricultural crops. Flor. Rev. 146:19-20, 70-73.

DAHLQUIST, R. L. and J. W. KNOLL. 1978. Inductively coupled plasma-atomic emission spectroscopy: analysis of biological materials and soil for major, trace, and ultra-trace elements. Appl. Spec. 32:1-30.

DAHNKE, W. C. and E. H. VASEY. 1973. Testing soils for nitrogen. p. 97-114. In: L. M. Walsh and J. B. Beaton (eds.), Soil testing and plant analysis (rev. ed.). Soil Sci. Soc. Amer., Madison, Wisconsin.

DAVIDESCU, D. and V. DAVIDESCU. 1982. Evaluation of fertility by plant and soil analysis. Abacus Press, Kent, England.

DEBOLT, D. C. 1980. Multi-element emission spectroscopic analysis of plant tissue using DC agron plasma source. J. Assoc. Offic. Anal. Chem. 63:802-805.

DORSHEIMER, W. T. and R. A. ISAAC. 1982. Application of NIR analysis. Amer. Lab. 14:58-63.

DUDAL, R. 1976. Inventory of the major soils of the world with special reference to mineral stress hazards. p. 3-14. In: M. J. Wright (ed.), Plant adaptation to mineral stress in problem soils. Cornell University, Ithaca, New York.

EL-SHEIKH, A. M., A. EL-HAKAM, and A. ULRICH. 1970. Critical nitrate levels for squash, cucumber and melon plants. Commun. Soil Sci. Plant Anal. 1:63-74.

EMBLETON, T. W., W. W. JONES, and R. G. PLATT. 1978. Leaf analysis as a guide to citrus fertilization. p. 4-9. In: H. M. Reisenaur, (ed.), Soil and plant-tissue testing in California. Bull. 1879 (rev. June 1978). California Coop. Ext. Serv., Davis, Calif.

ERICKSON, T., K. A. KELLING, and E. E. SCHULTE. 1982. Predicting alfalfa nutrient needs through DRIS. p. 233-246. In: Proc. 21st Wisconsin Fertilizer AgLime and Pest Management Conf., Madison.

FITTS, J. W. and W. L. NELSON 1956. The determination of lime and fertilizer requirements of soils through chemical test. Adv. Agron. 8:242-282.

FLANNERY, R. L. and D. K. MARKUS. 1971. Determination of phosphorus, potassium, calcium, and magnesium simultaneously in North Carolina ammonium acetate and Bray Pl soil extracts by AutoAnalyzer. p. 97-112. In: L. M. Walsh (ed.), Instrumental methods for analysis of soils and plant tissue. Soil Sci. Soc. Amer., Madison, Wisconsin.

FOLLETT, R. H., L. S. MURPHY, and R. L. DONAHUE. 1981. Fertilizers and soil admendments. Prentice-Hall, Englewood Cliffs, NJ.

GALLAHER, R. N. and J. B. JONES, JR. 1976. Total, extractable, and oxalate calcium and other elements in normal and mouse ear pecan tree tissues. J. Amer. Soc. Hort. Sci. 101:692-696.

GALLAHER, R. N., C. O. WELDON, and J. G. FUTRAL. 1975. An aluminum block digestor for plant and soil analysis. Soil Sci. Soc. Amer. Proc. 39:803-806.

GALLAHER, R. N., C. O. WELDON, and F. C. BOSWELL. 1976. A semi-automated procedure for total nitrogen in plant and soil samples. Soil Sci. Soc. Amer. Proc. 40:887-889.

GERALDSON, C. M. 1977. Nutrient intensity and balance. p. 75-84. In: T. R. Peck, J. T. Cope, Jr., and D. A. Whitney (eds.), Soil testing: correlating and interpreting the analytical results. Amer. Soc. Agron., Madison, Wisconsin.

GERALDSON, C. M., G. R. KLACAN, and O. A LORENZ. 1973. Plant analysis as an aid in fertilizing vegetable crops. p. 365-380. In: L. M. Walsh and J. D. Beaton (eds.), Soil testing and plant analysis (rev. ed.). Soil Sci. Soc. Amer., Madison, Wisconsin.

GLEIT, C. E. and W. D. HOLLAND. 1962. Use of electrically excited oxygen for the low temperature decomposition of organic substances. Anal. Chem. 34:1454-1457.

GOODALL, D. W. and F. G. GREGORY. 1947. Chemical composition of plants as an index of their nutritional status. Tech. Publ. 17. Imperial Bureau of Horticulture and Plantation Crops, Penglais, Aberystwyth, Wales.

GORSUCH, T. T. 1970. Destruction of organic matter. Vol. 39, International Series in Monographs in Analytical Chemistry. Pergamon Press, New York.

GORSUCH, T. T. 1976. Dissolution of organic matter. p. 491-508. In: P. D. Lafleur (ed.), Accuracy in trace analysis: sampling, sample handling, analysis. Vol. 1, Special Publ. 422. National Bureau of Standards, Washington, DC.

GRAHAM, E. R. 1959. An explanation of theory and methods of soil testing. Missouri Agr. Expt. Sta. Bull. 734.

GRIGG, J. L. 1953. Determination of available soil molybdenum. New Zealand News 3:37-40.

GROVE, J. H. and M. E. SUMNER. 1982. Yield and leaf composition of sunflower in relation to N, P, K and lime treatments. Fert. Res. 3:367-378.

HANNA, W. J. and T. B. HUTCHESON, JR. 1968. Soil-plant relationships. p. 141-162. In: L. B. Nelson (ed.), Changing patterns in fertilizer use. Soil Sci. Soc. Amer., Madison, Wisconsin.

HALVIN, J. L. and P. N. SOLTANPOUR. 1980. A nitric acid plant tissue digest method for use with inductively couple plasma spectrometry. Commun. Soil Sci. Plant Anal. 11:969-980.

HEADY, E. O., J. T. PESEK, W. G. BROWN, and J. L. DOLL. 1961. Crop response surfaces and economic options in fertilizer use. p. 475-525. In: E. O. Heady and J. L. Doll (eds.), Agricultural production functions. Iowa State Univ. Press, Ames.

HERN, J. L. 1984. Determination of total sulfur in plant materials using an automated sulfur analyzer. Commun. Soil Sci. Plant Anal. 15:99-107.

HOEFT, R. G., L. M. WALSH and D. R. KEENEY. 1973. Evaluation of various extractants for available soil sulfur. Soil Sci. Soc. Amer. Proc. 37:401-404.

HOEFT, R. G. 1982. Diagnosis of sulfur deficiency. In: Proceedings Soil and Plant Analysis Seminar, Anaheim, CA. Council on Soil Testing and Plant Analysis, Athens, Georgia.

HOOD, S. L., R. Q. PARKS, and C. HURWITZ. 1944. Mineral contamination resulting from grinding plant samples. Ind. Eng. Chem. Anal. Ed. 16:202-205.

HORWITZ, W. (ed.). 1980a. Official methods of analysis of the association of official analytical chemists. Section 7.003. AOAC, Arlington, Virginia.

HORWITZ, W. (ed.). 1980b. Official methods of analysis of the association of official analytical chemists. Section 51.028. AOAC, Arlington, Virginia.

HORWITZ, W. (ed.). 1980c. Official methods of analysis of the association of official analytical chemists. Section 3.007. AOAC, Arlington, Virginia

HORWITZ, W. (ed.). 1980d. Official methods of analysis of the association of official analytical chemists. Sections 2.057-2.058. AOAC, Arlington, Virginia.

HORWITZ, W. (ed.). 1980e. Official methods of analysis of the association of official analytical chemists. Sections 7.025-7.032. AOAC, Arlington, Virginia.

HORWITZ, W. (ed.). 1980f. Official methods of analysis of the association of official analytical chemists. Sections 3.061-3.064. AOAC, Arlington, Virginia.

ISAAC, R. A. and W. C. JOHNSON. 1976. Determination of total nitrogen in plant tissue. J. Assoc. Offic. Anal. Chem. 59:98-100.

ISAAC, R. A. and W. C. JOHNSON. 1983. Determination of protein nitrogen in plant tissue using near infrared spectroscopy. J. Assoc. Offic. Anal. Chem. 66: (in press).

ISAAC, R. A. and J. B. JONES, JR. 1970. AutoAnalyzer systems for the analysis of soil and plant tissue extracts. p. 57-64. In: Advances in automated analysis, Technicon Cong. Proc. Technicon Corp., Tarrytown, New York.

ISAAC, R. A. and J. B. JONES, JR. 1972. Effects of various dry ashing temperatures on the determination of 13 nutrient elements in five plant tissues. Commun. Soil Sci. Plant Anal. 3:261-269.

ISAAC, R. A. and H. D. KERBER. 1971. Atomic absorption and flame photometry: techniques and uses in soil, plant, and water analysis. p. 17-38. In: L.M. Walsh (ed.), Instrumental methods for analysis of soils and plant tissue. Soil Sci. Soc. Amer., Madison, Wisconsin.

JACKSON, M. L. 1958. Soil chemical analysis. Prentice-Hall, Englewood Cliffs, New Jersey.

JOINER, J. N., R. P. POOLE, and C. A. CONOVER. 1983. Nutrition and fertilization of ornamental greenhouse crops. Hort. Rev. 5:317-413.

JONES, J. B., JR. 1963. Effect of drying on ion accumulation in corn leaf margins. Agron. J. 55:579-580.

JONES, J. B., JR. 1965. Soil testing: catalyst for change. Plant Food Rev. II:1-5.

JONES, J. B., JR. 1970a. Uniformity, standardization and terminology: our problems and progress. Commun. Soil Sci. Plant Anal. 1:311-316.

JONES, J. B., JR. 1970b. Distribution of fifteen elements in corn leaves. Comm. Soil Sci. Plant Anal. 1:27-33.

JONES, J. B., JR. 1972a. Plant tissue analysis for micronutrients. p. 319-346. In: J. J. Mortvedt, P. M. Giordano and W. L. Lindsay (eds.), Micronutrients in agriculture. Soil Sci. Soc. Amer., Madison, Wisconsin.

JONES, J. B., JR. 1972b. General fertilizer recommendations for pecans. p. 101-112. In: Proc. 65th Annu. Conv. Southeastern Pecan Growers Assoc., Starkville, Mississippi.

JONES, J. B., JR. 1974. Plant analysis handbook for Georgia. Georgia Coop. Ext. Bull. 735.

JONES, J. B., JR. 1976. Elemental analyses of biological substances by direct-reading spark emission spectroscopy. Amer. Lab. 8:15-20.

JONES, J. B., JR. 1977. Elemental analysis of soil extracts and plant tissue ash by plasma emission spectroscopy. Commun. Soil Sci. Plant Anal. 8:349-365.

JONES, J. B., JR. (ed.). 1980. Handbook on reference methods for soil testing. Council on Soil Testing and Plant Analysis, Athens, Georgia.

JONES, J. B., JR. 1983. Soil test works when used right, Part 2. Solutions 27:61-70.

JONES, J. B., JR. 1984. Soil testing and plant analysis in the United States. Soil-Plant Analyst Newsletter 2:4.

JONES, J. B., JR. and W. W. BUDZYNSKI. 1980. A professional seminar on agronomics. Agricultural Associates, Inc., West Lafayette, Indiana.

JONES, J. B., JR. and R. A. ISAAC. 1972. Determination of sulfur in plant material using a LECO sulfur analyzer. J. Agr. Food Chem. 20:192–194.

JONES, J. B., JR. and W. J. A. STEYN. 1973. Sampling, handling, and analyzing plant tissue samples. p. 249–270. In: L. M. Walsh and J. D. Beaton (eds.), Soil testing and plant analysis (rev. ed.). Soil Sci. Soc. Amer., Madison, Wisconsin.

JONES, J. B., JR., R. L. LARGE, D. B. PFLEIDERER, and H. S. KLOSKY. 1971. How to properly sample for a plant analysis. Crops & Soils 23:114–120.

JONES, J. B., JR., J. E. PALLAS, and J. R. STANSELL. 1980. Tracking the elemental content of leaves and other plant parts of the peanut under irrigated culture in the sandy soils of south Georgia. Commun. Soil Sci. Plant Anal. 11:81–92.

JONES, U. S. 1979. Fertilizers and soil fertility. Reston Publ. Co., Reston, Virginia.

KEENEY, D. R. and D. W. NELSON. 1982. Nitrogen—inorganic forms. p. 643–698. In: A. L. Page, R. H. Miller, and D. R. Keeney (eds.), Methods of soil analysis, Part 2. Chemical and microbiological (2nd ed.). Amer. Soc. Agron., Madison, Wisconsin.

KENWORTHY, A. L. 1961. Interpreting the balance of nutrient-elements in leaves of fruit trees. p. 28–43. In: W. Reuther (ed.), Plant analysis and fertilizer problems. Publ. 8. American Institute of Biological Sciences, Washington, DC.

KENWORTHY, A. L. 1973. Leaf analysis as an aid in fertilizing orchards. p. 381–392. In: L. M. Walsh and J. D. Beaton (eds.), Soil testing and plant analysis (rev. ed.). Soil Sci. Soc. Amer., Madison, Wisconsin.

KIRKBY, E. A. 1979. Maximizing calcium uptake by plants. Commun. Soil Sci. Plant Anal. 10:89–113.

KOO, R. C. J. 1968. Potassium nutrition of tree crops. p. 469–488. In: V. J. Kilmer, S. E. Younts, and N. C. Brady (eds.), The role of potassium in agriculture. Amer. Soc. Agron., Madison, Wisconsin.

KORCAK, R. F. 1980. The importance of calcium and nitrogen source in fruit tree nutrition, p. 267–277. In: D. Atkinson, J. E. Jackson, R. O. Sharples, and W. M. Waller (eds.), Mineral nutrition of fruit trees. Butterworths, London.

KRANTZ, B. A., W. L. NELSON, and L. F. BURKHART. 1948. Plant-tissue tests as a tool in agronomic research. p. 137–156. In: H. B. KITCHEN (ed.), Diagnostic techniques for soils and crops. American Potash Institute, Washington, DC.

KUBOTA, J. and V. A. LAZAR. 1971. X-ray emission spectrograph: techniques and uses for plant and soil studies. p. 67–82. In: L. M. Walsh (ed.), Instrumental methods of analysis of soils and plant tissue. Soil Sci. Soc. Amer., Madison, Wisconsin.

LANGENEGGER, W. and B. L. SMITH. 1978. An evaluation of the DRIS system as applied to pineapple leaf analysis. p. 263–273. In: A. R. Ferguson, R. L. Bieleski, and I. B. Ferguson (eds.), Plant nutrition 1978. Information Series 134. New Zealand Depart. of Scientific & Industrial Res., Wellington, New Zealand.

LETZCH, W. S. and M. E. SUMNER. 1983. Computer program for calculating DRIS indices. Commun. Soil Sci. Plant Anal. 14:811–815.

LIEBHARDT, W. C. 1982. Testing . . . testing: nitrogen. Part 1. The New Farm. Jan.:29–36.

LIEBHARDT, W. C. 1982. Testing . . . testing: phosphorus. Part 2. The New Farm. Feb.:30–35.

LIEBHARDT, W. C. 1982. Testing . . . testing: potassium. Part 3. The New Farm. Mar.-Apr.:31–34.

LIEBHARDT, W. C. 1982. Testing . . . testing: lime. Part 4. The New Farm. May-June:21–24.

LIEBHARDT, W. C. 1982. Testing . . . testing: secondary & micronutrients. Part 5. The New Farm. July-Aug.:17–19.

LIEBHARDT, W. C. and M. CULIK. 1983. Testing . . . testing. Part 7. The New Farm. Feb.:21–25.

LINDSAY, W. L. and W. A. NORVELL. 1969. Development of a DTPA micronutrient soil test. Agronomy Abstr. (Amer. Soc. Agron.) p. 84.

LOCKMAN, R. B. 1970. Plant sample analysis as affected by sample decomposition prior to laboratory processing. Commun. Soil Sci. Plant Anal. 1:13-19.

LORENZ, O. A. and J. F. BARTZ. 1968. Fertilization for high yields and quality of vegetable crops. p. 327-352. In: L. B. Nelson (ed.), Changing patterns in fertilizer use. Soil Sci. Soc. Amer., Madison, Wisconsin.

LORENZ, O. A. and D. N. MAYNARD. 1980. Knott's handbook for vegetable growers. 2nd ed. John Wiley & Sons, New York.

LORENZ, O. A. and K. B. TYLER. 1978. Plant tissue analysis of vegetable crops. p. 21-24. In: H. M. Reisenaur (ed.), Soil and plant-tissue testing in California. Bull. 1879 (rev. June 1978). California Coop. Ext. Serv., Davis.

LUCAS, R. E. 1968. Potassium nutrition of vegetable crops. p. 489-498. In: V. J. Limer, S. E. Younts, and N. C. Brady (eds.), The role of potassium in agriculture. Amer. Soc. Agron., Madison, Wisconsin.

LUNDEGARDH, H. 1951. Leaf analysis (Translated by R. L. Mitchell). Hilger and Watts, Ltd., London, England.

MAYNARD, D. N., A. V. BARKER, P. L. MINOTTI, and N. H. PECK. 1976. Nitrate accumulation in vegetables. Adv. Agron. 28:71-118.

McLEAN, E. O. 1973. Testing soils for pH and lime requirement. p. 77-96. In: L. M. Walsh and J. D. Beaton (eds.), Soil testing and plant analysis (rev. ed.). Soil Sci. Soc. Amer., Madison, Wisconsin.

McLEAN, E. O. 1977a. Contrasting concepts in soil test interpretation: sufficiency levels of available nutrients versus basic cation saturation ratios. p. 39-54. In: T. R. Peck, J. T. Cope, Jr., and D. A. Whitney (eds.), Soil testing: correlating and interpreting the analytical results. Amer. Soc. Agron., Madison, Wisconsin.

McLEAN, E. O. 1977b. Fertilizer and lime recommendations based on soils tests: good, but could they be better? Commun. Soil Sci. Plant Anal. 8:441-464.

McLEAN, E. O. 1982a. Soil pH and lime requirement. p. 199-224. In: A. L. Page (ed.), Methods of soil analysis, Part 2. Chemical and microbiological properties (2nd ed.). Amer. Soc. Agron., Madison, Wisconsin.

McLEAN, E. O. 1982b. Chemical equilibrations with soil buffer systems as bases for future soil testing procedures. Commun. Soil Sci. Plant Anal. 13:411-433.

McLEAN, E. O. and S. MOSTAGHIMI. 1983. Improved corrective fertilizer recommendations based on a two-step alternative usage of soil tests: III. The Bray P-1 test on soils with concretions. Soil Sci. Soc. Amer. Proc. J. 47:966-971.

McLEAN, E. O., T. O. OLOYA, and J. L. ADAMS. 1979. Soil tests to inventory the initially available levels and to assess the fates of added P and K as bases for improved fertilizer recommendations. Commun. Soil Sci. Plant Anal. 10:623-630.

McLEAN, E. O., J. L. ADAMS, and R. C. HARTWIG. 1982. Improved corrective fertilizer recommendations based on a two-step alternative usage of soil tests: II. Recovery of soil-equilibrated K. Soil Sci. Soc. Amer. Proc. J. 46:1198-1201.

MEHLICH, A. 1953. Determination of P, Ca, Mg, K, Na, and NH_4. North Carolina Soil Testing Division (mimeo).

MEHLICH, A. 1972. Uniformity of expressing soil test results: a case for calculating results on a volume basis. Commun. Soil Sci. Plant Anal. 3:417-424.

MEHLICH, A. 1973. Uniformity of soil test results as influenced by volume weight. Commun. Soil Sci. Plant Anal. 4:475-486.

MEHLICH, A. 1978. New extractant for soil test evaluation of phosphorus, potassium, magnesium, calcium, sodium, manganese and zinc. Commun. Soil Sci. Plant Anal. 9:477-492.

MEHLICH, A. 1982. Comprehensive methods in soil testing. In: Proc. Soil and Plant Analysis Seminar, Anaheim, CA. Council on Soil Testing and Plant Analysis, Athens, Georgia.

MELDAL-JOHNSEN, A. 1975. An analysis of some factors influencing potato production. M. S. Thesis, Univ. of Natal, Peitermatitzburg, South Africa.

MELSTED, S. W. and T. R. PECK. 1977. The Mitscherlich-Bray growth function. p. 1-18. In: T. R. Peck, J. T. Cope, Jr. and D. A. Whitney (eds.), Soil testing: correlating and interpreting the analytical results. Amer. Soc. Agron., Madison, Wisconsin.

MILLAWAY, R. M. and L. WIERSHOLM. 1979. Calcium and metabolic disorders. Commun. Soil Sci. Plant Anal. 10:1-28.

MITCHELL, R. L. 1956. The spectrographic analysis of soils, plants, and related materials. Tech. Commun. 44. Commonwealth Bureau of Soils, Harpenden, Herts, England.

MORTVEDT, J. J. 1977. Micronutrient soil test correlations and interpretations. p. 99-117. In: T. R. Peck, J. T. Cope, Jr. and D. A. Whitney (eds.), Soil testing: correlating and interpreting the analytical results. Amer. Soc. Agron., Madison, Wisconsin.

MORRIES, P. 1983. A century of Kjeldahl (1883-1983). J. Assoc. Public Analysts 21:53-58.

MUNSINGER, R. A. and R. MCKINNEY. 1982. Modern Kjeldahl systems. Amer. Lab. 14:76-79.

MUNSON, R. D. and W. L. NELSON. 1973. Principles and practices in plant analysis. p. 223-248. In: L. M. Walsh and J. D. Beaton (eds.), Soil testing and plant analysis (rev. ed.) Soil Sci. Soc. Amer., Madison, Wisconsin.

MUNTER, R. C. and R. A. GRANDE. 1981. Plant tissue and soil extract analysis by ICP-atomic emission spectrometry. p. 653-672. In: R. M. Barnes (ed.), Developments in atomic plasma spectrochemical analysis. Heyden & Son, Ltd., London.

MUNTER, R. C., T. L. HALVERSON and R. D. ANDERSON. 1984. Quality assurance for plant tissue analysis by ICP-AES. Commun. Soil Sci. Plant Anal. 15:1285-1322.

NELSON, P. B. and J. W. BOODLEY. 1965. An error involved in the preparation of plant tissue for analysis. Proc. Amer. Soc. Hort Sci. 86:712-716.

NELSON, W. L. and C. M. HANSEN. 1968. Methods and frequency of fertilizer applications. p. 85-118. In: W. L. Nelson (ed.), Changing patterns in fertilizer use. Soil Sci. Soc. Amer., Madison, Wisconsin.

NELSON, D. W. and L. E. SOMMERS. 1980. Total nitrogen analysis of soil and plant tissues. J. Assoc. Offic. Anal. Chem. 63:770-778.

NELSON, D. W. and L. E. SOMMERS. 1982. Total carbon, organic carbon, and organic matter. p. 539-579. In: A. L. Page (ed.) Methods of soil analysis, Part 2. Chemical and microbiological properties (2nd ed.). Amer. Soc. Agron., Madison, Wisconsin.

NELSON, W. L., J. W. FITTS, L. T. KARDOS, W. T. MCGEORGE, R. Q. PARKS, and J. FIELDING REED. 1951. Soil testing in the United States. National Soil & Fertilizer Research Committee. Publ. 0-979953. U. S. Government Printing Office, Washington, DC.

NESMITH, J. and E. W. McELWEE. 1974. Soil reaction (pH) for flowers, shrubs, and lawn. Univ. of Florida Cir. 352A.

NEUBERT, P., W. WRAZIDLO, H. P. VIELEMEYER, I. HUNDT, F. GULLMICK, and W. BERGMANN. 1969. Tabellen zur pflanzenanalzre—erste orientierende ubersicht. Institut for Pflanzenernahrung Jerna, Berlin, Germany.

NICHOLAS, D. J. D. 1957. An appraisal of the use of chemical tissue tests for determining the mineral status of crop plants. p. 119-139. In: T. Wallace (ed.), Plant analysis and fertilizer problems. Institut de Recherches pour les Huiles et Oléagineux, Paris.

OLSEN, S. R. 1972. Micronutrient interactions. p. 243-264. In: J. J. Mortvedt, P. M. Giordano, and W. L. Lindsay (eds.). Micronutrients in agriculture. Soil Sci. Soc. Amer., Madison, Wisconsin.

OLSEN, S. R., C. V. COLE, F. S. WATANABE, and L. A. DEAN. 1954. Estimation of available phosphorus in soils by extraction with sodium bicarbonate. USDA Cir. 939.

OLSEN, S. R. and L. E. SOMMERS. 1982. Phosphorus. p. 403-430. In: A. L. Page (ed.), Methods of soil analysis, Part 2. Chemical and microbiological properties (2nd ed.). Amer. Soc. Agron., Madison, Wisconsin.

OWENS, H. I. 1980. Soil testing and plant analysis report. USDA-SEA-Extension, Washington, DC (mimeo).

PEARSON, R. W. and FRED ADAMS (eds.). 1967. Soil acidity and liming. Amer. Soc. Agron., Madison, Wisconsin.

PECK, T. R. 1980. Standard soil scoop. p. 3-4. In: W. C. Dahnke (ed.), Recommended chemical soil test procedures for the north central region. North Dakota Agr. Expt. Sta. Bull. 499 (rev.).

PECK, T. R. and S. W. MELSTED. 1973. Field sampling for soil testing. p. 67-76. In: L. M. Walsh and J. D. Beaton (eds.), Soil testing and plant analysis (rev. ed.). Soil Sci. Soc. Amer., Madison, Wisconsin.

PECK, T. R., J. T. COPE, JR., and D. A. WHITNEY (eds.). 1977. Soil testing: correlating and interpreting the analytical results. ASA Special Publ. 29. Amer. Soc. Agron., Madison, Wisconsin.

PETERSON, J. C. 1982. Effects of pH upon nutrient availability in a commercial soilless root medium utilized for floral crop production. Ohio Agr. Res. Cir. 268.

PIPER, C. S. 1942. Soil and plant analysis. Hassell Press, Adelaide, Australia.

PRASAD, M. and T. M. SPIERS. 1982. Evaluation of a simple sap nitrate test for some ornamental crops. p. 474-479. In: A. Scaife (ed.), Plant nutrition 1982: Proc. Ninth Intern. Plant Nutrition Colloquium. Commonwealth Agricultural Bureaux, Slough, England.

PRASAD, M., R. E. WIDMER, and R. R. MARSHALL. 1983. Soil testing horticultural substrates for cyclamen and poinsettia. Commun. Soil Sci. Plant Anal. 14: 553-573.

PREVOT, P. and M. OLLAGNIER. 1956. Methode d'utilisation du dianostic folaire. p. 177-194. In: T. WALLACE (ed.), Plant analysis and fertilizer problems. Institut de Recherches pour les Huiles et Oléagineux, Paris.

REISENAUER, H. M. (ed.). 1978. Soil and plant-tissue testing in California. Bull. 1879 (rev. June 1978). California Coop. Ext. Serv., Davis.

REITZ, H. J. and W. C. STILES. 1968. Fertilization of high producing orchards. p. 353-378. In: L. B. Nelson (ed.), Changing patterns in fertilizer use. Soil Sci. Soc. Amer., Madison, Wisconsin.

RHOADES, J. D. 1982. Soluble salts. p. 167-180. In: A. L. Page (ed.), Methods of soil analysis, Part 2. Chemical and microbiological properties (2nd ed.). Amer. Soc. Agron., Madison, Wisconsin.

SAYRE, J. D. 1952. Accumulation of radio-isotopes in corn leaves. Ohio Agr. Expt. Sta. Res. Bull. 723.

SCAIFE, A. and K. L. STEVENS. 1983. Monitoring sap nitrate in vegetable crops: comparison of test strips with electrode methods, and effects of time of day and leaf position. Commun. Soil Sci. Plant Anal. 14:761-771.

SCHOFIELD, R. K. and A. W. TAYLOR. 1955. The measurement of soil pH. Soil Sci. Soc. Amer. Proc. 19:164-167.

SCHOLLENBERGER, C. J. and R. H. SIMON. 1945. Determination of exchange capacity and exchangeable bases in soil-ammonium acetate method. Soil Sci. 59:13-24.

SHEAR, C. B. 1975. Calcium-related disorders of fruits and vegetables. HortScience 10:361-365.

SHEAR, C. B. and M. FAUST. 1980. Nutritional ranges in deciduous tree fruits and nuts. Hort. Rev. 2:142-163.

SHOEMAKER, H. E., E. O. McLEAN, and P. G. PRATT. 1962. Buffer methods for determination of lime requirement of soils with appreciable amount of exchangeable aluminum. Soil Sci. Soc. Amer. Proc. 25:274-277.

SMITH, P. F. 1962. Mineral analysis of plant tissues. Annu. Rev. Plant Physiol. 13:81-108.

SMITH, J. D., D. L. CARTER, M. J. BROWN, and C. L. DOUGLAS. 1968. Differences in chemical composition of plant sample fractions resulting from grinding or screening. Agron. J. 60:149-151.

SOLTANPOUR, P. N. and S. P. SCHWAB. 1977. A new soil test for simultaneous extraction of macro- and micro-nutrients in alkaline soils. Commun. Soil Sci. Plant Anal. 8:195-207.

SOLTANPOUR, P. N., J. B. JONES, JR., and S. M. WORKMAN. 1982. Optical emission spectrometry. p. 29-65. In: A. L. Page (ed.), Methods of soil analysis, Part 2. Chemical and microbiological properties (2nd ed.). Amer. Soc. Agron., Madison, Wisconsin.

SONNEVELD, C. J., VAN DEN ENDE, and P. A. VAN DIJK. 1974. Analysis of growing media by means of a 1:1½ volume extract. Commun. Soil Sci. Plant Anal. 5:183-202.

SPENCER, K., JR., R. FRENEY, and M. B. JONES. 1978. Diagnosis of sulphur deficiency in plants. p. 507-513. In: A. R. Ferguson, R. L. Bigleski, and I. B. Ferguson (eds.), Proc. 8th Intern. Colloquium—Plant Analysis and Fertilizer Problems. New Zealand Depart. of Scientific & Industrial Res., Wellington, New Zealand.

STECKEL, J. E. and R. L. FLANNERY. 1971. Simultaneous determinations of phosphorus, potassium, calcium, and magnesium in wet digestion solutions of plant tissue by AutoAnalyzer. p. 83-96. In: L. M. Walsh (ed.), Instrumental methods of analysis for soils and plant tissue. Soil Sci. Soc. Amer., Madison, Wisconsin.

STEENBJERG, F. 1954. Manuring and plant production. p. 31-34. In: W. Reuther (ed.), Plant analysis and fertilizer problems. Institut de Recherches pour les Huiles et Olèagineux, Paris.

STEYN, W. J. A. 1961. The errors involved in the sampling of citrus and pineapple plants for leaf analysis purposes. p. 409-430. In: W. Reuther (ed.). Plant analysis and fertilizer problems. American Institute of Biological Sciences, Washington, DC.

SUMNER, M. E. 1977a. Use of the DRIS system in foliar diagnosis of crops at high levels. Commun. Soil Sci. Plant Anal. 8:251-268.

SUMNER, M. E. 1977b. Effect of corn leaf sampled on N, P, K, Ca and Mg content and calculated DRIS indices. Commun. Soil Sci. Plant Anal. 8:269-280.

SUMNER, M. E. 1977c. Use of the DRIS system in foliar diagnosis of crops at high yield levels. Commun. Soil Sci. Plant Anal. 8:251-268.

SUMNER, M. E. 1982. The diagnosis and recommendation integrated system (DRIS). In: Proceedings Soil and Plant Analysis Seminar, Anaheim, CA. Council on Soil Testing and Plant Analysis, Athens, Georgia.

SUMNER, M. E., R. RENEAU, J. O. AROGUN, and E. E. SHULTE. 1983. Foliar diagnostic norms for sorghum. Commun. Soil Sci. Plant Anal. 14:817-825.

SUNG, J. F. C., A. E. NEVISSI and F. B. DEVALLE. 1984. Simple sample digestion of sewage sludge for multi-element analysis. J. Envir. Sci. Health A19:959-972.

SYLTIE, P. W., S. W. MELSTED, and W. M. WALKER. 1972. Rapid tissue tests as indicators of yield, plant composition, and soil fertility for corn and soybeans. Commun. Soil Sci. Plant Anal. 3:37-49.

THOMAS, G. W. 1982. Exchangeable cations. p. 159-166. In: A. L. Page (ed.), Methods of soil analysis, Part 2. Chemical and microbiological properties (2nd ed.). Amer. Soc. Agron., Madison, Wisconsin.

THOMAS, G. W. and D. E. PEASLEE. 1973. Testing soils for phosphorus. p. 115-132. In: L. M. Walsh and J. D. Beaton (eds.), Soil testing and plant analysis (rev. ed.). Soil Sci. Soc. Amer., Madison, Wisconsin.

TOLG, G. 1974. The basis of trace analysis. p. 698-710. In: F. KORTE (ed.), Methodicium chimicum. Vol. 1, Analytical method. Part B, Micromethods, biological methods, quality control, automation. Academic Press, New York.

TRIERWEILER, J. R. and W. L. LINDSAY. 1969. EDTA-ammonium bicarbonate soil test for zinc. Soil Sci. Soc. Amer. Proc. 33:49-54.

TUCKER, M. P. 1984. Volumetric soil measures for routine soil testing. Commun. Soil Sci. Plant Anal. 15:833-840.

ULRICH, A. 1948. Plant analysis—methods and interpretation of results. p. 157-198. In: H. B. Kitchen (ed.), Diagnostic techniques for soils and crops. American Potash Institute, Washington, DC.

ULRICH, A. 1952. A physiological bases for assessing the nutritional requirements of plants. Annu. Rev. Plant Physiol. 3:207-228.

ULRICH, A. 1961. Plant analysis in sugar beet nutrition. p. 190-211. In: W. Reuther (ed.), Plant analysis and fertilizer problems. Publ. 8. American Institute of Biological Sciences, Washington, DC.

ULRICH, A. and F. J. HILLS. 1973. Plant analysis as an aid in fertilizing sugar crops: Part 1, Sugar beets. p. 271-288. In: L. M. Walsh and J. D. Beaton (eds.), Soil testing and plant analysis. Soil Sci. Soc. Amer., Madison, Wisconsin.

VIGLER, M. S., A. W. VARNES and H. A. STRECKER. 1980. Sample preparation techniques for AA and ICP spectroscopy. Amer. Lab. 12:31-34.

WALKLEY, A. and C. A. BLACK. 1934. An examination of the Degtjareff method for determining soil organic matter and a proposed modification of the chromic acid titration method. Soil Sci. 37:29-38.

WALL, L. L., C. W. GEHRKE, and J. SUZUKI. 1980. Automated turbidiation of determination of sulfate sulfur in soils and fertilizers and total sulfur in plant tissues. J. Assoc. Offic. Anal. Chem. 63:845-853.

WALLACE, T. 1961. The diagnosis of mineral deficiencies in plants by visual symptoms. Chemical Publ. Co., New York.

WARNCKE, D. D. 1980. Recommended test procedure for greenhouse growth media. p. 31-33. In: W. C. Dahnke (ed.), Recommended chemical soil test procedures for the north central region. North Dakota Agr. Expt. Sta. Bull. 499 (rev.).

WICKSTROM, G. A. 1967. Use of tissue testing in field diagnosis. p. 109-112. In: G. W. Hardy (ed.), Soil testing and plant analysis, Part II. SSSA Special Publ. 2. Soil Sci. Soc. Amer., Madison, Wisconsin.

WIERSUM, L. K. 1979. Effects of environment and cultural practices in calcium nutrition. Commun. Soil Sci. Plant Anal. 10:259-278.

WOLF, B. 1982a. An improved universal extracting solution and its use for diagnosing soil fertility. Commun. Soil Sci. Plant Anal. 13:1005-1033.

WOLF, B. 1982b. A comprehensive system of leaf analyses and its use for diagnosing crop nutrient status. Commun. Soil Sci. Plant Anal. 13:1035-1059.

WOODSON, W. R. and J. W. BOODLEY. 1983. Petiole nitrate concentration as an

indicator of geranium nitrogen. Commun. Soil Sci. Plant Anal. 14:363–372.
WORLEY, R. E. 1969. Pecan leaf analysis service summary, 1967. Georgia Expt. Sta. Res. Rpt. 55.
ZASOSKI, R. J. and R. G. BURAU. 1977. A rapid nitric-perchloric acid digestion method for multi-element tissue analysis. Commun. Soil Sci. Plant Anal. 8:425–436.

2

Carbohydrate Partitioning and Metabolism in Crops[1]

Jaleh Daie[2]
Department of Biology, Utah State University,
Logan, Utah 84322

[1] Paper No. 2953, Utah Agricultural Experiment Station. The author wishes to thank Drs. R. Wyse and J. Bennett for reading the manuscript and Scott Barber for secretarial assistance. The literature search was completed in June 1984.
[2] Present address: Department of Soils and Crops, Cook College, Rutgers University, New Brunswick, NJ 08903.

Horticultural Reviews, Volume 7

ISBN 0-87055-492-1

I. INTRODUCTION

Since the domestication of plants as a reliable source of food, there has been intense interest in improving crop productivity. Crop yields have been increased largely through efforts to increase plant adaptability to environmental conditions, resistance to diseases and pests, and genetic yield potential, and efforts to improve agronomic and horticultural practices. In spite of these efforts, reports by national and international organizations suggest that agricultural productivity is on a yield plateau (Wittwer 1978). This conclusion is most applicable to certain species in specific regions. No strong evidence exists, however, for a worldwide yield plateau for any crop that is receiving adequate research attention. As the amount of land available for agriculture decreases through urbanization, and as energy and production costs increase, yield per unit land area must be improved to keep up with increasing populations and to improve standards of living. Additional basic research is needed to provide new knowledge that eventually will provide the basis for increasing the yield potential of crop plants. Since plant scientists have yet to define adequately the specific physiological criteria for higher yields, there is great hope for further improvement of crop productivity.

To date, increases in crop yield have been achieved primarily by empirical selection for plants' ability to produce more harvestable organs under progressively improved systems of agricultural input. The potential for modifying carbohydrate allocation among plant parts is the basis for current interest in the mechanisms that regulate carbon partitioning and allocation patterns in crops. In this chapter, I categorize and describe factors that are important in determining allocation patterns and will consider the present state of knowledge regarding each major factor.

II. BIOLOGICAL DETERMINANTS OF YIELD

Crop productivity is affected by genetic as well as environmental variables. Among the environmental factors are water and temperature, nutrient availability, disease and pest resistance, and various cultural practices. Although description of all yield-limiting factors is beyond the scope of this review, two genetic determinants are briefly discussed: photosynthetic efficiency and harvest index. Of all the biological yield determinants, photosynthesis continues to be considered a high-priority area of research. While there seem to be many alternative ways to enhance photosynthetic efficiency (National Academy of Sciences 1977), some may be more practical than others. These include manipulation of the mechanisms responsible for the distribution of assimilates, which in turn regulate yield (Jain 1975).

A. Photosynthetic Efficiency

Since more than 90% of a plant's dry matter is the result of photosynthesis, it seems likely that a predictable correlation should exist between net photosynthesis and dry matter content of the whole plant. Both net photosynthesis and dry matter production vary among cultivars (Evans 1975; Ozbun 1978). However, the relationship between photosynthesis and yield is very complex, and for many crops a positive correlation is not obtained (Wittwer 1980). The reasons for a poor correlation between leaf photosynthesis and yield are varied. Before a crop canopy achieves full interception of light, leaf area is more important than the photosynthetic rate per unit leaf area in determining crop growth rate. After canopy closure, the carbon exchange rate (CER) per unit leaf area becomes the most important factor (Gifford and Jenkins 1983). No evidence is available, however, to show that CER per unit leaf area has increased during the domestication of important crops such as wheat, maize, sugarcane, and cowpea. Actually, the highest measured CER for wheat, cotton, and pearl millet are found in the wild relatives, not in modern cultivars (Gifford and Evans 1981). However, a correlation seems to exist between yield of forage crops and their respective CER.

High CER has always been considered a desirable characteristic. Several attempts have been made to select for a specific photosynthetic component that would result in a higher CER. In general, these attempts have not been very successful. For example, increasing chlorophyll content does not seem to have much impact on CER (McCashin and Canvin 1979). No good correlation has been shown between Hill reaction activity or phosphorylation per unit leaf area and CER (Hanson and Grier 1973). Nevertheless, a correlation between ribulose bisphosphate (RuBP) carboxylase activity and CER does seem to exist for some crops (Randall *et al.* 1977). However, since this enzyme constitutes 30–50% of total soluble protein in chloroplasts, selecting for it may result in reduced amounts of other important proteins. Therefore, breeding for higher specific activities of RuBP carboxylase seems to be a better approach. In this regard, not much success has been achieved except in isolated cases. Efforts to breed out photorespiration have failed so far, i.e., there is no clear evidence that a C_3 plant can have low oxygenase activity with normal carboxylase activity (Gifford and Evans 1981). Considering the low success rate for all of these approaches, an alternative might be to breed directly for overall CER, rather than for any one specific photosynthetic component.

One difficulty in trying to breed for overall CER is that different results may be obtained depending on whether photosynthetic rate is measured on a leaf area or canopy basis. Although a positive correlation may exist between yield and the CER of a single leaf, the results will not be the same on a canopy basis (Evans 1975). Peet *et al.* (1977) found a highly significant correlation between photosynthetic rate and yield for eight dry bean genotypes during podfill. Kueneman *et al.* (1979) measured photosynthetic rates for the upper leaf canopy of five cultivars of *Phaseolus vulgaris.* They reported that photosynthetic rates were not directly related to seed yields and proposed that CER measurements made at one stage of crop development or on a small portion of the total canopy are not particularly useful selection criteria for breeding programs.

Sharma *et al.* (1982) observed a positive correlation between net assimilation rate (NAR) and soybean yield when measurements were completed during the time of pod development and physiological maturity. Rodriguez and Lambeth (1976) measured photosynthesis for individual leaves of tomato and observed a positive correlation between economic yield, yield components, and photosynthesis. Bhagsari (1981) reported no correlation between root yield and net photosynthesis of individual leaves for sweet potato.

Based on the current literature, no final conclusion can be stated about a correlation between yield and photosynthesis. A more standardized system for measuring photosynthetic variables is needed. More important, at the whole plant level, a lack of correlation between photosynthetic rates and economic yield may result from the chemical partitioning of carbon between starch and sucrose, as evidenced in the work of Marini and Barden (1981). They showed a linear correlation between net photosynthetic rates and specific leaf weight (a possible indicator of leaf starch content) of apple leaves. Since the relationship between CER and final yield is weak, this area of research does not appear to be overly promising.

B. Harvest Index

Two major determinants of high productivity in a crop are its ability to produce high levels of photosynthates over a wide range of environmental conditions and to efficiently transport and partition a high proportion of those assimilates into economically important organs. If a crop plant is not able to allocate a major portion of the fixed carbon into yields, high photosynthetic rates would not translate into increased food production. Therefore, bioregulation for more efficient allocation of fixed carbon to harvestable parts may hold the key to assuring adequate world food supplies in the future.

Because agricultural yield (the economic product) is not necessarily synonymous with biological yield (total biomass), the concept of an "efficiency" index was first proposed by E. S. Beaven in 1914 (Donald and Hamblin 1976). In 1962, Donald proposed the more explicit term harvest index (HI), which is defined as the ratio of economic yield to total biomass production (HI = Economic Yield/Biological Yield).

Harvest index, by definition, has a value less than unity, and a range between 0 and 55% for most crops. Although HI indicates how the products of photosynthesis are partitioned among different plant parts, a positive correlation between economic yield and HI is not, by itself, unequivocal evidence of the value of the harvest index as a selection criterion (Donald and Hamblin 1976; Charles-Edwards 1982). Charles-Edwards (1982) argued that HI integrates phenological, physiological, and environmental factors and, unless each factor is well defined, HI does not provide much aid in understanding crop performance. Therefore, direct analysis of the main parameters of economic yield is a more realistic means of evaluating crop yield than is analysis of HI. This argument is consistent with previous findings that HI values of many crops are sensitive to environmental factors such as crop density, water status, and nitrogen availability (Donald and Hamblin 1976). They showed, in cereal grains, that up to a point as the population density was increased, a greater grain yield per unit area was achieved, but HI declined before the maximum yield was attained. Likewise, under water stress conditions, the harvest indices of several grain species dropped (Poostchi *et al.* 1972). Nitrogen application usually increases biomass production but reduces HI for some crops (Donald and Hamblin 1976). On the other hand, McNeal *et al.* (1971) reported that the increase they noted in biomass production greatly exceeded the decline in HI, the net result being increased grain yield.

Of the two alternatives to further improve crop yield—increasing photosynthetic rates and/or increasing harvest index—the potential for eventual success seems to be much greater if HI is manipulated by modifying carbon partitioning and allocation patterns. Although some description of how assimilates are partitioned in plants is emerging, no single aspect of it is fully understood. Crop plants are highly integrated systems consisting of multiple sources and sinks. Presumably, the whole system is so integrated that it maintains a balance of growth. Although this balance determines the harvestable yield, its mechanism is poorly understood. Partitioning within the whole plant involves production of carbohydrates in photosynthetic organs (*sources*), phloem loading, and subsequent translocation and unloading at regions of growth or storage (*sinks*). The remainder of this review will focus on the factors involved in efforts to achieve higher harvest indices through understanding carbon allocation patterns.

III. REGULATION OF CARBON PARTITIONING

Assimilate partitioning can be regulated at several sites within the plant. Possible regulatory points in the source regions will be discussed before those in sink organs. Regulation that occurs in one region, however, is not isolated from the other parts of the plant. Consideration of the integrated system is essential to interpret these phenomena. Therefore, source–sink communication also is discussed.

As already mentioned, high net photosynthetic rates do not necessarily contribute to high HI because a large part of the fixed CO_2 may be diverted into starch or nonharvestable biomass. The yield at any given level of photosynthesis thus depends on (1) the proportion of carbon available for export, which depends on carbon end-product formation and phloem loading (possible regulatory points in source areas), and (2) the capacity of yield organs to import carbon, which is subject to regulation within the sink region and is a function of the sink strength.

A. Regulation at the Source

The initial product of CO_2 fixation in C_3 plants is phosphoglycerate, which is then reduced to glyceraldehyde phosphate or triosephosphate. Triosephosphate, which is a substrate for several alternative pathways, is involved in (1) regeneration of ribulose bisphosphate (required for CO_2 fixation), (2) the photorespiration process (a substantial waste of fixed carbon), and (3) sucrose or starch formation (carbon end-product formation).

Triosephosphates are transported from the chloroplast into the cytoplasm, where they may be oxidized to phosphoglycerate or condensed to hexose phosphate, a precursor for sucrose, fructosans, and cell wall polysaccharides (Walker and Robinson 1978). Coupled with triosephosphate transport, the free inorganic phosphate produced by subsequent metabolism in the cytoplasm is exchanged back into the chloroplast by the triose-P translocator. During periods of high net CO_2 fixation, triosephosphates still in the chloroplasts are partially converted to starch and stored there. Starch reserves can be utilized at later times when photosynthetic rates are low.

1. Chemical Partitioning of Carbon within the Leaf. The first step in carbon allocation occurs within source leaves and is termed chemical partitioning or carbon end-product formation. The major end products are sucrose, which is translocated, and starch, which is temporarily stored in the chloroplasts.

a. Sucrose Synthesis. Sucrose is the principal sugar translocated in most crops. The primary site of sucrose synthesis in source leaves is the cytoplasm of mesophyll cells. Little if any sucrose is synthesized within the chloroplast, and since the inner membrane of the chloroplast envelope is impermeable to sucrose, any synthesis within the chloroplast would not contribute to the sucrose pool available for export from leaves. The rate of sucrose synthesis is a function of the carbon fixation rate, chemical partitioning of carbon between starch and sucrose, and the rate of sucrose export from the leaf.

Figure 2.1 illustrates the path of carbon from the chloroplast into the cytoplasm, cell wall free space, and phloem. The triosephosphate/phosphate translocator on the chloroplast membrane plays a crucial role in determining the relationship between sucrose and starch synthesis. Triosephosphate allocation for sucrose or starch synthesis is regulated by the ratio of phosphoglycerate to free phosphate (Heldt *et al.* 1977). The level of free phosphate in the leaf is influenced by the rate of sucrose synthesis (Geiger 1979), which depends on sucrose phosphate synthase activity (Huber and Israel 1982). The activity of sucrose phosphate synthase is, in turn, partially regulated by the sucrose pool in the leaf and is correlated with the rate of sucrose transport out of the leaves (Silvius *et al.* 1979). Recently, Huber and Bickett (1984) reported that high contents of fructose-2,6-bisphosphate in spinach leaves caused a reduction in sucrose formation, while enhancing starch production.

Sucrose levels in the cytoplasm depends directly on four enzymes that are involved in its synthesis and degradation. They are sucrose phosphate synthase (SPS), with its associated sucrose phosphate phosphatase (SPP), sucrose synthase (SS), and invertase (I). These enzymes catalyze the following reactions:

$$\text{UDP-glucose} + \text{fructose-6-P} \xrightarrow{\text{SPS}} \text{UDP} + \text{sucrose-P}$$

$$\text{sucrose-P} \xrightarrow{\text{SPP}} \text{sucrose} + P_i$$

$$\text{UDP-glucose} + \text{fructose} \overset{\text{SS}}{\rightleftharpoons} \text{UDP} + \text{sucrose}$$

$$\text{sucrose} \overset{\text{I}}{\longrightarrow} \text{fructose} + \text{glucose}$$

Two of these enzymes, SPS and SPP, are found only in leaf tissue, whereas SS and invertase are present in all types of tissue. While two important regulatory points in sucrose synthesis are SPS and fructose-1,6-bisphosphatase (FBPase) activity (Fig. 2.1), the roles of SS and invertase in leaves are less clear. Claussen (1983) observed a positive correlation between the distribution of assimilates and SS activity in different tissues of eggplant and suggested a causal relationship between the two factors. Claussen and Lenz (1983) reported that removal of eggplant fruits caused a reduction in leaf SPS activity concurrent with increased leaf starch levels, implying a sink regulation of sucrose formation.

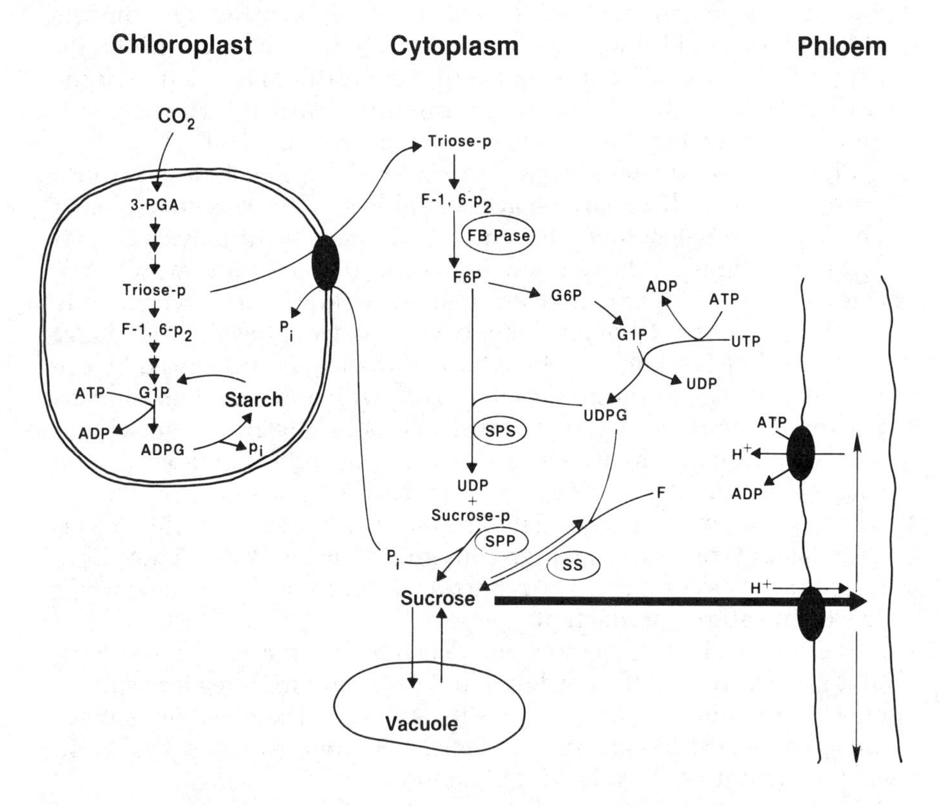

FIG. 2.1. Schematic presentation of carbon partitioning in source leaves. FBPase—Fructose-1,6-bisphosphatase. SPS—Sucrose phosphate synthase. SPP—Sucrose phosphate phosphatase. SS—Sucrose synthase. ●—At the chloroplast membrane, triose-P translocator; at the sieve tube membrane, sugar carrier protein or ATPase.

UDP-glucose and fructose-6-phosphate are allosteric activators of SPS, while UDP, other nucleotides, P_i, and fructose-1,6-bisphosphate inhibit SPS activity (Doehlert and Huber 1983). In some plants, sucrose causes a feedback inhibition of SPS (Huber and Israel 1982). Allocation of larger amounts of triosephosphate to sucrose rather than starch is related to SPS activity in leaves (Huber 1981). Reduced sucrose synthesis results in less free phosphate available for exchange with triosephosphate across the chloroplast membrane, thereby favoring starch synthesis in chloroplasts (Geiger 1979). Reaction rates for *in vivo* sucrose formation may be proportional to the cytoplasmic concentration of UDP-glucose, which in turn depends on the supplies of triosephosphate and ATP to regenerate UTP (Harbran *et al.* 1981).

Recent results by Huber (1981) support the hypotheses outlined here. Huber observed a good correlation between starch synthesis and sucrose-induced inhibition of SPS in plant tissues that exhibit different capacities for starch accumulation.

Sucrose levels within leaf cytoplasm are influenced not only by export from the source leaf, but by compartmentation within the vacuole. Recent work has shown that sucrose is transiently stored in leaf vacuoles as a buffer against short-term changes in photosynthesis and export from the mesophyll cells (Fisher and Outlaw 1979; Fondy and Geiger 1983). The mechanism and control of sucrose transport into these vacuoles, therefore, constitutes a potentially important control point for carbon allocation within source leaves.

b. Starch Formation. The net efflux of fixed carbon from chloroplasts during photosynthesis often is less than half of that assimilated during a light period because part of the fixed carbon is sequestered as starch within the chloroplast (Challa 1976; Clauss *et al.* 1964; Ho and Shaw 1977; Walker and Herold 1977). Starch synthesis does not seem to reflect fixation of "excess" carbon; rather, it may represent a preprogrammed portion of the total carbon fixed (Chatterton and Silvius 1980, 1981). If this interpretation is true, then starch synthesis would compete with sucrose synthesis and consequently with phloem loading for available carbon. Moreover, when starch levels are high in the leaf, its further synthesis may be impaired owing to a feedback process. The result is increased soluble carbohydrates, which may promote a reduction in net photosynthetic rates (Mauney *et al.* 1979; Nafziger and Koller 1976). Understanding starch metabolism within leaves may be an important lead for developing ways to enhance carbon partitioning into yield organs.

Starch deposition may be beneficial because the stored starch is mobilized during the dark period. Starch metabolism, however, is a dynamic process; synthesis and catabolism occur concomitantly throughout the light period. In view of the amount of energy required for the synthesis and breakdown of starch, its turnover during the day appears to be a wasteful process. Starch synthesis may also represent a costly process to the plant, in light of recent findings by Huber (1983) with vegetative plants. He showed that remobilized starch may be used preferentially for growth of shoots relative to roots (sink organs). How starch contributes to vegetative and reproductive yield organs during darkness is not known, and more work in this area is warranted.

Two critical enzymes required for starch synthesis in leaves are ADP-glucose pyrophosphorylase (ADPG-PP) and starch synthase (St.S), which catalyze the following reactions:

$$\text{ATP} + \text{glucose-1-P} \xrightarrow{\text{ADPG-PP}} \text{ADP-glucose} + \text{PP}_i$$
$$\text{ADP-glucose} + \alpha\text{-glucan} \xrightarrow{\text{St.S}} \alpha\text{-1,4-glucosyl-glucan} + \text{ADP}$$

The ADPG-PP step as well as the St.S reaction are likely to be control points in the starch biosynthesis pathway (Preiss and Levi 1979).

ADP-glucose pyrophosphorylase activity is regulated by the ratio of P_i to 3-phosphoglycerate in the chloroplast: 3-phosphoglycerate activates the synthesis of ADP-glucose, whereas inorganic phosphate is a potent inhibitor (Preiss 1984; Preiss and Boyer 1980). Because ADP-glucose is the sole glycosyl precursor for starch synthesis in leaf tissue, starch synthesis may be regulated by controlling ADP-glucose levels (Sanwal *et al.* 1967). Nomura *et al.* (1967), however, did not observe any stimulation of phosphorylase activity by 3-phosphoglycerate in bean and rice leaves.

2. End-Product Inhibition of Photosynthesis. Whether starch build-up and/or sucrose formation regulates photosynthetic rates in source leaves remains equivocal. Some investigators have concluded that starch accumulation has no direct effect on photosynthetic rates (Claussen and Biller 1977; Crookston *et al.* 1974; Forde *et al.* 1975). Numerous reports, however, state that photosynthetic rates decline as leaf starch content increases (Chatterton *et al.* 1972; Mauney *et al.* 1979; Thorne and Koller 1974; Nafziger and Koller 1976). Nafziger and Koller (1976) suggested that as long as starch levels remain low, no significant inhibition of photosynthesis occurs. Their data on soybean indicated that only a 10% reduction in photosynthetic rate was associated with a leaf starch content of 28% (dry weight basis); in order for a sharp drop in photosynthetic rates to occur, leaves must have at least 35% starch (dry weight).

Many of the findings cited have been the result of drastic manipulations, such as major source–sink alterations, that could cause secondary effects in addition to modifying photosynthetic rates. In an attempt to cause minimal alteration to the plant system, Potter and Breen (1980) subjected sunflower and soybean plants to 52 hr or more of continuous light. Starch accumulation in sunflower and, to a lesser extent in soybean, was not associated with reduced photosynthetic rates. Although the older leaves of sunflower accumulated less starch than did young leaves, photosynthesis declined to a greater extent than in the young leaves. Leaf maturity thus was a significant determinant in the photosynthetic response to continuous light and its associated starch accumulation. Furthermore, Potter and Breen (1980) suggested that starch inhibition of photosynthesis is not likely to be due to interference with CO_2 diffusion or light absorption processes.

The contrasting data available reflect the complexity of the plant system, which involves interacting factors that need to be considered as an integrated phenomenon. If starch formation is not merely a preprogrammed process, its buildup may result from reduced demand for sucrose export. If so, then sink demand will play a regulatory role in photosynthetic rate fluctuations. Furthermore, plant growth regulators, possibly originating from sinks or elsewhere in the plant, may also be involved in starch formation (Daie 1984a).

3. Phloem Loading. Phloem loading is the accumulation of sucrose in the sieve tube elements and may be an important control point within the source. As first proposed by Giaquinta (1977), it is thought to be an active, proton-sugar cotransport process. Sucrose availability for loading is an important factor controlling the rate of export from leaves and translocation. Therefore, understanding how sucrose exits from mesophyll cells remains central to understanding phloem loading and translocation.

a. Sucrose Transfer to Phloem. The movement of sucrose from mesophyll cells to sieve tubes may be symplastic, apoplastic, or a combination of the two. Plasmodesmatal connections are required for symplastic movement. However, solutes could diffuse apoplastically from mesophyll cells to sieve tubes along a concentration gradient (Ziegler 1974). If symplastic transport is involved, ^{14}C assimilates are expected to be retained by isolated mesophyll cells. Kaiser *et al.* (1979) showed that very little leakage of ^{14}C assimilate occurred from leaf protoplasts isolated from *Papaver* and *Spinacia*. Retention of radioactivity by the cells was also observed in the intact cells.

In maize leaves, extensive plasmodesmatal connections exist between the mesophyll cells and bundle sheath cells, as well as among mesophyll cells (Evert *et al.* 1978). This permits the symplastic movement of assimilates through leaf laminae. Apoplastic sucrose transfer, however, has also been reported for maize (Heyser *et al.* 1976, 1978). Francheschi and Giaquinta (1983) showed that in some species (soybean, mungbean, and wingbean) there is a unique cell layer called paraveinal mesophyll (PVM). Plasmodesmata frequently exist between mesophyll-PVM, PVM-PVM, and PVM-bundle sheath, indicating symplastic movement of solutes through leaf laminae.

In regard to mechanisms of sucrose efflux, Doman and Geiger (1979), based on calculations of the rate of sucrose movement and membrane surface properties, suggested a facilitated transfer mechanism for sugar beet efflux from mesophyll cells. They also showed a K^+-induced sucrose efflux from mesophyll cells of sugar beet (sucrose efflux coupled to K^+ influx). This suggestion is plausible if sucrose release is near the site of, and coupled to, the process of sucrose loading. Phloem loading of sucrose is a proton cotransport system, with K^+ ions moving out in response to the membrane potential. If this occurs *in vivo*, then K^+ in the free space would increase, which in turn would result in an increased sugar efflux from mesophyll cells near the phloem region. Huber and Moreland (1980, 1981) have described a sucrose–K^+ cotransport efflux system in tobacco and wheat protoplasts. The discrepancy between the results obtained in these studies may be from the differences in intact tissue versus protoplasts.

Giaquinta (1983), in a recent review of the literature, suggested that the most likely path for assimilate movement from mesophyll cells to the phloem regions is a symplastic transfer, since most prevailing evidence indicates that adequate symplastic connections exist between mesophyll cells. Additionally, an entirely apoplastic transfer of sucrose to the phloem region would be impeded by the transpiration stream.

b. Symplastic or Apoplastic Phloem Loading. Assimilate transfer into the sieve tube element/companion cell (the step immediately prior to loading) can also occur via the symplast or apoplast. Sucrose most likely enters the apoplast in the proximity of the sieve tube companion cell complex prior to loading. As evidenced by high resolution autoradiography of bean (Giaquinta and Geiger 1977), and also in work on *Vicia* (Delrot 1981), tobacco (Moss and Rasmussen 1969), and sugar beet (Giaquinta 1983), phloem has a significantly higher concentration of sucrose than do mesophyll cells. In order for this concentration to occur symplastically, there should be adequate continuity between the sieve elements and the surrounding mesophyll cells. Plasmodesmatal connections in this area are rare in *Vicia faba*, maize, *Tussilago*, and sugar beet (Evert *et al.* 1978; Geiger *et al.* 1973; Gunning 1976; Gunning *et al.* 1974). Furthermore, plasmodesmata do not seem to function as concentrating elements.

Several investigators have presented evidence supporting mesophyll unloading of sugars into the apoplast prior to loading into sieve tubes. Kursanov and Brovchenko (1970) showed that more than 20% of the sugar is present in the apoplast of sugar beet leaves. Geiger and co-investigators, using various elaborate techniques, have presented evidence that apoplastically applied sucrose is readily available for loading. In $^{14}CO_2$ gassing experiments with sugar beets, they reported that under conditions of high photosynthesis, [^{14}C] sucrose increased in a solution that perfused the apoplast and, when symplastic continuity was disrupted using 800 m*M* mannitol to cause plasmolysis, translocation rates could be restored by supplying 20 m*M* sucrose to the apoplast (Geiger 1975; Geiger *et al.* 1974; Sovonick *et al.* 1974). Heyser (1980) also showed that [^{14}C] sucrose was the only sugar present in xylem exudate after exposure to $^{14}CO_2$ and that its concentration increased with increasing photosynthetically active radiation. Assuming that the nonpermeating sulfhydryl-reacting inhibitor *p*-chloromercuribenzoylsulfonic acid (PCMBS) did not affect plasmodesmata function, Giaquinta (1976) concluded that sucrose entered the apoplast prior to loading into the phloem.

However, entry of sucrose into the apoplast before sieve tube loading is not a universal phenomenon. Madore and Webb (1981), for example, concluded, based on the absence of loading of exogenously applied [^{14}C] stachyose, that the major route of sugar loading in *Cucurbita* was symplastic. Nevertheless, the bulk of the evidence in the literature is consistent with apoplastic entry of sucrose before its loading into sieve tube elements. The following features of the loading system should be considered in deciding whether a symplastic or apoplastic pathway, prior to loading, is most consistent with the evidence.

- The selectivity of the loading process argues against a symplastic transfer system, which is unlikely to be selective.
- The presence of plasmodesmata is not final proof of a symplastic transport system. They may have developed there during cell division for the purpose of unloading (leaf sink stage), rather than loading (Thorne and Giaquinta 1982; Giaquinta *et al.* 1983).
- As a leaf progresses through sink-to-source transition, biochemical and structural changes occur in the phloem (Giaquinta 1983)

and membrane transport system for loading from the apoplast is developed (Giaquinta 1980). This is evidenced by the acquisition of ATPase activity in the phloem membrane with the onset of export capability (Cronshaw 1981).

- A concentrating step occurs at the sieve elements. Unless plasmodesmata act as "one-way valve" systems, the alternative explanation that an active influx of sugars occurs across the phloem membrane from the apoplast seems more likely.
- Phloem loading has been shown to be a proton-sugar cotransport process that is consistent with an active, carrier-mediated membrane transport system, a process that can only occur across a membrane.

*c. **Sucrose Loading.*** Active membrane transport of sucrose is coupled to the transport of protons across the membrane of the sieve tube element. This hypothesis has been substantiated by several investigators (Komor 1977; Malek and Baker 1978; Delrot and Bonnemain 1979; Delrot *et al.* 1980; Delrot 1981; Humphreys 1980). Indirect experimental evidence and the ion gradients known to exist across the plasmalemma of the sieve elements are consistent with a proton-sucrose cotransport system. According to the model, sucrose loading occurs via a sucrose-specific carrier system. The driving force for the sugar transport is a proton gradient from the acidic free space (pH 5.5) to the alkaline sieve element (pH 8.0). The electrochemical gradient is established by a proton-pumping ATPase (See Fig. 2.1). The evidence to date for this mechanism of uptake is circumstantial, and will remain so unless a relatively pure preparation of phloem cells can be studied. The proton cotransport mechanism of sucrose loading has received considerable attention, but the uptake kinetics have not been well studied.

Sucrose loading exhibits biphasic kinetics composed of a high-affinity, saturable component (low K_m) and a low-affinity, linear component (high K_m) (Maynard and Lucas 1982; Daie and Wyse 1983a). Maynard and Lucas (1982) suggested that the two transport systems operate in parallel, with the saturable component being associated with proton cotransport. Although the contribution of the linear component is significant, its nature remains obscure. The linear kinetics suggest that it is a physical process such as diffusion, but Maynard and Lucas (1982) have shown that it is sensitive to metabolic inhibitors, temperature, and anoxia.

4. Regulation of Carbon Export and Translocation. Factors involved in controlling sucrose export from leaves include chemical partitioning of carbon between starch and sucrose, sucrose compartmentation within cells, sucrose release from mesophyll cells for loading, membrane transport of sugar into phloem, and hormonal regulation of partitioning and transport.

Recent findings suggest that the role of a source leaf is to control the timing and supply of sucrose for export as well as to provide the energy for sucrose loading into the phloem (Fondy and Geiger 1983). Once sucrose is loaded, there is no evidence that the source controls its final destination (Gifford and Evans 1981).

Sucrose synthesized in the cytoplasm of mesophyll cells is first transferred to vein boundary cells and then into the sieve tubes for export from the leaf. Sieve tubes and companion cells are high in solute content (Geiger and Fondy 1980), which results in decreased water potential of the sieve tubes, leading to water movement into these cells. A water potential gradient is then established between source and sink (low solute content), along which phloem sap moves *en masse* (Fondy and Geiger 1983). Sucrose is the major contributor to the high solute content of sieve tubes in most species. As solutes are unloaded into various sink regions along the path of translocation, osmotic potential gradients between source and sink are regulated not only by the loss of solutes but also in response to frictional pressure potential losses along the pathway.

Translocation rates appear to be coupled directly to the sugar pool available for export. Decreased photosynthesis usually results in reduced translocation, and the ratio between the two rates changes linearly (Fondy and Geiger 1983). Servaites and Geiger (1974) observed that even when the rate of photosynthesis was high (high CO_2 and saturating light intensity), translocation did not slow. Thorne and Koller (1974) also showed that translocation increased in response to increased sink demand. Collectively, these studies, as well as studies on the kinetics of sucrose uptake, strongly indicate that translocation rates in the phloem do not limit the movement of assimilates, and that the presence of a rate-limiting step at sieve tube or companion cell membranes is not a likely barrier to rates of export (Fondy and Geiger 1983).

The availability of sucrose for export and turgor changes in the phloem appear to be the important determinants controlling translocation rates. Changes in sink demand also can modify sucrose loading and hence translocation rates by causing changes in turgor at the site of loading (Giaquinta 1983). Consistent with a sink-regulated phloem loading, Smith and Milburn (1980) have shown that when sink demand was increased by incisions along the phloem path, positive turgor was maintained at the site of loading. Maintenance of turgor was achieved by increased sucrose loading. These authors did not propose a mechanism for how enhanced rates of sucrose uptake occurred. However, Daie and Wyse (1985) provided evidence showing that under reduced turgor conditions, uptake kinetics are modified, resulting in the activation of the active component of sucrose transport. The active component is inhibited at high cellular turgor conditions, and sucrose uptake kinetics are linear.

B. Partitioning at the Sink

Growing sinks play a key role in determining the rate of sucrose export out of leaves. The sink's ability to absorb assimilates affects the rate and direction of carbon end-product formation. For example, sucrose loading and export are directly related to sucrose concentration in the leaf apoplast (Geiger 1976). Flux rates to the sinks, however, are regulated by the sucrose concentration gradient between source and sink (Ho 1979; Milburn 1974; Walker and Thornley 1977). This mobilizing ability of sinks may play a role in controlling sucrose flux rates and ultimately potential photosynthetic rates (Thorne and Koller 1974; Habeshaw 1973). Furthermore, starch accumulation during periods of high rates of photosynthesis may indicate that sucrose transport out of leaves may be lagging behind the rate of photosynthesis. This may be the result of reduced sink demand for assimilates or preferential control toward starch synthesis.

As discussed previously, the source does not seem to control the final distribution patterns of translocated sucrose. Carbohydrate allocation patterns ultimately depend on the competitive ability of various sinks and their mobilizing abilities. Phloem unloading, membrane transport, sugar metabolism, and storage within the sink region all influence partitioning in the sink.

1. Mobilizing Ability. The eventual carbohydrate distribution pattern in a plant is related to the relative competitive ability of the various sink regions within the whole plant. Increasing the harvest index of crops requires development of cultivars with sinks that are efficient importers of assimilates. A high sink import rate is a function of its mobilizing ability. In turn, the sink mobilizing ability, or strength, is a function of sink size and activity, as described by the following equation:

$$\text{Sink Strength (g/day)} = \text{Sink Size (g)} \times \text{Sink Activity (g/g/day)}$$

The factors that contribute to sink activity are not well understood, but they involve phloem unloading, uptake of assimilates by sink tissue, and metabolism and carbohydrate storage in sink regions.

Although the biochemical nature of the control factors remains unclear, sink activity influences sucrose gradients between various source and sink regions (Ho 1979). These sucrose gradients are hypothesized to control carbon flux to sink regions (as more assimilates are taken up by sink cells, a steeper gradient is established between source and sink, which leads to more carbon flux). Therefore, the ability of a sink to maintain a low apoplastic sucrose concentration, either by metabolism or compartmentalization, should enhance its mobilizing ability relative to other sinks.

The ultimate determinant of a sink's import rate is subject to debate. Depending on experimental approach and conditions, sink size or activity may appear to be the predominant factor. Walker and Ho (1977a) stated that the sink strength of tomato fruit appeared to depend more on sink activity than on size. They cited declines in the rate of carbon import by young tomato fruits that were 20–90% of their final size, as fruit size increased. Smaller fruits imported twice as much carbon on an individual fruit basis than did larger fruits. Hewitt *et al.* (1982) observed different fruit solid contents in two tomato genotypes and attributed the high fruit solids content of one genotype to greater sink strength and larger leaf area. Dinar and Stevens (1981) found that the genotypic differences in content of starch and soluble solids in tomato fruits were due to variations in sink activity and that starch accumulation in fruits appeared to be the result of greater sink activity. Using $^{11}CO_2$ as a tracer, Moorby *et al.* (1974) presented evidence for the effect of sink activity on translocation. They showed that translocation in tomato plants was reduced 15–30% when sink activity was reduced by cooling fruits. Walker and Ho (1976) observed similar results and suggested that reduced growth of tomato fruit when the fruit was exposed to low temperatures may have been a result of reduced sink activity.

In a study designed to determine the influence of bean pod size on competition for assimilates, Olufajo *et al.* (1982) found that, following the removal of major pod sinks, assimilates became available for general distribution in the plant. The final destination of assimilates thus was regulated by the developmental stage of alternative sinks. Their results indicated the advantage of larger sinks.

Cook and Evans (1983) designed an experimental system with wheat in which access to reserve carbohydrates was minimal and the distribution patterns were mainly from the use of current carbohydrate supplies. Larger sinks were better competitors for assimilates, and the partitioning between sinks of the same size was proportional to $1/d^2$ (d = distance from source). Melon fruits place a strong demand on the assimilate supply of nearby leaves but have much smaller effect on leaves at a distance from the fruit (Hughes *et al.* 1983), implying that the proximity of source and sink influences sink strength.

These examples illustrate that the relative importance of sink size, proximity to source, and sink activity as determinants of sink strength are not well defined. Complications notwithstanding, the important question seems to center on which factors allow a sink to become bigger relative to other, competing sinks in the first place. Sink activity and its proximity to the source certainly will be important, but sinks that grow at faster rates (i.e., are larger) than other sinks will undoubtedly be at an advantage.

Under either source- or sink-limiting conditions, sink strength is a major determinant of translocation rate, so that similar-sized fruits have similar import rates (Ho 1979). To further complicate the issue, depending on the developmental stage of the plant, nonstorage, actively growing regions may be the strongest sinks for assimilates. Gawronska *et al.* (1984) reported that the developing green organs of potato at an early stage have greater sink strength than do the small

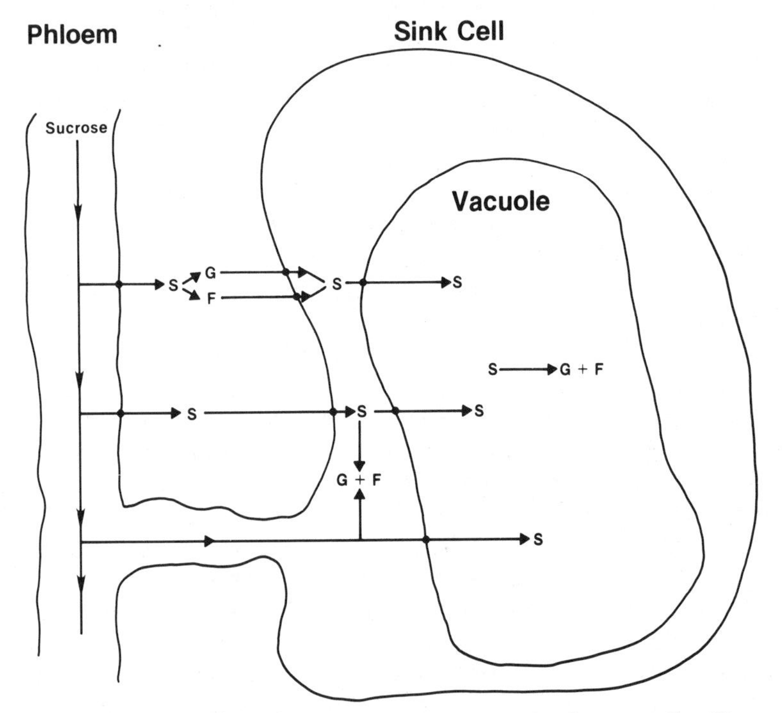

FIG. 2.2. Possible pathways of sucrose unloading. S—Sucrose. G—Glucose. F—Fructose.
From Giaquinta et al. (1983).

tubers. This obviously is the result of sink activity, since the meristem regions are much smaller than tubers.

The basis for high sink activity remains to be resolved. Ho (1979) reported that constant sink demand caused heavy loss of carbon even from leaves with minimal rates of carbon fixation. He concluded that sink demand can override assimilate availability in regulating translocation. Likewise, Gifford and Evans (1981) concluded that sink strength, rather than assimilate availability (source activity) or vascular connections between the source and sink, determines assimilate allocation. A rapidly growing sink creates a steeper gradient between source and sink, and it therefore competes at an advantage relative to a weaker sink. Hormones interact to modify the movement of assimilates towards active sink regions, but their involvement and mechanism of action are not clear. The possible role of hormones is discussed in Section III.D.1.

2. Phloem Unloading. Sucrose unloading from the phloem may be an important control point in determining carbon allocation patterns, but progress in understanding this process has been limited owing to the inaccessibility of unloading sites. Circumstantial evidence about whether the pathway of movement is apoplastic or symplastic can be gleaned from information about the structure of the unloading sites and the frequency of plasmodesmatal connections between phloem and sink cells. Sugars may be transported to the sink cells via a symplastic route through plasmodesmata, or sucrose may be unloaded into the apoplast prior to transport into sink cells (Giaquinta *et al.* 1983). Figure 2.2 is a schematic depiction of possible pathways for sucrose unloading. Based on the experimental evidence available, Ho and Baker (1982) classified different sinks according to the path of unloading (Table 2.1). No universal mechanism or pathway exists for sucrose unloading in all plants; unloading pathways should therefore be determined for specific tissues.

TABLE 2.1 Classification of Sink Organs

Sink	Unloading path	Control point
Growth sinks (apical meristems and young leaves)	Symplastic	Metabolic activity (rate of growth)
Storage sinks sugar sink sucrose (sugarcane stem, taproot) hexoses (grape berry) starch sink (cereal grains, tubers) starch–sugar sink (tomato fruit)	Apoplastic at some point, but not necessarily at step out of sieve element	Hydrolysis of sucrose; active uptake of sucrose; hexose or starch synthesis

Source: Ho and Baker (1982).

Sucrose is unloaded in the sink apoplast of several species, including maize (Felker and Shannon 1980), soybean (Thorne 1982), wheat (Jenner 1974), and sugarcane (Glasziou and Gayler 1972). Apoplastic unloading is required in some reproductive tissues (e.g., developing bean and soybean seeds) because there are no plasmodesmatal connections between the maternal phloem strands and developing endosperm (Thorne 1982). However, strong evidence supports the presence of symplastic sucrose unloading in young (sink) sugar beet leaves (Giaquinta 1977), pea roots (Dick and ApRees 1975), and maize roots (Giaquinta *et al.* 1983).

If an apoplastic step mediates unloading, the concentration of sucrose in the free space becomes important in controlling rates of unloading (Patrick 1983). Low apoplastic sucrose concentrations are achieved in two ways. Sucrose moves into the storage tissue of crops such as soybean (Thorne 1982), mature sugar beet root (Wyse 1979), and mature carrot (Daie 1984b) without hydrolysis. In these cases, the rate-determining step is sucrose transport across the plasmalemma of storage cells. In other crops such as maize (Shannon and Dougherty 1972) and sugarcane (Bowen and Hunter 1972), low apoplastic sucrose concentrations are maintained through sucrose hydrolysis by a cell wall invertase. Glucose and fructose then are the major transport sugars present in the apoplast. The presence of acid invertase activity at the site of unloading enhances unloading not only by ensuring low apoplastic sucrose levels but also by preventing sucrose from reloading and increasing the osmotic concentration in the free space. Cell wall invertase activity in sink regions has been suggested as a biochemical indicator of sink strength (Eschrich 1980).

The mechanism of unloading into the sink apoplast could involve passive, facilitated, or active transport. Gifford and Evans (1981) suggested that unloading can occur by leakage (passive efflux) into areas where phloem loading is impaired, or that sinks at some regions prevent loading. Bennett *et al.* (1984), however, reported results contrary to this hypothesis. Using soybean embryoless seed coat "cups," they showed that no reloading of [^{14}C] sucrose back into the shoot system occurred from the site of unloading.

Recent findings indicate that unloading may involve a carrier-mediated transport system. Thorne (1982, 1983) has shown that unloading of sucrose into developing soybean seeds is sensitive to temperature, anoxia, nonpenetrating inhibitors, metabolic inhibitors, and divalent metal chelators. He proposed that the energy dependence of phloem unloading may reflect the energy metabolism in the companion cell–sieve element complex, which is consistent with a facilitated mode of transport involving essential sulfhydryl groups on membrane-bound proteins. Recently, Patrick (1983) reported an energized, facilitated unloading system in the seed coats of *Phaseolus vulgaris* where use of PCMBs and other metabolic inhibitors resulted in a significant reduction of assimilate efflux. He also observed stimulation of sucrose efflux when apoplastic sucrose concentrations were low. Later, he reported (Patrick 1984) that the sucrose inhibition of unloading was not the result of transinhibition but rather a response to solution osmolality, i.e. cellular turgor. The rate of unloading depended on cell turgor potential and was stimulated under high turgor conditions. Patrick (1984) proposed that a turgor-sensitive unloading process provided the appropriate mechanism for sink control of assimilate unloading in the developing bean ovules. Similarly, Griffith and Wyse (1983) reported that sucrose efflux is carrier-mediated in hollow stems of *Vicia faba*. If unloading is carrier mediated, then high apoplastic sucrose concentrations are expected to inhibit efflux. This was assumed to be the case for *Vicia faba* stems. However, this conclusion may be incorrect, since these authors did not distinguish between sucrose concentration and solution osmolality as Patrick (1984) had done. Wolswinkel and Ammerlaan (1983), working with developing seeds of *Vicia faba*, also observed inhibition of sucrose and amino acid unloading in the presence of metabolic inhibitors. They did not, however, conclude that a carrier was present.

Phloem unloading simply may be a reversal of the loading process. When the proton-motive force is too low to maintain the sucrose gradient across a membrane, unloading occurs. In this context, unloading can be viewed as a change in the balance between sucrose influx and efflux.

3. Membrane Transport of Sugars. Uptake, metabolism, and storage of sugars and carbohydrates by sink cells are the final steps in carbon partitioning. Like phloem loading, sugar transport into many sink types occurs via a proton cotransport system (Komor 1982 and references therein). An active transport system is considered to be a universal feature of tissues that show apoplastic unloading. Such a process would serve as a control mechanism at the sink region and explains the influence of the sink on assimilate partitioning within the whole plant (Gifford and Evans 1981).

Biphasic sugar transport kinetics have been observed in several sink tissues including carrot root (Daie 1984b), sugar beet root (Saftner *et al.* 1983), and soybean seed (Thorne 1982; Lichtner and Spanswick 1981). At low apoplastic sucrose concentrations, an active, saturating transport component operates. At higher concentrations, transport is by a nonsaturating, linear system. Schmitt *et al.* (1984) showed that the linear portion of the uptake in protoplasts of soybean cotyledons is a combination of energy-dependent [(*p*-trifluoromethoxy) carbonyl cyanide, FCCP-sensitive] transport and diffusion. The linear component is a significant portion of total uptake and may represent transport via a system that utilizes a different energy source than is used by the active transport system. Alternatively, the biphasic kinetics of sugar transport in the sink can be explained by two forms of a carrier: a high-affinity (low K_m), saturable type, and a low-affinity (high K_m), nonsaturable type. Obviously, more research is needed to clarify the role and mechanism of the linear component.

Membrane transport of sugars is particularly complex in sinks that store simple sugars, as occurs in sugarcane, sugar beet, and carrot. Sucrose is stored in the vacuoles of parenchyma cells in these tissues at concentrations as high as 800 m*M* (e.g., sugar beet). Therefore, in addition to transport across the plasmalemma, a tonoplast transport step is also required. The transport mechanism at the tonoplast may be quite different from that across the plasmalemma since the pH gradient and electrical potentials are in opposite directions across plasmalemma and tonoplast membranes (Komor *et al.* 1982; Bennett and Spanswick 1983). Relative to the outside of the cell, a negative electrical potential exists across the plasmalemma with the inside pH being higher. A lower pH and more positive potential are found inside the vacuole relative to the cytoplasm. The pH differential is maintained by a proton–ion-pumping ATPase (Bennett and Spanswick 1983).

Sugar transport across the plasmalemma is considered to be a sugar-proton cotransport process. If so, it follows that, due to reversal of the gradients across the tonoplast, the vacuolar transport step must be a proton antiport system. Saftner and Wyse (1980) provided indirect evidence that sucrose uptake into the vacuoles of sugar beet taproot was a K^+-stimulated, proton antiport system. Recently, R. E. Wyse (personal communication) observed a sucrose-proton antiport using membrane vesicles derived from sugar beet tonoplasts. These data suggest that the tonoplast system differs mechanistically from the sink plasmalemma and phloem-loading systems in source tissue and may present a control point for carbon allocation in some species.

C. Establishment of Sink Load

The final yield is a function of the number of yield organs present and the size they attain. Usually an inverse relationship exists between these two factors. In the case of seed and fruit crops, the final number of yield organs depends on flower formation, abortion, and fruit set vs abortion. Another important factor affecting the potential yield of some perennial crops is alternate bearing. A full discussion of this subject is beyond the scope of this article. However, because establishment of specific sink load is relevant to horticultural production, a thorough review of the literature in this area is warranted.

Although alternate bearing by trees is still largely unexplained (Monselise and Goldschmidt 1982), it is widely believed to be related to the carbohydrate status of the plant. If carbohydrate reserves are exhausted during an "on" year, there will be an adverse effect on flower bud formation leading to a subsequent "off" year. Another suggested cause of alternate bearing may be fruit-produced inhibitor(s). Galliani *et al.* (1975) suggested that alternate bearing can be reduced by plant growth regulator–induced thinning. Since growth and abscission of some fruits is highly dependent on seed development (Valpuesta and Bukovac 1983), seed-produced hormones are believed to be crucial in this relationship (Crane 1964). Fruit abscission may be hormonally controlled for fruit crops such as peach (Martin and Nishijima 1972). Flower abscission in crops such as pistachio is also believed to be due to hormonal changes (Crane *et al.* 1976).

The cause(s) of flower and fruit formation, abortion, and abscission remain equivocal. Evidence favoring a hormonal vs a carbohydrate reserve hypothesis has been suggested for several crops. Obviously, persisting flowers and fruits are stronger sinks for photosynthates than are abscising ones. The higher sink activity of persisting fruits may be subject to hormonal regulation. In such cases, advanced and strong competition for assimilates (by persisting fruits) would be a secondary result of hormone action. The crucial question thus seems to be identifying the initial stimulus that triggers an organ to become a strong sink, ensuring its persistence. In the final analysis, the presence of yield organs is most likely due to interactions between hormones and the carbohydrate status of the plant.

D. Source–Sink Communication

Plants are highly integrated organisms requiring signal transmission and regulatory communication among their parts. The nature of such signal(s) is not known, but it may be a discrete entity or the combined result of several physiological processes. Source–sink communication has been the subject of much discussion, but little hard evidence is available to describe a mechanism. A simple hypothesis has been developed based on turgor pressure in the phloem. According to this hypothesis, a sink-induced reduction in turgor pressure at the site of loading will enhance loading and thereby facilitate assimilate movement toward sites of unloading.

Plant growth regulators have also been suggested as likely candidates for signal transmission because they are present in all plant parts and may be translocated throughout the plant. Furthermore, plant growth regulators may be involved in regulating turgor and thus may have a secondary effect on source–sink communication.

1. Hormonal Regulation of Assimilate Partitioning. Partitioning clearly is a function of the competitive ability of various growth centers and must be studied in the context of growth and differentiation. Plant growth regulators are involved in all aspects of growth and differentiation and therefore may play a key role in sink strength. In addition to their effect on sink growth, plant growth regulators also exert a mobilizing influence (Patrick 1982; Wyse *et al.* 1980; Saftner and Wyse 1984).

For almost half a century, plant growth regulators have been implicated in the regulation of assimilate translocation. More recently, a role has been proposed for some growth regulators in photosynthate allocation (Lis and Antoszewski 1982; Patrick and Wareing 1978; Mulligan and Patrick 1979). Although the mechanism for so-called hormone-directed transport remains unclear, the effect of plant growth regulators on assimilate transport generally is accepted not to be solely the result of hormone-induced growth (Patrick 1979). Since growth regulators are present in all parts of a plant, they may modify partitioning patterns within any organ or tissue.

a. Effects at the Source. Most current information on hormonal control of partitioning in the leaf is confined to the effects of hormones on photosynthesis and phloem loading. Wareing *et al.* (1968) have shown that gibberellic acid (GA_3) and kinetin enhance photosynthetic activity, whereas ABA generally is believed to lead to a reduction of photosynthesis through its effects on stomatal closure; ABA, however, may inhibit photosynthesis directly (Cornic and Miginiac 1983). Furthermore, phytohormones may modify carbon end-product formation. For example, GA and indoleacetic acid (IAA) enhanced and ABA inhibited SPS activity in sugar beet and bean leaves (Daie 1984a). SPS activity is hypothesized to determine sucrose availability for export.

Abscisic acid (ABA), IAA, and GA all have been shown to influence phloem loading. ABA significantly reduced phloem loading in bean (Daie and Wyse 1983a) and castor bean (Malek and Baker 1978). In both species, IAA enhanced phloem loading. Daie and Wyse (1983a) also observed enhancement of phloem loading by GA_3 and 6-

benzylamino purine (BA). Sturgis and Rubery (1982) reported stimulation of phloem loading by IAA. Lepp and Peel (1970) showed that IAA and kinetin increased the rate of loading of sugars in willow and suggested that these hormones increase the source capacity.

The mechanism by which plant growth regulators control phloem loading is unclear. Malek and Baker (1978) suggested that IAA and ABA regulate ATPase-mediated proton pumping responsible for the establishment of a proton gradient across the plasmalemma. Sturgis and Rubery (1982), however, did not believe that the IAA effect on loading was mediated through changes in net proton extrusion. Direct evidence supporting this hypothesis has yet to be presented.

b. Effects at the Sink. The potential capacity of a sink (sink strength) is a function of its size and activity (assimilate flux). Both of these characteristics are subject to hormonal regulation. The finding that growth-promoting hormones enhance assimilate transport within 1–2 hrs of application indicates that hormone-induced growth may not be the exclusive mechanism by which plant growth regulators affect partitioning (Patrick 1982). In contrast, the increased accumulation of ^{14}C-labeled assimilates in pea pods caused by pod warming was from both a direct effect on ovule growth and an influence on transport other than the warmed zone (Williams and Williams 1978). They suggested the latter to be hormonally mediated.

A sink-mediated control system is suggested by reports of a causal relationship between sink metabolism and hormone-directed transport. Regulation of sink strength could be mediated through direct hormone effects on sugar-proton cotransport at the membrane (Colombo *et al.* 1978). This situation can be unambiguous for the sinks that retrieve their assimilates from the apoplast (e.g., bean, carrot, sugar beet). *In vitro* studies of these sinks should provide valid evidence that the hormone is affecting sink strength. Indeed, Patrick and Turvey (1981) presented evidence for hormonal regulation of sink activity in bean stems. Wyse *et al.* (1980) and Saftner and Wyse (1984) have also shown that ABA stimulates and IAA inhibits membrane transport of sugar and therefore can influence sink activity in sugar beet taproot. ABA also enhanced sink activity of carrot (J. Daie, unpublished results). Likewise, sucrose uptake rate by bean stem was reduced by IAA, unaffected by kinetin, and stimulated by GA_3 (Patrick 1982). It was suggested that IAA and kinetin act principally on phloem transport, whereas GA_3 exerts a dual action on phloem transport and sink strength.

Sugar transport into cells from the apoplast is partially determined by the sugar concentration in the apoplast. Patrick and Wareing

(1980) observed a higher sugar concentration released from bean stem due to kinetin and GA_3. They attributed this to an effect on net phloem unloading. These effects, however, may be through hormone regulation of apoplastic pH (proton extrusion) to which phloem unloading is sensitive or due to the enhanced activity of a cell wall invertase. In a variety of sinks, assimilate import is positively correlated with cell wall invertase activity and high hormone levels (Morris 1982).

ABA acts as an antagonist of the fungal toxin fusicoccin, which hyperpolarizes plant cells and increases the pH gradient across membranes. Because of this and other information in the literature, Tanner (1980) postulated that ABA enhanced phloem unloading by decreasing the proton motive force responsible for maintaining sugar levels against a concentration gradient.

Setter *et al.* (1982) proposed that the accumulation of ABA in soybean seed sinks during rapid pod fill was the result of ABA translocation from the leaves along with other assimilates. A continuous flow of ABA to the sink regions would, however, result in excessively high ABA levels in those regions. Therefore, if plant growth regulators are to be effective signal transmitters in any specific tissue, their localized and overall levels must be tightly regulated in those tissues. In studies to determine the difference between source and sink tissues in regard to their transport and metabolism of ABA, Daie and Wyse (1983b) and Daie *et al.* (1984) reported that regulation of localized ABA concentration is not likely to be due to mechanisms involved in membrane transport of ABA in these tissues, but instead to the rate and direction of metabolism. Their data suggested that while source and sink tissue had similar mechanisms for ABA transport (Hartung and Dierich 1983), the tissues differed in their metabolic rates and in their preferred metabolic pathway for maintaining endogenous ABA levels, implying differential control mechanisms that are tissue specific. These mechanisms may be closely coupled to any communication that is present among different tissues.

c. ***Models for Action of Plant Growth Regulators.*** The exact mechanism(s) by which hormones regulate assimilate partitioning and transport remains unknown. Patrick (1982), however, has proposed two possible models for such control at the whole plant level (Fig. 2.3). In the "sink" hypothesis, control is mediated solely by assimilate transport within the sink, and changes in assimilate pool size provide the signal to coordinate supply to the sink. The assimilate pool at the sink region may be considered a "barometer" reflecting the balance between assimilate supply and utilization as well as possibly determining rates of these processes. In the "supply-sink" hypothesis,

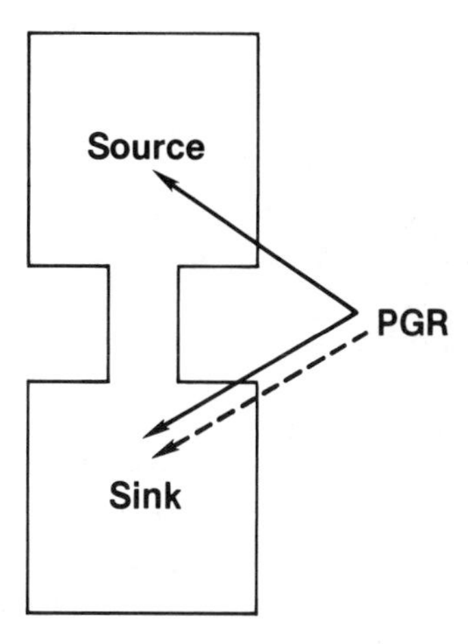

FIG. 2.3. Proposed model for plant growth regulator (PGR) action. Solid line represents "supply–sink" hypothesis; dashed line represents "sink" hypothesis. *Modified from Patrick (1982).*

sink-produced hormones integrate the assimilate supply with utilization by controlling both processes. "Sink" control could be an adequate strategy when assimilate concentrations are not limiting. Under conditions of limited assimilate concentration, the "supply-sink" strategy could ensure greater flow of assimilates to the sink region than that generated by "sink" control. In this case, sink-produced hormones would also mediate regulation of supply processes (Patrick 1982).

Finally, numerous reports exist regarding changes in crop yield caused by plant growth regulators (Doughty and Scheer 1975; Krewer *et al.* 1983; Bangerth 1980). The effects of exogenously applied plant hormones on yield and productivity, however, have been variable and unpredictable. Their effectiveness may be a function of genotype, environmental conditions, developmental stage of plant, and other factors that influence flower, fruit, and seed development (Morgan 1980).

2. Sink Demand and Photosynthetic Capacity. Substantial evidence shows that alterations in the source:sink ratio influence the rate of photosynthesis by a leaf. Partial or complete removal of sinks generally leads to a reduction in net photosynthesis. Associated with this phenomenon is the hypothesis that the concentration of assimilates in leaves alters the net photosynthetic rates of those leaves (Ho 1976); however, evidence for this is equivocal (Gifford and Evans 1981).

It is difficult to distinguish cause-and-effect relationships that affect assimilate movement from source to sink because the observed response depends upon the experimental approach. Rapid changes in the source:sink ratio initiated by selective leaf shading or defoliation cause a rapid, compensatory change in translocation from the remaining leaves without changing photosynthetic rates in short-term experiments (Geiger 1976; Borchers *et al.* 1979). In other studies, export rates were unchanged when source:sink ratios were altered (Fondy and Geiger 1983).

In the last decade, evidence has accumulated favoring the hypothesis that high sink demand stimulates photosynthetic rates. Hall and Milthorpe (1978) showed that removing rapidly growing pepper fruits caused a 30% reduction in net CO_2 uptake. Similarly, sink-related enhancement of photosynthetic rates has been reported for several tree crops, including sour cherry (Sams 1980), citrus (Lenz 1979), and apple (Monselise and Lenz 1979). Akie Fujii (1984) showed in an apple orchard study that fruiting spur leaves had a 20% higher rate of photosynthesis than did vegetative spur leaves during most of the growing season. When a similar experiment was repeated with potted plants in the laboratory, enhanced photosynthetic rates and increased carboxylation efficiency were observed for fruiting spur leaves compared with vegetative spur leaves. In contrast to Lenz's (1979) observation, Akie Fujii did not detect any effect of fruit on photorespiration. Likewise, Avery *et al.* (1979) suggested that increased fruit demand in apples is met by enhanced photosynthesis. Net photosynthesis by citrus and eggplant, as determined by gas exchange or growth analysis, was always higher for plants that had actively growing fruits compared with nonfruiting plants (Lenz 1979; Claussen and Lenz 1983). Hurewitz and Janes (1983) found that the CER of tomato leaves increased with rising root-zone temperature, which is a factor known to modify root metabolic activity as a sink. They speculated that enhanced phosphorus uptake by roots, or translocation of assimilates into the root zone, may be the mechanism by which the root-zone environment regulates CER.

Forney and Breen (1983) observed no significant differences in net photosynthesis of deblossomed or fruiting strawberries for the first 65 days after planting. After this time, photosynthetic rates of deblossomed plants decreased substantially, whereas that of fruiting plants was maintained or enhanced. Twofold to threefold increases in the photosynthetic rate in potato were reported by Moorby (1968) following tuber formation. Frier (1977), however, did not observe increased photosynthesis after the establishment of tubers; these results may have occurred because the latter experimental plants were grown at irradiation levels below saturation. Dewelle *et al.* (1983) observed a positive correlation between sink demand and photosynthetic rates in potato and suggested that photosynthetic rates may, in many cases, reflect sink demand.

In contrast to this evidence, other reports indicate there is no particularly useful correlation between sink demand and photosynthesis. For example, Crompton *et al.* (1981) reported that in *Vicia faba*, the presence of pods, although strong sinks, does not increase CO_2 uptake by adjoining leaves. Frier (1977) also disputes the apparent positive relationship between sink demand and photosynthesis in potato.

A photosynthetic rate response to altered sink activity is not necessarily a universal phenomenon. Certain conditions must be present for such feedback to be effective. Herold (1980) described some requirements for demonstrating when sink demand is communicated to the source. She proposed that, in order to demonstrate a positive correlation between photosynthetic rate and sink demand three conditions should be fulfilled: (1) the activity of the sink components (other than manipulated sinks) should remain constant; (2) the direct, vascular link between the supplying leaf and the receiving sink should be maintained; and (3) a potential must exist and be maintained for enhanced photosynthetic rate (dependent on environmental conditions). She further suggested that two kinds of sinks may be defined in studies of carbon partitioning: "primary" sinks, which are the cytoplasm of photosynthetic cells, and "secondary" sinks, which refer to distant, nonphotosynthetic cells such as roots or storage tissues. In this view, the chloroplast is considered to be the source; therefore, the ultimate barrier between source and sink is the chloroplast envelope, and changes in photosynthetic rate (given that environmental conditions are constant) must be initiated by some message that crosses the chloroplast membrane.

It appears that an increase in photosynthetic rate as sink demand is increased is not a typical response in short-term studies. However, since there is need for adjustment within the plant, this response is generally observed in longer-term experiments. Whether the adjustment is a direct result of sink demand or an incidental effect of associated changes in hormone synthesis or availability of crucial nutrients for the photosynthetic apparatus is difficult to establish (Herold 1980).

IV. YIELD: SOURCE OR SINK LIMITED?

Improving crop yields remains a highly desired objective for scientists involved in all aspects of plant research. Despite past yield increases, there still appears to be great potential for further yield improvements. Clearly, numerous processes occurring within plants have regulatory functions and some influence upon the final yield. What are the most important factors? Where do detrimental factors occur? Are these factors internal or external? Is yield limited by processes in the source or sink? At present, we cannot state clear conclusions. Furthermore, no single factor has been identified as universal for all crop species.

Field and laboratory manipulations of the plant and its environment can result in improvements in all processes affecting yield. In the context of monoculture crop production systems, however, environmental factors may be more limiting than the genetic makeup of the plant. For example, the potential yield of sugar beets has been estimated to be about 125 MT/ha, but the yield under the most advanced agricultural practices currently is about 80 MT/ha (Loomis and Williams 1963). From light and CO_2-enrichment studies, we know that crop yield most often is photosynthetically limited. Even under the most advanced systems of agriculture, these two environmental factors can remain limiting and therefore affect potential yield.

Although CO_2 and light are environmental factors, their effects on photosynthetic rates can be considered as causing source-induced yield limitations. If source is not limiting yield *per se*, it may be interacting at specific levels of organization that determine yield. For example, high contents of solids in tomato fruits are desirable. Stevens (1976) and Hewitt and Stevens (1981) showed that in tomato, low solids content of fruit was due to a low leaf area per fruit, suggesting that the source may have become the factor limiting solids production. Source limitation may be related in the short term to sucrose formation and rates of export from leaves. A long-term, source-related limitation may involve RuBP carboxylase activity.

While relatively few arguments for source-limited yield have been advanced for open-field agriculture, several lines of evidence support sink limitation of yield. Reciprocal grafting experiments using genotypes that differ in yield characteristics indicate that yield limitations may be due to the sink. Wyse (1979) showed that when roots of a low-sucrose fodder beet genotype were grafted to the tops of a high-sucrose sugar beet genotype (L_{19}), the sucrose content of the fodder root increased only 13%. Conversely, fodder tops reduced the sucrose content of the L_{19} root by only 5%. These experiments strongly indicate that control of sucrose storage was located in the sink.

Other evidence for sink-limited yields include enhanced photosynthetic rates when sink demand is increased in some species, and increased leaf starch under conditions of high photosynthetic rates. The following observations also suggest that sink demand may be lagging behind photosynthesis and therefore limiting yield: lack of a clear positive correlation between CER and yield (increased CER contributes to total dry matter production, but not necessarily yield, HI, or allocation patterns) and diminished mobilizing ability of the sink under conditions of high photosynthetic rates. For example, Saftner and Wyse (1982) showed that sugar uptake by root sink tissue of sugar beet was reduced under CO_2-enriched conditions. They suggested that the membrane ATPase responsible for establishing a pH gradient across the membrane in the sink could not be replenished as fast as sucrose was being provided from the source.

V. SUMMARY AND CONCLUDING REMARKS

Although yield usually is limited by photosynthetic rates under field conditions, genetic improvements in photosynthetic potential have not been achieved to date. Yield increases have resulted mainly from empirical selection for high harvest index. Therefore, increasing the harvest indices of crops seems to be the most effective means for future yield improvements. A high proportion of yield organs to total biomass should not, however, be obtained by jeopardizing leaf surface area or other structural and functional requirements of the plant.

The relative distribution of photosynthetic assimilates is of central importance to yield. It is evident that carbon allocation patterns may not be controlled as much by carbon fixation rates of mesophyll cells or by translocation rates (nonlimiting factor) as by carbon partitioning in different regions of the plant. These regions include the source tissue for regulation of available carbon for export and sink regions characterized by such important phenomena as sink formation and abortion, duration of sink growth, phloem unloading, and transport rates at the membrane level. These regulatory aspects are keys that may unlock the mystery of enhancing crop production in the future.

LITERATURE CITED

AKIE FUJII, J. A. 1984. Photosynthesis and photorespiration in apples. Ph.D. Thesis, Washington State Univ., Pullman.

AVERY, D.J., C.A. PRIESTLEY, and K.J. TREHAREN. 1979. Integration of assimilation and carbohydrate utilization in apple. p. 221–230. In: R. Marcelle, C. Clijsters, and M. Van Pouke (eds.), Photosynthesis and plant development. W. Junke, The Hague.

BANGERTH, F. 1980. Some effects of endogenous and exogenous hormones and growth regulators on growth and development of tomato fruits. p. 141–150. In: B.

Jeffcoat (ed.), Aspects and prospects of plant growth regulators. Monograph 6. British Plant Growth Regulator Group. Wessex Press, Wantage, Oxfordshire.

BENNETT, A.B. and R.M. SPANSWICK. 1983. Solubilization and reconstitution of an anion-sensitive H^+ ATPase from corn roots. J. Membrane Biol. 75:21–31.

BENNETT, A.B., B.L. SWEGER, and R.M. SPANSWICK. 1984. Sink to source translocation in soybean. Plant Physiol. 74:434–436.

BHAGSARI, A.S. 1981. Relation of photosynthetic rates to yield in sweet potato genotypes. HortScience 16:779–780.

BORCHERS, C.A., A.B.G. LAMM, and C.A. SWANSON. 1979. Source–sink patterns in relation to source strength. Plant Physiol. (Supp.) 64:34.

BOWEN, J.E. and J.E. HUNTER. 1972. Sugar transport in immature internodal tissue of sugarcane. II. Mechanism of sucrose transport. Plant Physiol. 49:789–793.

CHALLA, H. 1976. An analysis of the diurnal course of growth, carbon dioxide exchange and carbohydrate reserve content of cucumbers. Publ. 20. Centre Agr. Publ. Document, Wageningen, the Netherlands.

CHARLES-EDWARDS, D. A. 1982. Physiological determinants of crop growth. Academic Press, New York. p. 106-112.

CHATTERTON, N. J. and J. E. SILVIUS. 1980. Photosynthate partitioning into leaf starch as affected by daily photosynthetic period duration in six species. Physiol. Plant. 49:141-144.

CHATTERTON, N. J. and J. E. SILVIUS. 1981. Photosynthate partitioning in soybean leaves. II. Irradiance level and daily photosynthetic period duration effects. Plant Physiol. 67:257-260.

CHATTERTON, N. J., G. E. CARLSON, W. E. HUNGERFORD, and D.R. LEE. 1972. Effect of tillering and cool nights on photosynthesis and chloroplast starch in Pangola. Crop Sci. 12:206-208.

CLAUSS, H., D. C. MORTIMER, and P. R. GORHAM. 1964. Time-course study of translocation of products of photosynthesis in soybean plants. Plant Physiol. 39:269-273.

CLAUSSEN, W. 1983. Investigations on the relationship between the distribution of assimilates and sucrose synthase activity in *Solanum melongena*. Z. Pflanzenphysiol. 110:175-182.

CLAUSSEN, W. and E. BILLER. 1977. The significance of sucrose and starch contents of the leaves for the regulation of net photosynthetic rates. Z. Pflanzenphysiol. 81:189-198.

CLAUSSEN, W. and F. LENZ. 1983. Investigations on the relationship between sucrose phosphate synthase activity and net photosynthetic rates, and sucrose and starch content of leaves of *Solanum melongena* Z. Pflanzenphysiol. 109:459-468.

COLOMBO, R., M. I. DEMICHELIS, and P. LADO. 1978. 3-O-Methyl glucose uptake stimulation by auxin and fusicoccin in plant material and its relationship with proton extrusion. Planta 138:249-256.

COOK, M. G. and L. T. EVANS. 1983. The roles of sink size and location in the partitioning of assimilates in wheat ears. Aust. J. Plant Physiol. 10:313-327.

CORNIC, G. and E. MIGINIAC. 1983. Nonstomatal inhibition of net CO_2 uptake by (±) abscisic acid in *Pharbitis nil*. Plant Physiol. 73:529-533.

CRANE, J. C. 1964. Growth substances in fruit setting and development. Annu. Rev. Plant Physiol. 15:303-326.

CRANE, J. C., P. B. CATLIN, and I. AL-SHALAN. 1976. Carbohydrate levels in the pistachio as related to alternate bearing. Amer. J. Hort. Sci. 101:371-374.

CROMPTON, H. J., C.P. LLOYD-JONES, and D.G. HILL-COTTINGHAM. 1981. Translocation of labeled assimilates following photosynthesis of $^{14}CO_2$ by the field bean, *Vicia faba*. Physiol. Plant. 51:189-194.

CRONSHAW, J. 1981. Phloem structure and function. Annu. Rev. Plant Physiol. 32:465-84.

CROOKSTON, R. K., J. O'TOOLE, R. LEE, J. L. OZBUN, and D. H. WALLACE. 1974. Photosynthetic depression in beans after exposure to cold for one night. Crop Sci. 14:457-464.

DAIE, J. 1984a. Are phytohormones involved in carbon partitioning in leaves? p. 32-36. In: Proc. Plant Growth Regulator Soc. Amer.

DAIE, J. 1984b. Characteristics of sugar transport in storage tissue of carrot. J. Amer. Soc. Hort. Sci. 109:718-722.

DAIE, J. and R. E. WYSE. 1983a. Regulation of phloem loading by plant growth regulators in *Phaseolus vulgaris*. p. 139-149. In: Proc. Plant Growth Regulator Soc. Amer.

DAIE, J. and R. E. WYSE. 1983b. ABA uptake in source and sink tissues of sûgarbeet. Plant Physiol. 72:430-433.

DAIE, J. and R. E. WYSE. 1985. Evidence on the mechanism of enhanced sucrose uptake at low turgor in leaf discs of *Phaseolus coccinius*. Physiol. Plant. (In Press).

DAIE, J., R. E. WYSE, M. HEIN, and M. L. BRENNER. 1984. ABA metabolism in source and sink tissues of sugarbeet. Plant Physiol. 74:810-814.

DELROT, S. 1981. Proton fluxes associated with sugar uptake in *Vicia faba* leaf tissue. Plant Physiol. 68:706-711.

DELROT, S. and J. L. BONNEMAIN. 1979. Proton sugar cotransport in leaf tissues of *Vicia faba*. Plant Physiol. 63:35-44.

DELROT, S., J. P. DESPEGHEL, and J.L. BONNEMAIN. 1980. Phloem loading in *Vicia faba* leaves: effect of N-ethylmaleimide and parachloromercuribenzensulfonic acid on H^+ extrusion, K^+ and sucrose uptake. Planta 149:144-148.

DEWELLE, R. B., P. J. HURLEY, and J. J. PAVEK. 1983. Photosynthesis and stomatal conductance of potato clones (*Solanum tuberosum*). Plant Physiol. 72:172-176.

DICK, P. S. and T. APREES. 1975. The pathway of sugar transport in roots of *Pisum sativum*. J. Expt. Bot. 26:305-314.

DINAR, M. and M. A. STEVENS. 1981. The relationship between starch accumulation and soluble solids content of tomato fruits. J. Amer. Soc. Hort. Sci. 106:415-418.

DOEHLERT, D. C. and S. C. HUBER. 1983. Spinach leaf sucrose phosphate synthase: activation by G-6-P and interaction with inorganic phosphate. Fed. Europ. Biol. Sci. Lett. 153:293-296.

DOMAN, D. C. and D. R. GEIGER. 1979. Effect of exogenously supplied foliar potassium on phloem loading in *Beta vulgaris*. Plant Physiol. 64:528-533.

DONALD, C. M. 1962. In search of yield. J. Austral. Inst. Agr. Sci. 28:171-178.

DONALD, C. M. and J. HAMBLIN. 1976. The biological yield and harvest index of cereals as agronomic and plant breeding criteria. Adv. Agron. 28:361-405.

DOUGHTY, C. C. and W. P. A. SCHEER. 1975. Growth regulators increase yield and reduce length of harvest of highbush blueberries. HortScience 10:260-261.

ESCHRICH, W. 1980. Free space invertase, its possible role in phloem unloading. Ber. Deutsch. Bot. Ges. 93:363-378.

EVANS, L.T. 1975. Crop physiology. Cambridge Univ. Press, New York.

EVERT, R.F., W. ESCHRICH, and W. HEYSER. 1978. Leaf structure in relation to solute transport and phloem loading in *Zea mays*. Planta 138:279-294.

FELKER, F.C. and J.C. SHANNON. 1980. Movement of ^{14}C assimilates into kernels of *Zea mays*. Plant Physiol. 65:864-870.

FISHER, D.B. and W.H. OUTLAW. 1979. Sucrose compartmentation in the palisade parenchyma of *Vicia faba*. Plant Physiol. 64:481-483.

FONDY, B.R. and D.R. GEIGER. 1983. Control of export and of partitioning among

sinks by allocation of products of photosynthesis in source leaves. Proc. Ann. Plant. Bioch. Physiol. Symp. (Univ. Missouri-Colombia) 2:33–42.
FORDE, B.J., H.C.M. WHITEHEAD, and J.A. ROWLEY. 1975. Effect of light intensity and temperature on photosynthetic rate, leaf starch content and ultrastructure of *Paspalum dilatum*. Austral. J. Plant Physiol. 2:185–195.
FORNEY, C.F. and P.J. BREEN. 1983. Dry matter partitioning in fruiting and non-fruiting strawberry plants. HortScience 18:70.
FOYER, C., D. WALKER, and E. LATZKO. 1982. The regulation of cytoplasmic fructose-1, 6-bisphosphatase in relation to the control of carbon flow to sucrose. Z. Pflanzenphysiol. 107:457–465.
FRANCHESCHI, V. and R. GIAQUINTA. 1983. The paraveinal mesophyll of soybean leaves in relation to assimilate transfer and compartmentation. I. Ultrastructure and histochemistry during vegetative development. Planta 157:411–421.
FRIER, V. 1977. The relationship between photosynthesis and tuber growth in *Solanum tuberosum*. J. Expt. Bot. 28:999–1007.
GALLIANI, S., S.P. MONSELISE, and R. GOREN. 1975. Improving fruit size and breaking alternate bearing in mandarins by Ethephon and other agents. HortScience 10:27–28.
GAWRONSKA, H., R.B. DWELLE, J.J. PAVEK, and P. ROW. 1984. The partitioning of photoassimilates by four potato clones (*Solanum tuberosum*). Crop. Sci. 24:1031–1036.
GEIGER, D.R. 1975. Phloem loading. p. 395–450. In: I. M.H. Zimmerman and J. A. Milburn (eds.), Transport in plants. Springer-Verlag, New York.
GEIGER, D.R. 1976. Phloem loading in source leaves. In: I.F. Wardlaw and J.B. Passioura, (eds.), Transport and transfer processes in plants. Academic Press, New York.
GEIGER, D.R. 1979. Control of partitioning and export of carbon in leaves of higher plants. Bot. Gaz. 140:241–248.
GEIGER, D.R. and B.R. FONDY. 1980. Phloem loading and unloading: Pathways and mechanisms. What Is New in Plant Physiology 11:25–28.
GEIGER, D.R., R.T. GIAQUINTA, S.A. SOVANICK, and R.J. FELLOW. 1973. Solute distribution in sugar beet leaves in relation to phloem loading and translocation. Plant Physiol. 52:585–589.
GEIGER, D.R., S.A. SOVONICK, T.L. SHOCK, and R.J. FELLOW. 1974. Role of free space in translocation in sugarbeet. Plant Physiol. 54:892–898.
GIAQUINTA, R.T. 1976. Evidence for phloem loading from the apoplast: chemical modification of membrane sulfhydryl groups. Plant Physiol. 57:872–875.
GIAQUINTA, R.T. 1977. Possible role of pH gradient and membrane ATPase in the loading of sucrose into the sieve tubes. Nature 267:369–370.
GIAQUINTA, R.T. 1980. Sucrose/proton cotransport during phloem loading and its possible control by internal sucrose concentration. p. 273–282. In: R. M. Spanswick, W. J. Lucas, and J. Dainty (eds.), Plant membrane transport: current conceptual issues. North Holland-Elsevier, Amsterdam.
GIAQUINTA, R.T. 1983. Phloem loading of sucrose. Annu. Rev. Plant Physiol. 59:178–180.
GIAQUINTA, R.T. and D.R. GEIGER. 1977. Mechanism of cyanide inhibition of phloem translocation. Plant Physiol. 59:178–180.
GIAQUINTA, R.T., W. LIN, N.L. SADLER, and V.R. FRANCESCHI. 1983. Pathway of phloem unloading of sucrose in corn root. Plant Physiol. 72:362–367.
GIFFORD, R.M. and L.T. EVANS. 1981. Photosynthesis, carbon partitioning, and yield. Annu. Rev. Plant Physiol. 32:485–509.
GIFFORD, R.M. and C.L. JENKINS. 1983. Prospects of applying knowledge of

photosynthesis towards improving crop production. p. 419–434. In: Govindjee (ed.). Photosynthesis: CO_2 assimilation and plant productivity. Vol. 2. Academic Press, New York.

GLASZIOU, K.T. and K.R. GAYLER. 1972. Storage of sugars in stalks of sugar cane. Bot. Rev. 38:471–490.

GRIFFITH, S. and R.E. WYSE. 1983. Some characteristics of the phloem unloading system in stem segments of *Vicia faba*. Plant Physiol. 72:138.

GUNNING, B.E.S. 1976. The role of plasmodesmata in short distance transport to and from phloem. p. 203–227. In: B. E. S. Gunning and A. W. Robards (eds.), Intercellular communication in plants: studies on plasmodesmata. Springer, Berlin.

GUNNING, B.E.S. and J.S. PATE. 1974. Transfer cells. p. 441–480. In: A. W. Robards (ed.), Dynamic aspects of plant ultrastructure. McGraw-Hill, London.

HABESHAW, D. 1973. Translocation and the control of photosynthesis in sugarbeet. Planta 110:213–226.

HALL, A.J. and F.L. MILTHORPE. 1978. Assimilate source–sink relationships in *Capsicum annum*. II. The effects of fruit excision on photosynthesis and leaf and stem carbohydrate. Austral. J. Plant Physiol. 5:1–13.

HANSON, W.D. and R.E. GRIER. 1973. Rates of electron transfer and non-cyclic photophosphorylation for chloroplasts isolated from maize populations selected for juvenile productivity and leaf width. Genetics 75:247-257.

HARBRON, S., C. FOYER, and D.A. WALKER. 1981. The purification and properties of sucrose phosphate synthase from spinach leaves: The involvement of this enzyme and fructose bisphosphatase in the regulation of sucrose biosynthesis. Arch. Biochem. Biophys. 212:237-246.

HARTUNG, W. and B. DIERICH. 1983. Uptake and release of abscisic acid by runner bean root tip segments. L. Naturforsch. 38:719-723.

HELDT, H.W., C.J. CHAN, D. MARONDE, A. HEROLD, Z.S. STANKOVIC, D.A. WALKER, A. FRAMINER, M.R. KIRK, and U. HEBER. 1977. Role of orthophosphate and other factors in the regulation of starch formation in leaves and isolated chloroplasts. Plant Physiol. 59:1146-1155.

HEROLD, A. 1980. Regulation of photosynthesis by sink activity—The missing link. New Phytol. 86:131-144.

HEWITT, J.D. and M.A. STEVENS. 1981. Growth analysis of two tomato genotypes differing in total fruit solids. J. Amer. Soc. Hort. Sci. 106:723-727.

HEWITT, J.D., M. DINAR, and M.A. STEVENS. 1982. Sink strength of fruits of two tomato genotypes differing in total fruit solids content. J. Amer. Soc. Hort. Sci. 107:896-900.

HEYSER, W. 1980. Phloem loading in the maize leaf. Ber. Deutsch. Bot. Ges. 93:221-28.

HEYSER, W., R.F. EVERT, E. FRITZ, and W. ESCHRICH. 1978. Sucrose in the free space of translocating maize leaf bundles. Plant Physiol. 62:491-494.

HEYSER, W., R. HEYSER, W. ESCHRICH, and O.A. LEONARD. 1976. The influence of externally applied organic substances on phloem translocation in detached maize leaves. Planta 132:269-277.

HO, L.C. 1976. The relationship between the rates of carbon transport and photosynthesis in tomato leaves. J. Expt. Bot. 27:87-97.

HO, L.C. 1979. Regulation of assimilate translocation between leaves and fruits in the tomato. Ann. Bot. 43:437-448.

HO, L.C. and D.A. BAKER. 1982. Regulation of loading and unloading in long distance transport systems. Physiol. Plant. 56:225-230.

HO, L.C. and A.F. SHAW. 1977. Carbon economy and translocation of ^{14}C in leaflets of the seventh leaf of tomato during leaf expansion. Ann. Bot. 41:833-848.

HUBER, S.C. 1981. Interspecific variation in activity and regulation of leaf sucrose phosphate synthetase. Z. Pflanzenphysiol. 102:443-450.

HUBER, S.C. 1983. Relation between photosynthetic starch formation and dry weight partitioning between the shoot and root. Can. J. Bot. 61:2709-2716.

HUBER, S.C. and D.M. BICKETT. 1984. Evidence for control of carbon partitioning by fructose-2,6-bisphosphate in spinach leaves. Plant Physiol. 74:445-447.

HUBER, S.C. and D.W. ISRAEL. 1982. Biochemical basis for partitioning of photosynthetically fixed carbon between starch and sucrose in soybean leaves. Plant Physiol. 69:691-696.

HUBER, S.C. and D.E. MORELAND. 1980. Translocation. Efflux of sugars across the plasmalemma of mesophyll protoplasts. Plant Physiol. 65:560-562.

HUBER, S.C. and D.E. MORELAND. 1981. Co-transport of K^+ and sugars across plasmalemma of mesophyll protoplasts. Plant Physiol. 67:163–169.

HUGHES, D.L., J. BOSLAND, and M. YAMAGUCHI. 1983. Movement of photosynthates in muskmelon plants. J. Amer. Soc. Hort. Sci. 108:189–192.

HUMPHREYS, T.E. 1980. Sugar-proton cotransport and phloem loading. What Is New in Plant Physiology 11:9–12.

HUREWITZ, J. and H.W. JANES. 1983. Effect of altering the root-zone temperature on growth, translocation, carbon exchange rate, and leaf starch accumulation in tomato. Plant Physiol. 73:46–50.

JAIN, H.K. 1975. Breeding for yield and other attributes in grain legumes. Ind. Agr. Res. Inst., New Dehli.

JENNER, C.F. 1974. An investigation of the association between the hydrolysis of sucrose and its absorption by wheat grains. Austral. J. Plant Physiol. 1:319–329.

KAISER, W.M., J.S. PAUL, and A. J. BASSHAM. 1979. Release of photosynthates from mesophyll cells *in vitro* and *in vivo*. Plant Physiol. 64:377–385.

KOMOR, E. 1977. Sucrose uptake by cotyledons of *Ricinus communis*: Characteristics, mechanism and regulation. Planta 137:119–131.

KOMOR, E. 1982. Transport of sugar. 635–676. In: F. A. Loewus and W. Tanner (eds.), Encyclopedia of plant physiology. Plant carbohydrates I. Springer-Verlag, Berlin.

KOMOR, E., M. THOM, and A. MARETZKI. 1982. Vacuoles from sugarcane suspension cultures. III. Proton motive potential differences. Plant Physiol. 69:1326–1330.

KREWER, G.W., J.W., DANIEL, D.C. COSTON., G.A. COUVILLON, and S.J. KAYS. 1983. Transport of ^{14}C photosynthate into young peach fruits in response to CGA-IS281 and ethylene-releasing compounds. HortScience 18:476–478.

KUENEMAN, E.A., D.H. WALLACE, and P.M. LUDFORD. 1979. Photosynthetic measurements of field grown dry beans and their relations to selection for yield. J. Amer. Soc. Hort. Sci. 104:480–482.

KURSANOV, A.L. and M.I. BROVCHENKO. 1970. Sugars in the free space of leaf plates: Their origin and possible involvement in transport. Can. J. Bot. 48:1243–1250.

LENZ, F. 1979. Fruit effects on photosynthesis, light and dark respiration. p. 271–281. In: R. Marcelle, C. Clijsters, and M. Van Pouke (eds.), Photosynthesis and plant development. W. Junke, The Hague.

LEPP, N.W. and A.J. PEEL. 1970. Some effects of IAA and kinetin upon the movement of sugars in the phloem of willow. Planta 90:230–235.

LICHTNER, F.T. and R.M. SPANSWICK. 1981. Sucrose uptake by developing soybean cotyledons. Plant Physiol. 68:693–698.

LIS, E.K. and R. ANTOSZEWSKI. 1982. Do growth substances regulate the phloem as well as xylem transport of nutrients to the strawberry receptacles? Planta 156:492–495.

LOOMIS, R. S. and W. A. WILLIAMS. 1963. Maximum crop productivity: An estimate. Crop Sci. 3:67–72.

MADORE, M. and J.A. WEBB. 1981. Leaf free space analysis and vein loading in *Cucurbita pepo*. Can. J. Bot. 59:2550–2557.

MALEK, F. and D.A. BAKER. 1978. Effect of fusicoccin on proton cotransport of sugars in the phloem loading of *Ricinus communis*. Plant Sci. Lett. 11:233–239.

MARINI, R.P. and J.A. BARDEN. 1981. Seasonal correlations of specific leaf weight to net photosynthesis and dark respiration of apple leaves. Photosynth. Res. 2:251–258.

MARTIN, G.C. and C. NISHIJIMA. 1972. Levels of endogenous growth regulators in abscising and persisting peach fruits. J. Amer. Soc. Hort. Sci. 97:561–565.

MAUNEY, J.R., G. GUINN, K.E. FRY, and J.D. HESKETH. 1979. Correlation of photosynthetic CO_2 uptake and carbohydrate accumulation in cotton, soybean, sunflower and sorghum. Photosynthetica 13:260–266.

MAYNARD, J.W. and W. LUCAS. 1982. Sucrose and glucose uptake into *Beta vulgaris* leaf tissue: A case for general (apoplastic) retrieval systems. Plant Physiol. 70:1436–1443.

McCASHIN, B.G. and D.T. CANVIN. 1979. Photosynthetic and photorespiratory characteristics of mutants of *Hordeum vulgare*. Plant Physiol. 64:354–360.

McNEAL, F.H., M.A. BERG, P.L. BROWN, and C.F. McGUIRE. 1971. Productivity and quality response of five spring wheat genotypes, *Triticum aestivum* L., to nitrogen fertilizer. Agron. J. 63:908–910.

MILBURN, J.A. 1974. Phloem transport in *Ricinus*: Concentration gradients between source and sink. Planta 117:303–319.

MONSELISE, S.P. and E.E. GOLDSCHMIDT. 1982. Alternate bearing in fruit trees: A review. Hort. Rev. 4:128–173.

MONSELISE, S.P. and F. LENZ. 1979. Effect of fruit load on photosynthetic rates of budded apple trees. Gartenbauwissenschaft 45:220–224.

MOORBY, J. 1968. The influence of carbohydrate and mineral nutrient supply on the growth of potato tubers. Ann. Bot. 32:57–68.

MOORBY, J., J.H. TROUGHTON, and B.G. CURRIE. 1974. Investigations of carbon transport in plants. J. Expt. Bot. 25:937–944.

MORGAN, D.C. 1980. Factors affecting fruit and seed development in field beans and oil seed rape. p. 151–164. In: B. Jeffcoat (ed.), Aspects and prospects of plant growth regulators. Monograph 6. British Plant Growth Regulator Group. Wassex Press, Wantage, Oxfordshire.

MORRIS, D.A. 1982. Hormonal regulation of sink invertase activity: implications for control of assimilate partitioning. p. 659–668. In: P.F. Wareing (ed.), Plant growth substances. Academic Press, New York.

MOSS, D.N. and H.P. RASMUSSEN. 1969. Cellular localization of CO_2 fixation and translocation of metabolites. Plant Physiol. 44:1063–1068.

MULLIGAN, D.R. and J.W. PATRICK. 1979. Gibberellic acid promoted transport of assimilates in stems of *Phaseolus vulgaris*. Localized vs. remote sites of action. Planta 145:233–238.

NAFZIGER, E.D. and H.R. KOLLER. 1976. Influence of leaf starch concentration on CO_2 assimilation in soybean. Plant Physiol. 57:560–563.

NATIONAL ACADEMY OF SCIENCES. 1977. World food and nutrition study. The potential contribution of research. Natl. Res. Council, Washington, DC.

NOMURA, T., N. NAKAYAMA, T. MURATA, and T. AKAZAWA. 1967. Biosynthesis of starch in chloroplasts. Plant Physiol. 42:327–332.

OLUFAJO, O.O., R.W. DANIELS, and D.H. SCARISBRICK. 1982. The effect of pod removal on the translocation of ^{14}C photosynthates from leaves in *Phaseolus vulgaris*, c.v. Lochness. J. Hort. Sci. 57:333–338.

OZBUN, J.L. 1978. Photosynthetic efficiency and crop production. HortScience 6:678–679.

PATRICK, J.W. 1979. An assessment of auxin-promoted transport in decapitated stems and whole shoots of *Phaseolus vulgaris*. Planta 146:107–112.

PATRICK, J.W. 1982. Hormonal control of assimilate transport. p. 669–678 In: P.F. Wareing (ed.), Plant growth substances. Academic Press, New York.

PATRICK, J.W. 1983. Photosynthate unloading from seed coats of *Phaseolus vulgaris*. General characteristics and facilitated transfer. Z. Pflanzenphysiol. 111:9–18.

PATRICK, J.W. 1984. Photosynthate unloading from seed coats of *Phaseolus vulgaris* L. Control by tissue water relations. J. Plant Physiol. 115:297–310.

PATRICK, J.W. and P.M. TURVEY. 1981. The pathway of radial transfer of photosynthates in decapitated stems of *phaseolus vulgaris*. Ann. Bot. 47:611–621.

PATRICK, J.W. and P.F. WAREING. 1978. Auxin-promoted transport of metabolites in stems of *Phaseolus vulgaris*. J. Expt. Bot. 29:359–366.

PATRICK, J.W. and P.F. WAREING. 1980. Hormonal control of assimilate movement and distribution. p. 65–68. In: B. Jeffcoat (ed.), Aspects and prospects of plant growth regulators. Monograph 6. British Plant Growth Regulator Group. Wassex Press, Wantage, Oxfordshire, England.

PEET, M.M., A. BRAVO, D.H. WALLACE, and J.L. OXBUN. 1977. Photosynthesis, stomatal resistance, and enzyme activities in relation to yield of field grown dry bean cultivars. Crop Sci. 17:287–293.

POOSTCHI, I., I. ROUHANI, and K. RAZMI. 1972. Influence of levels of spring irrigation and fertility on yield of winter wheat (*Triticum aestivum* L.) under semi-arid conditions. Agron. J. 64:438–442.

POTTER, J.R. and P.J. BREEN. 1980. Maintenance of high photosynthetic rates during the accumulation of high leaf starch levels in sunflower and soybean. Plant Physiol. 66:528–531.

PREISS, J. 1982. Regulation of the biosynthesis and degradation of starch. Annu. Rev. Plant Physiol. 34:431–454.

PREISS, J. 1984. Starch, sucrose biosynthesis and partition of carbon in plants are regulated by orthophosphate and triose-phosphates. Trends Biol. Sci. 9:24–27.

PREISS, J. and C.D. BOYER. 1980. Evidence for independent genetic control of the multiple forms of maize endosperm branching enzymes and starch synthesis. p. 161–174. In: N.C. Brady (ed.), Mechanisms of saccharide polymerization and depolymerization. Academic Press, New York.

PREISS, J. and C. LEVI. 1979. Metabolism of primary products of photosynthesis. Metabolism of starch in leaves. p. 282–312. In: M. Gibbs and E. Latzk (eds.), Encyclopedia of plant physiology. New series. Springer-Verlag, New York.

RANDALL, D.D., C.J. NELSON, and K.H. ASAY. 1977. Ribulose bisphosphate carboxylase: Altered genetic expression in tall fescue. Plant Physiol. 59:38–41.

RODRIGUEZ, B.P. and V.N. LAMBETH. 1976. Photosynthetic rate and yield component relationship in winter greenhouse tomatoes. HortScience 11:430–431.

SAFTNER, R.A. and R.E. WYSE. 1980. Alkali cation/sucrose co-transport in the root sink of sugarbeet. Plant Physiol. 66:884–889.

SAFTNER, R.A. and R.E. WYSE. 1984. Effect of plant hormones on sucrose uptake by sugarbeet root tissue discs. Plant Physiol. 74:951–954.

SAFTNER, R.A., J. DAIE, and R.E. WYSE. 1983. Sucrose uptake and compartmentation in sugarbeet taproot tissue. Plant Physiol. 72:1–6.

SAMS, C. 1980. Factors affecting the leaf and shoot morphology and photosynthetic rate of sour cherry (*Prunus cerasus* L. Montmorency). Ph.D. Thesis, Michigan State Univ., East Lansing.

SANWAL, G.G., E. GREENBERG, J. HARDIE, E.C. CAMERON, and J. PREISS. 1967. Regulation of starch biosynthesis in plant leaves: Activation and inhibition of ADP-glucose pyrophosphorylase. Plant Physiol. 43:417–427.

SCHMITT, M.R., W.D. HITZ, W. LIN, and R.T. GIAQUINTA. 1984. Sugar transport into protoplasts isolated from developing soybean cotyledons. II. Sucrose transport kinetics, selectivity, and modeling studies. Plant Physiol. 75:941–946.

SERVAITES, J.C. and D.R. GEIGER. 1974. Effect of light intensity and oxygen on photosynthesis and translocation in sugarbeet. Plant Physiol. 54:575–578.

SETTER, T.L., W.A. BRUN, and M.L. BRENNER. 1980. Effect of obstructed translocation on leaf abscisic acid and associated stomatal closure and photosynthesis decline. Plant Physiol. 65:1111–1115.

SHANNON, J.C. and C.T. DOUGHERTY. 1972. Movement of ^{14}C-labelled assimilates into kernels of *Zea mays*. II. Invertase activity of the pedicel and placentachazal tissue. Plant Physiol. 49:203–206.

SHARMA, A.K., B.B. SINGH, and S.P. SINGH. 1982. Relationship among net assimilation rate, leaf area index and yield in soybean genotypes. Photosynthetica 16:115–118.

SILVIUS, J.E., N.J. CHATTERTON, and D.F. KREMER. 1979. Photosynthate partitioning in soybean leaves at two irradiance levels. Plant Physiol. 64:872–875.

SMITH, J. A. and J. A. MILBURN. 1980. Phloem turgor and the regulation of sucrose loading in *Ricinus communis* L. Planta 148:42–48.

SOVONICK, S.A., D.R. GEIGER, and R.J. FELLOWS. 1974. Evidence for active phloem loading in the minor veins of sugarbeet. Plant Physiol. 54:886–891.

STEVENS, M.A. 1976. Inheritance of viscosity potential in tomato. J. Amer. Soc. Hort. Sci. 101:152–155.

STURGIS, J.N. and P.H. RUBERY. 1982. The effects of indol-3-yl acetic acid and fusicoccin on the kinetic parameters of sucrose uptake by discs from expanded primary leaves of *Phaseolus vulgaris*. Plant Sci. Lett. 24:319–326.

TANNER, W. 1980. On the possible role of ABA on phloem unloading. Ber. Deutsch. Bot. Ges. 93:349–351.

THORNE, J.H. 1982. Characterization of the active sucrose transport system of immature soybean embryos. Plant. Physiol. 70:953-958.

THORNE, J.H. 1983. An *in vivo* technique for the study of phloem unloading in seed coats of developing soybean seeds. Plant Physiol. 72:268-271.

THORNE, J.H. and R.T. GIAQUINTA. 1982. Pathways and mechanisms associated with carbohydrate translocation in plants. In: D. H. Lewis (ed.), Physiology and biochemistry of storage carbohydrates in vascular plants. Soc. Expt. Bot. Symp. Series. Cambridge Univ. Press.

THORNE, J.H. and H.R. KOLLER. 1974. Influence of assimilate demand on photosynthesis, diffusive resistance, translocations and carbohydrate level of soybean leaves. Plant Physiol. 54:201-207.

VALPUESTA, V. and M. BUKOVAC. 1983. Cherry fruit growth: Localization of indoleacetic acid oxidase and inhibitors in the seed. J. Amer. Soc. Hort. Sci. 108:457-459.

WALKER, D.A. and A. HEROLD. 1977. Can the chloroplast support photosynthesis unaided? In: Y. Fujita, S. Katsh, K. Shibeta, S. Miyachi (eds.) Photosynthetic Organelles: Structure and function. (Special issue):1–7.

WALKER, A.J. and L.C. HO. 1974. Investigations of carbon transport in plants. J. Expt. Bot. 25:937-944.

WALKER, A.J. and L.C. HO. 1976. Young tomato fruits induced to export carbon by cooling. Nature 261:410–411.
WALKER, A.J. and L.C. HO. 1977a. Carbon translocation in tomato fruit: carbon import and fruit growth. Ann. Bot. 41:813–823.
WALKER, A.J. and L.C. HO. 1977b. Carbon translocation in tomato: Effect of fruit temperature on carbon metabolism and the rate of translocation. Ann. Bot. 41:825–832.
WALKER, D.A. and S.P. ROBINSON. 1978. A contemporary view of photosynthetic carbon assimilation. Ber. Deutsch. Bot. Gaz. 91:513–526.
WALKER, A.J. and J.H.M. THORNLEY. 1977. The tomato fruit: Import, growth, respiration, and carbon metabolism at different fruit sizes and temperatures. Ann. Bot. 41:977–985.
WAREING, P.F., M.M. KHALIFA, and C.J. TREHARNE. 1968. Rate-limiting processes in photosynthesis at saturating light intensities. Nature 220:453–457.
WILLIAMS, A.M. and K.R. WILLIAMS. 1978. Regulation of movement of assimilates into ovules of *Pisum sativum*. Austral. J. Plant Physiol. 5:295–300.
WITTWER, S.H. 1974. Maximum production capacity of food crops. BioScience 24:216–224.
WITTWER, S.H. 1978. The next generation of agricultural research. Science 199:375–378.
WITTWER, S.H. 1980. The shape of things to come. p. 413–459. In: P.S. Carlson (ed.), The biology of crop productivity. Academic Press, New York.
WYSE, R.E. 1979. Parameters controlling sucrose content and yield of sugarbeet roots. J. Amer. Soc. Sugarbeet Tech. 20:368–385.
WYSE, R.E. and R.A. SAFTNER. 1982. Reduction in sink mobilizing ability following periods of high carbon flux. Plant Physiol. 69:226–228.
WYSE, R.E., J. DAIE, and R.A. SAFTNER. 1980. Hormonal control of sink activity in sugarbeet. Plant Physiol. (Suppl.) 65:121.
WOLSWINKEL, P. and A. AMMERLAAN. 1983. Phloem unloading in developing seeds of *Vicia faba*. The effect of several inhibitors on the release of sucrose and amino acids by the seed coat. Planta 158:205–215.
ZIEGLER, H. 1974. Biochemical aspects of phloem transport. p. 43–62. In: M. A. Sleigh and D. H. Jennings (eds.), Transport at the cellular level. Cambridge Univ. Press, London.

3

Juvenility, Maturation, and Rejuvenation in Woody Plants

Wesley P. Hackett
Department of Horticultural Science and Landscape Architecture, University of Minnesota, St. Paul, Minnesota 55108

Horticultural Reviews, Volume 7

ISBN 0-87055-492-1

I. INTRODUCTION

In the development of all woody plants from seed there is a so-called juvenile phase lasting up to 30–40 years in certain forest trees, during which flowering does not occur and cannot be induced by the normal flower-initiating treatment or conditions (see Section II). In time, however, the ability to flower is achieved and maintained under natural conditions; at this stage, the tree is considered to have attained the adult or sexually mature condition. As will be discussed in Sections III. A and IV, the length of the juvenile period can be influenced by environmental and genetic factors. This transition from the juvenile to the mature phase has been referred to as phase change by Brink (1962), ontogenetic aging by Fortanier and Jonkers (1976), or meristem aging (cyclophysis) by Seeliger (1924) and Oleson (1978). Associated with this transition are progressive changes in morphological and developmental attributes, including leaf cuticular characteristics (Franich *et al*, 1977); bark characteristics (Oleson 1982); leaf shape and thickness, phyllotaxis, thorniness, and shoot orientation (Schaffalitzky de Muckadell 1959); branch number and branching pattern (Libby and Hood 1976); tracheid width (Oleson 1982) and length (Rumball 1963); shoot growth vigor (Goodin 1964; Sweet and Wells 1974); seasonal leaf retention and stem pigmentation (Schaffalitzky de Muckadell 1959); ability to form adventitious roots and buds (Burger 1980; Schaffalitzky de Muckadell 1959); partitioning of photosynthates into main stem or branches and disease resistance (W.J. Libby, personal communication); and cold resistance (Hood and Libby 1980). Changes in such characteristics during development vary from species to species. Most change gradually during the period preceding the mature phase, and usually no distinct change in any one characteristic is apparent at the time the ability to flower is attained.

Phase change is of considerable theoretical importance relative to morphogenetic control, differentiation, and determination in plant development. It has practical significance for the following reasons:

1. The length of the juvenile period is inversely related to the breeding efficiency of woody perennials and to the selection of improved cultivars (Hansche and Beres 1980; Hansche 1983; Sherman and Lyrene 1983).
2. The ease of cuttage propagation for all types of woody perennials is strongly affected by ontogenetic age (Heybroek and Visser 1976).

3. The quantity and quality of productivity of a forest tree species is related to its degree of maturity (Heybroek and Visser 1976).

This review will describe the characteristics of juvenility and maturation in order to establish a more precise understanding of these concepts, to analyze their genetic, physiological, and morphogenetic basis, and to describe and evaluate methods of manipulating phase change that have horticultural significance. Several reviews of juvenility and phase change in plants are available (Allsopp 1965, 1968; Brink 1962; Doorenbos 1965; Sax 1962; Zimmerman *et al.* 1985). Juvenility in perennial plants was the subject of a symposium held in 1975 and sponsored by the International Society for Horticultural Science (Zimmerman 1976).

II. DESCRIPTION AND DEVELOPMENTAL BASIS OF PHASE CHANGE

Attainment and maintenance of the ability or potential to flower is the only consistent criterion available to assess the termination of the juvenile period. Other characteristics known to change with development and/or age are not consistent from species to species, and none has been demonstrated to be causally related to sexual maturity. However, for individual species, a specific character—such as thorniness (Visser 1965) or anthocyanin formation (Romberg 1944)—may be correlated with ability to flower and thus be useful as an indicator of the termination of the juvenile period. For example, using thorniness as an indicator, Visser (1965) found a direct relationship between degree of juvenility and length of the juvenile period in apples. However, in most cases, the regular production of flowers is the only practical way to identify the end of the juvenile period. In a species such as *Ribes nigrum* L. that is sensitive to short photoperiods for flower initiation, it is possible to show that there is a transitional phase of development during which flowering potential is increasing. Robinson and Wareing (1969) showed that if seedlings of black currant were exposed to short days (SD) at different ages, there was a minimum age, size, and number of nodes at which seedlings could be induced to flower, but the intensity of flowering increased through a certain age–size span, indicating a transitional phase.

In assessing the ability to flower, it is important to use procedures that are known to induce flower initiation but do not promote the maturation process. In only a few woody species can day length or other environmental factors be used to induce flower initiation and assess flowering potential. In others such as apple, stem ringing can be used because it induces flower initiation but apparently does not influence maturation. However, a significant problem arises when the same factor(s) promotes attainment of the sexually mature condition and the flower induction and initiation process. In many of the experiments cited in this review, it is not possible to distinguish between induction of the mature condition and the induction of flowering.

The length of the juvenile period in woody plants is quite variable (Table 3.1). In *Rosa* species, flowering can occur in seedlings 20-30 days old, but in certain forest and landscape tree species the juvenile period can last 30–40 years. In a very approximate way, the length of the juvenile period is related to the ultimate size of the plant. In general, shrubs have a shorter juvenile period than trees, but as shown in Table 3.1, there are exceptions to this generalization. Although a juvenile phase occurs in herbaceous annual and perennial species, it is generally shorter in duration and the morphological and physiological changes associated with the phase transition are generally less distinct than in other species.

Once the sexually mature phase is attained, it is relatively stable. Reversion to the juvenile condition does not generally occur as a result

TABLE 3.1. Length of Juvenile Period in Some Woody Plant Species

Species	Length of juvenile period
Rosa (hybrid tea)	20–30 days
Vitis spp.	1 year
Prunus spp.	2–8 years
Malus spp.	4–8 years
Citrus spp.	5–8 years
Pinus sylvestris	5–10 years
Hedera helix	5–10 years
Betula pubescens	5–10 years
Pyrus spp.	6–10 years
Sequoia sempervirens	5–15 years
Pinus monticola	7–20 years
Larix decidua	10–15 years
Fraxinus excelsior	15–20 years
Acer pseudoplatanus	15–20 years
Thuja plicata	15–25 years
Pseudotsuga menziesii	20 years
Pinus aristata	20 years
Sequoiadendron giganteum	20 years
Picea abies	20–25 years
Tsuga heterophylla	20–30 years
Picea sitchensis	20–35 years
Quercus robur	25–30 years
Abies amabilis	30 years
Fagus sylvatica	30–40 years

Source: Clark (1983).

of asexual propagation such as cuttage or graftage involving a single bud and a small piece of stem, although flowering may be delayed to varying degrees by these techniques, depending on the species (Schaffalitzky de Muckadell 1954). Reversion to the juvenile condition does occur naturally during sexual and apomictic reproduction and can be induced by adventitious bud and embryo formation and by various nutritional, hormonal, or environmental treatments. In some plants (e.g., *Hedera helix* L. and *H. canariensis* L.), the juvenile phase can also be maintained in a stable morphological condition by cuttage propagation. This stability in characteristics associated with the juvenile-to-mature phase change contrasts with the changes that occur in other traits, such as reduced growth rate and type of branching, as the plant grows older. The reduced growth rate of older plants can often be reversed if an aged shoot is grafted onto a young seedling rootstock or if a cutting of an aged shoot is rooted. In some cases flowering may be delayed by such propagation, but not to the extent it would be by sexual reproduction.

To distinguish between these types of phenomena, Wareing (1959) has used the term *maturation* for the transition from the juvenile to the mature phase and the term *aging* to indicate the loss of vigor associated with increased complexity of the plant. Fortanier and Jonkers (1976) have referred to this loss of vigor as *physiological aging* in contrast to *ontogenetic aging*, or maturation. Zimmerman (1973) has pointed out that on the basis of these definitions, maturation occurs only in the development of seedling plants, while plants propagated vegetatively from sexually mature trees undergo the process of aging only. This does not, however, mean that seedlings do not undergo aging as well as maturation. It may be that some of the changes that occur during aging are related to, or an extension of, the processes involved in maturation (Borchert 1976). It may be that aging, as well as maturation, must take place before flowering can occur in some plants. Maturation, not aging, will be the main topic of consideration in this review.

The stability of the mature condition has a bearing on the definition of the juvenile phase. Once a plant attains the reproductively mature condition, flowering will continue as long as the requisite flower-inducing treatment is imposed or exists in nature. The age or size at which any one species or genotype attains this stable condition can vary with environmental factors (Section IV. A). However, examples are known where certain environmental conditions (Hield *et al.* 1966) or growth substance treatments (Pharis and Morf 1967, 1968) will cause transient precocious flowering in seedling plants. In these examples, the plants soon revert to a nonflowering condition and either do not flower again for several years as in citrus (Hield *et al.* 1966) or flower again only if treated with the appropriate growth substance as in conifers (Pharis and Morf 1967, 1968). Such examples point out the importance of defining reproductive maturation in terms of a *continuing* ability to flower when exposed to a normal (natural or imposed) flower-inducing treatment. Plants that flower transiently as the result of a treatment, but that do not maintain the ability to flower under natural or imposed environmental conditions, would not be considered reproductively mature.

Where it has been possible to analyze changes in attributes associated with phase change, it can be demonstrated that the upper and peripheral parts of a plant first obtain mature characteristics (e.g., flowering ability), while the basal and interior parts (including new growth) retain juvenile characteristics. For example, in seedling birch (*Betula verrucosa* Ehrh.) trees induced to flower very early under conditions of maximal growth (continuous light, high temperatures, high nutrition) in conjunction with stem girdling, the first-formed branches (nodes) near the base do not flower, whereas those less than 100 cm up the stem flower quite profusely (Arshad 1980; Longman 1976). Zimmerman (1973) also demonstrated with *Malus hupehensis* (Pamp.) Redh. that transition to the mature phase is closely correlated with node number.

Similarly, adventitious root and bud initiation potential, which are juvenile characteristics, seem to be related to ontogeny, but in an inverse fashion. In eucalyptus seedlings, it has been demonstrated that the cotyledonary node has a very high rooting potential, but by the fourth node in *Eucalyptus ficifolia* F. Muell. (Mazalewski 1978) and the fifteenth node in *E. grandis* (Paton *et al.* 1970), rooting capacity is nearly zero. It has also been observed that cuttings taken from shoots formed in the basal region have a higher capacity to form adventitious roots than those taken from shoots in the upper part of the plant (Roulund 1973; Achterberg 1959). With regard to adventitious bud formation, Burger (1980) has demonstrated in citrus seedlings that there is a strong negative gradient in bud-forming capacity from internode to internode of the epicotyl as distance from the cotyledon is increased. In *Carya illinoinensis* (Wangh.) Koch, where anthocyanin formation is a phase change-related character, pigment formation extends about the same distance from ground level by way of sap flow in the different branches of the same tree (Romberg 1944). Also, the trunk and basal portion of main branches of a seedling tree of citrus or *Gleditsia triacanthos* L. retains the ability to form thorns, whereas the upper and peripheral region of the tree is nearly thornless (Chase 1947; Soost and Cameron 1975). Similar observations with regard to leaf retention in *Fagus sylvatica* L. have been made (Schaffalitzky de Muckadell 1954, 1959). These findings suggest that juvenile characteristics such as rooting potential may be preserved at the base of plants in ontogenetically young tissue (meristems), while maturation occurs in the periphery of the plant in ontogenetically older but chronologically young tissues. This localization of characteristics is depicted in Fig. 3.1.

The observed relationships between phase-change characteristics and the ontogenetic age and node number just described suggests that phase change is closely related to changes in shoot apical meristem activity. Quite obviously, the ability to flower and at least some of the morphological differences between juvenile and mature shoots (e.g., phyllotaxis and thorn formation) are the result of changes in the structure and behavior of the shoot apical meristem. In *Hedera helix* L., for example, Stein and Fosket (1969) showed that mature apices have a larger meristematic area consisting of smaller cells than do juvenile apices. In contrast, the subapical region is larger in the juvenile shoot, and cell division continues longer in this region, resulting in longer internodes. The duration of the plastochron in the juvenile apex is longer than that in the mature apex.

Although this evidence for English ivy indicates that shoot apices of mature and juvenile shoots are quite different, it does not indicate anything about the nature of the events or processes that result in the

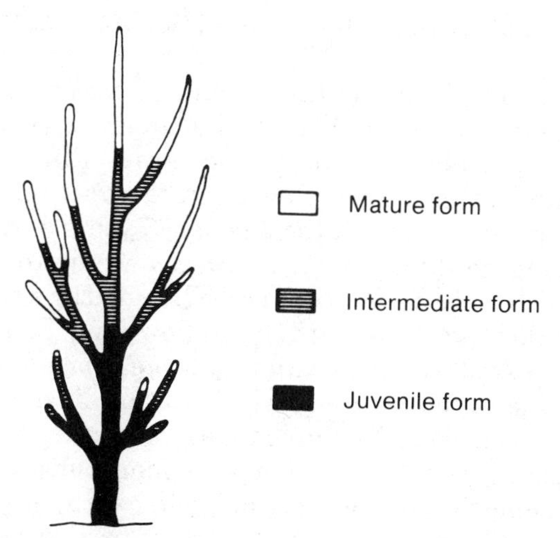

FIG. 3.1. Schematic representation of the localization of different maturation states on a mature seedling-grown plant.
After Kester (1976).

very different structure and behavior of mature and juvenile meristems. Do the changes in the meristems result from a single event or a series of events? Results of experiments involving GA_3-induced rejuvenation of mature English ivy plants demonstrate a differential sensitivity of various physiological and morphological characteristics to GA_3 dosage (Rogler and Hackett 1975a). Although indirect, this evidence suggests that maturation is a progressive transition, proceeding by several stages, and is not an abrupt single-event change.

III. GENETIC, PHYSIOLOGICAL, AND BIOCHEMICAL BASIS OF PHASE CHANGE

Wareing and Frydman (1976) have pointed out that it is necessary to make a distinction between two major aspects of the phase-change phenomenon:

- The factors controlling the transition from the juvenile to the mature stage
- The organizational, cellular, or molecular basis of the stable difference in juvenile and mature meristems which can be transmitted through many cell divisions

These two aspects of phase change will be discussed under separate headings.

A. Factors Controlling Attainment of the Mature Phase

1. Genetic. The length of the juvenile phase is genetically inherited. For example, Johnson (1940) found, among a very great number of seedlings, two plants of *Betula verrucosa* Ehrh. and one of *B. pubescens* Ehrh. that formed flowers in their second year, whereas seedlings normally flower at the age of 5–6 years or more. When the early-flowering seedlings of *B. verrucosa* were crossed with each other, early flowering appeared to be determined by a single dominant gene. In another genetic study of birch flowering, Stern (1961), using repeated controlled crossings with very large *B. pendula* Roth. populations, was able to select three early-flowering lines in three generations. These selected lines formed catkins during the first growth period from seed, even though they were morphologically juvenile.

In apple, some parents produce offspring with very short juvenile periods, many of which flower in 3–4 years from seed; other produce offspring with long juvenile periods of 10 or more years. The inheritance of this character in apple is quantitative and follows the usual pattern, except that it is not easy to measure the parental values because the actual length of the juvenile period of the parents is usually not known. By analysis of two incomplete half-diallel schemes of crosses involving 22 apple and 33 pear progenies, Visser (1976) showed a highly significant general combining ability and an insignificant specific combining ability variance for the juvenile period. This indicates that the inheritance of the juvenile period is additive in nature, a mode of inheritance that is a function of multigenic factors governing development.

Visser (1965) showed in apple and pear that there is a good correlation between the length of the vegetative phase (time from propagation to fruiting) of a cultivar and the contribution it makes to the length of the juvenile period of its progeny. In apple and in pear, as will be discussed in detail later, there is considerable evidence to show that seedlings must attain a certain size before they can flower. Therefore, within a progeny population, those seedlings that are most vigorous are likely to attain flowering size in the shortest time. Visser (1970b) suggested that by selecting the more vigorous seedlings, the breeder is also selecting for short juvenile period. However, Zimmerman (1977) concluded that vigor (stem diameter) was not a valid predictor of early flowering in pear seedlings; moreover, in apples (Saure 1970; Visser 1976), pears (Visser 1976), roses (de Vries 1976) and birch (Stern 1961), plants with short juvenile periods are smaller at time of flowering than plants with a long juvenile period.

2. Physiological and Biochemical. Little is known about the physiological or biochemical factors that cause maturation to occur. One approach to studying the phase-change phenomenon is to expose seedling plants to experimental treatments assumed to promote maturation and then to test the effect of such treatments by subjecting the treated plants to conditions known to favor flower initiation. By such an approach it is possible, in some species, to distinguish between conditions necessary for maturation and those necessary for flower initiation once the mature condition has been attained. Interpretation of such experiments may be something of a problem in species like apple, in which aging is the most significant factor known to influence flower initiation. However, both aging and flower initiation are promoted by scoring and growth-retarding chemicals, so these treatments can be used in maturation experiments to induce flowering, much as day length is used with day length–sensitive plants. Use of differences in node to first flower to assess the effect of a maturation-promoting treatment can also reduce the problem of interpretation since maturation is thought to occur at the shoot apical meristem.

This can be done by first establishing the node at which seedlings first initiate flower primordia. Zimmerman (1973) has shown that node to first flower is relatively constant under various environments that influence growth rate of *Malus hupehensis.*

Using such an approach, Higazy (1962), Longman and Wareing (1959), Doorenbos (1955), Zimmerman (1971), Robinson and Wareing (1969), Stephens (1964), Holst (1961), and others have shown that achieving a minimum size may be more important in completing the juvenile phase than is age or dormancy cycles. By using environments and treatments (fertilization) that induce continuous and/or vigorous growth, these workers reduced the chronological age of flowering in seedlings of *Lunaria biennis* L., *Malus hupehensis, Larix leptolepis* Gord., *Rhododendron, Ribes nigrum, Betula verrucosa, Pinus resinosa* Ait., and *Picea glauca* Voss. Longman and Wareing (1959) compared time to flower of birch seedlings grown continuously under long days in a greenhouse with those that went through cycles of growth, short day-induced dormancy, and chilling-to-break dormancy. Half the seedlings grown continuously in the greenhouse flowered when they were 2 m tall and less than 1 year old, whereas those exposed to cycles of growth and dormancy had not flowered after six cycles covering a period of more than 2 years. That rapid, continuous growth leads to a stable sexual maturation is indicated by the fact that grafts made from one of the continuously grown birch trees initiated both staminate and pistillate catkins in the following year (Longman 1976).

Visser (1964) has also demonstrated the importance of size in attaining the mature condition. Using 8- to 12-year-old apple and pear seedlings, he found an inverse correlation between tree size (stem diameter) and length of the juvenile period. He also found that stem diameter of apple seedlings before and after budding was positively correlated with percentage of seedlings flowering in a given year. He concluded that conditions that promote growth reduce the length of the juvenile period. However, as mentioned earlier, in apple, pear, and some other species, plants with a short juvenile period are smaller at time of flowering than are plants with a long juvenile period, indicating that rapid growth is not the only factor contributing to a short juvenile period.

That size or vigor *per se* is not critical to attaining maturity is suggested by observations of R. P. Pharis (personal communication) on age of flowering of sibling populations of *Pseudotsuga menziesii* under different climatic conditions. Plants grown in a low rainfall area of South Vancouver Island started flowering at 6–10 years, whereas those grown 50 miles away in a cool, wet region did not start to flower until 15–20 years of age. In this case, it would appear that water stress is the critical factor, although it is not possible to distinguish the apparent effects of water stress on maturation from those on induction of flowering.

Other treatments such as girdling, scoring, bark inversions, grafting onto dwarfing stocks, and application of growth regulators that slow growth and promote flowering of mature trees have, with few exceptions (Visser 1973; Kender 1974; Chacko *et al.* 1974; Longman 1976), little effect on or increase the length of the juvenile period.

Robinson and Wareing (1969) demonstrated that phase change is correlated with, but not dependent on, attainment of a certain size. This was demonstrated in *Ribes nigrum* seedlings in which a close correlation was found between size and response to short-day induction. Stem tip cuttings were taken from seedlings that had attained the minimum height (or distance from roots) required to respond to short-day induction. The rooted plants were grown to approximately the original height, then tips were again removed and rooted as cuttings. Even though the seedlings were prevented from ever reaching the minimum height or normal node number required for attainment of the mature condition by this treatment, the plants of the third and fourth decapitation cycle flowered in response to short days. The cumulative height of these plants, starting with the original seedling, was much greater than the previously determined minimum height. Robinson and Wareing (1969) interpreted these results as indicating that maturation is triggered by some mechanism or condition intrinsic to the shoot apex itself rather than by conditions prevailing within differentiated parts of the plant. They suggested that a minimum number of cell divisions must occur in the shoot apex for transition from the juvenile to the adult state.

Zimmerman's work with *Malus hupehensis* in which he found that transition to the adult phase is much more closely correlated with node number than with plant height indicates that cell division activity in the apical meristem rather than the subapical meristem is related to maturation (leaf initiation is a function of apical meristem activity, whereas internode length and height are functions of subapical meristem activity). On this basis he concluded that the stage of development at which maturation occurs is better measured by node number than by plant height.

In summary, the best strategy for obtaining rapid sexual maturation in many tree species is to grow the seedlings as rapidly as possible to a certain species- or genotype-dependent minimum size and then apply the flower-inducing treatment that is appropriate for the species.

It is not known why attainment of a minimum size is required for transition to the mature condition. However, the observations that the juvenile-to-mature phase change usually occurs at a predictable stage (size) in the developmental cycle of a plant and that changes occur at the shoot apical meristem raises the following questions about the organismal locus of the phase change: Does the plant body attain some critical size and then signal the meristem to initiate the developmental phase change response? or does the meristem behave independently of the remainder of the plant?

The best evidence indicating that meristems act as autonomous units and independently of the remainder of the plant comes from grafting experiments in which it has been found that grafting seedling scions onto mature trees capable of flowering has little or no effect on maturation. For example, Robinson and Wareing (1969) grafted juvenile and "near mature" scions of *Larix leptolepis* and *L. decidua* Mill. onto mature, bearing trees. One year after grafting, only one juvenile scion of 78 had cones, but 14 of 75 of the "near mature" scions had cones. None of the seedlings from which scions were taken had flowered at the time of observation. The authors concluded that grafting onto mature trees did not accelerate phase change in juvenile scions even though it did promote flowering of scions from "near mature" trees, suggesting that phase change is not primarily determined by nutritional conditions within the shoot system as the tree ages.

Contrary to these results with *Larix*, Singh (1959) found in an experiment with mango (*Mangifera indica* L) that 1½-year-old seedlings, approach-grafted to shoots of mature trees, flowered if the leaves were removed from the seedlings and the mature shoot was girdled below the union. Ungrafted control seedling plants did not flower. This suggests that meristems on seedlings of at least one woody species are capable of responding to stimuli from mature leaves and therefore that the meristem does *not* act autonomously but is influenced by surrounding tissues. However, these results do not imply that the mature stock caused maturation of the juvenile seedling since there is no evidence that it continued to flower if detached from the mature stock.

Other reports on the effects of grafting scions from seedlings onto mature flowering trees are conflicting (for a summary, see Doorenbos 1965; Zimmerman 1972). Some authors claim that flowering of grafted scions occurred 1-2 years earlier than in the parent seedlings, whereas others found no acceleration of flowering by grafting. These conflicting reports may be due to differences in the degree of maturity of the seedlings used or in the way the experiments were performed (Singh

1979; Lang 1965). Other researchers working with juvenility in herbaceous species have obtained results similar to those of Singh (1959) with mango. Lang (1965) cites three studies in which juvenile apices grafted onto mature plants flowered sooner than they would have without grafting onto donor plants. He concluded that leaves of juvenile plants were evidently incapable of producing the flowering stimulus (or were producing a flower-inhibiting stimulus) but that their growing points were entirely capable of responding to it.

Lang suggested that the lack of response of most woody plant juvenile apices to grafting onto mature plants may be more apparent than real because precautions (such as removal of metabolic sinks on the donor and removal of leaves on the receptor) necessary to obtain translocation from donor to receptor were not taken. The flower-promotive effect of defoliation of grafted seedling mangos in Singh's (1959) experiments suggests that transmission of the stimulus did not occur (or was not effective) unless leaves were removed from the receptor. These findings with mango point out the importance of manipulation of the donor and receptor in studies on maturation and flowering in other woody species. The importance of manipulating donor and receptor in grafting experiments also is emphasized by work of Clark and Hackett (1980), who demonstrated that translocation between juvenile and mature graft partners of *Hedera helix* under normal, unaltered conditions was very low or nonexistent when measured by detection of $^{14}CO_2$-labeled assimilates 24 or 48 hr after feeding. However, translocation of ^{14}C assimilates between graft partners in this system could be enhanced by increasing the leaf area treated with $^{14}CO_2$, girdling the stem of the donor plant, treatment of receptor shoot tips with the cytokinin benzylaminopurine, and selective defoliation of the receptor scion. In summary, the grafting evidence indicates that, depending upon the species, either the receptor shoot meristem or the leaf may be the site where maturation occurs.

Other evidence relevant to the question of the organismal locus of phase change has been obtained by grafting seedlings on various clonal rootstocks. Visser (1973) grafted apple seedlings on Malling 9 (M.9), a dwarfing rootstock, and M.16, an invigorating rootstock, and found that flowering of seedlings was hastened on M.9 and retarded on M.16 rootstocks, irrespective of the interstock and irrespective of the influence of rootstock or interstock on growth. *Malus sikkimensis* (Hook.f.) Koehne and *M. hupehenis* seedlings used as rootstocks reduced growth and retarded flowering. He concluded that vigor and precocity of a rootstock are independent properties, in that the ability of a rootstock to induce early flowering is determined by its specific precocity only. Several other workers have reported similar promotive effects of certain rootstocks on flowering in seedling apples (Tydeman 1961; Tydeman and Alston 1965; Kemmer 1953), indicating that roots can have some specific effect on the maturation processes in the shoot apices. Whether the effect of rootstocks on maturation and flowering is manifest through a promotive or inhibitory substance is not known, but it indicates that maturation or evocation of flowering at the apical meristem is being influenced by factors coming from other parts of the plant.

These experiments with clonal rootstocks and work by Frydman and Wareing (1973a,b; 1974) with *Hedera helix* and by Schwabe and Al-Doori (1973) with *Ribes nigrum* indicate that roots may be involved in control of phase change. Frydman and Wareing hypothesized that the juvenile phase is promoted by and dependent on gibberellins produced in the roots and that in *Hedera* the presence of aerial roots in close proximity to the apices causes them to remain juvenile. This hypothesis is supported by the following data: (1) apical buds of the juvenile phase contained higher levels of GA-like substances than their mature counterparts; (2) roots contained high levels of extractable GA-like substances; and (3) stems, apices, and leaves of derooted seedlings and cuttings contained lower levels of extractable GA-like substances than comparable organs of intact juvenile plants. Such a hypothesis is supported by the observation that exogenously applied GA_3 induces some juvenile characteristics not only in mature *Hedera helix* (Robbins 1957) and *H. canariensis* (Goodin and Stoutemyer 1961) but also in *Acacia melanoxylon* (Borchert 1965), *Citrus* (Cooper and Peynado 1958), pear (Griggs and Iwakiri 1961), *Cocos nucifera* (Schwabe 1976), *Morus nigra* (Trippi 1963c), and several *Prunus* species (Crane *et al.* 1960). Following this same line of reasoning, Wareing and Frydman (1976) suggested that as the height of a tree and the distance between shoot apices and roots increase, the levels of root-produced gibberellins in the shoot apices would be expected to decline, so that conditions for maintenance of the juvenile state no longer prevail and phase change to the mature condition can occur.

Results of investigations on the effects of growth-retardant chemicals on sexual maturation or flowering theoretically may have some bearing on the question of whether high gibberellin levels are inhibitory to the maturation process, since growth-retardant chemicals have been shown to inhibit gibberellin biosynthesis in the fungus *Gibberella fujikuroi* (Ninnemann *et al.* 1964) and to reduce gibberellin levels in some higher plant species (Zeevaart 1966). Early reports on the promotive effects of dwarfing rootstocks on flowering in apple seedlings suggested to investigators that retardation of growth might promote the maturation process. Even though this has been shown to be an erroneous conclusion (Visser 1973) because of the specificity of dwarfing stocks in promoting flowering, several investigators have attempted to promote flowering in seedlings by use of growth-retarding chemicals such as butanedioic acid mono-(2,2-dimethylhydrazide) (daminozide); (2-chloroethyl) trimethyl-ammonium chloride (chlormequat); and 2,4-dichlorobenzyl-tributylphosphonium chloride (chlorphonium chloride).

The reported effects of growth-retarding chemicals on maturation or flowering are conflicting and range from promotive to no effect to inhibitory. [In some cases, the growth retardants did not inhibit growth (Dennis 1968).] These differing results may be related to the age of the seedlings treated, as there is some indication that flowering of older seedlings is more likely to be promoted by growth retardants than is that of young seedlings that have never flowered previously. These results could also be interpreted to imply that (1) GA levels in seedlings are not readily controlled by growth retardants or (2) growth rates and the juvenile phase are not controlled by GAs. The results of Frydman and Wareing (1974) show that chlormequat can inhibit growth even while increasing the levels of GA-like substances in *Hedera helix*. The work of Arshad (1980) shows that chlormequat in conjunction with stem girdling increased the levels of GA-like substances in bark and buds of birch while promoting precocious flowering of seedlings (see later discussion).

The results of studies with growth retardants have not, so far, been very useful in determining whether declining levels of gibberellins in the shoot apices are required for transition from the juvenile to mature condition. Furthermore, if phase change in woody angiosperms depends solely on the reduced gibberellin levels in shoot apices accompanying increased distances between roots and shoot apices, it should be possible to induce phase change by grafting juvenile scions onto mature parts of trees. However, as already noted, such grafting treatments are seldom effective, and when effective, they require manipulation of the stock and scion to promote translocation from mature tissue into the seedling scion. It seems likely, as concluded by Wareing and Frydman (1976), that low gibberellin levels are a necessary but not sufficient condition for the juvenile-to-mature transition as evidenced by continued flowering.

This conclusion raises the question of what in addition to *low* gibberellin levels might be required for transition to the mature condition? The results of some experiments on control of flowering of birch seedlings carried out by Arshad (1980) may have some bearing on this question. Unfortunately, *Betula* is a genus in which it is difficult to distinguish between factors that influence maturation and those that influence induction of flower initiation. Birch seedlings normally begin to flower at 5–10 years of age, but time to flowering can be reduced to about 18 months by growing plants continually under long days in a greenhouse (Robinson and Wareing 1969), and such conditions appear to result in a stable phase change (Longman 1976). Time to flowering can be reduced even further by main stem or branch girdling (Longman 1976; Arshad 1980), and flowering occurs on the lowest branches in what might be expected to be the most juvenile part of the shoot system. Chlormequat at 10–25 mg/plant promoted flowering in girdled plants but had little or no effect on flowering of non-girdled plants (Arshad 1980). Arshad also found that stem or branch girdling of 6-month-old birch seedlings with or without chlormequat caused rapid, large increases in abscisic acid (ABA) and smaller increases in gibberellin-like substances in the bark and buds above the girdle. Exogenous applications of ABA promoted, and GA_3, GA_{4+7}, or GA_9 inhibited, flowering in girdled plants. As shown in these experiments and experiments with other species, many substances, both nutrients and hormones are likely to accumulate above a bark girdle. However, the fact that chlormequat enhanced the girdling effect on flowering and on elevation of ABA levels does implicate some aspect of terpene biosynthesis or metabolism in the response to girdling. These results suggest that ABA as an anti-gibberellin may be involved in the reproductive maturation process in birch and perhaps other species. This idea is supported by the findings of Rogler and Hackett (1975b) that ABA will counteract the rejuvenation influence of GA_3 in *Hedera helix*. Leaves of woody plants have been shown to synthesize inhibitors such as ABA (Eagles and Wareing 1964; Hillman *et al.* 1974) and the G inhibitor of eucalyptus, which is closely correlated with phase change–related loss of rooting potential in *Eucalyptus grandis* (Paton *et al.* 1970). However, Hillman *et al.* (1974) found that juvenile *Hedera helix* leaves contained five times as much ABA per unit fresh weight as did leaves of the mature form.

Inasmuch as ABA is intimately involved in stress phenomena (Walton 1980), the data with birch suggest that the relation of size and complexity to phase change might be mediated through a competition for water and the resultant stress, which would cause increases in ABA levels at meristems. Ross *et al.* (1981) have shown that GA_{4+7} promotes flower induction in 3-year-old seedlings of *Tsuga heterophylla* Sarg. only under conditions of girdling or water stress; others have shown that various stress treatments are synergistic with GA_{4+7} in this and other conifer species (Ross *et al.* 1983). Bonnet-Masimbert (1982) concluded from work with juvenile seedlings of *Pseudotsuga menziesii* that modest water stress, root flooding, or other treatment that inhibits root growth promotes flowering with or without a GA_{4+7} treatment. Unfortunately, in none of these studies were measurements made of the water status of the experimental plants.

Zimmermann (1978) has reported that water conductivity of the xylem decreases from base to top of the main stem, from main stem to branch, and from branch or twig to leaf due mainly to constriction of the xylem. The smaller the conductivity, the steeper the required pressure gradient to move water past that point. Lowest pressures are thus always found in leaves. According to Zimmermann, during periods of drought, pressures will always be lowest in the most peripheral parts of the tree. Such gradients in pressure should be greater for large trees than for small trees; therefore, stress conditions would be more likely in large than in small trees. However, because of technical difficulties, such measurements have not been made and the degree to which osmotic adjustment might counteract gradients in pressure is not known. Therefore, it is not clear whether the tensions generated are sufficient to induce synthesis and accumulation of large amounts of ABA or whether the difference in tensions in the periphery of large and small trees could be enough to implicate differential production of ABA as a mechanism for inducing maturation. Water potentials of about −10 bars are required to induce large amounts of ABA production in leaves (Walton 1980).

An alternate but complementary hypothesis concerning the relation of size to maturation is that assimilate accumulation or diversion to shoot meristems is modified as a result of size and complexity. Environmental conditions that enhance growth rate and early flowering of seedlings are the same ones that enhance photosynthesis. In addition, stem girdling, which causes early flowering in birch seedlings and some other species, increases the proportion of photosynthetic assimilate retained in the shoot. Hormonal control of partitioning of assimilates in the apical region might also be involved in phase change (Hackett 1976). It is well established that cytokinins, auxins, and gibberellins act to mobilize assimilates by creating metabolic sinks (Sachs 1977) and may create competition between sinks for assimilates (Hackett 1976). S. D. Ross (personal communication) has shown that GA_{4+7} enhances movement of ^{14}C assimilates to lateral long shoots within the bud of *Pinus radiata* at the expense of the apical meristem region. Recently, a similar role has been suggested for ABA (Brenner *et al.* 1982); such a role might relate to the possible involvement of ABA in promoting precocious flowering in young birch seedlings.

Allsopp (1968) and Franck (1976) have suggested that transition from juvenile to mature morphological character in several plants is correlated with an increase in the size of the shoot apical meristem. The anatomical studies of Stein and Fosket (1969), which show that the mature form of *Hedera helix* has an apical area twice as large as the juvenile, support this hypothesis. The transition from the smaller juvenile apex to the larger mature apex has been termed "apical strengthening" by Allsopp, who hypothesizes that nutritional factors cause stable alterations in the pattern of apical activity to become manifest in the juvenile-to-mature transition (Allsopp 1954, 1965, 1968).

In the water fern *Marsilea drummondii* A. Br., Allsopp (1953) demonstrated that increasing concentrations of glucose, fructose, or sucrose increased the rate of heteroblastic development with fewer leaves of each intermediate type being produced before the mature condition was obtained. Mannitol, a sugar that increases osmotic concentration but is poorly metabolized, was not effective (Allsopp 1954). These results indicate that the effects of sugar represent a response to nutrition rather than increased osmotic concentration in the external medium. Edwards and Allsopp (1956) found that low levels of nitrogen caused reversion from mature to juvenile and delayed appearance of the mature leaf. Allsopp (1965) presents evidence indicating that apical size closely follows changes in the nutritional status of several species of plants. He concluded that heteroblastic development appears to be largely dependent on changes in plant size and nutrition, which are reflected in changes in shoot apex size and morphological characteristics of the plant. Gaudet's work with *Marsilea vestita* Hook & Grev. generally supports Allsopp's nutritional hypotheses, but he has found that this species will form adult leaves in darkness, when supplied with an adequate carbon source, without an increase in apex size (Gaudet 1968; Gaudet and Malenky 1967).

A scanning electron microscope study of the apical meristem of mature *Hedera helix* shoots during GA_3-induced rejuvenation (Srinivasan and Hackett 1983) also brings into question the importance of dome size in relation to changes in phyllotaxis. In this study, phyllotaxis changed from the mature 2/5 to the juvenile 1/2 configuration before there was any significant reduction in the area of the apical meristem. At a much later time after treatment, there was a reduction in shoot apical meristem size as a result of GA_3-induced rejuvenation of mature plants. These results indicate that GA_3 does not act directly through changes in apical meristem size to influence phyllotaxis and other rejuvenation-related morphological characters. In *Acacia*, Kaplan (1980) has found that the increase in leaf length accompanied by an increased ratio of blade to petiole, which results in a short petiole, and the change to phyllode type of development are correlated with the developmental status of the shoot as a whole, and the size of the shoot apex in particular. However, the change from dissected to simple blade morphology shows no clear positional correspondence or relation to apex size.

There is indirect evidence that nutrition may be important in the control of phase change in other plants. It has been shown that factors such as low light intensity (Schaffalitzky de Muckadell 1959; Rogler and Hackett 1975b; Njoku 1956) and high temperature (Fisher 1954; Njoku 1957; Hudson and Williams 1955), which may reduce carbohydrate levels, cause rejuvenation or prolong the juvenile phase in *Hedera helix, Ipomoea caerulea* Koen. ex. Roxb., *Acacia melanoxylon* R. Br., *Fagus sylvatica, Ranunculus hirtus* B. & S., and *Rubus idaeus* L.

The experimental evidence discussed so far suggests that both hormonal and nutritional factors may be involved in the transition from the juvenile to the mature condition. However, what kind of mechanism(s) for control of phase change would be compatible with the evidence is not known. It is not even clear why attainment of minimum size is correlated with transition to the mature condition. Size most likely acts as an indicator of (1) cumulative cell divisions in the shoot apical meristem; (2) position of the shoot apex in relation to other parts of the plant (such as the roots) in terms of sources of, and translocation distance for, photosynthates, hormones, and water; and (3) the relative cumulative sizes and activities of various meristems (subapical and apical root and shoot meristems and the cambiums), as this influences competition for photosynthates, hormones, and water. It *is clear* that the meristem does not act autonomously and that its transition can be influenced by other tissues in the plant.

Several plant growth regulators besides ABA and growth retardants have been found to promote the juvenile-to-mature phase change. Kender (1974) reported that sprays of 1000 ppm (2-chloroethyl)phosphonic acid (ethephon) hastened and increased flowering of 3-, 4-, and 5-year-old apple seedlings, but control plants also flowered. Chacko *et al.* (1974) showed that 3-year-old mango trees could be induced to flower when treated with 1000 ppm ethephon (untreated controls did not flower). Mango seedlings do not normally flower until they are 6 or more years old. It is not clear from the report whether ethephon-treated seedlings flowered in subsequent years. Pharis and Morf (1967) induced flowering on very young seedlings. (3–12 months old) of several conifer species in the Taxodiaceae and Cupressaceae by application of various GAs. Pharis *et al.* (1976) also have stimulated flowering of young trees of species in the Pinaceae. With this family, nonpolar GAs (especially GA_{4+7}) are highly effective, whereas the polar GA_3 is virtually ineffective (Pharis *et al.* 1976; Ross *et al.* 1983).

Normally, reproductive structures would not be formed for 10–20 years. Pharis and Morf (1968) concluded that (1) GA applications do not terminate the juvenile phase since the seedlings revert to a non-flowering state after GA applications are stopped and (2) the juvenile state in these plants is one in which GA levels cannot be maintained at the level critical for floral induction.

The results and conclusions of Pharis and co-workers for conifers, coupled with numerous results indicating that gibberellins may be involved in maintaining the juvenile phase, rejuvenation, and/or inhibition of flowering in various woody angiosperm species, suggest that the hormonal basis of juvenility and maturation for conifers and for woody angiosperms may be quite different. However, as discussed previously, there is some direct and indirect evidence that terpene (GA or ABA) metabolism is involved in maturation in both groups of plants. Thus, it is possible that the apparent difference in the two groups of woody plants reflects an incomplete understanding of the results available rather than inherent differences.

B. Stability of Juvenile and Mature Meristems

The mature phase, once attained, is relatively stable and rejuvenation to the juvenile condition does not normally occur as a result of asexual propagation by cuttage or graftage. Rejuvenation does, of course, occur during sexual and apomictic (Frost 1938) reproduction and can be induced by other means (GA_3, high temperature, low light). Several explanations regarding the basis of this stability, which is transmitted through many cell divisions, can be hypothesized:

- The structure and properties of the apical meristem are controlled or stabilized by influences from preexisting, differentiated tissue located in more or less remote parts of the plant.
- The stable characteristic properties of the apical meristem are determined by the structure and organization of the shoot apex as a whole in terms of biophysical stability of the structure (organization) or diffusion-reaction schemes.
- There are intrinsic differences in the meristematic cells of juvenile and mature apices and their stable properties and organization arise from these intrinsic differences in the cells.

These three possible explanations can be termed the correlative, the structural and organizational, and the cellular. Evidence bearing on each type of explanation will be discussed in separate sections, starting with the third, for which the most data are available.

1. Cellular. The cellular explanation for the stability of the mature phase suggests that maturation involves a determination of cells, i.e., a process involving stable phenotypic change that persists in the absence of the initiating stimulus. Determination also involves a change in competence (the capacity of cells to react to developmental signals). Although the determined state is stable, the determination processes are not necessarily irreversible. Therefore, stability does not imply permanence. These characteristics of determination certainly fit the observations for phase change at the whole plant level. Do they also fit the observations for unorganized cells or a single cell derived from juvenile and mature tissue?

The findings of Stoutemyer and Britt (1965) that callus cultures derived from stem tissues of juvenile and mature *Hedera* have characteristically different and relatively stable growth rates over long periods of time indicate that there may be intrinsic differences in the unorganized cells derived from the two forms. Similar findings have been reported for *Castanea vulgaris* (Lam.), *Robinia pseudo-acacia* L., and *Aesculus hippocastanum* L. (Trippi 1963a,b). However, it should be borne in mind that in *Hedera* at least, these cultures were established from rather complex stem cross-sections, and therefore the observed differences could be from a number of complicating factors such as influences from preexisting differentiated tissues and medium selection of cell types with different growth rates. Polito and Alliata (1981) have recently shown that calli derived from shoot apical meristems of juvenile and mature *Hedera helix* also have characteristically different growth rates. It would, of course, be better to clone single cells derived from meristematic tissues of the shoot apices from juvenile and mature plants and determine if callus so derived has characteristic and stable growth rates.

There are a few other examples (Wardell and Skoog 1969; Meins and Binns 1979) of unorganized plant cells that retain rather stable but reversible characteristics through many subcultures, suggesting that differentiation (determination) is a replicative cellular state (i.e., a condition determined by intracellular control circuits that are not broken by cell division). The best example is tobacco pith explants, which normally require cytokinin for growth in culture but can be habituated for cytokinin by growing them at an elevated temperature (35°C) or on cytokinin-containing medium (Meins and Binns 1979).

The habituated phenotype is stable for many cell generations when returned to the lower temperature even when cultures are of single cell origin. The habituation process is reversible in that pith derived from adventitiously regenerated plants of habituated as well as nonhabituated cell origin always express the nonhabituated phenotype. Habituation is also reversible *in vitro* in that habituated cell clones grown at 16°C and on low cytokinin become nonhabituated (cytokinin requiring) after seven subcultures.

Whether the differential, stable growth rates of callus derived from juvenile and mature *Hedera* stems reflect the phenotypic differences of intact mature and juvenile English ivy plants has been studied by my colleagues and myself. We have been successful in regenerating plantlets from callus derived from stems of the two growth phases. Although the mode of adventive plantlet formation is different for callus of the two phases (embryo for mature callus, bud formation for juvenile callus), the resultant plantlets from callus of both phases have juvenile morphological characteristics (Banks 1979). The simplest interpretation of these observations is that the characteristic, stable differences in growth rate of callus from juvenile and mature stems do *not* reflect stable phenotypic characteristics of intact juvenile and mature *Hedera* plants; therefore, stable morphological differences are not likely to be the result of intrinsic differences in the cells of the meristems.

However, a number of observations spanning a long period of time (Mullins and Srinivasan 1976; Krul and Worley 1977; Barlass and Skene 1980; Schwabe 1976; Frost 1938; Baldini and Mosse 1956; Garner and Hatcher 1961) indicate that adventitious bud and embryo formation is a rejuvenating process. These observations span a number of species, organized and callus tissue, and natural and induced adventive processes. The most striking example is natural nucellar embryony in *Citrus*, which always results in complete rejuvenation (Frost 1938). Brink (1962) contends that the nucellus of citrus is juvenile tissue based on the juvenile character of nucellar seedlings. However, Brink's conclusion may be unwarranted based on the juvenile character of plants derived adventitiously from mature tissues. Schwabe (1976) cites the example of adventitious bud formation in apple roots in which rejuvenation occurs as measured by increased propensity to form adventitious roots, but time to flower is much less in plants derived from adventitious shoots than in those from seedlings. In grape, Barlass and Skene (1980) found that plantlets derived via adventitious buds from mature leaf primordia fragments had characteristics identical to seedlings. So perhaps a truly mature meristem cannot be formed *de novo*. To my knowledge, this statement holds true except for three cases. One case (Huhtinen 1976) involves adventitious bud formation on callus derived from very early flowering birch seedlings in which the only other clear evidence of the mature condition was the low propensity to form adventitious roots. In this case, plants derived adventitiously retained their ability to flower early, but there were some slight indications of rejuvenation such as the vegetative character of the terminal bud and a few of the uppermost lateral buds. Therefore, it can perhaps be postulated that some juvenile characteristics were induced even in these plants during callus growth and adventitious bud formation. In another case, Scorza and Janick (1980) found that callus derived from the upper parts of *Passiflora suberosa* L. vines produced shoots that flowered for several weeks but then lost the ability to flower. In a third case, embryoids derived from callus from roots of mature *Panax ginseng* plants flowered in the embryoid stage while still in culture (Chang and Hsing 1980). Whether plants that developed from these embryoids retained their ability to flower is not reported. These observations leave in doubt questions regarding the possibility that intrinsic cellular differences form the basis of stable juvenile and mature characteristics and whether a mature meristem can be formed adventitiously (*de novo*). However, the cytokinin-habituated state in tobacco pith callus is always reversed to the nonhabituated state when plants are regenerated adventitiously.

If there are intrinsic differences in the meristematic cells of juvenile and mature apices that form the basis of stable phenotypic characteristics, these might be expected to be reflected in quantitative or qualitative differences in deoxyribonucleic acid (DNA), ribonucleic acid (RNA), or proteins. The most data for these constituents in juvenile and mature cells are available for *Hedera helix*.

The DNA content of mature and juvenile *H. helix* cells has been determined by workers in several laboratories using various methods to examine different tissues. Kessler and Reches (1977) and Millikan and Ghosh (1971) found the DNA content of mature-phase leaf tissue to be lower than that of juvenile-phase leaves on a cellular or dry weight basis. Schaffner and Nagl (1979) analyzed cells of whole buds and leaves and concluded that nuclei of mature cells contained an average of 71% more DNA than those of comparable juvenile tissues. Wareing and Frydman (1976) Domoney and Timmis (1980), Polito and Alliata (1981), and Hackett *et al.* (1964) report no differences in DNA content per cell between juvenile and mature tissues taken from shoot apices, apical buds, apical meristems, and stem callus, respectively. The weight of evidence, then, suggests that the amount of DNA per 2C cell is not different for juvenile and mature tissues in *Hedera*. This conclusion has been confirmed recently in a cytological study by Polito and Chang (1984).

Fewer data are available for the RNA content of juvenile and mature *H. helix* tissues. Millikan and Ghosh (1971) reported less total, soluble, and ribosomal RNA on a dry weight basis in mature than in juvenile leaf tissue. Hackett *et al.* (1964) found higher levels of total RNA per cell in callus derived from juvenile than in callus derived from mature stems. Based on these findings, Domoney and Timmis (1980) did rRNA hybridization experiments with DNA prepared from juvenile and mature tissue and found no evidence for differences in redundancy in rRNA genes in DNA of the two forms. Rogler and Dahmus (1974) used DNA-RNA hybridization techniques to study the RNA populations of juvenile and mature *Hedera* apices and found no qualitative differences between species of RNA in the two forms. However, differences were observed in the frequency distribution of RNA species.

Fukasawa (1966) investigated the basic proteins from juvenile and mature stem callus of *Hedera* using electrophoretic separation techniques. He found several different protein bands, some of which were more dense in extracts from mature callus and one that was more dense in juvenile extracts.

2. Structural and Organizational. What evidence is available to support the idea that the stable phenotypic characteristics of juvenile and mature apices are determined by the structure and organization of the shoot apex? In woody plants that display distinct juvenile and mature characteristics, evidence to support this idea comes from grafting experiments in which mature *Citrus* apices comprised of the apical dome and only three leaf primordia were grafted onto juvenile seedling stocks (Navarro *et al.* 1975). The plants that resulted from these shoot apex grafts showed no rejuvenation and were completely adult in character. My co-workers and I have aseptically cultured on defined medium mature *Hedera* apices consisting initially of a dome and two leaf primordia. After 8 weeks in culture, they had as many as 12 unexpanded leaf primordia and had maintained the phyllotactic characteristics of mature *Hedera* plants. No root initiation had occurred. Other observations of excised mature shoot apices cultured *in vitro* verify the results with *Hedera helix* (Hackett 1980). However, as will be discussed later, multiple subculture of apices derived from axillary bud primordia on primary cultures of excised mature shoot apices tends to result in juvenile characteristics (Mullins *et al.* 1979; Gupta *et al.* 1981; Lyrene 1981; Franclet *et al.* 1980).

These observations also could be interpreted as evidence for intrinsic cellular characteristics forming the basis of stability. The likelihood of this interpretation is diminished, however, by the observations that *de novo*, adventitiously derived meristems from mature tissue almost invariably yield plants with juvenile characteristics. This combination of observations certainly suggests that there are restraints and controls imposed on cells by the organization and structure of the meristem and young primordia. The best available evidence from models and the results of surgical and growth regulator experiments (Steeves and Sussex 1972; Maksymowych and Erikson 1977) indicate that phyllotaxis is a cyclic process controlled by features near the meristem surface and near the prospective primordial site and not by long-range influences such as subjacent differentiating vascular tissue. Whether the cyclic conditions and responses are related to a morphogenic compound acting in a diffusion-reaction manner, or are related to space, contact, or pressure, is not clear. However, Green (1980) takes a biophysical perspective and presents evidence that cellulose microfibrillar alignment in the outer epidermal cell walls may vary in the plane of the apex and is pertinent to organ construction and may also be operative in determining the critical cyclic nature of phyllotaxis.

3. Correlative. The third possible explanation for the stability of plant growth phases has to do with influences from preexisting, differentiated tissues in parts of the plant more or less remote from the meristem. Some evidence for this possibility comes from grafting experiments with *Hedera helix* in which Doorenbos (1954) found that rejuvenation of mature shoots grafted to juvenile ones was promoted by the presence of juvenile leaves on stocks and inhibited by adult leaves on the scions. Other studies involving juvenile rootstock-induced rejuvenation of mature scions tend to support this possibility also (Muzik and Cruzado 1958; Martin and Quillet 1974).

From less direct evidence, it has also been suggested that reduced levels of gibberellins possibly coming from the roots are involved in the transition from the juvenile to the mature condition (i.e., the instability of the juvenile condition). However, based on the findings of Maksymowych and Erikson (1977) on the stability of gibberellin-induced phyllotactic changes in *Xanthium*, this third possible explanation for the stability of growth phases seems unlikely. In this work, changes in apex size and shape as well as phyllotaxis induced by GA_3 remained stable well after this growth regulator would be expected to be gone and, presumably, the normal hormonal supply to the shoot apices had resumed. The stability of these altered apices seems to argue against the idea that phyllotaxis (and perhaps other phase change-related morphological characteristics) depends on growth-regulating substances coming from substantially older tissue.

C. Future Research Directions

As the preceding discussion indicates, the factor(s) controlling the transition from the juvenile to the mature phase and the basis of the stability of the two phases are not well understood. Progress toward an understanding of phase change requires that consideration be given to the most important and fruitful areas of future research. Research on the following topics is important, is experimentally feasible, and is likely to be fruitful in terms of a better understanding of phase change:

1. Search for a woody plant experimental system, in which the transition to the mature phase is rapid, can be induced at will by a specific treatment, and is easily identified by rapid, controlled induction of flowering or associated morphological characteristics by a different treatment
2. Search for a biochemical or cytological marker(s) that is an early indicator of progress toward transition to the mature phase and an indicator of intrinsic differences in cells of the juvenile and mature phases by study of cells of apical meristems and single cell-cloned callus derived from them
3. Study surface cell characteristics of apical meristems that are in the process of rejuvenation as a result of gibberellin treatment and compare the surface cell characteristics of mature apical meristems with those that are derived adventitiously (*de novo*) from mature tissue

Discovery of a good experimental system, as described in item 1, is required if progress is to be made toward an understanding of the factors controlling transition from the juvenile to mature phase. The results of research on items 2 and 3 might help to distinguish between the most likely bases for the stability of the phases and might also help in following the progress of transition as influenced by various factors.

IV. PROCEDURES TO REDUCE LENGTH OF JUVENILE PERIOD

Because the delay in flowering caused by long juvenile periods is a major problem in the breeding and selection of tree species (Hansche and Beres 1980; Hansche 1983; Heybroek and Visser 1976), considerable effort has been made to find methods of reducing the length of the juvenile period. The effectiveness of some of the methods reported is in doubt because the experimental conditions were not well controlled and because it is difficult to determine whether the reported influence involved a reduction in the length of the juvenile period or a promotion of flower induction during the transitional period. The more frequently cited methods are evaluated in the following sections.

A. Environmental Treatments

As discussed in Section III.A.2, several workers have observed that seedlings require a minimum size to obtain the flowering condition. Therefore, the faster and longer the plant grows, the sooner it reaches reproductive maturation. Environmental conditions that promote rapid, continuous growth generally reduce the length of the juvenile period. It is common practice for plant breeders to use such conditions to obtain early flowering, thereby reducing generation time.

Various methods are used for promoting rapid, continuous growth, depending on the species and the feasibility of modifying the environment. Common treatments are those that prevent dormancy (long photoperiods), promote rapid growth (favorable temperatures, long photoperiods, high light intensity, optimal levels of water and mineral nutrition), and/or break dormancy (defoliation, low temperature, growth regulator application). Using such conditions in a greenhouse, Doorenbos (1955) was able to reduce the juvenile phase of *Rhododendron* seedlings to half its normal length. With apple seedling progenies grown continuously in optimal greenhouse conditions, Aldwinckle (1975) obtained flowering of 89–93% of the seedlings 26 months after germination by using defoliation to break dormancy. Seedlings from the same crosses grown in the field had not begun to flower 4 years after germination. Using conditions for rapid, continuous growth, other researchers have obtained equal or greater reductions in time to first flowering of seedlings of many other species (see Section III.A.2). In addition to optimal environmental conditions for growth, investigators stress the importance of a bare minimum of pruning or pruning to direct growth into a single main stem (Aldwinkle 1975; Visser 1970a).

Even when continuous growth in greenhouse conditions is not feasible, reduced time to flowering can be obtained by optimizing growth at the greenhouse, nursery, and orchard stages of growth. Visser (1970a) has demonstrated the importance of optimizing growth of apple seedlings at all stages. He has shown that the size of the seedlings transplanted into the orchard is very important for early flowering even when orchard environmental conditions are optimal in terms of soil moisture and fertility. Using optimal nursery and orchard conditions, he has obtained flowering of 90% of seedling progeny in 5 years with a mean time to flowering of 4.3 years. By improving growing conditions, time to flowering of genetically comparable groups of progeny has been reduced from about 7.5 to 4.5 years in apples and from about 9.0 years to 6.0 years in pears (Visser *et al.* 1976).

In contrast to the work cited thus far, investigations with conifers suggest that stress treatments (e.g., water stress, high temperature, low soil oxygen levels) induce precocious flowering in seedlings and that such treatments are synergystic with GA_{4+7} treatments (Brix and Portlock 1982; Bonnett-Masimbert 1982). As discussed earlier, Bonnett-Masimbert concludes that any treatment that inhibits root growth promotes flowering.

B. Grafting Seedlings on Clonal Rootstocks

As noted by Zimmermann (1972), the effectiveness of grafting seedling scions onto dwarfing rootstocks in order to shorten the juvenile period has been in dispute for many years because of conflicting reports in the literature. However, there is strong evidence that seedling scions grafted onto certain clonal apple dwarfing rootstocks flower 2–4 years earlier than the seedlings from which they are taken. In some cases the scions also show other mature characteristics—such as node of branching and leaf form—earlier than do parent seedlings (Tydeman 1937). The evidence for promotion of early flowering is most convincing for M.9, but M.27 also seems to show a similar effect. M.9 is one of the parents of M.27.

The extent to which earlier flowering is induced by the same rootstock appears to vary markedly in experiments carried out by different investigators. Visser (1973) suggests that the variation can be attributed to different growth conditions and that the better the growth conditions, the smaller the difference between seedlings on M.9 and seedlings on their own roots with respect to time of flowering. Visser also suggests that the budding or grafting operation limits the difference between grafted and ungrafted seedlings because the grafting operation checks growth. He concludes that under optimal conditions for growth, little is gained by grafting apple seedlings on precocious rootstocks such as M.9.

As discussed in Section III.A.2, Visser (1973) concluded from his grafting experiments with apple that vigor and precocity of a rootstock are independent properties, i.e., the ability of a rootstock to induce early flowering is determined by its specific precocity only. This work suggests that not all dwarfing rootstocks would be expected to promote early flowering of seedlings and that precocity and dwarfing should be evaluated as separate characters.

C. Grafting Seedlings on Mature Plants

Grafting of seedling scions into the top of reproductively mature plants has long been claimed to induce early flowering of the scion. Zimmerman (1972) concludes, based on a thorough review of the literature, that such grafting may induce early flowering when scions come from older seedlings, but that in these cases, the juvenile phase is already past and the scions were in the transitional phase. As noted earlier, the evidence for mango (Singh 1959) suggests that this species may be an exception to Zimmerman's general conclusion.

D. Treatments That Retard Growth

Many treatments that retard growth (ringing, scoring, bark inversion, root pruning, grafting on dwarfing interstocks, spraying with growth-retardant chemicals) have been demonstrated to promote flowering of reproductively mature trees of some species. However, there is no evidence that such treatments reduce the length of the juvenile phase. Zimmerman (1972) concludes that such treatments may promote earlier flowering in older seedlings, but where such treatments are effective, seedlings are probably in the transitional phase.

E. Treatments with Growth Regulators

There are a few instances where growth regulators such as gibberellins and ethephon will induce flowering in very young seedling plants. In the cases where gibberellins induce early flowering in seedlings of certain coniferous species, it is clear that GA treatments do not terminate the juvenile phase since after cessation of GA applications seedlings revert to a nonflowering state (Pharis and Morf 1968). In the report of ethephon-induced flowering of young mango seedlings by Chacko *et al.* (1974), it is not clear whether ethephon-treated seedlings flowered in subsequent years.

Even though GA treatments do not appear to cause stable changes in maturation state in conifers, they are very useful to plant breeders for induction of precocious flowering. For example, GA_3 is very effective for inducing strobilus formation when desired on very young seedlings of many species of the Cuppessaceae and Taxodioceae. Likewise, GA_{4+7} will induce precocious flowering on 16 species in five genera of the Pinaceae family (Ross *et al.* 1983). Flowering response of Pinaceae family seedlings induced by GA_{4+7} is often enhanced by cultural treatments such as girdling, water stress, high temperature, and nitrate treatment (Ross *et al.* 1983).

V. PROCEDURES TO OBTAIN JUVENILE MATERIAL FROM MATURE PLANTS

The adventitious root initiation potential of stem tissue of some species (particularly tree species) may decrease during ontogenetic aging and maturation. Cuttings from young trees and shrubs usually propagate easily from stem cuttings but, by the time they have been tested and selected for desirable characteristics, they lose much of their potential to root. Maturation state may also be important for other methods of clonal propagation, such as tissue culture, which require adventitious root and/or bud initiation. In addition, intraclonal variance in other maturation-related characteristics (e.g., growth rate and growth habit) can have important horticultural or silvicultural implications (Olesen 1973). In relation to clonal propagation and intraclonal variability, all of these characteristics are usually described by comparing them with the characteristics of a young seedling or the presumed mature part of a chronologically old seedling (ortet) or propagule (ramet) therefrom.

For these reasons, the ability to obtain relatively juvenile material from mature plants is very desirable. The most common methods for doing so are described and evaluated in the remaining sections, classified into two categories: use of juvenile parts of mature plants, and rejuvenation of mature parts of plants. Some of the methods cited are controversial because it is not clear whether the effect of the method and the characteristics observed are related to reversal of physiological age (invigoration) or to maturation (rejuvenation) (Fortanier and Jonkers 1976), and because it is not generally recognized that there may be various stages and separable characteristics of juvenility and maturation (Hackett and Rogler 1975a). This latter possibility suggests that rejuvenation has a quantitative aspect, which would result in varying degrees of rejuvenation dependent on the characteristic(s) being monitored and the method of rejuvenation being used.

A. Use of Juvenile Parts of Mature Plants

Selection and propagation of shoots from the part of a mature plant of seed origin that has retained all or part of its juvenile characteristics is a standard horticultural practice. Methods in this category are based on observations and experiments indicating that ontogenetically juvenile tissue generally persists at the base of seed-grown plants, particularly at the base of the main stem(s) or trunk(s) and the base of lower branches.

1. Material from Lower Part of the Crown. Cuttings taken from branches in the lower part of the crown have been shown to have higher rooting potential than those from the upper part of the crown of the same tree in *Picea abies* (L.) Karst. (Roulund 1973; Grace 1939), *Pseudotsuga menziesii* Mirb. Franco (Achterberg 1959), *Pinus radiata* D. Don. (Fielding 1969), *Pinus strobus* L. (Doran *et al.* 1940), and *Pinus* virginiana Mill. (Snow and May 1962). This is probably the result of unequal rate of advancement of maturation in the branches and main stem (Olesen 1978). However, Black (1972) was unable to detect differences in the rooting of cuttings from the upper and lower branches of the crown of *Pseudotsuga menziesii* in trees less than 24 years old, and Fielding (1954) was unable to detect crown position-related difference in rooting in *Pinus radiata* trees over 20 years old. Juvenile morphological characters (thorniness or leaf retention) are retained on the twigs of the trunk and lower main branches of citrus (Soost and Cameron 1975), *Gleditsia triacanthos* (Chase 1947), and *Fagus sylvatica* (Schaffalitzky de Muckadell 1954, 1959).

2. Epicormic Shoots on Trunk and Base of Primary Branches. Epicormic shoots growing naturally from latent buds on the base of trunks and primary branches and from adventitious buds on roots often have juvenile morphological characteristics and high rooting potential. Cuttings from such material have been used successfully for propagation of *Gleditsia triacanthos* L. (Stoutemyer *et al.* 1944), *Eucalyptus rostrata* Schlecht. and *E. polyanthemos* Schauer. (Fazio 1964), *Carya illinoinensis* (Smith and Chin 1980), and *Olea europaea* L. (Porlingis and Therios 1976). Shoots of adventitious origin on roots or shoots may be rejuvenated as a result of *de novo* meristem formation (see Section V.B.1) and not necessarily because they originate in the juvenile part of the plant.

3. Epicormic Shoots Obtained as a Result of Severe Pruning. Epicormic shoots can be forced into growth by severe pruning of the shoot system or treatment with the cytokinin benzylaminopurine (BA) (Mazalewski and Hackett 1979). This method has been used successfully to obtain shoots with higher rooting potential in apple (Stoutemyer 1937), *Pinus radiata* (Libby *et al.* 1972), and several *Eucalyptus* species (Mazalewski and Hackett 1970; Martin and Quillet 1974; Franclet 1979). Such results can be interpreted in several ways. Libby *et al.* (1972) interpreted their data to indicate that maturation had been arrested by hedging at the ontogenetic stage of the nodes from which the hedge shoots were formed. However, such results could also be interpreted in terms of a rejuvenation of tissue from a relatively more mature state. Others (Fortanier and Jonkers 1976) suggested that the increased rooting potential resulting from hedging or severe pruning is not related to maturation state (ontogenetic age) at all, but to a physiological invigoration or reversal of physiological aging. However, this seems unlikely with hedged *Pinus radiata* since Libby *et al.* (1972) have shown that plants resulting from cuttings of hedged plants have a growth habit very similar to that of plants grown from seed. Others have reported similar findings.

B. Rejuvenation of Mature Parts of Plants

Methods of rejuvenating mature parts of plants are based on the general conclusion, discussed earlier, that although maturation-related characteristics are relatively stable, they are reversible under certain conditions. It should be recognized that some characters appear to be more easily manipulated than others, and that a particular treatment or intensity or duration of treatment may influence one or some characters but not all characters (Rogler and Hackett 1975a).

1. Initiation of Adventitious Buds and Embryos. Most but not all investigations indicate that shoots or plantlets arising adventitiously are to some degree juvenile in appearance and exhibit delayed flowering and a greater rooting potential in comparison with shoots from the mature part of the same plant (Hackett 1980). Some of the observations are for adventitious buds arising from roots or adventive embryos from the nucellus. Unfortunately, the maturation status of roots is not known, and the maturation status of the nucellus has been questioned (Mullins and Srinivasan 1976; Brink 1962). Stoutemyer (1937) observed that shoots developing adventitiously on roots had juvenile morphological characteristics and high rooting potential, whereas cuttings from the mature mother tree failed to root. Robinson and Schwabe (1977) reported that shoots arising from the roots of the difficult-to-root 'Lord Lamborne' apple clone were juvenile in appearance and rooted easily but flowered within 2 years of cutting propagation. Obviously, this observation that flowering potential can be separated from rooting potential in rejuvenation is very important philosophically and horticulturally.

Wellensiek (1952) used adventitious shoots from sphaeroblasts of a number of species to obtain shoots with juvenile characteristics and high rooting potential. Garner and Hatcher (1962) showed that adventitious shoots arising from sphaeroblasts on stems of clonal apple have high rooting potential and the resulting plants have a disinclination to flower compared with similar aged (4 years) plants propagated from mature shoots. However, they concluded that the rooting potential of shoots of adventitious origin falls off in succeeding vegetative generations. In addition, they found that scions from 1-year-old sphaeroblast-originated trees grafted on M.7 and M.9 clonal rootstocks developed into trees that flowered only sparsely (and only at the top) after 4 years in comparison with similarly treated trees propagated from buds from mature trees.

Garner and Hatcher (1962) also observed that adventitious shoots originating from sphaeroblasts of *Ilex aquifolium* had "spiny" juvenile foliage and did not flower. Similar observations have been made with regard to the characteristics of olive trees developing from cuttings originating from sphaeroblasts (Baldini and Scaramuzzi 1957). Barlass and Skene (1980) found that plantlets originating adventitiously on leaf primordia from mature grape shoot apices were juvenile in morphological character and flowering ability. Recall that plants regenerated from callus originating from mature *Hedera helix* stems via adventive embryos have juvenile morphology and pigmentation (Banks 1979).

2. Grafting Mature Scions on Juvenile Rootstocks. The mature state is normally quite stable during vegetative propagation techniques, including grafting on seedling rootstocks. Generally such grafting causes an increase in vegetative vigor and some delay in flowering, but similar changes are also expected when such a scion is grafted on a mature clonal rootstock. Such responses are usually interpreted as changes in physiological rather than ontogenetic age.

Monselise (1973) reported greater vegetative vigor and delay of flowering when mature calamondin scions were grafted on young, rather than old, nucellar clone sour orange rootstocks. It is not clear from his report whether this response is a change in physiological or ontogenetic age. In contrast, Navarro *et al.* (1975) reported that *in vitro* micrografting of mature shoot apices onto virus-free apomictic seedlings had no influence on the maturation state of the resulting plants. However, there are reports that grafting on juvenile rootstocks in a few species causes a rejuvenation. The best documented example is with *Hedera helix* (Doorenbos 1954) in which such grafting causes complete rejuvenation based on morphological and physiological characteristics. With *Hedera canariensis*, greater rejuvenation occurred when scions consisted of a single bud with a small amount of stem than when scions consisted of a single bud with a larger amount of stem; successful rejuvenation in either case required that the temperature be above a certain minimum (Stoutemyer and Britt 1961). It was also shown by Doorenbos (1954) that the presence of leaves on the juvenile stock promotes rejuvenation, whereas leaves on the mature scion inhibit it.

In *Hevea brasiliensis* Muell. Arg., rooting potential of mature scions has been enhanced by repeated grafting in serial fashion at short intervals on successive young juvenile seedlings (Muzik and Cruzado 1958). Easy-to-root clones of *Eucalyptus platyphylla* and *Terminalia superba* were developed by grafting mature shoots to juvenile seedling rootstocks (Martin and Quillet 1974). With *Eucalyptus camaldulensis* Dehnhardt., Franclet (1979) reported that three serial micrografts onto seedling rootstocks with 2-month periods between were required to obtain increased rooting potential, whereas increased rooting potential was obtained after a single graft cycle followed by growth with *Cupressus dupreziana*. However, he noted that the *Cupressus dupreziana*-rejuvenated clone began to flower 8 years after rejuvenation, whereas a seedling plant would not be expected to flower until after 40 years.

In the grafting trials of Martin and Quillet (1974) and Franclet (1979), changes in foliage characteristics and habit and growth, as well as rooting potential, were observed. Chaperon (1979) stresses the importance of high vigor of the juvenile rootstock and proximity of the scion to the root of the seedling for successful rejuvenation of mature scions by juvenile rootstocks. In agreement with the findings of Stoutemyer and Britt (1961) and Doorenbos (1954) with *Hedera*, Franclet (1979) suggests that small size of the scion is important.

The grafting observations of Muzik and Cruzado (1958), Martin and Quillet (1974), and Franclet (1979) are difficult to interpret because results of control treatments in which a mature scion is grafted onto a mature rootstock are not provided. As a consequence, it is difficult to tell whether the effects observed are the result of rejuvenation or physiological invigoration.

The differences among species in response to grafting mature scions on juvenile stocks could be genetic or may occur because the conditions required for rejuvenation by grafting are not met. Furthermore, because it is difficult to distinguish between rejuvenation and reversal of physiological aging (invigoration) in some cases, some reports of rejuvenation may in reality be temporary invigoration related to reversal of physiological aging.

3. Treatment with Plant Growth Regulators. Mature *Hedera helix* can be rejuvenated to varying degrees by treatment with varying doses of GA_3 (Rogler and Hackett, 1975a); GA_{4+7} and GA_1 are also effective. The stability of the characteristics induced by GA_3 depends on dosage. For example, aerial root and anthocyanin formation are temporary characteristics when low doses of GA are used but stable characteristics when higher doses are used. Other growth substances such as auxins, cytokinins, and ethephon are ineffective (Rogler and Hackett 1975a). Application of GA_3 to mature pear (Griggs and Iwakiri 1961), *Citrus* (Cooper and Peynado 1958), *Acacia* (Borchert 1965), and some *Prunus* species (Crane *et al.* 1960) inhibits flowering and induces juvenile morphological characteristics, but the effects on rooting potential are not reported. In contrast, treatment of young seedlings of *Eucalyptus* with GA_3 speeds up the change from juvenile to mature type of foliage (Scurfield 1962); the effect on rooting potential is not reported.

4. Severe Pruning of Tree Crowns. Several workers have considered the effects on maturation characteristics of severe pruning of mature trees of seed origin and of hedging or stooling of mature trees of clonal origin. Hatcher (1959) has demonstrated with apple that severe pruning of rootstock clones delays flowering and increases vegetative vigor and rooting potential. Such severely pruned stool beds and hedges are routinely used to enhance rooting potential of apple rootstock clones, and cuttings taken from such sources have been utilized for the commercially successful propagation of clonal dwarfing rootstocks. Garner and Hatcher (1962) reported that even 1-year-old shoots on such stool beds can flower but still have high rooting potential. They consider such plants to be physiologically invigorated but not rejuvenated.

Franclet (1979) has used similar techniques with several conifer species to obtain cuttings with higher rooting potential from the mature crowns of trees of seed origin. In *Pseudotsuga menziesii*, such pruning yields invigorated shoot growth with juvenile morphological characteristics the first year and increased rooting potential the second year. Black (1972) has also used severe pruning of mature trees to enhance the rooting potential and vigor of *Pseudotsuga menziesii* cuttings. Franclet (1979) obtained similar results with *Picea abies*, in which rooted cuttings grew orthotropically. Franclet (1979) notes that in this case the "rejuvenation" of foliage characteristics and density of branches was very incomplete compared with the characteristics observed in 2- to 3-year-old seedlings. With pine species, Franclet (1979) found that pruning stimulated elongation of short shoots or brachyblasts. The resulting shoots had juvenile needle characteristics and a rooting potential characteristic of the species in the juvenile state and very much greater than shoots from areas of the tree that had not been severely pruned. However, with severely pruned *Pinus pinaster*, Franclet (1979) observed the general reappearance of female flowering the third year following pruning.

From the work of Garner and Hatcher (1962) and Franclet (1979), it would appear unlikely that severe pruning yields fully rejuvenated plants. In these studies it is not possible to distinguish between partial ontogenetic rejuvenation and reversal of physiological aging. Stability of the characteristics observed is crucial to distinguishing between the two possibilities. Franclet's (1979) observation that tissue from shoots of severely pruned trees is more amenable to graft and tissue culture rejuvenation suggests that physiological invigoration or partial ontogenetic rejuvenation (depending on interpretation) may be important for further ontogenetic rejuvenation.

5. Propagation by Cuttings. Maturation characteristics are generally stable during propagation by cuttings. However, Jacobs (1939) states that rooted cuttings of pine appeared to undergo a degree of rejuvenation and that in pines primary leaves are produced as well as occasional juvenile buds. Ooyama and Toyoshima (1965) showed that cuttings from second generation trees (*Pinus densiflora*) root more readily than original cuttings. Black (1972) reported similar results for *Pseudotsuga menziesii* and found that 3-year-old cutting-propagated trees had increased rooting potential, more juvenile morphological characteristics, and greater vigor than the original ortet. Morgan *et al.* (1980) reported that stem cuttings taken from rooted propagules of a 2-year-old *Quercus virginiana* seedling, maintained for 2 years in a greenhouse, rooted better than did cuttings taken from the original tree 2 years later. However, from the work of Morgan *et al.* it is difficult to distinguish between maturation and environmental effects on rooting potential. Franclet (1979) reported that environmental factors (e.g., nutrition, temperature, soil volume, and moisture) that promote root activity and shoot vigor of recently rooted cuttings and pruning influence the rooting potential and orientation of growth exhibited by shoots on recently rooted cuttings.

In the studies cited in this section, it appears that the treatments caused either partial rejuvenation or temporary physiological invigoration, but it is difficult to distinguish between these possibilities because the results of appropriate control treatments were not provided, the number of characteristics measured was not sufficient, and/or the stability of the characteristics was not well documented.

6. *In Vitro* Propagation. The results of recent studies indicate that even though mature meristems are quite stable both *in vivo* and *in vitro*, their phase-related characteristics can be modified as a result of *in vitro* culture; both the length of culture and the number of subcultures involved seem to be related to such changes. Such findings suggest the possible importance of the culture medium and other culture conditions with respect to rejuvenation *in vitro*.

David *et al.* (1978) have reported the *in vitro* rejuvenation of *Pinus pinaster* primary meristems. Elongating buds implanted on medium containing the cytokinin BA formed new buds from developing needle fascicles (brachyblasts or short shoots). The less elongated the buds implanted, the more juvenile the plantlets developed from the needle fascicle-derived buds in terms of foliage characteristics (Franclet 1979). This suggests that newly organized meristems are more plastic and subject to rejuvenation. High concentrations of BA and low concentrations of sucrose favored juvenile characteristics (David *et al.* 1978). Buds taken from cutting-propagated plants that had been subsequently hedged and given high levels of fertility exhibited the highest percentage of successful regeneration (Franclet 1979).

With *Sequoia sempervirens*, Boulay (1979) found that when the primary explant was of plagiotropic orientation, the number of orthotropic shoots produced increased with number of subcultures, and orthotropic shoot formation was enhanced by inclusion of activated charcoal in the medium. He also found that there was increasing root formation after a variable number of subcultures and suggested that conditions in the culture produce ontogenetic rejuvenation. Franclet (1979) indicated that *in vitro*-produced plantlets were impossible to distinguish from seedlings in the nursery. Mullins *et al.* (1979) found that the juvenile form of 'Cabernet Sauvignon,' an ancient grape clone, reappeared *in vitro* in shoots from serially subcultured mature shoot tips. Juvenility was indicated by lack of tendrils (the flower-bearing structures) and spiral phyllotaxy, which are both juvenile characters in grape. Likewise, thornless blackberries cultured from shoot tips produce monophyllous leaves once a few preformed trifoliate leaves have grown out (Broome and Zimmerman 1978), and production of monophyllous leaves in culture has been maintained by repeated subculturing for more than a year (Zimmerman 1981). Shoot cuttings with monophyllous leaves are easily rooted, and the resulting plants produce trifoliate leaves once established in the greenhouse. There is no delay in flowering.

Apple cultivars cultured from shoot tips also show some reversion in leaf shape in tissue culture (Sriskandarajah *et al.* 1982; Zimmerman 1981), although it is not as distinct as in the other cases described. This reversion is manifested by lobed leaves, deep, irregular serrations, and thinner leaf blades than normally found in shade leaves. This apparent reversion to the juvenile phase in apple might also explain the increased rooting of shoots after a certain number of subcultures (Sriskandarajah *et al.* 1982). Takeno *et al.* (1982/83) found that this increase in rooting ability with increasing numbers of subcultures was associated with a reduction in endogenous gibberellin and cytokinin levels. Although this evidence suggests a rejuvenation of shoots *in vitro*, apple trees propagated by these *in vitro* methods do not show delayed flowering in the field (Zimmerman 1981). Lyrene (1981), Gupta *et al.* (1981), and Economou (1982) have reported similar findings for blueberry, eucalyptus, and deciduous azalea, respectively. In all three cases, rejuvenation was measured in terms of increased rooting potential and changes in morphological characteristics.

B. McCown (personal communication) has also observed rapid changes to juvenile morphology in cultures of *Thuja occidentalis* and *Cryptomeria japonica*. *In vitro* propagation involving adventive bud or embryo formation would be expected to yield rejuvenated plants, but such rejuvenation in plantlets proliferated from preexisting meristems is less expected and, as indicated earlier, suggests that the small size of the mature explant and the conditions of the culture combine over time (subculture) to induce rejuvenation. It is also possible that over time in culture, adventitious bud formation contributes to the proliferation of plantlets and these would be expected to be more juvenile.

C. Conclusions and Commercial Application

The preceding discussion indicates that reliable methods are available to recover juvenile tissue from juvenile buds preserved at the base of mature seedling-grown plants of some species. Methods for rejuvenation of ontogenetically mature tissues are less reliable because our knowledge of the developmental basis of maturation and rejuvenation is incomplete, especially with regard to their relationship (if any) to physiological aging and invigoration. However, from the results of the studies described in the previous section, it appears that several methods can be used to induce partial rejuvenation and that a few methods induce complete rejuvenation.

Rejuvenation methods pioneered by the East Malling Research Station in England, which utilize cuttings from invigorated or partially rejuvenated hedges, have been used commercially for many years to propagate clonal, dwarfing apple rootstocks. Successful methods, however, will vary with species. The results also suggest that reversal of physiological aging may be an important first step for subsequent partial or complete rejuvenation and that environmental conditions for optimal growth are important for all methods. It seems that retreatment (or several cycles) using a single method (e.g., serial grafting) is likely to be more effective than a single application of a treatment. This might be considered a dose effect similar to the GA dose effect in *Hedera*. It also appears that the greatest degree of rejuvenation may come as a result of using several methods in succession, e.g., severe pruning, followed by serial grafting on juvenile rootstocks, followed by micropropagation *in vitro*. These latter observations strengthen previous suggestions that maturation and rejuvenation are progressive phenomena.

By recovery of existing juvenile material at the base of mature plants or by use of one or more rejuvenation methods, it seems possible (and has been accomplished with some species) to establish stock block plantings that will yield cuttings of higher rooting potential and less intraclonal variability than cuttings taken from the crown of a mature plant. Cuttings from such stock blocks should have, in addition to high rooting potential, the desirable juvenile characteristics of vegetative vigor, orthotropic growth habit, and juvenile branching pattern. Their potential to flower may be reduced, an undesirable side effect for some purposes. Plants in stock blocks would be severely pruned to increase cutting production and maintain juvenile characteristics. Periodically, it might be necessary to re-rejuvenate stock plants by the techniques originally used and to establish a new stock block from the resulting material.

LITERATURE CITED

ACHTERBERG, H.H. 1959. Zur vegetativein triebuer mehrung einigen fort geholze. Forst. V. Jagd. 9:59–60.

ALDWINCKLE, H.S. 1975. Flowering of apple seedlings 16–20 months after germination. HortScience 10:124–126.

ALLSOPP, A. 1953. Experimental and analytical studies of pteridophytes. XIX. Investigations on *Marsilea*. 3. The effect of various sugars on development and morphology. Ann. Bot. (London) 17:447–463.

ALLSOPP, A. 1954. Juvenile stages of plants and the nutritional status of the shoot apex. Nature 173:1032–1035.

ALLSOPP, A. 1965. Heteroblastic development in cormophytes. p. 1172–1221. In: W. Ruhland (ed.), Encyclopedia of plant physiology, XV/1. Springer, Berlin.

ALLSOPP, A. 1968. Heteroblastic development in vascular plants. Adv. Morphol. 8:127–171.

ARSHAD, N. L. 1980. Studies on endogenous growth substances in relation to flowering in seedlings of *Betula pendula*, Roth. Ph.D. Thesis, Univ. of Wales, Aberystwyth.

BALDINI, E. and B. MOSSE. 1956. Observations on the origin and development of sphaeroblasts in the apple. J. Hort. Sci. 31:156–162.

BALDINI, E. and F. SCARAMUZZI. 1957. Indagini et asservazioni sugli sferoblasti delli olivo. Ann. Sper. Agr. 11:723–740.

BANKS, M.S. 1979. Plant regeneration from callus from two growth phases of English ivy *Hedera helix* L. Z. Pflanzenphysiol. 92:349–353.

BARLASS, M. and K.G.M. SKENE. 1980. Studies on the fragmented shoot apex of grapevine. II. Factors affecting growth and differentiation *in vitro*. J. Expt. Bot. 31:489–495.

BLACK, D.K. 1972. The influence of shoot origin on the rooting of Douglas-fir stem cuttings. Proc. Intern. Plant Prop. Soc. 22:142–159.

BONNET-MASIMBERT, M. 1982. Influence de l'etat de activite des racines sur la floraison induite par des gibberellines 4 et 7 chez *Pseudotsuga menziesii* (Mirb.) Franco. Silvae Genetica 31:178–183.

BORCHERT, R. 1965. Gibberellic acid and rejuvenation of apical meristems in *Acacia melanoxylon*. Naturwissenschaften 52:65–66.

BORCHERT, R. 1976. The concept of juvenility in woody plants. Acta Hort. 56:21–36.

BOULAY, M. 1979. Multiplication et clonage rapide du *Sequoia sempervirens* par la culture *in vitro*. p. 49–56. In: Micropropagation d'arbres forestiers. Assoc. Foret-Cellulose, Domaine de l'Etancon, 77370 Nangis, France.

BRENNER, M.L., M.B. HEIN, J. SCHUSSLER, J. DAIR, and W. A. BRUN. 1982. Coordinate control: The involvement of abscisic acid, its transport and metabolism. p. 343–352. In: P. F. Wareing (ed.), Plant growth substances. Academic Press, London.

BRINK, R.A. 1962. Phase change in higher plants and somatic cell heredity. Quart. Rev. Biol. 37:1–22.

BRIX, H. and F.F. PORTLOCK. 1982. Flowering response of western hemlock seedlings to gibberellin and water-stress treatments. Can. J. For. Res. 12:76–82.

BROOME, O.C. and R.H. ZIMMERMAN. 1978. *In vitro* propagation of blackberry. HortScience 13:151–153.

BURGER, D.W. 1980. The use of cell and tissue culture systems to study juvenility and reproductive maturation in citrus. Ph.D. Thesis, Univ. of California, Davis.

CHACKO, E.K., R.R. KOHLI, R. DORE SWAMY, and G.S. RANDHAWA. 1974. Effect of (2-chloroethyl) phosphonic-acid on flower induction in juvenile mango (*Mangifera indica*) seedlings. Physiol. Plant. 32:188–190.

CHANG, W. and Y. HSING. 1980. *In vitro* flowering of embryoids derived from mature root callus of ginseng (*Panax ginseng*). Nature 284:341–342.

CHAPERON, H. 1979. Maturation et bouturage des arbres forestiers. p. 19–31. In: Micropropagation d'arbres forestiers. Assoc. Foret-Cellulose, Domaine de l'Etancon, 77370 Nangis, France.

CHASE, S.B. 1947. Propagation of thornless honey locust. J. For. 45:715–722.

CLARK, J.R. 1983. Age-related changes in trees. J. Arboriculture 9:201–205.

CLARK, J. and W.P. HACKETT. 1980. Assimilate translocation in juvenile-adult grafts of *Hedera helix*. J. Amer. Soc. Hort. Sci. 105:727–729.

COOPER, W.C. and A. PEYNADO. 1958. Effect of gibberellic acid on growth and dormancy in citrus. Proc. Amer. Soc. Hort. Sci. 72:284-289.

CRANE, J.C., P.E. PRIMER, and R.C. CAMPBELL. 1960. Gibberellin induced parthenocarphy in *Prunus*. Proc. Amer. Soc. Hort. Sci. 75:129-137.

DAVID, H., K. ISEMUKALI, and A. DAVID. 1978. Obtention de plants de pin maritime (*Pinus pinaster* Sol) a partir de brachyblastes ou d'apex caulinaires de tres jeunes sujets cultives *in vitro*. C. R. Acad. Sci., Ser. D287:245–248.

DENNIS, F.G. 1968. Growth and flowering response of apple and pear seedlings to growth retardants and scoring. Proc. Amer. Soc. Hort. Sci. 93:53-61.

DeVRIES, D.P. 1976. Juvenility in hybrid tea roses. Acta Hort. 56:235-239.

DOMONEY, C. and J.N. TIMMIS. 1980. Ribosomal RNA gene redundancy in juvenile and mature ivy (*Hedera helix*). J. Expt. Bot. 31:1093-1100.

DOORENBOS, J. 1954. Rejuvenation of *Hedera helix* in graft combinations. Proc. Koninkl. Ned. Akad. Wetenschap., Ser. C. 57:99-102.

DOORENBOS, J. 1955. Shortening of breeding cycle of rhododendron. Euphytica 4:141-146.

DOORENBOS, J. 1965. Juvenile and adult phases in woody plants. p. 1222–1235. In: W. Ruhland (ed.), Encyclopedia of plant physiology, XV/1. Springer, Berlin.

DORAN, W.L., R.P. HOLDSEVORTH, and A.D. RHODES. 1940. Propagation of white pine by cuttings. J. For. 38:817-823.

EAGLES, C.F. and WAREING, P.F. 1964. Dormancy regulators in woody plants. Nature 199:874-875.

ECONOMOU, A.S. 1982. Chemical and physical factors influencing *in vitro* propagation of hardy deciduous azaleas (*Rhododendron* spp.) Ph.D. Thesis, Univ. of Minnesota, St. Paul.

EDWARDS, P. and A. ALLSOPP. 1956. The effects of changes in inorganic nitrogen supply on the growth and development of *Marsilea* in aseptic culture. J. Expt. Bot. 7:194-202.

FAZIO, S. 1964. Propagating eucalyptus from cuttings. Proc. Intern. Plant Prop. Soc. 14:288-290.

FIELDING, J.M. 1954. Methods of raising Monterey pine cuttings in the open nursery. Bull. 32. Forest & Timber Bur., Canberra, Australia.

FIELDING, J.M. 1969. Factors affecting the rooting and growth of *Pinus radiata* cuttings in the open nursery. Bull. 45 Forest & Timber Bur., Canberra, Australia.

FISHER, F.J.J. 1954. Effect of temperature on leaf-shape in *Ranunculus*. Nature 1733:406-407.

FORTANIER, E.J. and H. JONKERS. 1976. Juvenility and maturity of plants as influenced by their ontogenetical and physiological ageing. Acta Hort. 56:37-44.

FRANCK, D.H. 1976. Comparative morphology and early leaf histogenesis of adult and juvenile leaves of *Darlingtonia californica* and their bearing on the concept of heterophylly. Bot. Gaz. 137:20-34.

FRANCLET, A. 1979. Rajeunissement des arbres adults en vue de leur propagation vegetative. p. 3-18. In: Micropagation d'arbres forestiers. Assoc. Foret-Cellulose, Domaine de l'Etancon, 77370 Nangis, France.

FRANCLET, A., A. DAVID, H. DAVID, and M. BOULAY. 1980. Premiere mise en evidence morphologique d'un rajeunissement de meristemes primaires caulinaires de pin maritime age *Pinus pinaster* Sol. C. R. Acad. Sci., Ser. D290:927-930.

FRANICH, R.A., L.G. WELLS, and J.R. BARNETT. 1977. Variation with tree age of needle cuticle topography and stomatal structure in *Pinus radiata* D. Don. Ann. Bot. 41:621-626.

FROST, H. B. 1938. Nucellar embryony and juvenile characters in clonal varieties of *Citrus*. J. Hered. 29:423-432.

FRYDMAN, V. M. and P. F. WAREING. 1973a. Phase change in *Hedera helix* L. I. Gibberellin-like substances in the two growth phases. J. Expt. Bot. 24:1131-1138.

FRYDMAN, V. M. and P. F. WAREING. 1973b. Phase change in *Hedera helix* L. II. The possible role of roots as a source of shoot gibberellin-like substances. J. Expt. Bot. 24:1139-1148.

FRYDMAN, V. M. and P. F. WAREING. 1974. Phase change in *Hedera helix* L. III. The effects of gibberellins, abscisic acid and growth retardants on juvenile and adult ivy. J. Expt. Bot. 25:420-429.

FUKASAWA, H. 1966. Disc electrophoresis of proteins from juvenile and adult specimens of ivy. Nature 212:516-517.

GARNER, R. J. and E. S. J. HATCHER. 1961. Delayed flowering of trees raised from adventitious scions. p. 52-53. In: Annu. Rpt. East Malling Res. Sta. for 1960.

GARNER, R. J. and E. S. J. HATCHER. 1962. Regeneration in relation to vegetative vigor and flowering. p. 105-111. In: Proc. 16th Intern. Hort. Cong., Brussels.

GAUDET, J. J. 1968. The correlation of physiological differences and various leaf forms of an aquatic plant. Physiol. Plant. 21:594-601.

GAUDET, J. J. and R. K. MALENKY. 1967. Changes in the shoot apex during early development of the fern *Marsilea vestita*. Nature 213:945-946.

GOODIN, J. R. 1964. Shoot growth rates as a factor in growth phase transitions in *Hedera*. Proc. Amer. Soc. Hort. Sci. 84:600–605.

GOODIN, J. R. and V. T. STOUTEMYER. 1961. Effect of temperature and potassium gibberellate on phases of growth of Algerian ivy. Nature 192:677–678.

GRACE, N. H. 1939. Vegetative propagation of conifers I. Rooting of cuttings taken from the upper and lower regions of a Norway spruce tree. Can. J. Res., Ser. C 17:178–180.

GREEN, P. B. 1980. Organogenesis—A biophysical view. Annu. Rev. Plant Physiol. 31:51–82.

GRIGGS, W. H. and B. T. IWAKIRI. 1961. Effects of gibberellin and 2,4,5-trichlorophenoxypropionic acid sprays on Bartlett pear trees. Proc. Amer. Soc. Hort. Sci. 77:73–89.

GUPTA, P. K., A. F. MASCARENHAS, and V. JAGANNATHAN. 1981. Tissue culture of forest trees: clonal propagation of mature trees of *Eucalyptus citriodora* by tissue culture. Plant Sci. Lett. 20:195–201.

HACKETT, W. P. 1976. Control of phase change in woody plants. Acta Hort. 56:143–154.

HACKETT, W. P. 1980. Control of phase change in woody plants. p. 257–272. In: C. H. A. Little (ed.), Control of shoot growth in trees. Proc. Inter. Union For. Res. Organ. Working Parties on Xylem Physiology and Shoot Growth Physiology, Fredericton, New Brunswick, Canada.

HACKETT, W. P., V. J. STOUTEMEYER, and O. K. BRITT. 1964. Some cellular characteristics of tissue cultures from various growth phases of *Hedera helix* L. Plant Physiol. (Suppl.) 39:LXIV.

HANSCHE, P. E. 1983. Response to selection. p. 154–171. In: J. N. Moore and J. Janick (eds.), Methods in fruit breeding. Purdue Univ. Press, West Lafayette, Indiana.

HANSCHE, P. E. and W. BERES. 1980. Genetic remodeling of fruit and nut trees to facilitate cultivar improvement. HortScience 15:710–715.

HATCHER, E. J. S. 1959. The propagation of rootstocks from stem cuttings. Ann. Appl. Biol. 47:635–639.

HEYBROEK, H. H. and T. VISSER. 1976. Juvenility in fruit growing and forestry. Acta Hort. 56:71–80.

HIELD, H. Z., C. W. COGGINS, JR., and L. N. LEWIS. 1966. Temperature influence on flowering of grapefruit seedlings. Proc. Amer. Soc. Hort. Sci. 89:175–181.

HIGAZY, M. K. M. T. 1962. Shortening the juvenile phase for flowering. Meded. Landbhogesch. (Wageningen) 62:1–53.

HILLMAN, J. R., I. YOUNG, and B. A. KNIGHT. 1974. Abscisic acid in leaves of *Hedera helix* L. Planta 119:263–266.

HOLST, M. J. 1961. Experiments with flower promotion in *Picea glauca* (Moench) Voss. and *Pinus resinosa*. Ait. Recent Adv. Bot. 2:1654–1658.

HOOD, J. V. and W. J. LIBBY, JR. 1980. A clonal study of intraspecific variability in radiata pine. I. Cold and animal damage. Austral. For. Res. 10:9–20.

HUDSON, J. P. and I. M. WILLIAMS. 1955. Juvenility phenomena associated with crown gall. Nature 175:814.

HUHTINEN, O. 1976. Early flowering of birch and its maintenance in plants regenerated through tissue cultures. Acta Hort. 56:243–249.

JACOBS, M. R. 1939. The vegetative reproduction of forest trees. 1. Experiments with cuttings of *Pinus radiata* Don. Commonwealth For. Bull. 25.

JOHNSSON, H. 1940. Hereditary precocious flowering in *Betula verrucosa* and *B. pubescens*. Hereditas 35:112–114.

KAPLAN, D. R. 1980. Heteroblastic leaf development in acacia. Morphological and morphogenetic implications. Cellule 73:137–203.

KEMMER, E. 1953. Uber das primare und das fertile Stadium bei Apfelgeholzen. Züuchter 23:122–127.

KENDER, W. J. 1974. Ethephon induced flowering in apple seedlings. HortScience 9:444–445.

KESSLER, B. and S. RECHES. 1977. Structural and functional changes of chromosomal DNA during ageing and phase change in plants. Chromosomes Today 6:237-246.

KESTER, D. E. 1976. The relationship of juvenility to plant propagation. Proc. Intern. Plant Prop. Soc. 26:71-84.

KRUL, W. R. and J. F. WORLEY. 1977. Formation of adventitious embryos in callus cultures of 'Seyval,' a French hybrid grape. J. Amer. Soc. Hort. Sci. 102:360-363.

LANG, A. 1965. Physiology of flower initiation. p. 1380-1536. In: W. Ruhland (ed.), Encyclopedia of Plant Physiology, XV/1. Springer, Berlin.

LIBBY, W. J., JR., A. G. BROWN, and J. M. FIELDING. 1972. Effects of hedging radiata pine on production, rooting and early growth of cuttings. New Zealand J. For. Sci. 2:263-283.

LIBBY, W. J., JR. and J. V. HOOD. 1976. Juvenility in hedged radiata pine. Acta Hort. 56:91-98.

LONGMAN, K. A. 1976. Some experimental approaches to the problem of phase change in forest trees. Acta Hort. 56:81-90.

LONGMAN, K. A. and P. F. WAREING. 1959. Early induction of flowering in birch seedlings. Nature 184:2037-2038.

LYRENE, P. M. 1981. Juvenility and production of fast-rooting cuttings from blueberry shoot cultures. J. Amer. Soc. Hort. Sci. 106:396-398.

MAKSYMOWYCH, R. and R. O. ERIKSON. 1977. Phyllotactic changes induced by gibberellic acid in *Xanthium* shoot apices. Amer. J. Bot. 64:33–44.

MARTIN, B. and G. QUILLET. 1974. Bouturage des arbres forestiers an Congo. Resultats des essais effectues a Pointe-Noire de 1969 a 1973. Rajeunissement des arbres plus et constitution du parc a bois. Bois et Forets des Tropiques 157:21-40.

MAZALEWSKI, R. L. 1978. The influence of plant growth regulators, hedging and other rejuvenating methods upon rooting of *Eucalyptus ficifolia* stem cuttings. M.S. Thesis, Univ. of California, Davis.

MAZALEWSKI, R. L. and W. P. HACKETT. 1979. Cutting propagation of *Eucalyptus ficifolia* using cytokinin-induced basal trunk sprouts. Proc. Intern. Plant Prop. Soc. 29:118–124.

MEINS, F., JR. and A. BINNS. 1979. Cell determination in plant development. BioScience 29:221–225.

MILLIKAN, D. F. and B. N. GHOSH. 1971. Changes in nucleic acids associated with maturation and senescence in *Hedera helix*. Physiol. Plant. 24:10–13.

MONSELISE, S. P. 1973. Recent advances in the understanding of flower formation in fruit trees and its hormonal control. Acta Hort. 34:157-166.

MORGAN, D. L., E. L. McWILLIAMS, and W. C. PARR. 1980. Maintaining juvenility in live oak. HortScience 15:493–494.

MULLINS, M. G. and C. SRINIVASAN. 1976. Somatic embryos and plantlets from an ancient clone of the grapevine (cv. Cabernet-Sauvignon) by apomixis *in vitro*. J. Expt. Bot. 27:1022–1030.

MULLINS, M. G., Y. NAIR, and P. SAMPET. 1979. Rejuvenation *in vitro*: Induction of juvenile characters in an adult clone of *Vitis vinifera* L. Ann. Bot. 44:623-627.

MUZIK, T. J. and H. J. CRUZADO. 1958. Transmission of juvenile rooting ability from seedlings to adults of *Hevea brasiliensis*. Nature 181:1288.

NAVARRO, L., C. N. ROISTACHER, and T. MURASHIGE. 1975. Improvement of shoot tip grafting *in vitro* for virus-free *Citrus*. J. Amer. Soc. Hort. Sci. 100:471–479.

NINNEMANN, H. J., A. D. ZEEVAART, H. KENDE, and A. LANG. 1964. The plant growth retardant CCC as inhibitor of gibberellin biosynthesis in *Fusarium moniliforme*. Planta 61:229–235.

NJOKU, E. 1956. Studies on the morphogenesis of leaves. XI. The effect of light intensity on leaf shape in *Ipomoea caerulea*. New Phytol. 55:91–110.

NJOKU, E. 1957. The effect of mineral nutrition and temperature on leaf shape in *Ipomoea caerulea*. New Phytol. 56:154–171.

OLESEN, P. O. 1973. On transmission of age changes in woody plants by vegetative propagation. p. 64–70. In: Yearbook, Royal Vet. & Agr. Univ., Copenhagen.

OLESEN, P. O. 1978. On cyclophysis and topophysis. Silvae Genetica 27:173–178.

OLESEN, P. O. 1982. The effect of cyclophysis on tracheid width and basic density in Norway spruce. For. Tree Improvement Arbor. (Horsholm) 15:1–80.

OOYAMA, N. and A. TOYOSHIMA. 1965. Rooting ability of pine cuttings and its promotion. Bull. Govt. For. Expt. Sta. (Tokyo) 179:100–125.

PATON, D. M., R. R. WILLING, W. NICHOLLS, and L. D. PRYOR. 1970. Rooting of stem cuttings of Eucalyptus: A rooting inhibitor in adult tissues. Austral. J. Bot. 18:175–183.

PHARIS, R. P. and W. MORF. 1967. Experiments on the precocious flowering of western red cedar and four species of *Cupressus* with gibberellins A_3 and A_4/A_7 mixture. Can. J. Bot. 45:1519–1524.

PHARIS, R. P. and W. MORF. 1968. Physiology of gibberellin induced flowering in conifers. p. 1341–1356. In: F. Wightman and G. Setterfield (eds.), Biochemistry and physiology of plant growth substances. Runge Press, Ottawa, Canada.

PHARIS, R. P., S. D. ROSS, R. L. WAMPLE, and J. N. OWENS. 1976. Promotion of flowering in conifers of *Pinaceae* by certain of the gibberellins. Acta Hort. 56:155–162.

POLITO, V. S. and V. ALLIATA. 1981. Growth of calluses derived from shoot apical meristems of adult and juvenile ivy (*Hedera helix* L.). Plant Sci. Lett. 22:387–393.

POLITO, V. S. and Y.-C. CHANG. 1984. Quantitative nuclear cytology of English ivy (*Hedera helix* L.). Plant Sci. Lett. 34:369–377.

PORLINGIS, I. C. and I. THERIOS. 1976. Rooting response of juvenile and adult leafy olive cuttings to various factors. J. Hort. Sci. 51:31–39.

ROBBINS, W. J. 1957. Gibberellic acid and the reversal of adult *Hedera* to a juvenile state. Amer. J. Bot. 44:743–746.

ROBINSON, J. C. and W. W. SCHWABE. 1977. Studies on the regeneration of apple cultivars from root cuttings. I. Propagation aspects. J. Hort. Sci. 52:205–220.

ROBINSON, L. W. and P. F. WAREING. 1969. Experiments on the juvenile-adult phase change in some woody species. New Phytol. 68:67–78.

ROGLER, C. E. and M. E. DAHMUS. 1974. Gibberellic acid-induced phase change in *Hedera helix* as studied by deoxyribonucleic acid-ribonucleic acid hybridization. Plant Physiol. 54:88–94.

ROGLER, C. E. and W. P. HACKETT. 1975a. Phase change in *Hedera helix*: Induction of the mature to juvenile phase change by gibberellin A_3. Physiol. Plant. 34:141–147.

ROGLER, C. E. and W. P. HACKETT. 1975b. Phase change in *Hedera helix*: Stabilization of the mature form with abscisic acid and growth retardants. Physiol. Plant. 34:148–152.

ROMBERG, L. D. 1944. Some characteristics of the juvenile and bearing pecan tree. Proc. Amer. Soc. Hort. Sci. 44:255–259.

ROSS, S. D., R. F. PIESCH, and F. F. PORTLACK. 1981. Promotion of cone and seed production in rooted ramets and seedlings of western hemlock by gibberellins and adjunct cultural treatments. Can. J. For. Res. 11:90–98.

ROSS, S. D., R. P. PHARIS, and W. D. BINDER. 1983. Growth regulators and conifers: their physiology and potential uses in forestry. p. 35–78. In: L. G. Nickell (ed.), Plant growth regulating chemicals. Vol. II. CRC Press, Boca Raton, Florida.

ROULUND, H. 1973. The effect of cyclophysis and topophysis on rooting ability of Norway spruce cutting. For. Tree Improvement Arbor. (Horsholm) 5:21–41.

RUMBALL, W. 1963. Wood structure in relation to heteroblastism. Phytomorphology 13:206–214.

SACHS, R. M. 1977. Nutrient diversion: An hypothesis to explain the chemical control of flowering. HortScience 12:220–222.

SAURE, M. 1970. Beitrage zur Fruhselektion beim Apfel. I. Beziehungen zwischen dem Eintritt der Bluhfahigkeit und der Wuchsleistung von Apfelsamlingen. Mitt. Obst-Versuchsringes, Jork 25:156–161.

SAX, K. 1962. Aspects of ageing in plants. Annu. Rev. Plant Physiol. 13:489–506.

SCHAFFALITZKY DE MUCKADELL, M. 1954. Juvenile stages in woody plants. Physiol. Plant. 7:782–796.

SCHAFFALITZKY DE MUCKADELL, M. 1959. Investigations on ageing of apical meristems in woody plants and its importance in silviculture. Forstl. Forsgsv. Danm. 25:310–455.

SCHAFFNER, K. H. and W. NAGL. 1979. Differential DNA replication involved in transition from juvenile to adult phase in *Hedera helix* (Araliaceae). Proc. Symp. on Genome and Chromatin: Organization, Evolution, Function. Plant Stematics and Evolution Suppl. 2:105–110.

SCHWABE, W. W. 1976. Applied aspects of juvenility and some theoretical considerations. Acta Hort. 56:45–56.

SCHWABE, W. W. and A. H. AL DOORI. 1973. Analysis of a juvenile-like condition affecting flowering in the black currant (*Ribes nigrum*). J. Expt. Bot. 24:969–981.

SCORZA, R. and J. JANICK. 1980. *In vitro* flowering of *Passiflora suberosa* L. J. Amer. Soc. Hort. Sci. 105:892–897.

SCURFIELD, G. 1962. Effects of gibberellic acid on woody perennials with special reference to species of eucalyptus. For. Sci. 8:168–179.

SEELIGER, R. 1924. Topophysis und Zyklophysis pflanzlicher organe und ihre bedeutung fur die Pflanzenkultur. Angew. Bot. 6:191–200.

SHERMAN, W. B. and P. N. LYRENE. 1983. Handling seedling populations. p. 66–73. In: J. N. Moore and J. Janick (eds.), Methods in fruit breeding. Purdue Univ. Press, West Lafayette, Indiana.

SINGH, L. B. 1959. Movement of flowering substances in the mango leaves (*Mangifera indica* L.). Hort. Adv. 3:20–28.

SMITH, M. W. and H. CHIN. 1980. Seasonal changes in rooting of juvenile and adult pecan cuttings. HortScience 15:594–595.

SNOW, A. G., JR. and C. MAY. 1962. Rooting of Virginia pine cuttings. J. For. 60:257–258.

SOOST, R. K. and J. W. CAMERON. 1975. Citrus. p. 507–540. In: J. N. Moore and J. Janick (eds.), Advances in fruit breeding. Purdue Univ. Press, West Lafayette, Indiana.

SRINIVASAN, C. and W. P. HACKETT. 1983. Changes in shoot apices during gibberellin induced phase change from mature to juvenile in *Hedera helix*. HortScience 18:606 (Abstr.)

SRISKANDARAJAH, D., M. G. MULLINS, and Y. NAIR. 1982. Induction of adventitious rooting *in vitro* in difficult-to-propagate cultivars of apple. Plant Sci. Lett. 24:1–9.

STEEVES, T. A. and I. SUSSEX. 1972. Patterns in plant development. Prentice-Hall, Englewood Cliffs, New Jersey.

STEIN, O. L. and E. B. FOSKET. 1969. Comparative developmental anatomy of shoots of juvenile and adult *Hedera helix*. Amer. J. Bot. 56:546–551.

STEPHENS, G.R., JR. 1964. Stimulation of flowering in eastern white pine. For. Sci. 10:28–34.

STERN, K. 1961. Uber den Erfolg einer uber drei Generationen gefuhrten Auslese auf fruhes Bluhen bei *Betula verrucosa* Silvae Genetica 10:48–51.

STOUTEMYER, V. T. 1937. Regeneration in various types of apple wood. Iowa Agr. Expt. Sta. Bull. 220. p. 307–352.

STOUTEMYER, V. T. and O. K. BRITT. 1961. Effect of temperature and grafting on vegetative growth phases of Algerian ivy. Nature 189:854–855.

STOUTEMYER, V. T. and O. K. BRITT. 1965. The behavior of tissue cultures from English and Algerian ivy in different growth phases. Amer. J. Bot. 52:805–810.

STOUTEMYER, V. T., F. L. S. O'ROURKE, and W. M. STEINER. 1944. Some observations on the vegetative propagation of honey locust. Amer. J. For. 42:32–36.

SWEET, G. B. and L. G. WELLS. 1974. Comparison of the growth of vegetative propagules and seedlings of *Pinus radiata*. New Zealand J. For. Sci. 4:399–409.

TAKENO, K., J. S. TAYLOR, S. SRISKANDARIJAH, R. P. PHARIS, and M. G. MULLINS. 1982/83. Endogenous gibberellin and cytokinin-like substances in cultured shoot tissues of apple, *malus pumila* cv. Jonathan, in relation to adventitious root formation. Plant Growth Reg. 1:261–268.

TRIPPI, V. S. 1963a. Studies on ontogeny and senility in plants. II. Seasonal variation in proliferative capacity *in vitro* of tissues from juvenile and adult zones of *Aesculus hippocastanum* and *Castanea vulgaris*. Phyton 20:146–152.

TRIPPI, V. S. 1963b. Studies on ontogeny and senility in plants. III. Changes in proliferative capacity *in vitro* during ontogeny in *Robinia pseudoacacia* and *Castanea vulgaris* and in adult and juvenile clones of *R. pseudoacacia*. Phyton 20:153–159.

TRIPPI, V. S. 1963c. Studies on ontogeny and senility in plants. V. Leaf-fall in plants of different age of *Robinia pseudoacacia* and the effect of gibberellic acid on *R. pseudoacacia* and *Morus nigra*. Phyton 20:167–171.

TYDEMAN, H. M. 1937. Experiments on hastening the fruiting of seedling apple. p. 92–99. In: Annu. Rpt. East Malling Res. Sta. for 1936.

TYDEMAN, H. M. 1961. Rootstock influence on the flowering of seedlings of apples. Nature 192:83.

TYDEMAN, H. M. and F. H. ALSTON. 1965. The influence of dwarfing rootstocks in shortening the juvenile phase of apple seedlings. p. 97–98. In: Annu. Rpt. East Malling Res. Sta. for 1964.

VISSER, T. 1964. Juvenile phase and growth of apple and pear seedlings. Euphytica 13:119–129.

VISSER, T. 1965. On the inheritance of the juvenile period in apple. Euphytica 14:125–134.

VISSER, T. 1970a. Environmental and genetic factors influencing the juvenile period in apple. p. 101–116. In: Proc. Angers Fruit Breeding Symp, Angers, France, 1970.

VISSER, T. 1970b. The relation between growth, juvenile period and fruiting of apple seedlings and its use to improve breeding efficiency. Euphytica 19:293–302.

VISSER, T. 1973. The effect of rootstocks on growth and flowering of apple seedlings. J. Amer. Soc. Hort. Sci. 98:26–28.

VISSER, T. 1976. A comparison of apple and pear seedlings with reference to the juvenile period. II. Mode of inheritance. Acta Hort. 56:215–224.

VISSER, T., J. J. VERHAEGH, and D. P. DE VRIES. 1976. A comparison of apple and pear seedlings with reference to the juvenile period. I. Seedling growth and yield. Acta Hort. 56:205–214.

WALTON, D. C. 1980. Biochemistry and physiology of abscisic acid. Annu. Rev. Plant Physiol. 31:453–489.

WARDELL, W. L. and F. SKOOG. 1969. Flower formation in excised tobacco stem segments. I. Methodology and effects of plant hormones. Plant Physiol. 44:1402–1406.

WAREING, P. F. 1959. Problems of juvenility and flowering in trees. J. Linn. Soc. (London) 56:282–289.

WAREING, P. F. and V. M. FRYDMAN. 1976. General aspects of phase change with special reference to *Hedera helix* L. Acta Hort. 56:57–68.

WELLENSIEK, S. J. 1952. Rejuvenation of woody plants by formation of sphaeroblasts. Proc. Kon. Ned. Akad. Wet. 567–573.

ZEEVAART, J. A. D. 1966. Reduction in gibberellin content of Pharbitis seeds by CCC and after effects on the progeny. Plant Physiol. 51:856–862.

ZIMMERMAN, R. H. 1971. Flowering in crabapple seedlings. Methods of shortening the juvenile phase. J. Amer. Soc. Hort. Sci. 96:404–411.

ZIMMERMAN, R. H. 1972. Juvenility and flowering in woody plants: a review. HortScience 7:447–455.

ZIMMERMAN, R. H. 1973. Juvenility and flowering in fruit trees. Acta Hort. 34:139–142.

ZIMMERMAN, R. H. 1976. Juvenility in perennial plants. Proc. Symp. Intern. Soc. Hort. Sci., 1975, Acta Hort. 56:1–317.

ZIMMERMAN, R. H. 1977. Relation of pear seedling size to length of the juvenile period. J. Amer. Soc. Hort. Sci. 102:443–447.

ZIMMERMAN, R. H. 1981. Micropropagation of fruit plants. Acta Hort. 120:217–222.

ZIMMERMANN, M. H. 1978. Structural requirements for optimal water conduction in tree stems. p. 517–532. In: P. B. Tomlinson and M. H. Zimmermann (eds.), Tropical trees as living systems. Cambridge Univ. Press, London.

ZIMMERMAN, R. H., W. P. HACKETT, and R. P. PHARIS. 1985. Hormonal aspects of phase change and precocious flowering, pp. 79–115. In: A. Pirson (ed.), Encyclopedia of Plant Physiology (N.S.), Vol. 11/IV. Springer-Verlag, Berlin and New York.

4

In Vitro Systems for Propagation and Improvement of Tropical Fruits and Palms[1,2]

Richard E. Litz
Tropical Research and Education Center, Institute of Food and Agricultural Sciences, University of Florida, Homestead, Florida 33031
Gloria A. Moore
Fruit Crops Department, Institute of Food and Agricultural Sciences, University of Florida, Gainesville, Florida 32611
Chinnathambi Srinivasan
Tropical Research and Education Center, Institute of Food and Agricultural Sciences, University of Florida, Homestead, Florida 33031

---

[1] The assistance provided to R. E. Litz by the Rockefeller Foundation and by the USDA Tropical and Subtropical Agricultural Research Program (Agreement No. 58-7B30-9-116) and to G. A. Moore by the USDA-SEA Competitive Grants Office (82-CRCR-1-1011) is gratefully acknowledged. Additional support has been provided by the State of Florida and by the Florida Mango Forum. Florida Agricultural Experiment Stations Journal Series No. 5691.

[2] Abbreviations: ABA, abscisic acid; BA, benzyladenine; CH, casein hydrolysate; pCPA, *p*-chlorophenoxyacetic acid; CW, coconut water; 2,4-D, 2,4-dichlorophenoxyacetic acid; GA, gibberellic acid; IAA, indoleacetic acid; IBA, indolebutyric acid; 2iP, isopentenyladenine; KIN, kinetin; MS, Murashige and Skoog (1962); NAA, naphthaleneacetic acid; 2,4,5-T, 2,4,5-trichlorophenoxyacetic acid.

Horticultural Reviews, Volume 7

ISBN 0-87055-492-1

I. INTRODUCTION

Conventional plant breeding has had little success in improving tropical fruits and palms. Despite the significance of perennial tropical plants for food and export income, little is known about the genetics of most of these plants. Most of the superior cultivars have resulted from selections made within seedling populations from uncontrolled crosses, or have been derived from useful somatic mutations within existing clones. Because most important horticultural characteristics are conferred by complexes of genes, the genetic integrity of many important tropical crop cultivars has been maintained by means of vegetative propagation. With some notable exceptions, including papaya, *Musa*, and pineapple, tropical fruits are generally produced on woody, perennial trees having long juvenile periods. Breeding programs for such crops can involve the professional lifetimes of several generations of scientists.

Recent advances in the culture of plant cells, tissues, and organs and in somatic cell genetics appear to have considerable potential for the improvement of many staple crops of the temperate zones and for tropical crops such as sugarcane and rice. The use of *in vitro* systems as alternative propagation techniques will quite likely have the greatest immediate impact on the productivity of many tropical crop plants. Many tropical fruit and spice trees, e.g., clove [*Eugenia caryophyllus* (Sprengel) Bullock & Harrison], are nearly impossible to propagate asexually. The *in vitro* clonal propagation of selected, high-yielding superior trees would greatly affect the efficiency and quality of production. Moreover, long-term storage and international exchange of tropical plant germplasm would be greatly facilitated by the development of *in vitro* systems for regeneration and propagation, in conjunction with effective disease-indexing techniques. The importance of such research for improving tropical perennial plants is obvious, particularly since disease, pest, and environmental stresses are constant threats to production of tropical food and export crops (Litz 1984a).

Occasionally, the outbreak of a new disease can eliminate commercial production of certain crops throughout an entire region. This occurred following the outbreak of Panama disease in the large-scale monoculture of 'Gros Michel' bananas in Central America earlier in this century. Currently, banana and plantain production are similarly threatened by black sigatoka disease in the same region. Epidemics of monilia, witch's broom, and swollen shoot diseases are threatening cacao production in various regions of the tropics. The use of *in vitro* systems for selection of mutant cells or for the induction of somaclonal variants in crop improvement schemes could lower the genetic vulnerability of many economically important perennial tropical plants and could increase the effectiveness of breeding programs. Clonal integrity would remain intact with the exception of the desirable mutant if efficient selection schemes could be developed.

The tropical fruit and palm species discussed in this review are listed in Table 4.1; the abbreviations used are listed in footnote 2.

II. TROPICAL FRUIT TREES

The most notable successes in the cell, tissue, and organ culture of tropical fruit trees have been with species in which there is naturally occurring polyembryony. Probably the greatest obstacle to developing *in vitro* systems for woody plant regeneration is the absence of juvenil-

TABLE 4.1. Common and Scientific Names of Tropical Crops Referred to in This Review

Plant family	Genus	Species	Common name
Anacardiaceae	*Mangifera*	*indica* L.	Mango
Annonaceae	*Annona*	*squamosa* L.	Custard apple
Arecaceae	*Bactris*	*gasipaes* H.B.K.	Peach palm
	Cocos	*nucifera* L.	Coconut palm
	Elaeis	*guineensis* Jacq.	Oil palm
	Phoenix	*dactylifera* L.	Date palm
Bromeliaceae	*Ananas*	*comosus* L. Merr.	Pineapple
Caricaceae	*Carica*	× *heilbornii* Badillo n.m. pentagona	Babaco
	Carica	*papaya* L.	Papaya
Ebenaceae	*Diospyros*	*kaki* Thunb.	Persimmon
Lauraceae	*Persea*	*americana* L.	Avocado
Moraceae	*Ficus*	*carica* L.	Fig
	Morus	*alba* L.	Mulberry
Musaceae	*Musa*	AAA group	Banana
	Musa	AAB, ABB groups	Plantains
Myrtaceae	*Eugenia*	*jambos* L.	Rose apple
	Eugenia	*malaccensis* Lam.	Malay apple
	Myrciaria	*cauliflora* D.C. Berg.	Jaboticaba
Oxalidaceae	*Averrhoa*	*carambola* L.	Carambola
Passifloraceae	*Passiflora*	*edulis* Sims	Purple passion fruit
		edulis var. flavicarpa	Yellow passion fruit
		mollisima (HBK) Bailey	Banana passion fruit
		suberosa L.	Corky passion fruit
Rutaceae	*Citrus*	*aurantiifolia* (Christm.) Swingle	Lime
		aurantium L.	Sour orange
		grandis (L.) Osb.	Pummelo
		limetoides Tanaka	Sweet lime
		limon (L.) Burns. f.	Lemon
		medica L.	Citron
		paradisi Macf.	Grapefruit
		reticulata Blanco	Mandarin
		sinensis (L.) Osb.	Sweet orange
	Poncirus	*trifoliata* L.	Trifoliate orange

ity in mature trees (Bonga 1982). The nucellus, a juvenile tissue, retains considerable morphogenetic potential in many plant species and the capability of producing adventitious embryos *in vivo* (Rangaswamy 1982). The regeneration of somatic embryos *in vitro* from the nucellus of cultured *Citrus* ovules was first reported by Maheshwari and Rangaswamy (1958). Many subsequent advances in the development of *in vitro* systems for *Citrus* and other tropical fruit species have been the result of this early observation. Other regeneration pathways, such as shoot-tip culture, have been described primarily for herbaceous tropical fruit trees or for seedlings.

A. Citrus

Citrus species and cultivars constitute the most important tropical fruits (FAO 1982). *Citrus* trees are generally propagated by grafting buds of a scion cultivar onto seedling rootstocks. This propagation method allows the transmission of virus and mycoplasma diseases, including exocortis, psorosis, tristeza, xyloporosis, greening, and stubborn (Button 1977). Since virus diseases of *Citrus* are very rarely seed-transmitted, virus elimination from polyembryonic clones can be accomplished by selection of true-to-type nucellar seedlings (Button and Kochba 1977). However, some *Citrus* clones are monoembryonic and many commercially desirable cultivars are seedless.

1. Shoot-Tip Culture and Virus Elimination. Nucellar plants of some monoembryonic cultivars have been obtained through somatic embryogenesis following *in vitro* culture of nucellar explants (Rangan *et al.* 1968, 1969), although this technique has not been successful in all cases (Button and Kochba 1977). Nucellar explants or entire unfertilized or abortive ovules have been cultured to produce nucellar plants of some seedless *Citrus* cultivars (Button and Bornman 1971; Kochba *et al.* 1972; Button and Kochba 1977). Although *Citrus* plants produced from somatic embryos *in vivo* or *in vitro* are free from most pathogenic viruses, they are juvenile and thus require an extended growth period before they become productive (Button 1977; Button and Kochba 1977).

The technique of meristem culture for the production of virus-free plants has been unsuccessful with *Citrus*, although multiple shoot formation from apical cultures of *Citrus* has been reported (Kitto and Young 1981; Barlass and Skene 1982). A modified technique, *in vitro* shoot-tip grafting, was developed by Murashige *et al.* (1972) and refined by Navarro *et al.* (1975). Aseptically grown, etiolated, 2-week-old nucellar seedlings are used as rootstocks. The seedlings are decapitated, their cotyledons are removed, and the root is shortened. Shoot tips containing the apical meristem and three leaf primordia are isolated from new shoots of a desired scion. An inverted-T incision is made on the epicotyl of the rootstock seedling and the shoot tip is placed in the incision. Grafted plants are allowed to develop aseptically in a nutrient solution for several weeks before being transfered to soil. Navarro *et al.* (1975) reported that the success rate for this technique usually ranged from 30 to 50% and sometimes was as high as 90%. Grafted plants are largely virus-free and are not juvenile. This technique is now being used in *Citrus* improvement programs throughout the world to obtain virus-free budwood for commercial propagation. In addition, shoot-tip grafting of imported bud material is also effective in reducing the risk of introducing diseases from one country to another (Navarro 1981a).

Several factors influence the rate of success of grafting or virus elimination. Successful grafts have been obtained using many different scion cultivars of the commercially grown *Citrus* species, and significant differences in grafting success among these different types have not been noted when appropriate rootstocks were used. Both good and poor rootstock–scion combinations have been identified, but it is not yet known if grafting success *in vitro* is related to graft compatibility *in vivo* (Navarro 1981a). The frequency of successful grafts increases with the size of the shoot tip, but the percentage of virus-free plants declines, although this depends on the pathogen. Therefore, the shoot-tip size chosen reflects a compromise between these two criteria (Navarro *et al.* 1975; Navarro 1981a). Rootstocks can be successfully grafted only during a rather narrow range of development (Navarro *et al.* 1975), and optimal conditions may differ for different rootstock cultivars (Navarro 1981a). The cultural conditions of stock plants and grafted plants also affect grafting success. For example, growing rootstock seedlings in continuous darkness appears to be beneficial (Navarro *et al.* 1975). Scion material taken from plants grown in a heated greenhouse yields a higher frequency of pathogen-free plants than scions taken from field-grown trees. Scions from *in vitro*-cultured buds produce fewer healthy plants than scions from whole trees (Navarro 1981a).

2. Organogenesis. Shoots have been produced in cultures of different *Citrus* species on a modification of MS medium (Chaturvedi and Mitra 1974; Raj Bhansali and Arya 1978). Explants have included stem sections (Grinblat 1972; Chaturvedi and Mitra 1974; Raj Bhansali and Arya 1978, 1979; Barlass and Skene 1982), root sections (Raj Bhansali and Arya 1978), root meristems (Sauton *et al.* 1982), and leaf sections (Chaturvedi and Mitra 1974). Maximum shoot production has been achieved with 0.25–5.0 mg/liter BA. The inclusion of exogenous auxin and malt extract in the medium is beneficial or essential for shoot production in some cases (Chaturvedi and Mitra 1974; Raj Bhansali and Arya 1978, 1979).

3. Somatic Embryogenesis. Among woody plants, somatic embryogenesis has been most thoroughly studied in *Citrus* species. Since many *Citrus* species are polyembryonic and adventive embryos are produced *in vivo* from nucellar tissue, most studies of *in vitro* somatic embryogenesis have involved the culture of adventitious embryos, isolated nucelli, or whole fertilized or unfertilized ovules. Embryos may be produced directly from the explant without an intermediate callus stage. Entire fertilized ovules, excised nucelli from fertilized ovules, or isolated nucellar embryos from polyembryonic *Citrus* types are responsive in culture, although embryos and calli obtained from these cultures may have been derived solely from embryos already present in the explant.

Rangan *et al.* (1968, 1969) first reported the direct induction of somatic embryos from nucellus of monoembryonic *Citrus*. Nucelli were extracted from fertilized ovules of several monoembryonic cultivars 100–120 days after pollination. Since monoembryonic types were used and the zygotic embryo was discarded prior to culture, there was no possibility that embryos initiated *in vivo* were present in the explants. The embryos could be germinated and grown directly into plants. There are other reports of somatic embryogenesis in nucellar explants from fertilized ovules of monoembryonic genotypes (Bitters *et al.* 1970; Deidda 1973; Navarro and Juarez 1977). All attempts to induce somatic embryogenesis in unfertilized ovaries and ovules, and in excised nucelli from such ovules, in monoembryonic genotypes have failed (Button and Kochba 1977; Kobayashi *et al.* 1982).

Direct induction of somatic embryos has been obtained in nucellar explants from abortive (Bitters *et al.* 1970) and unfertilized (Button and Bornman 1971) ovules of the seedless navel orange, as well as from entirely unfertilized ovules (Button and Bornman 1971; Button and Kochba 1977). Kochba *et al.* (1972) obtained somatic embryos from cultured unfertilized ovules and excised nucelli of 'Shamouti' and 'Valencia' sweet oranges and 'Marsh' grapefruit cultivars with a tendency to seedlessness. Mitra and Chaturvedi (1972) produced embryos from unpollinated ovaries of sweet orange and lime and from sliced unfertilized ovules of lime flower buds. Somatic embryos have also been produced from cultured unfertilized ovules of other *Citrus* species and cultivars (Kobayashi *et al.* 1982). In all these cases, the ovules or nucelli were obtained from flowers or immature fruit.

Starrantino and Russo (1980) cultured undeveloped ovules of several cultivars of navel oranges and lemons from ripe, mature, harvested fruits 8–15 months after anthesis. Embryos arose *in vitro* directly from already existing adventive embryos in the undeveloped ovules. Fruiting trees from embryos produced in these cultures have been obtained (Starrantino and Russo 1983). Embryos can be induced from undeveloped ovules from mature fruits of many polyembryonic *Citrus* types, including sweet oranges, grapefruits, mandarins, and lemons (G. A. Moore, unpublished data), so that embryogenic *Citrus* cultures can be initiated throughout much of the year.

There have been no reports of direct induction of *Citrus* somatic embryos *in vitro* from tissue that is neither nucellar nor ovular in origin. Matsumoto and Yamaguchi (1983) reported the induction of globular embryos and adventitious buds from the stems of *in vitro*-grown nucellar seedlings of trifoliate orange, a *Citrus* relative. Secondary embryos are produced from adventive embryos (Chaturvedi and Mitra 1972; Kochba *et al.* 1972; Juarez *et al.* 1976; Button and Kochba 1977), frequently from the hypocotyl regions of existing embryos.

Citrus calli or "pseudobulbils" (white or green, rounded or elongated bodies) have been obtained from cultures of entire fertilized ovules or nucelli from such ovules (Maheshwari and Rangaswamy 1958; Rangaswamy 1961; Sabharwal 1963). Calli and pseudobulbils appeared to originate directly from embryos already present in the explants. The callus was embryogenic and could be maintained for long periods of time.

Kochba *et al.* (1972) obtained embryogenic callus from unfertilized ovules of 'Shamouti' orange in which there was no evidence of nucellar or zygotic embryos (Spiegel-Roy and Kochba 1973a). When the callus was subcultured onto medium containing 1.0 mg/liter each of IAA and KIN, it proliferated rapidly and formed numerous green embryos that later developed leaves (Kochba and Spiegel-Roy 1977a). Malt extract and low concentrations of adenine enhanced somatic embryogenesis, whereas other KIN:IAA ratios and high adenine concentrations suppressed it (Kochba and Spiegel-Roy 1977b). Continued subculture of this callus yielded habituated callus lines that proliferated on medium devoid of any growth hormones (Kochba and Spiegel-Roy 1977b; Spiegel-Roy and Kochba 1973b). Light and electron microscope studies of the callus (Button *et al.* 1974) showed that it was entirely composed of small globular proembryos at various stages of development. Proembryos arose from single cells and subsequently developed into mature embryos from which plants could be established (Kochba and Spiegel-Roy 1977b).

Mitra and Chaturvedi (1972) found that embryogenic calli and embryos were initiated from sliced unfertilized ovules of polyembryonic limes. However, the addition of growth regulators to the medium was necessary for continued growth of this callus and embryo production declined after 1 year in culture. Starrantino and Russo (1980) obtained embryogenic calli and embryos in their cultures of navel oranges and lemons; proliferation of calli, embryos, and pseudobulbils was 70% from orange ovules and 50% from lemon ovules after 30 days. These cultures were maintained for more than 3 years, but no habituation of the calli to growth regulators was reported. Habituated embryogenic callus of navel orange (Button and Rijkenberg 1977) and of other *Citrus* species (Vardi *et al.* 1982) has also been obtained.

Embryogenic *Citrus* suspension cultures have been obtained generally from embryogenic calli produced on semisolid medium (Button *et al.* 1974; Harms and Potrykus 1980; Kochba *et al.* 1982a). There has been only one report of embryogenic callus induction from monoembryonic *Citrus*. Juarez *et al.* (1976) cultured excised nucelli from fertilized ovules of monoembryonic 'Clementine' from fruits 3–4 weeks after pollination. Embryogenic calli and embryos arose from the nucelli.

Embryogenic callus has been induced from other explant sources. Embryogenesis as well as organogenesis occurs in stem segment callus from sweet orange, lime, and sweet lime seedlings (Raj Bhansali and Arya 1978, 1979). Triploid somatic embryos and plants have been obtained from pummelo endosperm callus (Wang and Chang 1978). Mitra and Chaturvedi (1972) reported somatic embryogenesis from ovary wall tissue.

Citrus calli and callus-derived plants have been reported to be consistently diploid ($2n = 18$) or morphologically normal, but cytological evidence has not been provided (Spiegel-Roy and Kochba 1973c; Button 1977). In a 'Shamouti' suspension culture, cells with chromosome numbers between 36 and 72 were found, but the cultures were still embryogenic (Harms and Potrykus 1980). Chromosome counts on a small number of cells in nucellar callus lines of eight cultivars were made by Vardi (1981). In six of the cultivars, all of the counted cells were diploid. In counts of the sour orange callus line, 17 cells were diploid and 3 cells contained 19 chromosomes. In a lemon callus line, the cells were predominantly tetraploid, possibly because 2,4-D was used in the culture initiation medium. When 10 plants were regenerated from 'Shamouti' protoplasts, 9 were diploid and 1 was tetraploid (Vardi 1977).

Many factors affecting the induction of *in vitro* embryogenesis in *Citrus* species have been identified. Monoembryony in *Citrus* is recessive to polyembryony, and the trait appears to be governed by more than one gene (Parlevliet and Cameron 1959). The degree of polyembryony in a genotype may influence its success in culture. Esan (1973), in an attempt to determine the basis for the polyembryony in *Citrus*, found that naturally monoembryonic cultivars produced a graft-transmissible and diffusable repressor of embryogenesis. The factor or factors appeared to be concentrated in the chalazal region of the nucellus or ovule. Tisserat and Murashige (1977a) reported that the presence of monoembryonic citron ovules in the same culture vessel repressed embryogenesis in embryogenic callus of both polyembryonic 'Ponkan' mandarin and carrot (*Daucus carota* L.). Volatile inhibitors included CO_2, ethylene, and ethanol. A gas chromatographic analysis of citron and 'Ponkan' mandarin ovules indicated that the monoembryonic citron ovules contained substantially higher endogenous levels of IAA, ABA, and gibberellins (Tisserat and Murashige 1977b).

Mitra and Chaturvedi (1972) found a direct relationship between the degree of polyembryony and the potential for somatic embryogenesis. Highly polyembryonic sweet orange callus was more prolific than that of less polyembryonic limes, and calli of monoembryonic pummelo formed no embryos.

Various researchers have found that embryos can be produced from undeveloped or unfertilized ovules of many *Citrus* genotypes when they are removed from flower buds (Kobayashi *et al.* 1982), young fruit (Button and Bornman 1971; Kochba *et al.* 1972; Button and Kochba 1977), or mature fruit (Starrantino and Russo 1980). It seems likely that embryos can be obtained from undeveloped or unfertilized ovules from fruit of many *Citrus* genotypes at any stage of development. However, the relative efficiency of embryo production from ovules extracted at various stages of fruit development has not been determined.

Relatively high sucrose concentrations (5% w/v) have generally been found to be optimal for induction of embryogenic *Citrus* callus (Murashige and Tucker 1969; Kochba and Button 1974). Somatic embryogenesis has also been stimulated by other sugars such as galactose and lactose (Button 1978; Kochba *et al.* 1982b). Malt extract and other organic substances have frequently been added to *Citrus* cultures and have usually been found to be either essential or beneficial for embryogenesis.

Embryogenic cell cultures of *Citrus* species may have fairly high levels of endogenous auxin. Exogenous auxin does not appear to be essential for *Citrus* callus proliferation, although it may be stimulatory (Murashige and Tucker 1969). Many of the treatments that stimulate embryogenesis in habituated *Citrus* callus may do so by reducing the endogenous auxin level in the callus (Spiegel-Roy and Kochba

1973a; Epstein *et al.* 1977; Kochba and Spiegel-Roy 1977a; Kochba *et al.* 1978). Tisserat and Murashige (1977b) found that IAA and 2,4-D suppressed embryogenesis in 'Ponkan' mandarin nucellar cultures. Embryogenic lines of habituated 'Shamouti' callus were inhibited by IAA and NAA (Kochba and Spiegel-Roy 1977c), as was cell division of protoplasts from this callus (Harms and Potrykus 1980). Kochba and Spiegel-Roy (1977c) found that 5-hydroxy-nitro-benzylbromide and 7-aza-indole, inhibitors of auxin synthesis, greatly increased somatic embryogenesis in the habituated 'Shamouti' callus.

Cytokinins may inhibit embryo formation (Kochba and Spiegel-Roy 1977c; Tisserat and Murashige 1977b). Somatic embryogenesis in habituated 'Shamouti' orange callus is stimulated by 8-aza-guanine, which is believed to be an antagonist of cytokinin (Kochba and Spiegel-Roy 1977c). Kochba *et al.* (1978) found that low levels of ABA stimulate callus growth and somatic embryogenesis, whereas Tisserat and Murashige (1977b) reported that high concentrations of ABA inhibit somatic embryogenesis. Gibberellic acid stimulates the growth of *Citrus* callus (Murashige and Tucker 1969) and root formation in *Citrus* (Kochba *et al.* 1974). Both (2-chloroethyl) trimethylammonium chloride (Chromequat, CCC) and butanedioic acid (2,2-dimethylhydrazide) (daminozide, Alar), which are inhibitors of GA biosynthesis, increase somatic embryogenesis in habituated 'Shamouti' calli (Kochba et al. 1978). Tisserat and Murashige (1977b) reported that 3.0 mg/liter ethylene added as (2-chloroethyl) phosphoric acid (ethephon) significantly decreased somatic embyrogenesis in 'Ponkan' mandarin nucellar callus, while Kochba *et al.* (1978) found that 0.5 mg/liter ethepon stimulated embryogenesis in 'Shamouti' callus.

Mature *Citrus* embryos have been germinated and grown *in vitro* on basal medium with or without a carbohydrate source and with or without additives such as malt extract (Button and Kochba 1977). However, consistent germination of fully and partially developed *Citrus* embryos has most often been achieved on a medium containing GA (Kochba *et al.* 1974; Button and Kochba 1977). Adenine sulfate has also been reported to enhance rooting of *Citrus* embryos (Kochba *et al.* 1974). After root development is initiated, shoot development usually occurs spontaneously (Button and Kochba 1977). The germinated embryos are then transferred to a medium without hormones or with a low level of auxin for further growth and rooting (Button and Kochba 1977; Kobayashi *et al.* 1982). The concentration of iron in the medium is often elevated to prevent chlorosis of the plants (Button and Kochba 1977). Most plant losses occur when the plants are transferred to soil.

Citrus plants must be carefully hardened off, usually by enclosing them with plastic and growing them under low light intensities (Juarez *et al.* 1976; Button and Kochba 1977).

4. Embryo Rescue. Ohta and Furasato (1957) first reported that fully developed *Citrus* embryos could be successfully cultured. Less-developed *Citrus* embryos, including those at a globular stage, also can grow and develop *in vitro*, although often the presence of nucellar tissue or additives such as CH or CW are necessary for development (Maheshwari and Rangaswamy 1958; Rangaswamy 1958, 1959, 1961; Sabharwal 1962, 1963; Singh 1963).

5. Protoplast Isolation and Culture. Early protoplast research with *Citrus* involved the use of habituated callus of 'Shamouti' orange from callus originally obtained by Kochba *et al.* (1972). Button and Botha (1975) obtained single cells of this callus through the use of 2–3% pectinase (Macerase) in a basal medium containing 3% sucrose. Embryos and plants were regenerated on a medium supplemented with malt extract or CW.

Although Vardi *et al.* (1975) reported the isolation and culture of *Citrus* protoplasts from a nondifferentiating line of 'Shamouti' callus, this method for protoplast isolation in *Citrus* was later refined. A modified enzyme solution containing driselase and reduced levels of pectinase and cellulase was used together with a simplified one-step overnight enzyme treatment (Vardi and Raveh 1976; Vardi 1977). The culture of 'Shamouti' protoplasts above a feeder layer of X-ray–irradiated orange or *Nicotiana tabacum* L. protoplasts allowed cell regeneration at much lower plating densities (Vardi and Raveh 1976). Use of a feeder layer also enhanced somatic embryogenesis (Vardi 1977). When an embryogenic line of the habituated 'Shamouti' callus was used for protoplast isolation in place of the nondifferentiating line and the protoplasts were irradiated, approximately 80% of the colonies formed became embryogenic (Vardi 1977). Plants were obtained from both irradiated and nonirradiated protoplasts.

Embryogenic nucellar calli of nine other *Citrus* cultivars have been used for protoplast isolation (Vardi 1981; Vardi *et al.* 1982; Kobayashi *et al.* 1983). In all cases, calli, embryos, and morphologically normal plants have been regenerated from the protoplasts. Protoplasts from different species have required the same conditions for induction of cell division and plant regeneration, but have varied in their sensitivity to certain enzymes, growth regulators, and sorbitol in the isolation medium (Vardi 1981; Vardi *et al.* 1982).

Viable protoplasts have also been obtained from cotyledons of 'Valencia' orange (Burger and Hackett 1982). Callus cultures were produced from the protoplast colonies, but plants were not regenerated. Plants have been regenerated only from *Citrus* protoplasts isolated from nucellar calli or suspension cultures derived from such calli (Harms and Potrykus 1980).

There have been no reports of fusion between *Citrus* protoplasts. Attempts have been made to fuse protoplasts of 'Shamouti' orange and *N. tabacum* (Harms and Potrykus 1980; Harms *et al.* 1980). However, only *Citrus*-type colonies were obtained from the fusion mixture.

6. In Vitro Selection. Habituated 'Shamouti' orange callus tolerant to sodium chloride (NaCl) was selected by plating callus on medium containing 5.0 g/liter NaCl (Kochba *et al.* 1980; 1982). Some callus lines were gamma-irradiated prior to plating. The best-growing cultures were selected and then subcultured on NaCl-containing medium for 10 passages of 5 weeks each. Five irradiated-callus lines and five nonirradiated-callus lines were selected on the basis of their potential for growth or embryogenesis. All of the selected lines grew markedly better than a control line on medium containing up to 10.0 g/liter of NaCl. Individual lines differed in their salt tolerance, but the differences did not depend on whether the lines arose from irradiated or nonirradiated callus. Sodium chloride tolerance was maintained after up to four subcultures in salt-free medium, and at least one of the selected cell lines was considered to be a genetic variant. Further evidence for this was gathered from physiological and biochemical studies on the selected callus line (Ben-Hayyim and Kochba 1982, 1983). The salt-tolerant lines remained embryogenic, but some lines required the presence of NaCl in the medium for somatic embryogenesis (Kochba *et al.* 1982a). A salt-tolerant cell line of sour orange was selected by the same method from suspension cultures (Kochba *et al.* 1982), but salt-tolerant plants from either species were not recovered.

A similar selection scheme was used to isolate lines of 'Shamouti' callus that were tolerant to 2,4-D (Kochba *et al.* 1980; Spiegel-Roy *et al.* 1983). Six callus lines were selected on a medium containing 1×10^5 M 2,4-D. Only one callus line remained embryogenic on 2,4-D-containing medium, although other lines were able to grow on high 2,4-D concentrations. The 2,4-D tolerance was stable for up to three passages on 2,4-D–free medium. Production of plants tolerant to 2,4-D was not reported.

In vitro systems may eventually be of value in *Citrus* for selection or screening of disease-resistant plants. The presence of "mal secco" toxin in the culture medium inhibited growth of ovular callus of the susceptible 'Eureka' lemon, whereas callus of the resistant 'Shamouti' orange was not affected (Nachmias *et al.* 1977). The formation of callus and roots from internode stem pieces of 'Etrog' citron was inhibited by the presence of *Citrus* exocortis viroid in the plant tissue (Navarro 1981b).

7. In Vitro Physiological Studies. Altman and Goren (1974a) devised an *in vitro* system for the culture of *Citrus* buds in order to study the mechanism of bud development, particularly the hormonal control of growth and dormancy. The buds could develop and sprout on basal medium, and the dormancy and sprouting periods of the buds were the same *in vitro* as *in vivo*. An innate dormancy dependent upon a balance of auxin, cytokinin, and GA was found (Altman and Goren 1977). Exogenous hormones affected both the time required for sprouting and the type of growth produced, but the effects of the hormones varied at different times of the year, suggesting that there is an endogenous critical level of each hormone needed, above which there is a sprouting response, and that this level may vary during the annual growth cycle.

The *in vitro* production of callus in the abscission zone of *Citrus* bud cultures was also studied. Callus formation was promoted by the presence of ABA in the medium, especially if IAA was also present (Altman and Goren 1971, 1974b, 1977). A synergistic effect between GA and ABA on callus formation was observed; it was concluded that ABA induced the formation of callus and GA enhanced the growth of already formed callus cells (Altman and Goren 1974b). The omission of sucrose (Altman and Goren 1977) or the presence of ethylene (Altman and Goren 1977; Giladi *et al.* 1977) also enhanced callus production. The effect of ABA on callus formation was found to be correlated with ethylene evolution (Goren *et al.* 1979).

There have also been efforts to develop *in vitro* systems to examine the physiological behavior of *Citrus* flowers and fruits (Gulsen *et al.* 1981; Sipers and Einset 1982). Cultures of grapefruit peel have been used to study the metabolism of carotenoids and chlorophyll (Oberbacher 1967). Mitotic activity and cell proliferation *in vitro* of juice vesicle stalks of lemon have been studied by Kordan (1977).

8. Anther Culture. Chen *et al.* (1980) have obtained haploid plants from anther cultures of *C. microcarpa* Bunge. Haploid and aneuploid plants have also been produced from anthers of trifoliate orange (Hikada *et al.* 1979). Plants were obtained from sour orange anther cultures, but all of them were diploid (Hikada *et al.* 1982). The optimal stage of anther development, cultural conditions, and effect of hormones appeared to differ for each species.

B. Mango

Mango is one of at least 15 *Mangifera* species that produce edible fruit, although only mango is widely planted. Believed to have originated in eastern India (Knight 1980), it is an allotetraploid that arose after interspecific hybridization and subsequent doubling of the chromosome number (Mukherjee 1953). The annual production of mango is exceeded among fruit crops only by grape, *Citrus* spp., *Musa* (bananas and plantains), and apples (FAO 1982).

Both polyembryonic and monoembryonic mango cultivars are grown. According to Leroy (1947), adventive embryony is probably due to the effect of one or more recessive genes. Polyembryonic cultivars are important in southeast Asia and in tropical Latin America, whereas monoembryonic cultivars are the basis of commercial production in India and in Florida. Polyembryonic mango cultivars are seed-propagated and exhibit little variability in seedling populations. Monoembryonic mango cultivars are propagated either by inarching or by grafting budwood onto seedling rootstock.

1. Disease Elimination. Virus diseases have not been limiting factors in mango production, eliminating the need for *in vitro* systems for the recovery of virus-free plants. On the other hand, fungal and bacterial pathogens (e.g., *Colletotrichum gloeosporiodes*, the cause of anthracnose; *Pseudomonas mangiferae*; and *Erwinia mangiferae*) can cause considerable damage to fruit and young vegetative growth. Mango decline is believed to be caused by a systemic pathogen of an unknown nature.

Because of the absence of vascular connections between the nucellus and other maternal tissues, mango plants derived from the nucelli of polyembryonic cultivars are generally free of infections that might have affected the parent plant. Plantlets derived via somatic embryogenesis from nucellar callus of either polyembryonic or monoembryonic mango cultivars would therefore also be relatively free of pathogens. There have been no reported attempts of shoot-tip or meristem-tip culture of mango, possibly because there is a rapid activation of oxidative enzymes during the excision of explants, leading to the eventual death of excised, cultured tissues.

2. Somatic Embryogenesis. Although Maheshwari and Rangaswamy (1958) were aware of the *in vitro* embryogenic potential of the nucellus from polyembryonic mango cultivars, they did not stimulate *in vitro* development of already formed adventitious embryos. Somatic embryogenesis from nucellar tissues of polyembryonic mango cultivars was reported by Litz *et al.* (1982). Entire fertilized ovules from 40- to 60-day-old fruitlets were initially cultured on modified MS medium containing half strength major salts, ascorbic acid, glutamine, CW, and other additives. The ovules were dissected 2–3 weeks later; proliferating, globular somatic embryos were removed and transferred to the original medium or to liquid medium. The *in vitro* response was cultivar dependent. Somatic embryos were more likely to develop in cultures of highly polyembryonic cultivars (e.g., 'Ono', 'Chino', and 'Sabre') than in cultures of mango cultivars in which relatively few adventitious embryos occur (e.g., 'Heart' and 'Carabao'). Because the cultures were derived from fertilized ovules, somatic embryos could have developed directly from adventitious, nucellar embryos. It is less likely that somatic embryos developed from the zygotic embryos present in the ovules at the time of culturing because the zygote aborts at an early stage in several polyembryonic mango cultivars, including 'Carabao', 'Pico', 'Strawberry', 'Olour', and 'Cambodiana' (Singh 1960). Therefore, all of the embryos in ovules of these cultivars should be adventitious. The direct formation of somatic embryos from hypocotyls of germinating mango somatic embryos has also been reported (Litz *et al.* 1984).

Rao *et al.* (1982) induced calli from cotyledons of mango seeds in the presence of 5.0 mg/liter NAA and 2.5–5.0 mg/liter KIN; they observed only adventitious root formation from this callus. The induction of embryogenic callus from mango nucellar tissue derived from fertilized ovules of polyembryonic cultivars can occur on modified MS medium containing 1.0–2.0 mg/liter 2,4-D (Litz *et al.* 1984). Mango nucellar callus is very similar to *Citrus* nucellar callus in appearance and is initially globular and white. Sabharwal (1963) indicated that *Citrus* pseudobulbils arise directly from adventitious embryos that are present in the ovules at the time of culturing. Similar observations have been made with mango (R. E. Litz, unpublished data). Somatic embryogenesis occurs in the presence of 2,4-D, although the development of embryos is arrested at the globular stage. After subculture onto medium without 2,4-D, somatic embryos develop to maturity. Embryogenic polyembryonic mango callus has been subcultured for more than a year without any noticeable loss of embryogenic potential.

Somatic embryogenesis from unfertilized mango ovules has not been reported. However, using a variation of the procedure described by Rangan *et al.* (1968) for induction of somatic embryos from isolated nucellus of monoembryonic *Citrus*, Litz (1984b) has regenerated somatic embryos from monoembryonic mango nucellar callus derived from ovules from which the zygotic embryo has been removed. Whereas Rangan *et al.* (1968) demonstrated that somatic embryogenesis occurs directly from the cultured nucellus without an intermediate callus stage, induction of a nucellar callus in the presence of 1.0–2.0 mg/liter 2,4-D is necessary in mango. Somatic embryogenesis has occurred from this callus in the presence of 2,4-D, but more efficiently following subculture to medium without 2,4-D. Monoembryonic mango cultivars respond *in vitro* in different ways; only three monoembryonic cultivars ('Tommy Atkins', 'Ruby', and 'Irwin') have so far regenerated from mango callus.

During the maturation of mango somatic embryos in the absence of 2,4-D, the size of the embryos increases substantially from small, 0.1- to 0.2-cm globular embryos to 3- to 4-cm–long mature embryos, 4 or 5 months after embryo induction. Maturation of somatic embryos has often been accompanied by gradual necrosis of the hypocotyls and cotyledons, which has been difficult to control. Secondary embryos frequently develop from the hypocotyls of germinating somatic embryos. Until these developmental anomalies can be controlled, the efficient regeneration of whole mango plants from somatic embryos will be impeded.

C. Miscellaneous Tropical Fruit Trees

Several reports have described tissue culture studies with different perennial tropical fruit trees, e.g., carambola (Litz and Conover 1980a; Rao *et al.* 1982), custard apple (Nair *et al.* 1984), avocado (Skene and Barlass 1983), and mulberry (Oka and Ohyama 1981; Ohyama and Oka 1982). However, most of these investigations have involved embryonic or seedling tissues as explants for callus induction and occasionally organogenesis and therefore do not represent clonal propagation. Fruit mesocarp tissues have also been used as explants for callus induction in avocado (Schroeder 1961) and sapodilla (Bapat and Narayanaswamy 1977), although plantlet regeneration from callus of this nature has never been reported.

In addition to mango and *Citrus*, many tropical fruit trees that originated in the mature rain forests of South America and southeast Asia are naturally polyembryonic. Litz (1984c) reported somatic embryogenesis from callus derived from immature adventitious embryos of the jaboticaba, a polyembryonic fruit tree from South America. Embryogenic callus was induced from immature embryos from ovule explants (0.6–0.8 cm) on modified MS medium with 1.0–2.0 mg/liter 2,4-D, 6% sucrose, and other additives. Somatic embryogenesis occurred either directly from adventitious embryos or from the callus, but was particularly likely to occur following subculture of callus into liquid medium without 2,4-D. Stimulation of axillary bud proliferation from cultured immature adventitious embryos on medium with BA was also possible.

Similar regeneration pathways have been observed for two naturally polyembryonic *Eugenia* (*Syzygium*) species native to southeast Asia, which produce fruit of minor importance, i.e., the rose apple and the Malay apple (Litz 1984d). Embryogenic callus was obtained from immature adventitious embryos from polyembryonic ovules of both species that were cultured on modified MS medium containing 1.0–2.0 mg/liter 2,4-D, 6% sucrose, and other additives. Globular somatic embryos that were induced in the presence of 2,4-D matured and germinated on medium without growth regulators. Stimulation of axillary bud proliferation occurred from immature adventitious embryos on medium with 3.0–10.0 mg/liter BA.

The regeneration of jaboticaba, Malay apple, and rose apple via somatic embryogenesis represents clonal propagation of these species because adventitious embryos are derived either from the nucellus (jaboticaba and rose apple) or from the inner integument of the ovule (Malay apple). Although these fruit tree species are of minor importance, they are closely related to other tropical fruit and spice trees (e.g., clove) that are monoembryonic and often difficult to propagate asexually. Based on *in vitro* studies with *Citrus*, it should also be possible to induce somatic embryogenesis from the related monoembryonic *Eugenia* species.

The regeneration of fig plants from cultured shoot tips has been reported (Muriithi *et al.* 1982), although proliferating cultures were not obtained. A modified MS medium containing 0.18 mg NAA, 0.1 mg BA, and 0.03 mg GA per liter was used. Rooting was induced with 0.5 mg/liter NAA, and regenerated plants appeared to be free of fig mosaic virus. Shoot-tip culture has not been effectively used for other woody tropical fruit trees.

Promising results have been reported for the Japanese persimmon 'Tsurunoko' by Yokoyama and Takeuchi (1981). Callus was induced from stem explants from mature trees on MS medium containing 1.0 mg/liter NAA. With this medium, root formation occurred from the callus, and shoot primordia differentiated from the callus in the presence of NAA and 1.0 mg/liter KIN. Normal plantlets have apparently not been recovered.

Nair *et al.* (1983) described the recovery of haploid custard apple from anther cultures following the dissection of flower buds in a suspension of activated charcoal and sucrose. Cultures were maintained in darkness in medium containing 5% sucrose; they transferred to the light and onto medium containing lower sucrose concentrations in order to induce haploid plant differentiation.

III. Herbaceous Tropical Fruits

A. Papaya

Papaya is an important fruit crop with worldwide production comparable to that of avocado or strawberry (FAO 1982). Papaya selections and inbred cultivars are generally seed-propagated, although vegetative propagation by rooted cuttings has been developed for dioecious papaya cultivars (Allan 1964). Papayas are herbaceous plants and have a relatively short juvenile period of 4–5 months; however, plants can remain in production for several years in the absence of environmental and disease stresses. Some papaya production is devoted to the recovery of papain, a proteolytic enzyme with several commercial applications.

Serious papaya diseases are caused by viruses (e.g., papaya ringspot, papaya mosaic, papaya apical necrosis), by fungi (e.g., *Phytophthora infestans*), and by a mycoplasma-like organism. Papaya production has been restricted in many tropical regions because of one or more disease threats. Papaya viruses are not seed-transmitted; since most papaya cultivars are seed-propagated, maintenance of virus-free stock plants is not essential.

1. Shoot-Tip Culture. The regeneration of single papaya plants from seedling shoot tips was reported by Mehdi and Hogan (1976); proliferation was not observed. Yie and Liaw (1977) subsequently described the proliferation of excised seedling shoot tips and obtained eightfold increases in the number of shoots between each subculture. Litz and Conover (1977, 1978a) described the *in vitro* propagation of mature papaya selections by shoot-tip culture. Explant establishment on MS medium containing 10.0 mg KIN and 2.0 mg NAA per liter is strongly influenced by clone, sex type, and season (Litz and Conover 1981a). Shoot tips from staminate plants respond more rapidly on establishment medium and attain greater proliferation rates following transfer to MS proliferation medium containing 0.5 mg BA and 0.1 mg NAA per liter. Systemic bacterial infections are difficult to control in field-grown stock plants, and contamination problems have been observed to increase dramatically during the subtropical dry season. Seven- and eightfold increases in shoot numbers between subcultures have been obtained, although these rates cannot be sustained for more than 8 to 13 subcultures, depending on the cultivar (Yie and Liaw 1977; Litz and Conover 1981a). Rooting and reestablishment of apical dominance occur equally well on medium containing 1.0–3.0 mg/liter IBA and NAA, although the success rate has been observed to decline in proportion with increased time on proliferation medium.

2. Organogenesis. The regeneration of papaya by organogenesis from callus from seedling stem segments (DeBruijne *et al.* 1974; Yie and Liaw 1977; Arora and Singh 1978) and from cotyledons (Litz *et al.* 1983b) has been demonstrated. Although Yie and Liaw (1977) obtained callus on medium containing 1.0 mg NAA and 0.1 mg KIN per liter, Arora and Singh (1978) used higher concentrations of the same growth regulators. Adventitious shoots were formed following subculture of callus onto medium containing 0–0.2 mg IAA and 1.0–2.0 mg KIN per liter, whereas root formation required a higher auxin to cytokinin ratio (Arora and Singh, 1978). Callus induction from papaya cotyledons has been observed from the lamina and the midrib (Litz *et al.* 1983b). Optimum growth regulator levels for callus induction from the midrib were 0.3–2.0 mg/liter BA and 0.5–3.0 mg/liter NAA; callus induction from the lamina required slightly higher concentrations of both regulators. Shoot formation from cotyledon callus occurred on media containing 0.05–2.0 mg/liter BA and 0–2.0 mg/liter NAA, and root formation occurred on media with auxin concentrations as high as 15.0 mg/liter NAA.

3. Somatic Embryogenesis. Somatic embryogenesis from papaya callus has been achieved. Generally, this has occurred from callus on the same medium that induces adventitious shoot formation (Yie and Liaw 1977; Debruijne *et al.* 1974). Activated charcoal (0.5%) stimulates somatic embryogenesis from peduncle callus of a related species, *Carica stipulata* Badillo (Litz and Conover 1980b). *In vitro* polyembryony can occur in papaya ovules resulting from interspecific crosses with *C. cauliflora* Jacq. (Litz and Conover 1981b), and a high frequency of somatic embryogenesis from ovular callus occurs (Litz and Conover 1982) on MS medium with 20% CW. In medium with 1.0–2.0 mg/liter 2,4-D, globular somatic embryos multiply by budding (Litz and Conover 1983). Neither Debruijne *et al.* (1974) nor Yie and Liaw (1977) were able to germinate papaya somatic embryos, although plants have been regenerated from somatic embryos derived from ovular callus either by subculture to medium containing 0.02–0.2 mg BA and 0.05–2.0 mg NAA per liter (Litz and Conover 1982) or by transfer from medium with 2,4-D to medium that is free of growth regulators (Litz and Conover 1983). Somatic embryogenesis in papaya may be partly dependent on osmotic factors, as Litz and Conover (1981b, 1982, 1983) have found that relatively high concentrations of sucrose (6%) have stimulated very efficient somatic embryogenesis.

4. Embryo Rescue. One of the major challenges in papaya breeding has been to transfer characteristics such as virus resistance and cold tolerance from wild *Carica* species to papaya. Unfortunately, many of these wild species (e.g., *C. cauliflora, C. stipulata*, and *C. pubescens*) are sexually incompatible with papaya. Ovule culture is a highly efficient approach to embryo rescue for hybridization involving *C. cauliflora* and papaya. Moore and Litz (1984) demonstrated by the use of biochemical markers that the somatic embryos derived from papaya ovular callus resulting from *C. papaya* × *C. cauliflora* crosses are in fact interspecific hybrids.

Early studies with excised papaya embryos (Phadnus *et al.* 1970) indicated that embryo culture could also be utilized for rescue of interspecific embryos from immature papaya seeds in which failure of endosperm formation had occurred. Culture of excised embryos as a means for effecting embryo rescue has also been reported by Khuspe *et al.* (1980), Phadnus *et al.* (1970), and Rojkind *et al.* (1982). Hybrid embryos excised from papaya ovules resulting from *C. papaya* × *C. cauliflora* crosses have been successfully cultured on modified White's medium 60 days (Khuspe *et al.* 1980) and 90 days (Rojkind *et al.* 1982) after pollination. The hybrid was fertile and has been backcrossed with papaya. Litz and Conover (1983) reported that the ability to induce callus and somatic embryogenesis from *C. papaya* × *C. cauliflora* crosses was correlated with fruit shape. Crosses involving pistillate papaya selections with long or elliptical fruit and *C. cauliflora* were generally successful, although those with round fruit were unsuccessful.

5. Protoplast Isolation and Culture. The isolation and culture of papaya protoplasts has been attempted in order to facilitate wide interspecific crosses and induction of somaclonal variants. Since plants have been regenerated from papaya cotyledons (Litz *et al.* 1983b), efforts have concentrated on isolation of cotyledon protoplasts (Litz and Conover 1979; Liu and Yang 1983). Mesophyll protoplasts are easily recovered by digesting papaya cotyledons in modified MS medium containing 2% cellulase, 0.5% pectinase, and 0.3 *M* mannitol. Plant regeneration from callus derived from papaya leaf protoplasts has not been reported (Litz 1984e; Liu and Yang 1983).

6. Anther Culture. One of the chief problems in programs to improve dioecious papaya types is the difficulty in determining the genetic contribution from the staminate parent. Therefore haploid plant production and the recovery of homozygous, diploid male plants by endomitosis would theoretically result in plants that yield homozygous pollen. Litz and Conover (1978b) attempted to develop *in vitro* systems for recovery of haploid plants. They reported a low recovery frequency of haploid plants directly from cultured anthers on liquid MS medium with 0.5% activated charcoal, 0.5 mg/liter BA, and 0.1 mg/liter NAA; however, the efficiency of regeneration was so low that it may have been spontaneous and not a direct effect of *in vitro* conditions. The development of systems for haploid plant induction would have a great impact on papaya breeding.

B. Babaco

A relative of papaya, babaco is believed to be a natural hybrid between *C. stipulata* Badillo and *C. pubescens* Lenne and Koch. It originated in the Andes and is grown commercially in that region and in areas with a Mediterranean type of climate. It produces parthenocarpic fruit and therefore must be propagated vegetatively by rooted stem cuttings. Attempts to utilize the *in vitro* shoot-tip propagation procedure developed by Litz and Conover (1977, 1978a) for papaya have been unsuccessful with babaco (Cohen and Cooper 1982), although both the putative parental species could be propagated by this method. A variant medium for shoot-tip propagation that includes Knop's macronutrients, MS micronutrients, adenine sulfate, CH, and 1.0 mg/liter BA has been developed by Cohen and Cooper (1982). Rooting of shoots could be induced on medium containing 1.0 mg/liter IBA.

Apparently, Cohen and Cooper (1982) have also regenerated vegetative buds from babaco leaf tissue; however, details of this procedure have not been reported.

C. Bananas and Plantains

The annual worldwide production of plantains and bananas is exceeded among fruit crops only by that of grapes (FAO 1982). Bananas are particularly important as an export crop and are crucial to the economies of many tropical countries. The plantain, or cooking, banana is generally grown as a starchy staple crop and is therefore unique among fruit crops.

Both bananas and plantains are propagated vegetatively by removal of suckers from established plants. Bananas are sterile triploids and produce parthenocarpic fruit. They are entirely derived from *Musa acuminata*, a diploid. All banana cultivars are assumed to have originated via somatic mutation and persistent selection. Plantains are also sterile triploids, but are derived from interspecific hybridization between *M. acuminata* and *M. balbisiana*. Because of the relatively narrow genetic base of cultivated *Musa* types, they appear to be highly susceptible to new diseases or mutated pathogenic microorganisms (Rowe 1984). Consequently, banana and plantain production has been severely threatened in different parts of the world at different times.

1. Shoot-Tip Culture and Virus Elimination. Many of the most important banana clones are infected with cucumber mosaic virus (CMV), which is thought to limit production and is spread easily by the use of vegetative propagation methods. Berg and Bustamante (1974) reported that clones of 'Cavendish' banana could be freed of CMV by culturing the meristems from young suckers following heat treatment of the donor plant. A simple medium consisting of Knudsen's salts was supplemented with Berthollet's minor elements, 10% CW, CH, and other organic additives. This procedure does not stimulate proliferation of axillary buds, and only a single shoot develops from a cultured shoot tip.

Ma and Shii (1972) and Doreswamy *et al.* (1983) demonstrated that decapitation of the terminal bud is necessary in order to release the axillary buds from apical dominance. This procedure results in the proliferation of lateral shoots. According to Ma and Shii (1972), shoot proliferation is dependent on the presence of 340 mg adenine sulfate, 2.0 mg KIN, and 100 mg tyrosine per liter of medium. High rates of shoot proliferation can also occur in liquid culture even in the absence of growth regulators (Ma and Shii 1974). De Guzman *et al.* (1980) demonstrated that rapid proliferation rates were dependent on a high concentration of cytokinin (15% CW and 5.0 mg/liter BA). This was confirmed by Doreswamy *et al.* (1983), who cultured axillary buds on MS medium containing 15% CW and 10.0 mg/liter BA. Krikorian and Cronauer (1984) and Cronauer and Krikorian (1984a,b) have used a similar medium and approach. The rooting of regenerated plants has been reported to occur on medium containing 5.0 mg/liter IBA (Doreswamy *et al.* 1983) or 1.0 mg/liter NAA, IBA, or IAA together with 0.25% (w/v) activated charcoal (Cronauer and Krikorian 1984a,b). It has been reported that tissue culture–derived 'Giant Cavendish' plants are more vigorous than plants propagated by conventional means (Hwang *et al.* 1984).

2. Somatic Embryogenesis. Induction of embryogenic *Musa* 'ABB', 'Pelipita', and 'Saba' callus was reported by Cronauer and Krikorian (1983) from subcultured, multiplying shoot-tip cultures. An MS medium supplemented with 5% CW and 1.0 mg/liter 2,4,5-T and

1.0 mg/liter BA was used. Somatic embryogenesis occurred from the shoot bases and along the leaf sheaths. Although germination was not achieved, the somatic embryos were morphologically similar to zygotic embryos from seeds of *M. acuminata* and *M. balbisiana*.

3. Embryo Rescue. Embryo culture has been used to hasten the germination of some *Musa* clones that have the ability to set seed and to rescue the zygote resulting from pollinations between diploid and tetraploid *Musa* clones or that involve seed-fertile triploids (Cox *et al.* 1960). Excised, immature embryos can grow on a modification of Knudsen's medium supplemented with 4% sucrose.

4. *In Vitro* Selection. Because virtually all dessert and cooking bananas have arisen by somatic mutations within existing clones, there has been considerable interest in adapting *in vitro* systems to hasten the rate of mutation by means of ionizing irradiation, chemical mutagens, or somaclonal variation, particularly within clones that are threatened by disease pressure (De Langhe 1969; Broertjes and van Harten 1978). This approach to *Musa* improvement is dependent on the ability to induce calli with high regenerative potential. Unfortunately, this has not yet been possible (Tongdee and Boon-Long 1973; de Guzman 1975). The regeneration of dessert bananas from calli derived from leaf protoplasts has not been demonstrated (Krikorian and Cronauer 1984).

D. Passion Fruit

The genus *Passiflora* contains a number of edible tropical and subtropical fruits, including yellow passion fruit, purple passion fruit, granadilla, bell apple, banana passion fruit, sweet granadilla, etc. Yellow passion fruit is widely grown as a fresh fruit and as a juice ingredient (Knight 1980). Recent cultivars are hybrids that are generally seed-propagated, although some vines are propagated by rooted cuttings. However, serious diseases caused by a virus (*Passiflora* woodiness virus) and several fungi (e.g., *Phytophthora parasitica, Fusarium oxysporum, Pythium* spp., and *Rhizoctonia solani*) severely affect production and limit the usefulness of clonal propagation, particularly of susceptible cultivars.

1. Shoot-Tip Culture. Robles (1978) developed a clonal propagation method for yellow passion fruit and banana passion fruit by culturing shoot tips of vines on MS medium containing 2.0 mg/liter KIN; however, proliferation of axillary buds did not occur. Using a similar medium supplemented with 170 mg $NaH_2PO_4 \cdot H_2O$, 80 mg adenine sulfate, 2.0 mg KIN, and 2.0 mg IAA per liter. R.E. Litz and R.J. Knight (unpublished data) have observed proliferation of cultured axillary buds of a hybrid of *P. edulis* and *P. edulis* var. flavicarpa. Apical dominance can be restored and plant regeneration can occur on medium without growth regulators. Rooting has been achieved in medium containing 2.0 mg/liter IAA (Robles 1978) and 1.0–2.0 mg/liter IBA or NAA (R.E. Litz and R.J. Knight, unpublished data). Meristem-tip culture has not been attempted for eliminating *Passiflora* fruit woodiness virus.

2. Organogenesis. Regeneration of plantlets from seedling *Passiflora* callus (Scorza and Janick 1976) and from internodal segments from mature vines (Robles 1979) has been demonstrated. Leaf disks from immature leaves of several *Passiflora* species form calli on MS medium supplemented with 1.0 mg BA and 1.0 mg NAA per liter (Scorza and Janick 1976). Callus responded to different growth regulator concentrations in different ways. Both shoots and flowers were induced from *P. suberosa* callus on medium with 1.0 mg BA and 1.0 mg NAA per liter. Only shoots were formed from *P. suberosa* and *P. caerula* calli on medium with 1.0 mg/liter NAA. Rooting was stimulated from *P. foetida* callus by 1.0 mg/liter BA and from purple passion fruit callus by 1.0 mg BA and 1.0 mg NAA per liter. Organogenesis was never induced from yellow passion fruit callus. Tendril, stem, and leaf explants that originated near the shoot apex of *P. suberosa* could be induced to flower in the presence of 0.1 mg/liter BA, whereas juvenile and basal explants produced only vegetative growth (Scorza and Janick 1980).

E. Pineapple

Pineapple is grown in many tropical and subtropical regions, not only for local consumption, but also as an export crop. It is anticipated that the annual, worldwide production of pineapple will soon exceed that of peaches (Knight 1980). Pineapples have normally been propagated by suckers that develop from axillary buds on the stem. Cultivars are self-incompatible and are capable of producing parthenocarpic fruit. Consequently, when grown in isolation from other cultivars, the fruits are seedless. Although pineapple selections have been clonally propagated for many centuries, there are relatively few serious disease problems. However, the rootknot nematode can reduce yields considerably.

1. Shoot-Tip Culture. The culture of excised buds of pineapple was first described by Aghion and Beauchesne (1960). They stimulated the growth of single plantlets from each cultured axillary bud on Knop's medium with 15% CW. Similar observations were made by Sita *et al.* (1974). The proliferation of lateral buds from pineapple axillary bud explants was reported by Pannetier and Lanaud (1976), who used modified MS medium containing 2.0 mg BA and 1.0 mg NAA per liter, and by Mathews *et al.* (1976), who used a medium containing 1.8 mg NAA, 2.0 mg IBA, and 2.1 mg KIN per liter. A similar approach was used by Zepeda and Sagawa (1981). Both Mathews and Rangan (1979) and Zepeda and Sagawa (1981) demonstrated that liquid culture medium appeared to be most suitable for obtaining rapid proliferation. Rooting has been accomplished in the presence of 1.0 mg BA and 1.0 mg IBA per liter (Mathews and Rangan 1979).

2. Organogenesis. Mapes (1973) described a procedure for inducing callus from pineapple axillary buds on modified MS medium containing 20 mg/liter adenine following subculture of buds from medium with adenine and 20% CW. Shoots developed from protocorm-like structures and formed plants that were normal in appearance. Wakasa *et al.* (1978) also reported the differentiation of plants from protocorm-like callus. Differentiation of meristemoids from hybrid embryo callus occurs on MS medium with 2.0 mg NAA, 2.0 mg IBA, and 2.5 mg BA per liter (Srinivasa Rao *et al.* 1981). Leaf explants have also produced callus with regenerative potential (Mathews and Rangan 1979) on MS medium containing 1.8 mg NAA, 2.0 mg IBA, and 11.0 mg BA per liter.

According to Wakasa (1979), considerable variation occurs in pineapple plants that have been differentiated from calli. Morphological characteristics such as foliage density, wax secretion, spines, and leaf color are all affected. He reported that in cultures established from slips or syncarp, there was considerable variability (98–100%), whereas in plants regenerated from the crown or from axillary buds, the variation in the regenerated population was less (7% for crown-derived plants and 34% for axillary bud-derived plants). This level of variability in regenerated populations is unacceptable for clonal propagation, but it could be an important breeding tool.

IV. PALMS

The palm family comprises about 212 genera and 2650 species. Palms are, for the most part, unbranched, arborescent woody monocotyledons, although some species produce offshoots or suckers from axillary buds. Several palm species have been used as ornamental or landscape plants, but four palm species (coconut palm, date palm, oil palm, and peach palm, or pejibaye) are particularly important in the tropics and subtropics. Heterozygosity in seedling populations, prolonged juvenile periods, and limited vegetative axillary meristems are serious obstacles to conventional approaches for improving palm and to clonal propagation. The major problems in palm tissue culture in general are the difficulty in acquiring explants, microbial contamination, browning and slow growth of callus and somatic embryos, and poor germination of plantlets (Reynolds 1982).

A. Coconut Palm

The most important nut crop in the world, coconut is widely grown in the tropics as a source of food, oil, and fiber. According to Woodroof (1979), coconut growing is the largest industry in the Philippines. Coconut palms are seed-propagated, although considerable variation exists in seedling populations. Air layering has also been utilized, but mainly to reduce the height of seed trees (Davis 1962).

Coconut production is seriously threatened by cadang-cadang, a viroid disease in southeast Asia and by lethal yellowing disease in the New World. Palms throughout south Florida and the Caribbean have been destroyed by lethal yellowing disease in the past three decades (McCoy 1983). Some coconut palm cultivars are tolerant or resistant to lethal yellowing disease; these include 'Malayan Dwarf' and 'Maypan', an F_1 hybrid between 'Malayan Dwarf' and the susceptible 'Panama Tall'. However, there is a shortage of available 'Maypan' seed.

1. Organogenesis. Since coconuts do not normally branch, induction of adventitious buds or embryony and conversion of flower primordia to vegetative shoots are possible approaches to clonal multiplication of elite coconut palm selections. Occasionally, branched coconut palms have been observed both *in vivo* and *in vitro* (Davis 1969; Balaga 1975; Fisher and Tsai 1979). The branches are adventitious in origin (Balaga 1975). Spontaneous reversion of flowers into shoots has been observed in palms whose terminal shoots have been destroyed by natural causes (Davis 1969). Since coconuts produce enormous numbers of flowers, conversion of potential flowers into shoots would be a potential method of clonal multiplication. Transformation of coconut flower primordia into vegetative shoots has been achieved *in vitro* (Schwabe 1973; Eeuwens 1976; Blake and Eeuwens

1981). The ninth or tenth immature inflorescence (counting down from the mature unopened inflorescence as the first) has the greatest potential to form shootlets *in vitro* regardless of the composition of medium or type of auxin or cytokinin used (C. Srinivasan, R.E. Litz, and K. Norstog, unpublished data). Some shoots have rooted, but the vestigial leaves do not expand.

2. Somatic Embryogenesis. Callus can be induced from several types of explant. Eeuwens (1976) used a modified MS medium (Y_3 medium) to induce callus from leaf and inflorescence explants. The original Y_3 medium had less ammonium and higher potassium and iodine than MS medium; it was later improved by the addition of glutamine, arginine, and asparagine (Eeuwens 1978). Callus formation and somatic embryogenesis have been achieved from leaf explants of seedling and mature trees on Y_3 minerals, Morel's vitamins, activated charcoal, 2,4-D, and BA by Pannetier and Buffard-Morel (1982), but details of hormone levels and procedures were never reported by these workers.

Recently, Branton and Blake (1983a) also obtained somatic embryo-like structures from leaf and inflorescence callus that developed on medium containing MS macronutrients, Y_3 micronutrients, Blake's (1972) vitamins, 0.25% activated charcoal plus 300 mg CH, 1.0 mg BA, 22.5 mg 2,4-D, and 1.0 mg 2iP per liter. The cultures were incubated in the dark at 30°C, and the calli were subcultured onto the same medium every 4–6 weeks. Somatic embryogenesis occurred when the nodular callus was subcultured on medium with 0.002–0.02 mg/liter 2,4-D. These somatic embryos formed only spongy haustoria and did not germinate. Normal plants grew from inflorescence-derived somatic embryos when the auxin concentration was gradually reduced over an 8-month period (Branton and Blake 1983b).

Calli and somatic embryos have also been induced from leaf primordia and shoot tips of seedlings and mature plants by culturing explants initially on Gamborg's B_5 protoplast medium (Gamborg *et al.* 1968) supplemented with 100–200 mg/liter 2,4-D and 0.3% activated charcoal.

3. Embryo Rescue. Coconut water has been used extensively for culture of zygotic embryos and for inducing embryogenic callus in numerous plant species (Raghavan 1976). Unautoclaved CW from immature fruits stimulates germination of excised coconut embryos (Cutter and Wilson 1954; Abraham and Thomas 1962). Embryo culture has been used to rescue the normally abortive 'Makapuno' coconut embryo (de Guzman and del Rosario 1964, 1969; de Guzman *et al.*

1979). 'Makapuno' embryos produce shoots and some roots on agar, but healthy bipolar seedlings develop in liquid culture, mainly due to reduced browning (Balaga and de Guzman 1972). Activated charcoal also reduces browning and improves the growth of coconut palm embryos (Fisher and Tsai 1978).

4. Protoplast Isolation and Culture. Protoplasts have been isolated from coconut palm calli and from young inflorescences, but regeneration has not been observed (Eeuwens 1976; Halibou and Kovoor 1981).

5. Anther Culture. Embryos have differentiated directly from microspores of cultured anthers that were excised from flowers 4–5 weeks before anthesis (Thanh-Tuyen and de Guzman 1983). Neither the tapetum nor pollen grains produced callus. Embryogenesis occurred 6–9 weeks after culturing on modified Blaydes (1966) medium supplemented with 6–9% sucrose, 15% CW, 0.5% activated charcoal, and 2.0 mg/liter NAA. The ploidy level of anther-derived embryos was not determined. These somatic embryos failed to germinate.

B. Date Palm

Date palms were among the earliest cultivated plants (Zohary and Spiegel-Roy 1975). They are dioecious, and seedling populations segregate equally into staminate and pistillate plants. The pistillate plants in seedling populations are heterozygous (Knight 1980). Consequently, date palms are usually propagated by offshoots, although fewer than a dozen suckers generally are produced during the life of a plant (Tisserat 1983). Air layering has been used occasionally in order to rejuvenate date palms (Reynolds 1982). Lethal yellowing disease has been reported to affect date palms (McCoy 1983), and resistant selections have not been identified.

1. Shoot-Tip Culture. Although the vegetative propagation of elite date palm selections by shoot-tip culture has not been achieved, Tisserat (1984) has reported a procedure using 2-year-old stock plants. Following establishment of 0.5-mm shoot tips on MS medium supplemented with 10 mg/liter NAA and 0.3% activated charcoal, axillary buds were induced to proliferate on MS medium containing 0.1 mg NAA and 10.0 mg BA per liter. Rooting occurred in the absence of BA.

2. Somatic Embryogenesis. Reuveni and Lilien-Kipnis (1974) obtained callus and root formation from cotyledonary sheath explants of date palm embryos *in vitro*. Embryogenic callus is formed when embryos from naturally polyembryonic date palm ovules are cultured on MS medium containing 5% sucrose, 1% activated charcoal, and 250 mg CH, 5.0 mg KIN, 2.0 mg IBA, 2.0 mg NAA, and 0.5 mg 2,4-D per liter. Somatic embryogenesis occurs following subculture on medium containing 0.1 mg KIN and 0.1 mg NAA per liter with 0.5% activated charcoal (Reuveni 1979). Embryogenic callus was also induced from immature zygotic embryos on MS medium supplemented with 100 mg/liter 2,4-D, 3.0 mg/liter 2iP, and 0.3% activated charcoal. Somatic embryos developed from callus on hormone-free medium (Reynolds and Murashige 1979). Tisserat (1979, 1981) subsequently refined the physical and chemical parameters for cloning date palms from several explants.

Efficient embryonic callus production has been achieved from lateral buds on medium with 0.1 or 1.0 mg/liter pCPA or 2,4-D or on medium with 100 mg/liter 2,4-D, 3.0 mg/liter 2iP, and 0.3% activated charcoal (Tisserat 1982). Somatic embryogenesis occurred from callus that was transferred to 0.1 mg/liter NAA, whereas other auxins were less effective (Tisserat 1982). The morphogenetic potential of date palm callus depends on the source of the explants. Shoot-tip explants from suckers and seedlings usually produce embryogenic callus, whereas leaf callus differentiates only roots. There have been no reports of organized structures developing from root calli. Mature inflorescences and leafless shoot apices rarely produce calli, although roots are occasionally produced from mature inflorescences (Zaid and Tisserat 1983). In contrast, primordial inflorescences near shoot tips produce embryogenic calli after several months (Tisserat 1981; Zaid and Tisserat 1983). Explants from staminate and pistillate palms behave similarly.

Somatic embryogenesis occurs from the periphery as well as from the interior of friable date palm callus masses. Somatic embryos develop in the presence of 2,4-D and 2iP, but full development of bipolar embryos occurs only in a hormone-free medium (Tisserat and De Mason 1980). Although somatic embryogenesis is relatively efficient, only 1–4% of mature somatic embryos germinate (Tisserat 1982). The germination of somatic embryos is very slow due to the lack of endosperm and the presence of a vestigial haustorium.

Date palm somatic embryos germinate and grow best on solid medium containing 0.01 mg/liter NAA and 3% sucrose. Axillary shoots develop *in vitro* in 30–40% of 2- to 4-month-old plantlets, grown in slowly rotating (1 rpm) liquid medium with 0.01–1.0 mg BA and 0.001–0.01 mg IAA or 0.1 mg NAA per liter (Tisserat 1982). Offshoots are also produced on solid medium in the presence of NAA, 2iP, KIN, and BA. Fewer offshoots are initiated from plants derived from zygotic embryos and shoot tips. As many as 15 to 20 offshoots can be obtained from each plantlet within 8 weeks.

In 5% of the plants grown from zygotic embryos, differentiation of inflorescences occurred in the presence of 5.0–15.0 mg/liter of KIN, BA or 2iP with or without 1.0 mg/liter NAA (Ammar and Benbadis 1977). These inflorescences are short-lived.

Date palm plantlets can be successfully transferred into soil when they are 10 cm in length or longer. The plants have to be protected from desiccation by a plastic cover and sprayed with 0.5% benomyl every week to reduce fungal infection. The clonal basis of date palm plantlets has been confirmed by means of isozyme markers for six enzymes (Torres and Tisserat 1980); however, palms have yet to be grown to maturity in order to assess their yield and quality.

C. Oil Palm

Oil palm is cultivated on a large scale in the tropics for palm oil and palm kernel oil, which are used as cooking oils and in the manufacture of soaps (Hartley 1977). It produces a greater oil yield per unit area than any other crop. The oil palm does not produce suckers, and propagation has been by seed. As a result, there is considerable variability in seedling populations. There have been reports of numerous diseases of oil palms caused by bacteria and fungi, although the palms are resistant to lethal yellowing disease.

1. Shoot-Tip Culture. Staritsky (1970) described a procedure for stimulating plantlet formation from cultured apices of oil palms on medium containing 100 mg CH, 0.1 mg KIN, and 5.0 mg NAA or IAA per liter. Only one plantlet developed from each shoot tip, and removal of the shoot tip caused the death of the donor plant.

2. Somatic Embryogenesis. Embryogenic callus has been established from zygotic embryos, leaves, and roots of seedlings and from inflorescences (Smith and Jones 1970; Rabechault *et al.* 1970; Jones 1974; Corley *et al.* 1977, 1981).

Hanower and Pannetier (1982) induced embryogenic callus from young leaves of oil palm seedlings and adult palms on medium containing MS macronutrients, Nitsch's (1969) micronutrients, 100 mg/liter sodium ascorbate, and 2% glucose. However, no information was provided regarding growth regulator concentrations in order for this work to be verified.

Most oil palm tissue culture has been limited to *tenera* and *dura* cultivars, both of which have a stony endocarp. *Tenera* has good pulp, but *dura* has very little pulp and oil. *Tenera* is a hybrid of *dura* and *pisifera*, a shell-less palm with considerable oil-rich pulp. The *pisifera* palm is used as a pollinator. The seeds of *pisifera* are mostly sterile and even fertile seeds germinate poorly (Nwankwo 1981). To perpetuate *pisifera* germplasm, Nwankwo and Krikorian (1983a) induced somatic embryogenesis from zygotic embryos (the embryonic axes of ripe fruits) and from leaves and roots of seedlings on half strength MS medium with 5.0–10.0 mg NAA or 2,4-D per liter. The concentration of auxin could be as high as 70 mg/liter provided that 0.3% activated charcoal was also present. Calli formed in 2–3 months, and quickly became nodular. Nodular calli were subcultured at 2-week intervals under 1.5–3.0 klx light intensity and 30°C and on medium with a low auxin concentration. On solid medium containing 30.0 mg NAA and 1.0 mg GA per liter and 0.3% activated charcoal, roots and shoots were produced. Seeds of *pisifera* rapidly lose viability from desiccation and microbial infections due to the absence of a stony endocarp. Nwankwo and Krikorian (1983b) succeeded in extending embryo viability up to 30 days by aseptic *in vitro* storage.

Transplanting oil palm plantlets into sterile soil has been difficult (Wooi *et al.* 1982). In nature, vesicular-arbuscular mycorrhizae are normally associated with oil palm roots. Therefore, oil palm plantlets survive better in nonsterile soil or in soil inoculated with mycorrhizae (Wooi *et al.* 1982). Thousands of *in vitro*-cloned oil palms are currently being field-tested in Malaysia. These oil palms are reported to be genetically very stable and uniform in vegetative growth and kernel and oil content (Wooi *et al.* 1982).

In vitro cloning of elite selections of oil palms will almost certainly have an immediate economic impact because oil palms cannot be propagated asexually by other methods.

3. Embryo Rescue. The seeds of oil palm have a dormancy requirement, and germination is a complicated process (Reynolds 1982; Rabechault and Cas 1974). During germination, the haustorium elongates and penetrates the endosperm. The petiole of the haustorium enlarges and produces a shoot and root simultaneously (Reynolds 1982). *In vitro*-grown embryos produce either shoots or roots, but rarely both. Embryos excised from dormant seeds germinate more slowly *in vitro* than those from nondormant seeds. Coconut water improves the rate of germination, particularly of embryos extracted 45–60 days after harvesting. Gibberellic acid is ineffective; 2,4-D and NAA increase the size of the haustorium but suppress root and shoot development (Rabechault *et al.* 1973; Bouvinet and Rabechault 1965).

D. Peach Palm (Pejibaye)

Peach palm is widely grown in Central and South America as a source for heart of palm and for the fruit, which is a rich source of protein, oil, carbohydrates, vitamins, and minerals. Normally, the plants are propagated by seed or by removal of suckers from elite selections. The peach palm is considered to be one of the major, under-exploited tropical plants (Anon. 1975).

1. Somatic Embryogenesis. Arias and Huete (1983) were able to clonally propagate the peach palm, or pejibaye, from cultured shoot apices of suckers from mature trees. Callus was induced in the dark on modified MS medium containing either 20.0–50.0 mg 2,4-D and 3.0–6.0 mg BA per liter or 1.25–10.0 mg NAA and 0.5–2.0 mg KIN per liter. Differentiation of plantlets, possibly by somatic embryogenesis, occurred following subculture of this callus onto medium without growth regulators in the light. Germination of regenerated plantlets has not been reported.

V. CONCLUSION

The development of *in vitro* systems for the propagation and improvement of many tropical fruits and palms has been hindered by problems inherent in the regeneration of woody plants from tissue cultures. The feasibility of plant regeneration via somatic embryogenesis from tissue cultures of the most important palm species and from nucellar explants of some of the most important tropical tree fruit species (e.g., *Citrus*, mangos, and some of the Myrtaceous fruit trees) has been demonstrated. Thus, for many woody, tropical fruit and palm species, there already exist procedures for *in vitro* regeneration that could be utilized for storage and international exchange of sterile, disease-indexed material. Unlike the woody, tropical fruit trees, herbaceous tropical fruit plants are well suited for *in vitro* propagation procedures and for the development of *in vitro* systems that can complement conventional plant breeding approaches. The exploitation of *in vitro* propagation techniques may lead to the upgrading of large commercial plantings not only with newly introduced and disease-indexed cultivars, but also with difficult-to-propagate fruit and palm species. When large-scale plantings of field crops have been attempted to replace the natural forest cover in many tropical regions, severe problems have resulted from the lack of available nutrients and from the disruption of the sensitive ecology of these regions. *In vitro* propagation of elite selections of perennial fruit and palm trees may be one solution for economic exploitation of these sensitive regions.

A significant potential benefit of *in vitro* approaches to tropical tree improvement exists in the use of mutant cell selection techniques and somaclonal variation as means for lowering the genetic vulnerability of clonally propagated tropical fruit and palm trees without altering the unique character of elite selections. This would greatly enhance the effectiveness of tropical fruit and palm improvement schemes, particularly in the face of constant disease, pest, and environmental stresses that exist under tropical conditions. Research in the future must concentrate on improving the efficiency of plant recovery from cell cultures, extending the range of tropical tree species that can be regenerated *in vitro*, and actually demonstrating the effectiveness of these techniques in plant improvement schemes. The main obstacles to progress are unfamiliarity with tropical crop plants in developed countries, where biotechnology has otherwise been well supported by government agencies, and the perceived inaccessibility of major tropical plant germplasm resources. As human populations continue to increase rapidly in the tropics, greater attention must be focused on the research needs of the staple foods, export, and fruit crops of this region. Clearly, more studies are needed to realize the potential of *in vitro* techniques.

LITERATURE CITED

ABRAHAM, A. and K.J. THOMAS. 1962. A note on the *in vitro* culture of excised coconut embryos. Ind. Coconut J. 15:84–87.

AGHION, D. and G. BEAUCHESNE. 1960. Utilisation de la technique de culture sterile d'organes pour obtenir des clones d'ananas. Fruits d'Outre Mer 15:444–446.

ALLAN, P. 1964. Pawpaws grown from cuttings. Farming South Africa 101:1–6.

ALTMAN, A. and R. GOREN. 1971. Promotion of callus formation by abscisic acid in citrus bud cultures. Plant Physiol. 47:844–846.

ALTMAN, A. and R. GOREN. 1974a. Growth and dormancy cycles in *Citrus* bud cultures and their hormonal control. Physiol. Plant. 30:240–245.

ALTMAN, A. and R. GOREN. 1974b. Interrelationship of abscisic acid in the promotion of callus formation in the abscission zone of citrus bud cultures. Physiol. Plant. 32:55-61.

ALTMAN, A. and R. GOREN. 1977. Horticultural and physiological aspects of citrus bud culture. Acta Hort. 78:51-59.

AMMAR, S. and A. BENBADIS. 1977. Multiplication vegetative du palmier dattier (*Phoenix dactylifera* L) par la culture de tissue de jeunes plantes de semis. C. R. Acad. Sci., Ser. D. 284:1789–1792.

ANON. 1975. Underexploited tropical plants with promising economic value. National Academy of Sciences, Washington, DC.

ARIAS, O. and F. HUETE. 1983. Propagacion vegetativa *in vitro* de pejibaye (*Bactris gasipaes* H.B.K.). Turrialba 33:103–108.

ARORA, I.K. and R.N. SINGH. 1978. *In vitro* plant regeneration in papaya. Curr. Sci. 47:867-868.

BALAGA, H.Y. 1975. Induction of branching in coconut. Kalikasan, Philip. J. Biol. 4:135-140.

BALAGA, H.Y. and E.V. DE GUZMAN. 1972. The growth and development of coconut 'Makapuno' embryos *in vitro*. II. Increased root incidence and growth in response to media composition and to sequential culture from liquid to solid medium. Philip. Agr. 53:551-565.

BAPAT, V.A. and S. NARAYANASWAMY. 1977. Mesocarp and endosperm culture of *Achras sapota* Linn. *in vitro*. Ind. J. Expt. Biol. 15:294.

BARLASS, M. and K.G.M. SKENE. 1982. In vitro plantlet formation from *Citrus* species and hybrids. Scientia Hort. 17:333-341.

BEN-HAYYIM, G. and J. KOCHBA. 1982. Growth characteristics and stability of tolerance of citrus callus cells selected to NaCl stress. Plant Sci. Lett. 27:87-94.

BEN-HAYYIM, G. and J. KOCHBA. 1983. Aspects of salt tolerance in a NaCl-selected stable cell line of *Citrus sinensis*. Plant Physiol. 72:685-690.

BERG, L.A. and M. BUSTAMANTE. 1974. Heat treatment and meristem culture for the production of virus-free bananas. Phytopathology 64:320-322.

BITTERS, W.P., T. MURASHIGE, T.S. RANGAN, and E. NAUER. 1970. Investigations on established virus-free plants through tissue culture. Calif. Citrus Nursery. Soc. 9:27-30.

BLAKE, J. 1972. A specific bioassay for the inhibition of flowering. Planta 103:126-128.

BLAKE, J. and C.J. EEUWENS. 1981. Culture of coconut palm tissue with a view to vegetative propagation. p. 145-148. In: A.N. Rao (ed.), Tissue culture of economically important plants. COSTED, Singapore.

BLAYDES, D.F. 1966. Interaction of kinetin and various inhibitors in the growth of soybean tissue. Physiol. Plant. 19:748-753.

BONGA, J.M. 1982. Vegetative propagation in relation to juvenility, maturity, and rejuvenation. p. 387-412. In: J.M. Bonga and D.J. Durzan (eds.), Tissue culture in forestry. Martinus Nijhoff/Dr. W. Junk Publishers, The Hague.

BOUVINET, J. and H. RABECHAULT. 1965. Effects de l'acide gibberellique sur les embryons de palmier a huile (*Elaeis guineensis* Jacq. var. dura) en culture *in vitro*. C. R. Acad. Sci., Ser. D. 260:5336-5338.

BRANTON, R. L. and J. BLAKE. 1983a. Development of organized structures in callus derived from explants of *Cocos nucifera* L. Ann. Bot. 52:673-678.

BRANTON, R. L. and J. BLAKE. 1983b. A lovely clone of coconuts. New Scient. 98:554-557.

BROERTJES, C. and A.M. VAN HARTEN. 1978. Application of mutation breeding in the improvement of vegetatively propagated crops. Elsevier, Amsterdam.

BURGER, D.W. and W.P. HACKETT. 1982. The isolation, culture and division of protoplasts from citrus cotyledons. Physiol. Plant. 56:324-328.

BUTTON, J. 1977. International exchange of disease-free citrus clones by means of tissue culture. Outlook on Agr. 9:155-159.

BUTTON, J. 1978. The effects of some carbohydrates on the growth and organization of *Citrus* ovular callus. Z. Pflanzenphysiol. 88:61-68.

BUTTON, J. and C.H. BORNMAN. 1971. Development of nucellar plants from unpollinated and unfertilized ovules of the Washington navel orange *in vitro*. J. South Afr. Bot. 37:127-134.

BUTTON, J. and C.E.J. BOTHA. 1975. Enzymic macerations of *Citrus* callus and the regeneration of plants from single cells. J. Expt. Bot. 26:723-729.

BUTTON, J. and J. KOCHBA. 1977. Tissue culture in the citrus industry. p. 70-92. In: J. Reinert and Y.P.S. Bajaj (eds.), Applied and fundamental aspects of plant cell, tissue and organ culture. Springer-Verlag, Berlin.

BUTTON, J. and F.H.J. RIJKENBERG. 1977. The effect of subculture interval on organogenesis in callus cultures of *Citrus sinensis*. Acta Hort. 78:225-236.

BUTTON, J., J. KOCHBA, and C.H. BORNMAN. 1974. Fine structure of and embryoid development from embryogenic ovular callus of 'Shamouti' orange (*Citrus sinensis* Osb.). J. Expt. Bot. 25:446–457.

CHATURVEDI, H.C. and G.C. MITRA. 1974. Clonal propagation of citrus from somatic callus cultures. HortScience 9:118–120.

CHEN, Z., M. WANG, and H. LIAO. 1980. The induction of citrus pollen plants in artificial media (in Chinese). Acta Genet. Sin. 7:6-9.

COHEN, D. and P.A. COOPER. 1982. Micropropagation of babaco—a *Carica* hybrid from Ecuador. p. 743-744. In: A. Fujiwara (ed.), Plant tissue culture 1982. Japan. Assoc. Plant Tissue Culture, Tokyo.

CORLEY, R.H.V., J.N. BARRETT, and L.H. JONES. 1977. Vegetative propagation of oil palm via tissue culture. p. 1-8. In: D.A. Earp and W. Newell (eds.), International development in oil palm. Proc. Malaysian Intern. Agr. Oil Palm Conf., Intern. Soc. Planters, Kuala Lumpur, Malaysia.

CORLEY, R.H.V., C.Y. WONG, K.C. WOOI, and L.H. JONES. 1981. Early results from the first oil palm clone trials. Proc. The oil palm in agriculture in the eighties. Kuala Lumpur, Malaysia.

COX, E.A., G. STOTZKY, and R.D. GOOS. 1960. *In vitro* culture of *Musa balbisiana* Colla embryos. Nature 185:403-404.

CRONAUER, S.S. and A.D. KRIKORIAN. 1983. Somatic embryos from cultured tissues of triploid plantains (*Musa* 'ABB'). Plant Cell Rpt. 2:289–291.

CRONAUER, S.S. and A.D. KRIKORIAN. 1984a. Multiplication of *Musa* from excised stem tips. Ann. Bot. 53:321-328.

CRONAUER, S.S. and A.D. KRIKORIAN. 1984b. Rapid multiplication of bananas and plantains by *in vitro* shoot tip culture. HortScience 19:234-235.

CUTTER, JR., V.M. and K.S. WILSON. 1954. Effects of coconut endosperm and their growth stimulants upon the development *in vitro* of embryos of *Cocos nucifera*. Bot. Gaz. 115:234-240.

DAVIS, T.A. 1962. Rejuvenation of coconut palms. World Crops 14:254-259.

DAVIS, T.A. 1969. Clonal propagation of the coconut. World Crops 21:253-255.

DEBRUIJNE, E., E. DE LANGHE, and R. VAN RIJCK. 1974. Action of hormones and embryoid formation in callus cultures of *Carica papaya*. Intern. Symp. Fytofarm. Fytiat. 26:637-645.

DE GUZMAN, E.V. 1975. Project on production of mutants by irradiation of *in vitro* cultured tissues of coconut and bananas and their mass propagation by the tissue culture technique. p. 53-76. In: Improvement of vegetatively propagated plants through induced mutations. Tech. Soc. 173. IAEA, Vienna.

DE GUZMAN, E.V. and A. DEL ROSARIO. 1964. The growth and development of *Cocos nucifera* L. 'Makapuno' embryo *in vitro*. Philip. Agr. 48:82–94.

DE GUZMAN, E.V. and A. DEL ROSARIO. 1969. Growth and development of coconut 'Makapuno' embryo *in vitro*. I. The induction of rooting. Philip. Agr. 53:65–78.

DE GUZMAN, E.V., A.G. DEL ROSARIO, and E.M. UBALDE. 1979. Proliferative growths and organogenesis in coconut embryo and tissue cultures. Philip. J. Coconut Studies 7:1-10.

DE GUZMAN, E.V., A.C. DECENA, and E.M. UBALDE. 1980. Plantlet regeneration from unirradiated and irradiated banana shoot tip tissue cultured *in vitro*. Philip. Agr. 63:140-146.

DEIDDA, P. 1973. *In vitro* nucellar embryogenesis in monoembryonic Clementine seeds. p. 33–35. In: O. Carpena (ed.), Congreso Mundial de Citricultura, Vol. 2. Min. de Agr., Murcia, Spain.

DE LANGHE, E. 1969. Bananas, *Musa* spp. p. 53–78. In: F.P. Ferwerda and F. Wit (eds.), Outlines of perennial crop breeding in the tropics. Misc. Papers 4. Landbouwhogeschool, Wageningen, the Netherlands.

DORESWAMY, R., N.K.S. RAO, and E.K. CHACKO. 1983. Tissue culture propagation of banana. Scientia Hort. 18:247–252.

EEUWENS, C.J. 1976. Mineral requirements for growth and callus initiation of tissue explants excised from mature coconut palms (*Cocos nucifera*) and cultured *in vitro*. Physiol. Plant. 36:23–28.

EEUWENS, C.J. 1978. Effects of organic nutrients and hormones on growth and development of tissue explants from coconut (*Cocos nucifera*) and date (*Phoenix dactylifera*) palms cultured *in vitro*. Physiol. Plant. 42:173–178.

EPSTEIN, E., J. KOCHBA, and H. NEUMANN. 1977. Metabolism of indoleacetic acid by embryogenic and non-embryogenic callus lines of 'Shamouti' orange (*Citrus sinensis* Osb.). Z. Pflanzenphysiol. 85:263–268.

ESAN, E.B. 1973. A detailed study of adventive embryogenesis in the Rutaceae. Ph.D. Thesis, Univ. of California, Riverside.

FAO (Food and Agricultural Organization of the U.N.). 1982. FAO Production Yearbook. FAO, Rome.

FISHER, J.B. and J.H. TSAI. 1978. *In vitro* growth of embryos and callus of coconut palm. In Vitro 14:307–311.

FISHER, J.B. and J.H. TSAI. 1979. A branched coconut seedling in tissue culture. Principes 23:128–131.

GAMBORG, O.L., R.A. MILLER, and K. OJIMA. 1968. Nutrient requirements of suspension cultures of soybean root cells. Expt. Cell Res. 50:151–158.

GILADI, I., A. ALTMAN, and R. GOREN. 1977. Differential effects of sucrose, abscisic acid, and benzyladenine on shoot growth and callus formation in the abscission zone of excised citrus buds. Plant Physiol. 59:1161–1164.

GOREN, R., A. ALTMAN, and I. GILADI. 1979. Role of ethylene in abscisic acid-induced callus formation in citrus bud cultures. Plant Physiol. 63:280–282.

GRINBLAT, U. 1972. Differentiation of citrus stem *in vitro*. J. Amer. Soc. Hort. Sci. 97:559–603.

GULSEN, Y., A. ALTMAN, and R. GOREN. 1981. Growth and development of *Citrus* pistils and fruit explants in vitro. Physiol. Plant. 53:295–300.

HALIBOU, T.K. and A. KOVOOR. 1981. Regeneration of callus from coconut protoplasts. p. 149–151. In: A.N. Rao (ed.), Tissue culture of economically important plants. COSTED, Singapore.

HANOWER, J. and C. PANNETIER. 1982. In vitro vegetative propagation of the oil palm, *Elaeis guineensis* Jacq. p. 745–746. In: A. Fujiwara (ed.), Plant tissue culture 1982. Japan. Assoc. Plant Tissue Culture, Tokyo.

HARMS, C.T. and I. POTRYKUS. 1980. Hormone-inhibition of *Citrus* protoplasts released by co-culturing with *Nicotiana tabacum* protoplasts—Its significance for somatic hybrid selection. Plant Sci. Lett. 19:611–614.

HARMS, C.T., J. KOCHBA, and I. POTRYKUS. 1980. Fusion of citrus and tobacco protoplasts—A new system for somatic hybridization studies with remote species. p. 321–326. In: L. Ferenczy and G. L. Farkas (eds.), Advances in protoplast research. Pergamon Press, Oxford.

HARTLEY, C.W.S. 1977. The oil palm. Longman Group Ltd., London.

HIKADA, T., Y. YAMADA, and T. SCHICHIJO. 1979. *In vitro* differentiation of haploid plants by anther culture in *Poncirus trifoliata* (L.). Japan. J. Breed. 29:248–254.

HIKADA, T., Y. YAMADA, and T. SHICHIJO. 1982. Plantlet formation by anther culture of *Citrus aurantium* L. Japan. J. Breed. 32:247-252.

HWANG, S.C., C.L. CHEN, J.C. LIN, and H.L. LIN. 1984. Cultivation of banana using plantlets from meristem culture. HortScience 19:231-233.

JONES, L.H. 1974. Propagation of clonal oil palm by tissue culture. Oil Palm News 17:1-9.

JUAREZ, J., L. NAVARRO, and J.L. GUARDIOLA. 1976. Obtention de plantes nucellaires de divers cultivars de clementiners au moyen de la culture de nucelle "in vitro." Fruits d'Outre Mer 31:751-761.

KHUSPE, S.S., R.R. HENDRE, A.F. MASCARENHAS, and V. JAGANNATHAN. 1980. Utilization of tissue culture to isolate interspecific hybrids in *Carica* L. p. 198-205. In: P.S. Rao, M.R. Heble, and M.S. Chadha (eds.), Plant tissue culture, genetic manipulation, and somatic hybridization of plant cells. BARC, Bombay, India.

KITTO, S.L. and M.J. YOUNG. 1981. *In vitro* propagation of Carrizo citrange. HortScience 16:305-306.

KNIGHT, R. 1980. Origin and world importance of tropical and subtropical fruit crops. p. 1-120. In: S. Nagy and P.E. Shaw (eds.), Tropical and subtropical fruits: composition, properties, and uses. AVI Publ. Co., Westport, Connecticut.

KOBAYASHI, S., I. IKEDA, and N. NAKATANI. 1982. Studies on nucellar embryogenesis in citrus. III. On the differences in ability to form embryoids in *in vitro* culture of ovules from poly- and mono-embryonic cultivars (in Japanese). Bull. Fruit Tree Res. Sta. E. (Japan) 4:21-27.

KOBAYASHI, S., H. UCHIMIYA, and I. IKEDA. 1983. Plant regeneration from 'Trovita' orange protoplasts. Japan. J. Breed. 33:119-122.

KOCHBA, J., G. BEN-HAYYIM, P. SPIEGEL-ROY, S. SAAD, and H. NEUMANN. 1982a. Selection of stable salt-tolerant callus cell lines and embryos in *Citrus sinensis* and *C. aurantium.* Z. Pflanzenphysiol. 106:111-118.

KOCHBA, J. and J. BUTTON. 1974. The stimulation of embryogenesis and embryoid development in habituated ovular callus from the 'Shamouti' orange (*Citrus sinensis*) as affected by tissue age and sucrose concentration. Z. Pflanzenphysiol. 73:415-421.

KOCHBA, J. and P. SPIEGEL-ROY. 1977a. Cell and tissue culture for breeding and developmental studies of citrus. HortScience 12:110-114.

KOCHBA, J. and P. SPIEGEL-ROY. 1977b. Embryogenesis in gamma-irradiated habituated ovular callus of the 'Shamouti' orange as affected by auxin and by tissue age. Environ. Expt. Bot. 17:151-159.

KOCHBA, J. and P. SPIEGEL-ROY. 1977c. The effects of auxins, cytokinins and inhibitors on embryogenesis in habituated ovular callus of the 'Shamouti' orange (*Citrus sinensis*). Z. Pflanzenphysiol. 81:283-288.

KOCHBA, J., P. SPIEGEL-ROY, and H. SAFRAN. 1972. Adventive plants from ovules and nucelli in citrus. Planta 106:237-245.

KOCHBA, J., J. BUTTON, P. SPIEGEL-ROY, C.H. BORNMAN, and M. KOCHBA. 1974. Stimulation of rooting of *Citrus* embryoids by gibberellic acid and adenine sulfate. Ann. Bot. 38:415-421.

KOCHBA, J., P. SPIEGEL-ROY, H. NEUMANN, and S. SAAD. 1978. Stimulation of embryogenesis in citrus ovular callus by ABA, ethephon, CCC and Alar and its suppression by GA_3. Z. Pflanzenphysiol. 89:427-432.

KOCHBA, J., P. SPIEGEL-ROY, and S. SAAD. 1980. Selection for tolerance to sodium chloride (NaCl) and 2,4-dichlorophenoxyacetic acid (2,4-D) in ovular callus lines of *Citrus sinensis* p. 187-192. In: F. Sala, B. Parisis, R. Cella, and O. Ciferri (eds.), Plant cell cultures: results and perspectives. Elsevier/North-Holland Biomedical Press, Amsterdam.

KOCHBA, J., G. BEN-HAYYIM, P. SPIEGEL-ROY, S. SAAD, and H. NEUMANN. 1982a. Selection of stable salt-tolerant callus cell lines and embryos in *Citrus sinensis* and *C. aurantium*. Z. Pflanzenphysiol. 106:111–118.

KOCHBA, J., P. SPIEGEL-ROY, H. NEUMANN, and S. SAAD. 1982b. Effect of carbohydrates on somatic embryogenesis in subcultured nucellar callus of *Citrus* cultivars. Z. Pflanzenphysiol. 105:359–368.

KORDAN, H.A. 1977. Mitosis and cell proliferation in lemon fruit explants incubated on attenuated nutrient solutions. New Phytol. 79:673–678.

KRIKORIAN, A.D. and S.S. CRONAUER. 1984. Banana. p. 327–348. In: W.R. Sharp, D.A. Evans, P.V. Ammirato, and Y. Yamada (eds.), Handbook of plant cell culture. Vol. II. Crop species. Macmillan, New York.

LEROY, J.F. 1947. La polyembryonie chez les *Citrus* son interêt dans la culture et l'amelioration. Rev. Intern. Bot. Appl. (Paris) 27:483–495.

LITZ, R.E. 1984a. Tissue culture for the improvement of tropical fruits. Fla. Agr. Res. 3:26–28.

LITZ, R.E. 1984b. *In vitro* somatic embryogenesis from nucellar callus of monoembryonic *Mangifera indica* L. HortScience 19:715–717.

LITZ, R.E. 1984c. *In vitro* somatic embryogenesis from callus of jaboticaba, *Myrciaria cauliflora*. HortScience 19:62–64.

LITZ, R.E. 1984d. *In vitro* responses of adventitious embryos of two polyembryonic *Eugenia* species. HortScience 19:720–722.

LITZ, R.E. 1984e. Papaya. p. 349–368. In: W.R. Sharp, D.A. Evans, P.V. Ammirato, and Y. Yamada (eds.), Handbook of plant cell culture. Vol. II. Crop species. Macmillan, New York.

LITZ, R.E. and R.A. CONOVER. 1977. Tissue culture propagation of papaya. Proc. Fla. State Hort. Soc. 90:245–246.

LITZ, R.E. and R.A. CONOVER. 1978a. *In vitro* propagation of papaya. HortScience 13:241–242.

LITZ, R.E. and R.A. CONOVER. 1978b. Recent advances in papaya tissue culture. Proc. Fla. State Hort. Soc. 91:181–184.

LITZ, R.E. and R.A. CONOVER. 1979. Development of systems for obtaining parasexual *Carica* hybrids. Proc. Fla. State Hort. Soc. 92:180–182.

LITZ, R.E. and R.A. CONOVER. 1980a. Partial organogenesis in tissue cultures of *Averrhoa carambola*. HortScience 15:735.

LITZ, R.E. and R.A. CONOVER. 1980b. Somatic embryogenesis in cell cultures of *Carica stipulata*. HortScience 15:733–735.

LITZ, R.E. and R.A. CONOVER. 1981a. Effect of sex type, season, and other factors on *in vitro* establishment and culture of *Carica papaya* L. explants. J. Amer. Soc. Hort. Sci. 106:792–794.

LITZ, R.E. and R.A. CONOVER. 1981b. *In vitro* polyembryony in *Carica papaya* L. ovules. Z. Pflanzenphysiol. 104:285–288.

LITZ, R.E. and R.A. CONOVER. 1982. In vitro somatic embryogenesis and plant regeneration from *Carica papaya* L. ovular callus. Plant Sci. Lett. 26:153–158.

LITZ, R.E. and R.A. CONOVER. 1983. High frequency somatic embryogenesis from *Carica* suspension cultures. Ann. Bot. 51:683–686.

LITZ, R.E., R.L. KNIGHT, and S. GAZIT. 1982. Somatic embryos from cultured ovules of polyembryonic *Mangifera indica* L. Plant Cell Rpt. 1:264-266.

LITZ, R.E., S.K. O'HAIR, and R.A. CONOVER. 1983. In vitro growth of *Carica papaya* L. cotyledons. Scientia Hort. 19:287-293.

LITZ, R.E., R.J. KNIGHT, JR., and S. GAZIT. 1984. In vitro somatic embryogenesis from *Mangifera indica* L. callus. Scientia Hort. 22:233-240.

LIU, C. and J. YANG. 1983. Isolation and culture of papaya protoplasts. Plant Physiol. (Suppl.) 72:144 (Abstr.)

MA, S. and C. SHII. 1972. *In vitro* formation of adventitious buds in banana shoot apex following decapitation (in Chinese). J. Hort. Sci. (China) 18:135-142.

MA, S. and C. SHII. 1974. Growing banana plants from adventitious buds (in Chinese). J. Hort. Sci. (China) 20:6-12.

MAHESHWARI, P. and N.S. RANGASWAMY. 1958. Polyembryony and *in vitro* culture of embryos of *Citrus* and *Mangifera*. Ind. J. Hort. 15:275-282.

MAPES, M.O. 1974. Tissue culture of Bromeliada. Intern. Plant Prop. Comb. Soc. 23:47-55.

MATHEWS, V.H. and T.S. RANGAN. 1979. Multiple plantlets in lateral bud and leaf explant in vitro cultures of pineapple. Scientia Hort. 11:319-328.

MATHEWS, V.H., T.S. RANGAN, and S. NARAYANASWAMY. 1976. Micropropagation of *Ananas sativus in vitro*. Z. Pflanzenphysiol. 79:450-454.

MATSUMOTO, K. and H. YAMAGUCHI. 1983. Induction of adventitious buds and globular embryoids on seedlings of trifoliate orange (*Poncirus trifoliata*). Japan. J. Breed. 33:123-129.

McCOY, R.E. 1983. Lethal yellowing of palms. Institute of Food and Agr. Sciences, Univ. of Florida, Gainesville.

MEHDI, A.A. and L. HOGAN. 1976. Tissue culture of *Carica papaya*. HortScience 11:311 (Abstr.).

MITRA, G.C. and H.C. CHATURVEDI. 1972. Embryoids and complete plants from unpollinated ovaries and from ovules of in vivo-grown emasculated flower buds of *Citrus* spp. Bull. Torrey Bot. Club 99:184-189.

MOORE, G.A. and R.E. LITZ. 1984. Biochemical markers for *Carica papaya*, *C. caulifora* and plants from somatic embryos of their hybrid. J. Amer. Soc. Hort. Sci. 109:213-218.

MUKHERJEE, S.K. 1953. The mango—its botany, cultivation, uses and future improvements, especially as observed in India. Econ. Bot. 7:130-162.

MURASHIGE, T. and F. SKOOG. 1962. A revised medium for rapid growth and bioassays with tobacco tissue cultures. Physiol. Plant. 15:473-497.

MURASHIGE, T. and D.P.H. TUCKER. 1969. Growth factor requirements of citrus tissue culture. p. 1155-1161. In: H. D. Chapman (ed.), Proc. 1st Intern. Citrus Symp., Vol. 3. Univ. of California, Riverside.

MURASHIGE, T., W.P. BITTERS, T.S. RANGAN, E.M. NAUER, C.N. ROISTACHER, and P.B. HOLLIDAY. 1972. A technique of shoot apex grafting and its utilization towards recovering virus-free *Citrus* clones. HortScience 7:118-119.

MURIITHI, L.M., T.S. RANGAN, and B.H. WAITE. 1982. *In vitro* propagation of fig through shoot tip culture. HortScience 17:86-87.

NACHMIAS, A., I. BARASH, Z. SOLEL, and G.A. STROBEL. 1977. Translocation of mal secco toxin in lemons and its effect on electrolyte leakage, transpiration, and citrus callus growth. Phytoparasitica 5:94-103.

NAIR, S., P. GUPTA, and A.F. MASCARENHAS. 1983. Haploid plants from *in vitro* anther culture of *Annona squamosa* Linn. Plant Cell Rpt. 2:198-200.

NAIR, S., P.K. GUPTA, M.V. SHIRGURKAR and A.F. MASCARENHAS. 1984. *In vitro* organogenesis from leaf explants of *Annona squamosa* Linn. Plant Cell Tiss. Org. Cult. 3:29-40.

NAVARRO, L. 1981a. Citrus shoot-tip grafting *in vitro* (STG) and its applications: a review. p. 452-456. Proc. Intern. Soc. Citriculture.

NAVARRO, L. 1981b. Effect of citrus exocortis viroid (CEV) on root and callus formation by stem tissue of Etrog citron (*Citrus medica* L.) cultured *in vitro*. p. 437-439. Proc. Intern. Soc. Citriculture.

NAVARRO, L. and J. JUAREZ. 1977. Tissue culture techniques used in Spain to recover virus-free citrus plants. Acta Hort. 78:425-435.

NAVARRO, L., C.N. ROISTACHER, and T. MURASHIGE. 1975. Improvement of shoot-tip grafting *in vitro* for virus-free citrus. J. Amer. Soc. Hort. Sci. 100:471-479.

NITSCH, J.P. and C. NITSCH. 1969. Haploid plants from pollen grains. Science 163:85-87.

NWANKWO, B.A. 1981. Facilitated germination of *Elaeis guineensis* Jacq. var. *pisifera*. Ann. Bot. 48:251-254.

NWANKWO, B.A. and A.D. KRIKORIAN. 1983a. Morphogenetic potential of embryo and seedling-derived callus of *Elaeis guineensis* Jacq. var *pisifera* Becc. Ann. Bot. 51:65-76.

NWANKWO, B.A. and A.D. KRIKORIAN. 1983b. Aseptic storage of *Elaeis guineensis* form *pisifera* seeds. Principes 27:34-37.

OBERBACHER, M.F. 1967. Citrus tissue culture as a means of studying the metabolism of carotenoids and chlorophyll. Proc. Fla. State Hort. Soc. 80:254-257.

OHTA, Y. and K. FURUSATO. 1957. Embryo culture in *Citrus*. Seiken Ziho 8:49-54.

OHYAMA, K. and S. OKA. 1982. Multiple shoot formation from mulberry (*Morus alba* L.) hypocotyls by N-(2-chloro-4-pyridyl)-N^1-phenylurea. p. 149-150. In: A. Fujiwara (ed.), Plant tissue culture 1982. Japan. Assoc. Plant Tissue Culture, Tokyo.

OKA, S. and K. OHYAMA. 1981. *In vitro* initiation of adventitious buds and its modification by high concentrations of benzyladenine in leaf tissues of mulberry (*Morus alba*). Can. J. Bot. 59:68-74.

PANNETIER, C. and J. BUFFARD-MOREL. 1982. Production of somatic embryos from leaf tissues of coconut, *Cocos nucifera* L. p. 755-756. In: A. Fujiwara (ed.), Plant tissue culture 1982. Japan. Assoc. Plant Tissue Culture, Tokyo.

PANNETIER, C. and C. LANAUD. 1976. Divers aspects de l'utilisation possible des cultures "*in vitro*" pour la multiplication vegetative de l'*Ananas comosus* L. Merr., variété 'Cayenne lisse'. Fruits d'Outre Mer 31:739-750.

PARLEVLIET, J.E. and J.W. CAMERON. 1959. Evidence on the inheritance of nucellar embryony in citrus. Proc. Amer. Soc. Hort. Sci. 74:252-259.

PHADNUS, N.A., N.D. BUDRUKKAR, and S.N. KAULGUD. 1970. Embryo culture technique in papaya (*Carica papaya* L.). Poona Agr. Coll. Mag. 60:101-104.

RABECHAULT, H. and S. CAS. 1974. Recherches sur la culture *in vitro* des embryons de palmier à huile (*Elaeis guineensis* Jacq. var. *dura* Becc). Oleagineux 29:73-78.

RABECHAULT, H.,G. GUENIN, and J. AHEE. 1970. Recherches sur la culture *in vitro* des embryons de palmier á huile (*Elaeis guineensis*. Jacq. var. *dura* Becc.) VII. Comparison de divers milieux mineraux. Oleagineux 25:519-524.

RABECHAULT, H., G. GUENIN, and J. AHEE. 1973. Recherches sur la culture *in vitro* de embryons de palmier a huile (*Elaeis guineensis* Jacq. var. *dura* Becc.) IX. Activation de la sensibilite au lait de coco par une rehydration des graines. Oleagineux 28:333-336.

RAGHAVAN, V. 1976. Experimental embryogenesis in vascular plants. Academic Press, London.

RAJ BHANSALI, R. and H.C. ARYA. 1978. Tissue culture propagation of citrus trees. p. 135-140. Proc. Intern. Soc. Citriculture.

RAJ BHANSALI, R. and H.C. ARYA. 1979. Organogensis in *Citrus limetoides* Tanaka (sweet lime) callus culture. Phytomorphology 28:97-100.

RANGAN, T.S., T. MURASHIGE, and W.P. BITTERS. 1968. *In vitro* initiation of nucellar embryos in monoembryonic *Citrus*. HortScience 3:226–227.

RANGAN, T.S., T. MURASHIGE, and W.P. BITTERS. 1969. In vitro studies of zygotic and nucellar embryogenesis in citrus. p. 225–229. In: H. D. Chapman (ed.), Proc. 1st Intern. Citrus Symp., Univ. of California, Riverside.

RANGASWAMY, N.S. 1958. Culture of nucellar tissue of *Citrus* in vitro. Experientia 14:111–112.

RANGASWAMY, N.S. 1959. Morphogenetic response of *Citrus* ovules to growth adjuvants in culture. Nature 183:735–736.

RANGASWAMY, N.S. 1961. Experimental studies on female productive structures of *Citrus microcarpa* Bunge. Phytomorphology 11:109–127.

RANGASWAMY, N.S. 1982. Nucellus as an experimental system in basic and applied tissue culture research. p. 269–286. In: A.N. Rao (ed.), Tissue culture of economically important plants. COSTED, Singapore.

RAO, A.N., Y.M. SIN, N. KOTHAGODA, and J. HUTCHINSON. 1982. Cotyledon tissue culture of some tropical fruits. p. 124–137. In: A.N. Rao (ed.), Tissue culture of economically important plants. COSTED, Singapore.

REUVENI, O. 1979. Embryogenesis and plantlet growth of date palm (*Phoenix dactylifera* L.) derived from callus tissue. Plant Physiol. (Suppl.) 63:138 (Abstr.).

REUVENI, O. and H. LILIEN-KIPNIS. 1974. Studies of the in vitro culture of date palm (*Phoenix dactylifera* L.) tissues and organs. Pamphlet 145. Volcani Inst. of Agr. Res., Bet Dagan, Israel.

REYNOLDS, J.F. 1982. Vegetative propagation of palm trees. p. 182-207. In: J.M. Bonga and D.J. Durzan (eds.), Tissue culture in forestry. Martinus Nijhoff/Dr. W. Junk Publishers, The Hague.

REYNOLDS, J.F. and T. MURASHIGE. 1979. Asexual embryogenesis in callus cultures of palms. In Vitro 5:383-387.

ROBLES, M.J.M. 1978. Multiplication vegetative, *in vitro*, des bourgeons axillaires de *Passiflora edulis* var. *flavicarpa* Degener et de *P. mollissima* Bailey. Fruits d'Outre Mer 33:693-699.

ROBLES, M.J.M. 1979. Potential morphogenetique des entrenoeuds de *Passiflora edulis* var. *flavicarpa* Degener. et *P. mollissima* Bailey en culture *in vitro*. Turrialba 29:224.

ROJKIND, C., N. QUEZADA, and G. GUTIERREZ. 1982. Embryo culture of *Carica papaya, Carica cauliflora* and its hybrids *in vitro*. p. 763-764. In: A. Fujiwara (ed.), Plant tissue culture 1982. Japan. Assoc. Plant Tissue Culture, Tokyo.

ROWE, P. 1984. Breeding bananas and plantains. Plant Breed. Rev. 2:135-155.

SABHARWAL, P.S. 1962. *In vitro* culture of nucelli and embryos of *Citrus aurantiifolia* Swingle. p. 239- 243. In: P. Maheshwari (ed.), Plant embryology—A symposium. Council Sci. and Ind. Res., New Delhi.

SABHARWAL, P.S. 1963. *In vitro* culture of ovules, nucelli and embryos of *Citrus reticulata* var. *Nagpuri*, p. 265-274. In: P. Maheshwari and N.S. Rangaswamy (eds.), Plant tissue and organ culture—A symposium. Intern. Soc. Plant Morphologists, Delhi.

SAUTON, A., A. MOURAS, and A. LUTZ. 1982. Plant regeneration from citrus root meristems. J. Hort. Sci. 57:227-231.

SCHROEDER, C.A. 1961. Some morphological aspects of fruit tissues grown *in vitro*. Bot. Gaz. 122:198-204.

SCHWABE, W.W. 1973. The long, slow road to better coconut palms. Spectrum 103:9-10.

SCORZA, R. and J. JANICK. 1976. Tissue culture in *Passiflora*. Proc. Trop. Reg., Amer. Soc. Hort. Sci. 20:179-183.

SCORZA, L.B. and J. JANICK. 1980. *In vitro* flowering of *Passiflora suberosa* L. J. Amer. Soc. Hort. Sci. 105:892-897.

SINGH, L.B. 1960. The mango. Leonard Hill Ltd., London.

SINGH, U.P. 1963. Raising nucellar seedlings of some Rutaceae *in vitro*. p. 275–277. In: P. Maheswari and N.S. Rangaswamy (eds.), Plant tissue culture and a organ culture—A symposium. Intern. Soc. Plant Morphologists, Delhi.

SIPES, D.L. and J.W. EINSET. 1982. Role of ethylene in stimulating stylar abscission in pistil explants of lemons. Physiol. Plant. 56:6-10.

SITA, G.L., R. SINGH, and C.P.A. IYER. 1974. Plantlets through shoot tip cultures of pineapple. Curr. Sci. 43:724-725.

SKENE, K.G.M. and M. BARLASS. 1983. *In vitro* culture of abscissed immature avocado embryos. Ann. Bot. 52:667- 672.

SMITH, W.K. and L.H. JONES. 1970. Plant propagation through cell cultures. Chem. Ind. 44:1399.

SPIEGEL-ROY, P. and J. KOCHBA. 1973a. Mutation breeding in *Citrus*. p. 91-103. In: Induced mutations in vegetatively propagated plants. IAEA, Vienna.

SPIEGEL-ROY, P. and J. KOCHBA. 1973b. Tissue culture techniques in citrus. p. 29-32. In: O. Carpena (ed.), Congreso Mundial de Citricultura, Vol. 2. Min. de Agr., Murcia, Spain.

SPIEGEL-ROY, P., J. KOCHBA, and S. SAAD. 1983. Selection for tolerance to 2,4-dichlorophenoxyacetic acid in ovular callus of orange (*Citrus sinensis*). Z. Pflanzenphysiol. 109:41–48.

SRINIVASA RAO, N.K., R. DORESWAMY, and E.K. CHACKO. 1981. Differentiation of plantlets in hybrid embryo callus of pineapple. Scientia Hort. 15:235-238.

STARITSKY, G. 1970. Tissue culture of the oil palm (*Elaeis guineensis* Jacq.) as a tool for its vegetative propagation. Euphytica 19:288-292.

STARRANTINO, A. and F. RUSSO. 1980. Seedlings from undeveloped ovules of ripe fruits of polyembryonic citrus cultivars. HortScience 15:296-297.

STARRANTINO, A. and F. RUSSO. 1983. Reproduction of seedless orange cultivars from undeveloped ovules raised '*in vitro*'. Acta Hort. 131:253–258.

THANH-TUYEN, N.T. and E.V. DE GUZMAN. 1983. Formation of pollen embryos in cultured anthers of coconut (*Cocos nucifera* L.). Plant Sci. Lett. 29:81–88.

TISSERAT, B. 1979. Propagation of date palm (*Phoenix dactylifera* L.) *in vitro*. J. Expt. Bot. 30:1275–1283.

TISSERAT, B. 1981. Date palm tissue culture. p. 1-50. In: Adv. Agr. Tech., AAT-WR-17. USDA Agr. Res. Serv.

TISSERAT, B. 1982. Factors involved in the production of plantlets from date palm callus cultures. Euphytica 31:201-214.

TISSERAT, B. 1983. Tissue culture of date palms—A new method to propagate an ancient crop—and a short discussion of the California date industry. Principes 27:105-111.

TISSERAT, B. 1984. Propagation of date palms by shoot tip cultures. HortScience 19:230–231.

TISSERAT, B. and D.A. DE MASON. 1980. A histological study of development of adventive embryos in organ cultures of *Phoenix dactylifera* L. Ann. Bot. 46:465–472.

TISSERAT, B. and T. MURASHIGE. 1977a. Probable identity of substances in citrus that repress asexual embryogenesis. In Vitro 13:785–789.

TISSERAT, B. and T. MURASHIGE. 1977b. Repression of asexual embryogenesis *in vitro* by some plant growth regulators. In Vitro 13:799–805.

TONGDEE, S.C. and S. BOON-LONG. 1973. Proliferation of banana fruit tissues grown *in vitro*. Thailand J. Agr. Sci. 6:29–33.

TORRES, A.M. and B. TISSERAT. 1980. Leaf isozymes as genetic markers in date

palm. Amer. J. Bot. 67:162–167.
VARDI, A. 1977. Isolation of protoplasts in citrus. p. 575–578. Proc. Intern. Soc. Citriculture.
VARDI, A. 1981. Protoplast derived plants from different citrus species and cultivars. p. 149–152. Proc. Intern. Soc. Citriculture.
VARDI, A. and D. RAVEH. 1976. Cross feeder experiments between tobacco and orange protoplasts. Z. Pflanzenphysiol. 78:350–359.
VARDI, A., P. SPIEGEL-ROY, and E. GALUN. 1975. Citrus cell culture: Isolation of protoplasts, plating densities, effect of mutagens, and regeneration of embryos. Plant Sci. Lett. 4:231–236.
VARDI, A., P. SPIEGEL-ROY, and E. GALUN. 1982. Plant regeneration from Citrus protoplasts: variability in methodological requirements among cultivars and species. Theor. Appl. Genet. 62:171–176.
WAKASA, K. 1979. Variation in the plants differentiated from tissue culture of pineapple. Japan. J. Breed. 29:13–22.
WAKASA, K., K. YOSHIAKI, and K. MASAAKI. 1978. Differentiation from *in vitro* cultures of *Ananas comosus*. Japan. J. Breed. 28:113–121.
WANG, T. and C. CHANG. 1978. Triploid citrus plantlet from endosperm culture. p. 463–467. In: Proc. Symp. on Plant Tissue Culture, Peking. Science Press, Peking.
WOODROOF, J.G. 1979. Coconuts: Production, processing, products, 2nd ed. AVI Publ. Co., Westport, Connecticut.
WOOI, K.C., C.Y. WONG, and R.H.V. CORLEY. 1981. Tissue culture of palms—a review. p. 138–144. In: A.N. Rao (ed.), Tissue culture of economically important plants. COSTED, Singapore.
YIE, S. and S.I. LIAW. 1977. Plant regeneration from shoot tips and callus of papaya. In Vitro 13:564–567.
YOKOYAMA, T. and M. TAKEUCHI. 1981. The induction and formation of organs in callus cultures from twigs of mature Japanese persimmon (*Diospyros kaki* Thunb.). J. Japan. Soc. Hort. Sci. 49:557–562.
ZAID, A. and B. TISSERAT. 1983. Morphogenetic responses obtained from a variety of somatic explant tissues of date palm. Bot. Mag. (Tokyo) 96:67–73.
ZEPEDA, C. and Y. SAGAWA. 1981. *In vitro* propagation of pineapple. HortScience 16:495.
ZOHARY, D. and P. SPIEGEL-ROY. 1975. Beginnings of fruit growing in the old world. Science 189:319–327.

5

Cold Hardiness in Citrus

George Yelenosky
U.S. Department of Agriculture, Agricultural Research Service, Florida 32803

I. INTRODUCTION

Citrus is generally classified as a cold-tender evergreen with a tropical and subtropical origin (Webber *et al.* 1967), and its capacity to survive freezing temperatures does not approach that of northern woody plants, some of which exhibit −40°C and lower killing temperatures (Becwar and Burke 1982; George *et al.* 1974; Quamme *et al.* 1982; Sakai 1982). However, the continued existence of the citrus industry in the United States, with an annual on-tree value of more than $1 billion, attests to the potential of citrus to survive freeze situations (Cooper *et al.* 1969; Uphof 1938; Vasil'yev 1956; Webber 1895; Yelenosky 1977; Young 1963a). The manner and nature of this survival are so interwoven with and yet so different from those of other plants (Alden and Hermann 1971; Alleweldt 1969; Burke *et al.* 1976; Chandler 1913; Johansson 1970; Kaperska-Palacz 1978; Kenefick 1964; Larcher *et al.* 1973; Levitt 1968, 1980a; Luyet and Gehenio 1939; Olien 1981; Mayland and Cary 1970; Mazur 1969; Meryman 1966; Parker 1963; Proebsting 1978; Sakai 1962; Siminovitch *et al.*

Horticultural Reviews, Volume 7

ISBN 0-87055-492-1

1968; Weiser 1970; Weiser *et al.* 1979) that questions posed by Molisch (1897) on the causes and moment of lethal injury are still relevant to citrus frost hardiness. From the initial investigations by Molisch to present-day discussions on the freezing adaptation of plants (Siminovitch and Cloutier 1983), role of cell organelles (Steponkus *et al.* 1983), osmoregulation (Hendrix and Pierce 1983), and ice nucleation (Anderson *et al.* 1982; Franks *et al.* 1983; Lindow *et al.* 1982), the frost hardiness of citrus remains largely unexplained.

II. EXTENT OF FROST HARDINESS

Citrus species are vulnerable to freeze injury at −2.2°C and below. Yet, some of the most valued citrus crops are grown in relatively high-risk freeze areas throughout the world. In the moist, marine climatic zone of southern Japan, the freeze hazard is virtually continuous all winter, with minimum temperatures as low as −9.1°C (Ikeda *et al.* 1977). Considerable freeze risks also develop at intervals during the winter in the Caspian littoral, along the coast of the Black Sea, in the French Riviera, along the Adriatic seacoast in Yugoslavia, in the Italian Lyguria, and in parts of Turkey, Spain, Israel, Lebanon, and Greece—all important citrus-growing regions. However, some of the most severe freezes occur in the United States. The subtropical, arid to semiarid areas in California, Arizona, and the Lower Rio Grande Valley in Texas periodically experience devastating freezes to citrus (Cooper 1973), with temperatures as low as −10°C (Cooper *et al.* 1962; Young and Peynado 1963).

Severe freezes in Florida have the greatest economic impact on production of citrus and citrus products in the United States. Florida is in the humid semitropics, as are parts of China, India, South Africa, Australia, Argentina, and Brazil. Periodic severe freezes have threatened the very existence of commercial citriculture in Florida. The historic freezes of 1894–1895 caused extensive losses that essentially changed the northern limits of citrus growing in the United States (Webber 1895).

Since 1895, Florida has experienced temperatures of −5°C and lower on the average of once every 7 years; as a rule of thumb, Florida experiences a severe freeze of −6.7°C once every 10 years on the average. Despite past averages, severe freezes occurred in 1977, 1981, 1982, 1983, and 1985. There is, in fact, a 50% probability of freezing temperatures in central Florida from December to February (Bradley 1975).

In December 1962, prolonged temperatures of −6.7°C and minimums as low as −10.6°C cost Florida one-fourth of its 52 million citrus trees, with total losses exceeding $500 million (Hearn *et al.* 1963; Johnson 1963). Two highly unusual back-to-back severe freezes developed on January 12 and 13, 1981, and on January 11 and 12, 1982, with minimum temperatures as low as −13°C in 1981 (Yelenosky *et al.* 1981) and −10°C in 1982 (Buchanan *et al.* 1982), causing extensive damage to Florida citrus. It was largely the excellent condition of the trees and timely use of cold-protection cultural practices that kept losses below 40% of total annual production. However, the losses in fruit and damaged trees caused Florida to lose its world leadership role in citrus production to Brazil (Riemenschneider 1983).

Production problems were increased in Florida by another severe freeze situation on December 24, 25, and 26, 1983. Damage in certain areas equaled and may have even exceeded that caused by the devastating freeze of 1962. Much of the damage was related to the warm prefreeze temperatures, which kept trees in a nonhardened and freeze-vulnerable condition. The time of the freeze (Christmas Eve) and delays in freeze-warning forecasts due to unusual atmospheric conditions added to the usual difficulties in protecting groves. Little could be done to prevent devastating citrus losses in Marion, Lake, Orange, Volusia, Pasco, Hillsborough, Brevard, Polk, and Hardee counties, located above 28°N latitude. Revised Florida orange production figures in January 1984 indicated an overall decrease of 20–25% from the December 1, 1983, forecast of 168 million boxes. Production recovery from the 1983 freeze was severely set back during −8°C and lower temperatures that occurred in January 1985 with losses expected to equal those of 1983.

Many growers in Florida used water as a cold protectant rather than more costly heating (petroleum fuels) practices during the 1983 freeze. Water, which protected trees so well during the radiation-type severe freezes of 1981 and 1982 (Buchanan *et al.* 1982; Parsons, *et al.* 1982), was less effective during the advective Christmas freeze of 1983. My own observations of the 1983–1984 damage revealed that some protection was provided by sprinkler-irrigation systems, but that problems continue to exist in adapting sprinkler-irrigation practices to freeze protection during windy conditions. Major problems have occurred in supplying and maintaining critical distribution patterns and application rates. These difficulties also contributed to the ineffectiveness of overhead sprinklers during the 1962 freeze in Florida (Gerber and Martsolf 1965). In addition to the erratic effectiveness of sprinkler irrigation for cold protection, water applied in furrows under citrus trees is an effective method in California (Brewer 1978). The application of freeze-protection technology to orchard management draws upon basic concepts in atmospheric conditions, heat transfer, and tree physiology (Turrell 1973).

Protection methods are extremely valuable in the short term but are very sensitive to changing economic and social conditions. Development of cold-hardy citrus cultivars, through either classical breeding approaches (Barrett 1981) or nontraditional genetic manipulations (Kochba and Spiegel-Roy 1977), offers the possibility of long-term, stable protection against cold. The intent is to provide a few additional degrees of protection through inherent tree factors, and not to extend the northern limits of commercial citriculture. An increase of only 1°–2°C in cold hardiness, compared with that expressed in present commercial cultivars, would reduce tree and fruit losses by at least 10–20% in freeze-prone areas, and would often delay or eliminate the need for costly protection systems. The economic benefits of cold-hardy cultivars would be substantial, especially in high-risk freeze regions.

III. CULTIVAR DIFFERENCES

Essentially all commercial citrus trees are scion–rootstock combinations. Scion–rootstock interactions, along with differences in ambient environment, cultural practices, mass relationships, planting sites, and the health of the tree, complicate comparative ratings of citrus frost hardiness within, as well as among, freezes of different intensities and durations. Extrapolation of any given genetic potential to any given freeze situation is not yet satisfactory in citriculture. Generalities and broad guidelines have resulted from numerous observations over many years under various freeze situations and from partially controlled environment screening of potted citrus seedlings and young trees (Cooper 1959; Young and Hearn 1972; Young *et al.* 1977).

There are relatively large differences in frost hardiness among scion cultivars. In broad terms, the mandarin (*C. reticulata* Blanco) types are the most cold hardy, followed by sweet orange [*C. sinensis* (L.) Osbeck] and grapefruit (*C. paradisi* Macf.); lemons [*C. limon* (L.) Burm. f.] and limes (*C. aurantifolia* Christm. Swingle) are the least cold hardy. Extensive seedling screening was conducted in Texas after the 1961–1962 freezes (Cooper 1963; Furr *et al.* 1966; Young 1963a,b; Young and Olson 1963a,b), and controlled-environment screening followed shortly thereafter (Young 1969a; Young and Peynado 1967b).

Similar studies have been done on seedling rootstocks that were classified into six horticultural groups (Young 1970). Trifoliate orange [*Poncirus trifoliata* (L.) Raf.], a deciduous relative of *Citrus*, is a superior cold-hardy rootstock; sour orange (*C. aurantium* L.) is intermediate; and rough lemon (*C. jambhiri* Lush.) is one of the most cold-sensitive rootstocks. Different strains of trifoliate orange, as seedling trees, all seem to possess excellent cold hardiness under exceptionally cold (−16°C) field conditions (Yelenosky *et al.* 1968). Differences in freeze survival occur when different strains of trifoliate orange are used as rootstocks (Yelenosky *et al.* 1973). Seedling trees of *P. trifoliata* hybrids tend to retain but do not equal the cold hardiness of trifoliate orange seedlings. The citradia (*C. aurantium* × *P. trifoliata*) and citrumelo (*C. paradisi* × *P. trifoliata*) seedlings were more cold hardy than the citrange (*C. sinensis* × *P. trifoliata*) during three consecutive winters with minimum temperatures of −8.9°, −16.1°, and −7.8°C, respectively (Yelenosky *et al.* 1973). One of the selections, citrumelo 4475, was released to the industry in 1974 as 'Swingle' citrumelo rootstock (Hutchison 1974) and performed well as a cold-hardy rootstock for 'Valencia' [*C. sinensis* (L.) Osbeck] orange in controlled-environment studies (Yelenosky 1976a) and during the 1981 severe freeze in Florida (Yelenosky 1981b). During the 1981 freeze in Florida, when temperatures fell to −13°C, exceptional frost hardiness was observed in experimental selections of *Eremocitrus*, in hybrids of *P. trifoliata*, in progeny from hybridization of *C. reticulata* types, and in an open-pollinated, off-type seedling from citrumelo germplasm.

Only small to moderate changes can be expected in the frost hardiness of trees due to differences in rootstocks, and the influence of rootstocks on frost hardiness is most evident in older trees (Cooper 1952; Cooper *et al.* 1963a,b; Hearn *et al.* 1963; Gardner and Horanic 1963; Yelenosky and Young 1977; Young *et al.* 1960; Young and Olson 1963a). In general, the grafting of a cold-hardy scion, such as 'Satsuma' mandarin (*C. reticulata*), on the very cold-hardy trifoliate orange rootstock results in one of the most cold-hardy citrus trees in commercial citriculture. A cold-hardy scion is less cold hardy on a sensitive rootstock. Cold-sensitive 'Bearss' lemon (*C. latifolia*) on sensitive rough lemon rootstock is one of the more cold-sensitive trees (Castle 1983). The cold hardiness of citrus trees does not always directly relate to the cold hardiness of the rootstock. For example, trees on trifoliate orange seemingly express better freeze tolerance when cool temperatures prevail during 2 or more weeks immediately before a freeze. Otherwise, trees on trifoliate orange may not be any hardier than trees on rough lemon rootstock (Ziegler and Wolfe 1961).

Differences in cold acclimation and temperature-inducing bud dormancy are important factors in cold hardiness (Young 1970), as are differences in root system responses and development. The relatively small root system of trifoliate orange (Castle and Youtsey 1977) was considered to be a partial cause of the higher freeze mortality of mandarin trees on trifoliate orange, compared with the mortality of mandarin on rough lemon rootstock, during recurring freezes in one winter season (Yelenosky and Hearn 1967). Yoshimura (1967) suggested the use of rootstocks with deep-spreading root systems to increase cold protection in Japan. Temperature-dependent and diurnal root conductivity in citrus rootstocks (Wilcox and Davies 1981) is yet another variable that may partially determine the rootstock influence on citrus frost hardiness.

Considerable variation in frost hardiness among citrus cultivars has also been observed in Corsica, northern India, Japan, and the Georgia region of the USSR. In Corsica, a likely cold-sensitive rootstock hybrid, *C. volkameriana* ('Volkamer' lemon), was found (Blondel 1977) to be a more cold-hardy rootstock for citron scions (*C. medica* L.) than either the cold-hardy rootstock sour orange or the relatively cold-sensitive 'Alemow' (*C. macrophylla* Wester). 'Yuzu' (Ikeda *et al.* 1980; Kawase *et al.* 1982; Yoshida 1981), in Japan, and *Citrus juko* (Kokaya 1977), in the USSR, are probably cold-hardy *C. junos* Sieb. ex Tan. rootstocks resulting from cold-hardy *C. ichangensis* Swingle × *C. reticulata* (Wutscher 1979). Much of the work done in northern India (Gupta *et al.* 1978) has involved cultivars used in the United States, and similar observations on comparative cold hardiness have generally been obtained. Although many observations have been made in the USSR, they are not widely available (Alekseev 1978; Ivanov 1939; Leysle 1948; Vasil'yev 1956).

In summary, citrus cultivars can be ranked in the following order (Larcher *et al.* 1973) from the very cold hardy to the very cold sensitive: trifoliate orange (*P. trifoliata*), kumquat (*Fortunella* sp.), sour orange (*C. aurantium*), mandarin (*C. reticulata*), sweet orange (*C. sinensis*), grapefruit (*C. paradisi*), lemon (*C. limon*), and citron (*C. medica*).

IV. NATURE OF FROST HARDINESS

Frost hardiness in citrus is largely a function of frost avoidance and frost tolerance, concepts presented by Levitt (1980a) to provide some common ground of acceptable terminology for describing plant responses to environmental stresses. The words *frost* and *freeze* are used interchangeably in this review; freeze seems to be preferred in the United States and frost outside the United States (Larcher *et al.* 1973). Turrell (1973) discusses "frost-freeze" hardiness and helps to clarify points of contention that Levitt (1980a) hoped to circumvent by use of a common terminology relating to plant-environmental stresses. It is in this context that Levitt stated: "Whatever the stress, a plant may achieve resistance to it by either an avoidance or a tolerance mechanism. These two mechanisms of resistance may be developed at any one of three levels—the stress level, the elastic strain level, or the plastic strain level."

A. Frost Avoidance

Frost protection in citrus crops, as in other crops, largely reflects continuing progress and development in the science and technology of passive frost avoidance (Buchanan *et al.* 1982; Ikeda 1982; Gerber and Martsolf 1965; Parsons *et al.* 1982; Martsolf 1982; Turrell 1973; Yelenosky 1981a). Less important is active frost avoidance, or supercooling, in plants (Camp 1965; Luyet and Gehenio 1940; Marcellos and Single 1979; Quamme 1978; Rajashekar *et al.* 1982). The amount and duration of supercooling during natural freezes is not well documented in citriculture.

For practical purposes, citrus remains uninjured if ice does not form in the tissues. This does not include the fruit peel, which is chill sensitive or injured at temperatures above 0°C (Eaks 1960; Purvis 1980). Thus, any mechanism that prevents or delays ice formation in citrus tissues essentially increases frost hardiness in commercial citriculture. On the basis of observations during natural freezes, citrus leaves are classified as chill resistant (Graham and Patterson 1982), although chill injury can be induced with prolonged exposure to 1.7°C and continuous light (Yelenosky 1982a). However, such conditions are unknown in major citrus-growing areas. Ice avoidance as a result of freezing point depression (due to solute accumulation and/or decreases in tissue water) is considered less of a factor in the cold hardiness of citrus than either passive frost-protection methods or active supercooling; ice avoidance because of freezing point depression probably does not account for more than a 1°C added protection.

The merits of passive frost protection in limiting freeze damage are readily apparent in the viable citrus industries of the United States and Japan, which continue to be major world suppliers of citrus products regardless of periodic severe freezes. In contrast, the significance of supercooling in world citrus production is not so apparent. Indeed, the relevance of supercooling may lie in its relationship to evolutionary trends in frost acclimation of vascular plants (Larcher 1982) and to the distribution of woody plants (George *et al.* 1982; Kaku and Iwaya 1978; Quamme *et al.* 1982). It is largely differences in the depth of supercooling that separate subtropical citrus from temperate woody plants. Most citrus species rarely supercool below −10°C, although *Poncirus* may supercool appreciably below −10°C (Table 5.1). In contrast, temperate woody plants frequently approach homogeneous nucleation temperatures for plant tissue water, about −40°C±, a phenomenon commonly referred to as "deep" supercooling or undercooling of plants (George *et al.* 1982; Rajashekar *et al.* 1982a,b; Sakai 1982).

Supercooling in citrus is of utmost importance since the inherent ability to supercool is sufficiently great to avoid major freezing in areas where citrus is commercially grown. Flowers and fruit probably are the tissues that supercool the least in citrus (excluding succulent new leaves) during natural freezes. In controlled-temperature studies, citrus flowers supercooled to −4.3°C (G. Yelenosky, unpublished data) and fruit to −5°C (Hendershott 1962b). Woody tissues (stems) probably supercool the most in citrus; they have been shown to supercool as low as −8.9°C in 10-month-old sour orange seedlings (Yelenosky and Horanic 1969). In this instance, differences in supercooling were not adequate to index cultivar differences in frost hardiness among unhardened citrus seedlings. The amount of supercooling in citrus tissues at different stages of development is unresolved.

Much of the information on supercooling in citrus comes from freez-

TABLE 5.1. Apparent Supercooling in Citrus Types

Selection	Minimum temperature (°C)	Reference
C. medica	−4	Larcher 1971
C. limon	−5	Larcher 1971
C. paradisi	−5	Larcher 1971
cv. 'Redblush'	−3.9 for 4 hr	Young and Peynado 1965
C. sinensis	−5	Larcher 1971
cv. 'Pineapple'	−3.3 to −4.4 for 10 hr	Hendershott 1962
cv. 'Valencia' stems	−6.2 for 3 hr	Yelenosky 1976b
C. aurantium	−6	Larcher 1971
C. reticulata	−6	Larcher 1971
cv. 'Dancy'	−6.1	Young 1963a
cv. 'Satsuma'	−7.8	Magness and Traub 1941
cv. 'Satsuma'	−9.4	G. Yelenosky, unpublished data
Fortunella margarita	−7	Larcher 1971
Poncirus trifoliata × *C. grandis*	−7.8 for 8 hr	G. Yelenosky, unpublished data
P. trifoliata × *C. sinensis*	−11	Young 1963a
P. trifoliata		
leaves	−15	G. Yelenosky, unpublished data
stems	−20	Larcher 1971

Source: Yelenosky (1977).

ing point determinations in which citrus leaves supercooled to −8.3°C (Jackson and Gerber 1963) and detached lemon vesicles supercooled to −12°C (Lucas 1954). Differences in methodology are critical in comparative determinations (Young 1966).

Regardless of the many observations of supercooling in citrus during controlled-temperature scans, very little is known about the mechanism of supercooling in citrus. It tends to be a random event with a bell-shaped distribution pattern (Yelenosky and Horanic 1969). Supercooling probably occurs in all parts of a citrus tree and can be increased with water stress (Fig. 5.1) and temperature cold-hardening regimes (Young and Peynado 1965). Before 1975, supercooling was largely considered to be a relatively unimportant factor in citrus freeze survival. It was generally accepted that advective parameters characteristic of most severe freezes in citrus orchards largely precluded any significant amount of supercooling in citrus trees, which were mostly observed after damaging freezes had occurred. But there were occasional reports and observations that gave more credence to supercooling as a significant factor in citrus frost hardiness.

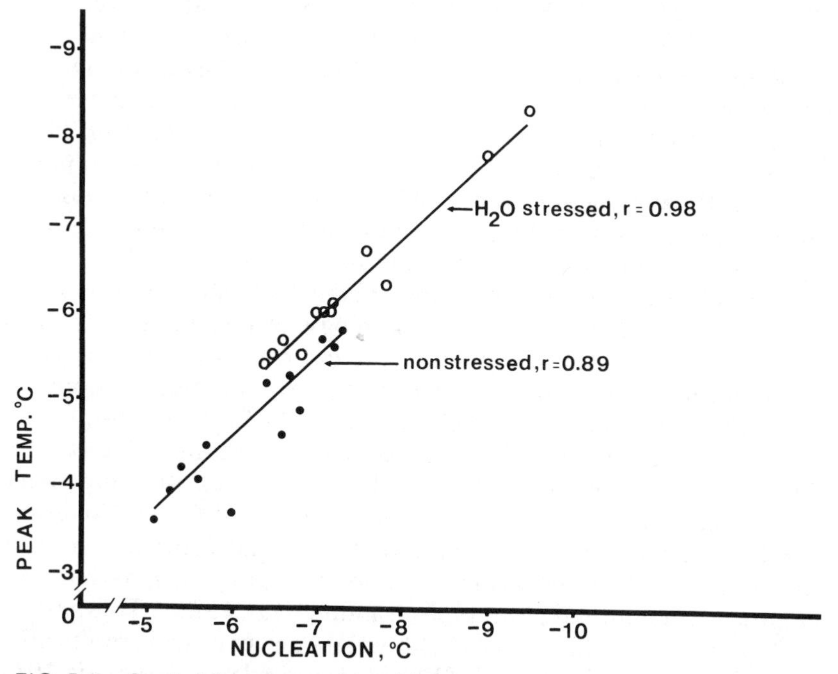

FIG. 5.1. Start of freezing and resulting heat evolution shown in exotherms of citron (*Citrus medica*) seedlings during freeze avoidance tests.
From Yelenosky (1979c).

The concept of ice-nucleation-active (INA) bacteria brought supercooling to the forefront in the discussions of frost hardiness in citrus, as well as in many other agricultural crops. Major national and international attention was soon focused on INA bacteria as the primary factor that induced ice formation in plants during freeze conditions. Lindow (1982a) was the catalyst in proposing that INA bacteria, especially species of *Pseudomonas* and *Erwinia*, were not only necessary but were present in sufficient number to account for the frost sensitivity of agricultural crops, especially in the temperate climatic zones. Some of the most active INA bacteria thought to act as heterogeneous ice nuclei that efficiently "triggered" the transition of water to ice at −1° to −2°C included strains of *P. syringae* pv. *syringa* van Hall, *P. fluorescens, E. herbicola* (Lohnis) Dye, and *E. stewartii* (Smith) Dye.

From the National Collection of Plant-Pathogenic Bacteria (Harpenden, England) Hirano *et al.* (1978) tested nine different strains of *P. syringae*; five strains showed ice nucleation activity at −5.5°C, one strain at −10°C, and one strain was inactive. Isolations from citrus plants in Israel resulted in two distinct bacteria with ice nucleation activity at −2.5°C (Yankofsky *et al.* 1981). Additional positive isolations of INA bacteria continued to be made from citrus and other major agricultural plants (Lindow 1982b). These results, along with observations of increased freezing injury induced by INA bacteria on tomato and soybeans (Anderson *et al.* 1982), maize (Wallin *et al.* 1979), snap beans and corn (Lindow *et al.* 1978), and peach, cherry, plum, and pear (Proebsting and Andrews 1982), helped to rekindle industry and scientific interest in the bioregulation of freezing stress, not just in citrus but in all major agricultural plants.

The implications of these findings about INA bacteria were enormous both to agriculture and meteorology. Atmospheric scientists were concerned about atmospheric ice nuclei from decomposing vegetation (Schnell and Vali 1972). The apparently abundant supply of freezing nuclei in the tropics (Schnell and Vali 1973), and the demonstration of biogenic ice nuclei obtained from marine as well as terrestrial sources (Schnell and Vali 1976), helped to focus attention on ice nuclei in agriculture. Leaf washings obtained in California, Colorado, Florida, Louisiana, and Wisconsin indicated that conifers were less likely to harbor INA bacteria than woody angiosperms, herbaceous perennials, and annual plant types (Lindow *et al.* 1978b). This survey also found low populations of INA bacteria on citrus trees in Florida. Researchers at the University of Florida initiated new surveys in 1984 to determine trends in seasonal populations in different citrus groves.

Lindow (1982a) raised many questions of concern to citriculture in the early reports of epiphytic INA bacteria in agriculture. The nucleation frequency (the number of active ice nuclei per total population of bacterial cells) was especially disturbing, since it was not an invariant and intrinsic property, but varied with many factors identifiable in cell culture (Lindow *et al.* 1982a). It was also relatively certain that high populations of INA bacteria must occur on citrus leaves in the field because nucleation frequency was as low as 1 ice nucleus per 5×10^3 bacterial cells (Lindow 1982b). Although definitive data were lacking, the assumption was made that critical populations of INA bacteria existed on citrus trees during natural freezes. Thus, the methods and cost of controlling INA bacteria became relevant to citriculture.

Three general approaches currently are available to control INA bacteria (Lindow 1983): (1) the use of bactericides, such as streptomycin, to reduce bacterial populations on trees; (2) the culturing and spraying of non-INA antagonistic bacteria on trees to occupy sites that otherwise would harbor INA bacteria; and (3) the use of inhibitors that stop the yet-unclear mechanisms of bacterial ice nucleation. A fourth approach involves the use of viruses to reduce INA bacteria populations (R. C. Schnell, personal communication). Whether one of these approaches, some combination, or a new approach will succeed in citriculture remains speculative since research on INA bacteria as a frost hazard to citrus trees is still in the initial stages (Anderson and Buchanan 1983; Yelenosky 1983).

In 1983, a *Pseudomonas syringae* bacterium, genetically altered in Lindow's laboratory (personal communication) at the University of California at Berkeley, was proposed for field tests on potato plants (*Solanum tuberosum* L.), which are seriously damaged at −2.5°C (Li and Palta 1978; Rajashekar *et al.* 1983). The altered bacterium shows no ice nucleation activity to about −5°C, and, in theory, it will successfully compete for sites on plants otherwise available to INA strains. (Since citrus trees need to avoid −6.7°C and colder freezes, which periodically devastate the industry, this particular bacterium would not be too helpful in citrus.) At the time of this writing, the field tests on potato were stopped through court petitions because of concerns of the petitioner about introducing gene-engineered bacteria into the ambient environment. The eventual outcome of this legal action, probably will set legal precedents concerning the utilization of altered life forms in agriculture.

In the meantime, bactericides may be the most practical approach for the control of INA bacteria in the citrus industry, although the importance of INA bacteria has not yet been adequately established in major freeze-prone, citrus-growing areas.

B. Frost Tolerance

Frost tolerance in citrus is considered synonymous with ice tolerance in this review. The moment and temperature at which ice crystals form in citrus tissues, the rate of ice spread, the duration of freezing temperatures, and the minimum temperature are considered to be significant variables.

The onset of freezing is generally determined under laboratory-type or controlled-temperature conditions using differential thermal or exothermic analyses. Such analyses essentially profile supercooling and the heat of crystallization of tissue water (latent heat of fusion). The exothermic increase in temperature can be as much as 3°C in stems of freezing citrus seedlings (Horanic and Yelenosky 1969; Yelenosky and Horanic 1969). In some instances, greater sensitivity in detecting the onset of freezing can be achieved by profiling electrical potential (EP) in stems of citrus seedlings (Fig. 5.2). Such techniques are not well suited nor easily adapted to highly unstable freeze conditions in the field. However, they are useful in estimating freezing points and lethal temperatures at which 50% of the total kill occurs (LT_{50}) in citrus plants (Gerber and Hashemi 1965; Rouse and Wiltbank 1972; Young 1966), as well as in estimating extremely low exotherms in woody plants (George *et al.* 1974).

The appearance of water-soaked tissue is probably the most acceptable practical alternative to differential thermal analyses (DTA) for determining the onset of freezing in citrus trees during natural freezes. It is often used as an indicator of freeze severity and is based on visual observations of darkened areas that develop when citrus tissue water freezes (Young and Peynado 1967a). Water-soaking seemingly is unique to freeze situations, but photographs of water-soaked tissue closely resemble photographs of windburn injury of orange leaves (Reed and Bartholomew 1930). Citrus water soaking is usually first observed adjacent to the midvein of leaves; freezing in interveinal areas of citrus leaves is largely the result of multiple nucleation sites that coalesce with time and minimum temperature. Eventually, ice propagates throughout the entire leaf. The breakdown in supercooling usually occurs above −6.7°C. The apparent multiple nucleation sites and the delay in lateral ice spread suggest that external heterogeneous INA bacteria are present and that there are momentary barriers to ice spread in citrus leaves. It also is unclear whether ice forms first in the leaves of citrus and spreads into the wood and fruit, or whether leaves, wood, and fruit nucleate individually with a meeting of ice fronts during severe freezes to kill entire trees above ground level.

Lucas (1954) found that the rate of ice spread from one lemon fruit to

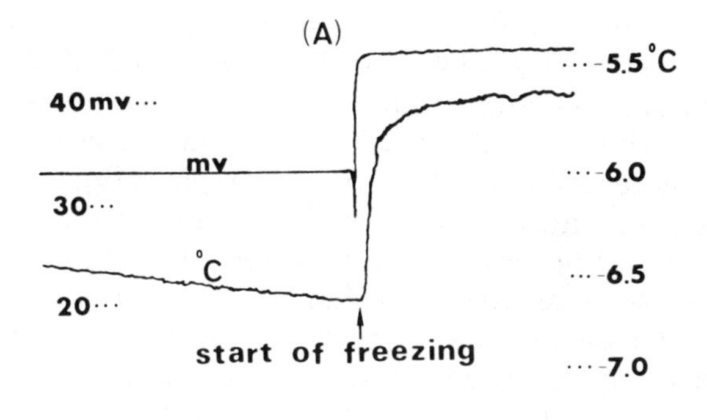

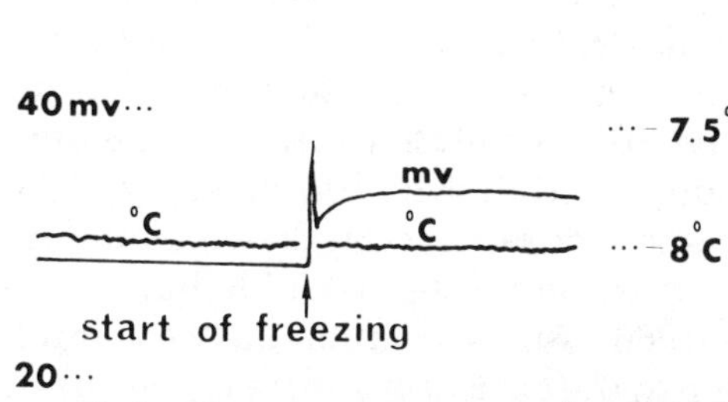

FIG. 5.2. Tracings of simultaneous measurements of electrical potential signals (mv) and heat of fusion (° C) to determine ice nucleation in citrus seedlings. A—Unhardened stems during a temperature decrease of 4° C/hr (chart speed of 2.54 cm/min). B—Cold-hardened stems during a temperature decrease of 1° C/hr; note that thermocouple technique (° C) failed to detect freezing indicated by electrical potential.
From Yelenosky et al. (1970b).

another fruit on the same branchlet was greater than 15 cm/min (about 30 m in 3.3 hr) at −5.2°C. A "domino" effect from a single nucleation site, as postulated by Lucas, has not yet been substantiated in the field. However, his suspicion that faster rates of crystallization can occur was supported in freeze tests of 2-year-old 'Marsh' grapefruit trees at −6.4°C (Yelenosky 1975). In this instance, rate of ice spread was estimated at about 74 cm/min in stems of unhardened trees. This rate was reduced to about one-third in cold-hardened trees. Even these limited attempts to target the onset of freezing in citrus provide sufficient data to support the presence of rapid rates of crystallization, multiple nucleation sites, and natural mechanisms that decrease the rate of ice spread in citrus trees.

Freezing in citrus tissues is largely confined to a single exotherm, with lethal injury usually occurring between −2.2° and −6.7°C (Young

1969). Lethal injury in other woody plants may not occur until the second (McLeester *et al.* 1969), third (Graham and Mullin 1976), or fourth exotherm (Quamme *et al.* 1972), at temperatures as cold as −30°C. These and other observations (Rajashekar *et al.* 1982a) suggest partial segregation of cold hardiness in plant types according to the lethal exotherm and a readily available water-continuum in citrus. In general, freezing profiles (Yelenosky 1975) reflect many unanswered questions in citrus frost hardiness. The abrupt increase in temperature "triggered" by a yet-unresolved mechanism and followed by different rates of net heat loss suggests continuing freezing of additional tissue water after initial crystallization. Whether continued freezing causes massing of ice in citrus stems sufficient to split the bark (Fig. 5.3) is not known. This massing of ice seemingly can be reduced with imposed water stress (Yelenosky 1979c), as well as by cold-hardening temperature regimes (Yelenosky 1975). The relevance, if any, of the duration of exotherm to final lethal injury in citrus is largely speculative, as is its relationship to such plant-freezing factors as (1) the imposing of lethal dehydration stresses (Olien 1981; Williams 1981), (2) tissue separation (Hatakeyama and Kato 1965; Single and Marcellos 1981), and (3) the risk of lethal intracellular freezing (Meryman 1956, 1966). An association has been implied between apparent ice buildup in the bark of citrus trees and bark splitting during freezes (Yelenosky 1979b).

It is suspected that once ice has formed in citrus tissues, a critical amount of time, inversely related to minimum temperature, is required before irreversible injury is induced. Lethal injury may be induced within 3–5 min after the initial surge of ice crystallization in unhardened citrus seedlings (Yelenosky and Horanic 1969). In other instances, leaves and wood killed in young citrus trees at −6°C can be discerned 10 min after ice forms (Yelenosky 1976b). Among different tissues, succulent flowers and new growth with higher than 85% tissue water content, on a fresh-weight basis, are the most vulnerable to ice damage. Increasing tolerance to ice is exhibited, in order, by leaves, buds, cambium, and xylem, respectively (Larcher *et al.* 1973). Within tissues, it is largely a matter of cold hardening that determines ice tolerance at different minimum temperatures and duration. The assumption that most of the ice formed in citrus tissues during natural freezes is largely the result of "equilibrium" freezing is based on observations of (1) relatively slow rates of temperature decrease, usually less than 2°C/hr, in the field and (2) instances where citrus tissues are not visibly injured by ice (Yelenosky 1976b; Young and Peynado 1967a). Such observations are not compatible with intracellular freezing, which is considered lethal almost instantaneously

FIG. 5.3. Bark-splitting on the trunk and limbs of a 4-year-old Hamlin' orange tree after −4.4° C for 1 hr near Dade City, Florida, on January 3, 1979. Splits were covered with a protective asphalt paint to aid healing.
From Yelenosky (1979b).

after nucleation (Meryman 1956, 1966). However, intracellular freezing can be implicated in the severe freezing injury of citrus flowers and new leaves that occurs virtually instantaneously with ice nucleation. In severely frozen grapefruit leaves, photosynthesis (measured by CO_2 uptake) virtually stops after injury at −6.7°C for 4 hr, and severe cellular disorganization in sour orange leaves ensues (Young 1969c; Young and Mann 1974).

Protective mechanisms against ice-imposed injury seemingly have evolved in citrus and contribute, along with protective cultural practices, to the survival of citrus trees during severe episodic freezes. Although these natural mechanisms are not well understood, they apparently involve adaptive changes in citrus physiology and metabolism that result in changes in cellular composition and physical relationships; these changes are collectively referred to by the term *cold hardening* (*frost hardening*).

Cold hardening in citrus is minimal relative to that in northern woody plants, which can survive at −25°C and colder; commercial citrus does well to survive at −6.7°C for 4 hr without devastating damage. Commercial citrus trees are generally considered cold-sensitive, subtropical evergreens. Their ability to survive minimum winter temperatures of −8.9°C in 1977 (Yelenosky and Young 1977) and −13.3°C in 1981 (Yelenosky *et al.* 1981), and still be rated commercially productive in 1983, is impressive. Less severe freeze conditions in the winter of 1962 (Hearn *et al.* 1962) were devastating.

Citrus trees are most cold hardy during the winter when growth is "arrested" or quiescent, provided the trees are in good health (Garnsey 1982). For practical purposes, citrus trees seemingly stop growing when ambient air temperatures are below 12°C (Bain 1949; Young and Peynado 1962). Based on temperature–growth differences, Young (1962, 1970) separated 24 cultivars of citrus into six horticultural groups, using bud dormancy–inducing temperatures between 15° and 15.6°C for seedlings of trifoliate orange; 8.9° and 10°C for citrumelos, limes, and lemons; and 8.9° and 15.6°C for mandarins and citranges. The relative brief stage of arrested or quiescent growth during winter conditions helps to predispose citrus trees to cold-hardening temperatures.

The relationship of climate (Young 1963b; Cooper *et al.* 1963b) and winter temperatures (Cooper and Peynado 1959) to citrus cold hardiness suggests that this phenomenon involves a three-stage process.

1. Predisposition of the trees to cold-hardening mechanisms optimized through good tree health
2. Prehardening of the trees induced largely by cool but nonfreezing temperatures through favorable weather changes preceding damaging freezes
3. Cold hardening of the trees induced by cellular responses to cooler temperatures through favorable genetic traits in temperature-sensitive systems in tree physiology and metabolism.

Since each stage of the phase process apparently feeds into the stage immediately following, the depth of citrus cold hardening will largely reflect the weakest link in the sequence of events. The importance of each stage in practical citriculture is dramatically illustrated by the survival of budded 'Satsuma' orange trees on trifoliate orange rootstock located near 31°15′ N latitude (Yelenosky 1978b). Trees survived −10°C the first winter after budding and −8.9°C the following winter. But, during the third winter, the trees were all killed, although minimum winter temperatures did not fall below −6.7°C. In this situation, the weakest link in the cold-hardening process was thought to be the prehardening stage during which prefreeze temperatures averaged as much as 6.1°C higher than they did in the first and second winters. Differences in winter-hardening temperatures tend to be reflected in the spring bud break in citrus and related genera, with the quickest bud break occurring after cold winter-hardening temperatures of at least 752 hr at or below 7.2°C (Young 1981). Thus, a favorable cold-hardening winter could conceivably increase the risk of late winter or early spring freeze injury to early flushes of flowers and leaves.

The associations of citrus cold hardiness with temperature in the field, as well as in controlled-environment facilities, indicate that citrus cold hardening is largely a function of temperature. However, light is also important (Niijar and Sites 1959; Yelenosky 1971a; Young 1961, 1969b); water relations have an effect (Yelenosky 1979c); and anatomical differences may be significant (Salazar 1966). The prehardening temperature effects that lead to different cold-hardened states are associated with increases in solute accumulation coupled with decreased water uptake. This suggests more closely knit liquid-solute interfaces. All of these events are considered temperature and time dependent. Presumably, smaller and less numerous ice loci would form and tissue dehydration would be less. Also, one might expect greater compartmentalization or isolation of ice crystals from critical cell sites. It remains to be determined how such generalizations will eventually be related to concepts about the protective systems that have evolved in citrus and such biophysical phenomena as the free energy of ice growth in plants (Olien and Smith 1981). In addition, there are osmotic relationships in frost desiccation that take into account the properties of colligative cryoprotection, water binding, mechanical resistance, and critical minimum cell-volume (Williams 1981).

Nuclear magnetic resonance (NMR) analyses are providing new and significant information about citrus frost hardiness. The suspicion that water binding occurs during cold hardening is supported by NMR analyses that indicated cold-hardened 'Satsuma' mandarin leaves had larger percentages of unfrozen water than nonhardened leaves during initial freezing and subsequent durations (Rouse *et al.* 1982). In a more recent NMR study, leaves from detached terminal shoots of lemon, grapefruit, orange, and mandarin cultivars ranged in cold hardiness from −4° to −11°C with no apparent differences in either water content or melting point depression (Anderson *et al.*

1983). These workers observed differences among cultivars in the amount of ice formed in the leaves that could not be accounted for by osmotic effects. Departures from ideal freezing curves indicated the presence of negative pressure potentials, which in plant-water relations expresses cell resistance to collapse or cell wall rigidity. It was concluded that differences in citrus frost hardiness can best be explained by differences in the amount of frozen water tolerated at critical temperatures. A similar conclusion was reached in another NMR study on the cold hardiness of *Solanum* species (Chen *et al.* 1976).

Stout (1981) has suggested that frozen water fractions are a preferred indicator in determining lethal dehydration strains in plants if cell injury is due to solution effects. In such instances, simplified calculations are proposed for determining cell water content during freezing (Farhy 1981). Negative pressure potential may very well be the most definitive factor yet to be considered in citrus frost hardiness. Previously, citrus frost hardiness has been viewed largely from associated rather than causal factor relationships.

In one way or another, all of the major categories of plant components (carbohydrates, proteins, lipids, and water) have been implicated in citrus cold hardiness (Young 1970; Nordby and Yelenosky 1982). The association of carbohydrates with citrus frost hardiness is based largely on the rapid accumulation of carbohydrates in the leaves and wood citrus during temperature cold hardening. The most rapid rates of accumulation seemingly occur between 15° and 5°C (Yelenosky and Guy 1977), thus coinciding with temperature regimes used to screen cold-hardy citrus selections (Young and Hearn 1972). The sugar fraction, especially sucrose, is principally associated with citrus cold hardiness. The association between sucrose and cold hardiness is virtually universal in plants (Levitt 1980b), and sucrose has been targeted as a possible translocatable promoter of frost hardiness in *Hedera helix* (Steponkus and Lanphear 1967). But, like many other suggested relationships, increases in sugar concentrations do not always correlate well with degrees of frost hardiness in citrus trees.

However, the relationship between sugar increases and citrus frost hardiness is supported by several lines of evidence. High sugar concentrations found in citrus tissues are not injurious, although they may have impact on growth regulator–enzyme mechanisms via "osmotic repression." Sugars also readily enter into water-binding osmotic roles that have impact on colligative properties in freezing point depression and influence the influx and efflux of cellular water, which is critical in dehydration stress concepts. Estimated dehydration to a preset level in citrus leaves, as a function of sucrose concentrations at different temperatures, is within the cold hardiness range of citrus (Yelenosky 1978a,b). Also, there are cryobiological data to support contentions that sugars protect cell membranes during freeze stress (Heber 1965, 1968; Santarius and Bauer 1983), that cell membranes help to prevent lethal intracellular freezing in the presence of extracellular ice at temperatures as low as −10°C (Mazur 1965), that the molecular weight of sugars influences the rate of ice spread (Yelenosky 1971b), and that changes in solutes induce changes in water potential, which in turn influence activation of enzyme systems (Darbyshire and Steer 1972). In studies on the distribution of ^{14}C photosynthetic assimilates, a greater retention of ^{14}C was found in the sugar fraction in the leaves of 'Valencia' orange seedlings at 10°C than at 25°C (Guy *et al.* 1981). One-year-old 'Valencia' trees that had 20:1 sugar:starch ratios in the leaves and 17:1 in the wood withstood −6.7°C without injury, whereas trees with 2:1 and 1:1 sugar:starch ratios were killed (Yelenosky 1978a). In this study, proline was also found to accumulate in 'Valencia' orange leaves at 10°C and decreases in water potential were observed. Both sugars and proline are also implicated in preventing chill injury in grapefruits, which acquire characteristics of cold-hardening concomitant with vegetative tissues (Purvis and Yelenosky 1982). Sakai and Yoshida (1968) vividly demonstrated the association of carbohydrates (sugars and starch) as well as proteins in the seasonal cold hardiness of woody plants.

The association of proteins with frost hardiness has not been as well documented in citrus as it has in other species. For example, in black locust, rapid increases in protein concentrations occur during the winter season (Siminovitch *et al.* 1968) and the structure of ribosomal patterns is altered during induction of cold hardiness (Bixby and Brown 1975); in apple, nucleic acids increase during cooler temperatures (Li and Weiser 1969); and Levitt's sulfhydryl hypothesis on low-temperature protein aggregation because of the formation of intermolecular disulfide bonds has been supported in non-citrus species (Huner *et al.* 1982). In contrast, cold-hardening regimes did not appreciably change the concentration of water-soluble proteins in 'Redblush' grapefruit leaves (Young 1969b). However, water-soluble proteins did accumulate in leaves of 'Hamlin' sweet orange trees exposed to cool night temperatures, although protein concentrations did not correlate well with freezing point determinations in the leaves (Ghazaleh and Hendershott 1967). The appearance of new protein fractions associated with temperature cold hardening has not been demonstrated in citrus, but greater protein denaturation in unhardened than in cold-hardened 'Valencia' orange leaves after damaging freezes was detected by Yelenosky and Guy (1982a).

Free proline, often associated with citrus frost hardiness, is one of three amino acids found to increase during low-temperature cold hardening (Yelenosky 1978a). Partitioning of [^{15}N] ammonium uptake supports the formation of active free proline in citrus during low temperatures (Kato and Kuboto 1982), and free proline apparently is not involved in any hydroxylation during cold-hardening temperatures (Yelenosky and Gilbert 1974).

Proline was the most abundant amino acid found in the tracheal sap of orange trees during the entire year and was especially high in concentration during autumn and winter (Moreno and Garcia-Martinez 1983). Free proline concentrations significantly increased in leaves of citrus trees on a wide range of different rootstocks during temperature cold hardening, and proline accumulation is closely allied with sugar accumulation (Yelenosky 1979a) and water-stress-induced cold hardening of young citrus trees (Yelenosky 1979c). With more mature trees, proline concentrations were generally higher in nonirrigated than irrigated trees, but proline concentrations did not correlate with water stress in the leaves of fruit in short-term studies (Syvertsen and Smith 1983). During longer periods of water stress, proline does accumulate in lemon trees (Levy 1980). Like sugars, proline is not known to be membrane toxic (Heber *et al.* 1971), but whether proline concentrations are high enough to help protect citrus against frost injury remains unclear. Proline may also have some role in moderating salt stress (Hanson and Scott 1980; Kappen *et al.* 1978), which has been implicated in citrus freeze injury where trees are grown under high salt conditions (Peynado 1982).

Of all the main plant components implicated in citrus frost hardiness, probably the least is known about the role of lipids. However, the association of lipids with frost hardiness merits attention based on the involvement of lipid metabolism in the adaptation of plants to their environment (Kuiper 1980). The role of lipids, which is largely related to the structure and function of plant membranes, ranges from the apparent importance of augmentation (Horvath *et al.* 1980; Siminovitch *et al.* 1975) to the apparent nonimportance of membrane lipid fluidity and phase transition temperatures (Pomeroy and Raison 1981) in the freezing of plants. Initial investigations with citrus show characteristic increases in linoleic acid and increased levels of lipid nonsaturation associated with cold-hardened rootstock seedlings (Nordby and Yelenosky 1982) and 8-month-old Valencia' trees on sour orange rootstock (Nordby and Yelenosky 1984). The relatively large increases in triglycerides in citrus leaves during cold hardening were not expected, and now it is thought that triglycerides may have some role in maintaining membrane integrity during citrus freezing and may be major inclusions of osmiophilic cell wall deposits (Griffith and Brown 1982) or osmiophilic globuli found in citrus leaves (Young and Mann 1974). Degradation of lipids in freeze-damaged citrus has not yet been evaluated, but it has been suggested as a possible alternative for indexing plant cold hardiness in instances where lipid differences cannot be found in widely different cold-hardy cultivars (Willemot 1983).

Cold acclimation in citrus is also characterized by increases in reduced glutathione content in leaves of orange trees during low but nonfreezing temperatures (Guy and Carter 1982). However, the speculation that accumulation of reduced glutathione plays a significant role in citrus freeze tolerance has not been borne out in recent studies (Guy *et al.* 1984).

V. BREEDING FOR IMPROVED COLD HARDINESS

New developments in genetic engineering offer exciting prospects for developing cold-hardy citrus trees with commercially acceptable fruit, but there are considerable obstacles to adapting recombinant DNA techniques to practical citriculture. Thus, cultivar improvement for increased frost hardiness will most likely continue to depend on more traditional breeding approaches, utilizing new technology whenever possible. Techniques that are being used include identity of special cold-hardiness traits in citrus selections through low-temperature screening procedures (Young and Hearn 1972), the isolation and preservation of these traits through cell and tissue culture (Murashige 1974; Kitto and Young 1981; Kochba and Spiegel-Roy 1977), additional genomic alterations through colchicine-induced polyploidy (Barrett 1974), and X-ray mutations (Gregory and Gregory 1965). Broadening the genetic base by introducing new germplasm from field explorations is being actively pursued in the long-range development of frost hardiness in citrus.

Interspecific and intergeneric hybridization (Barrett 1977, 1981; Barrett and Rhodes 1976) of the highly heterozygous citrus is slowly producing desired recombinations with increased frost hardiness. Some of the more promising selections have resulted from crossing grapefruit with trifoliate orange and backcrossing to sweet orange [(*C. paradisi* × *P. trifoliata*) × *C. sinensis*]. *The cold hardiness of P. trifoliata* is thus combined with acceptable fruit quality from grapefruit and sweet orange types. Improvement in fruit quality is important since unacceptable fruit quality has been a major obstacle in developing cold-hardy citrus trees ever since the initial attempts were made in 1897 (Webber and Swingle 1904).

The initial intergeneric crosses of the very cold-hardy *P. trifoliata* with the more cold-tender, edible *C. sinensis* and *C. paradisi* resulted in numerous *P. trifoliata* hybrids, such as the citranges and the citrumelos, which were relatively cold hardy but had unacceptable fruit quality (Cooper *et al.* 1962b). Although these trifoliate orange hybrids were of little or no value as scion cultivars, a few have developed into important rootstocks, some more cold-hardy than others. One example is the 'Swingle' citrumelo rootstock, released by the USDA in 1974. This release was hybridized in 1907 and resulted from crossing grapefruit with trifoliate orange, *C. paradisi* × *P. trifoliata* (Hutchison 1974). The seedling, which was eventually named and released as the 'Swingle' rootstock, was originally bred to produce a cold-hardy scion cultivar, but this failed due to unacceptable fruit quality. In the mid-1940s, 'Swingle' was included in extensive rootstock trials in Florida, which lasted for several decades. Thus, an extraordinary span of 67 years elapsed from hybridization to commercial release, a process that normally takes no more than 30 years.

Delays in breeding improved cultivars also result from geographical differences, communication barriers, and nonrecognition of desired germplasm. For example, *C. juko* is considered a promising cold-hardy rootstock in subtropical regions of Georgia, USSR, but dates back to the cultivation of *C. junos* during the fifth century BC (Kokaya 1977). Unfortunately, the Russian scientific literature is largely unavailable, although considerable work on citrus frost hardiness in the USSR is evident in untranslated reports. Brief accounts from the USSR have described the crossing of an intergeneric orange × trifoliate hybrid with a tetraploid orange that resulted in a triploid with high frost resistance as well as improved fruit quality; tetraploids were inferior to diploids in most agriculturally useful features (Maisuradze 1978). Tetraploid rootstocks have not performed well during natural freezes in the United States (Yelenosky *et al.* 1981), although growth reductions were noted that may warrant testing tetraploids as tree size–controlling rootstocks (Barrett and Hutchison 1978).

One of the most recent and potentially promising sources of cold hardiness for breeding cold-hardy citrus trees is *Eremocitrus glauca* (Lindl.) Swing. (the Australian desert lime). This species easily survives −10°C freezes, and some hybrids perform well in both natural and controlled freezes (Yelenosky *et al.* 1978, 1981).

Mandarin hybrids probably have been the most frequently used in improving scion cold hardiness but have had mixed acceptance in the industry. The 'Clementine' mandarin × Orlando tangelo cross [*C. reticulata* × (*C. paradisi* × *C. reticulata*)] resulted in some of the earliest-ripening hybrids with good cold hardiness. Examples are

'Robinson', 'Osceola', and 'Lee' tangerine hybrids (Reece and Gardner 1959) and 'Nova' tangelo (Reece *et al.* 1964). 'Page' orange [(*C. paradisi* × *C. reticulata*) × *C. reticulata*] also exhibits good cold hardiness, but its small fruit is a problem, although eating quality is high (Reece *et al.* 1963). A cross of sibling tangerine hybrids, 'Robinson' × 'Osceola' [(*C. reticulata* × *C. paradisi*) × *C. reticulata*], resulted in the most recent USDA release, 'Sunburst', which exhibited good cold hardiness during initial trials (Hearn 1979, 1981). The earliness of fruit ripening in these mandarin- or tangerine-type hybrids largely ensures harvest before damaging freezes occur in the United States (Hearn 1973). In Japan, the cold-hardy Satsuma mandarin types 'Okitsu-Wase' and 'Miho-Wase' seemingly are replacing old-line citrus types and demonstrate that selection of nucellar seedlings for improved traits is viable in citriculture (Iwasaki *et al.* 1966). The range of cold-hardiness ratings generally assigned to different *C. reticulata* hybrids is illustrated in Table 5.2.

In all probability, the development of increased cold hardiness through citrus breeding will be a slow, long-term process. Some immediate relief from devastating freezes may be possible by application of chemicals that control growth and other physiological processes in citrus trees, as discussed in the next section.

VI. BIOREGULATION

Bioregulation of citrus frost hardiness through the application of growth regulators and other chemicals is largely in the infancy stages, although many hope this approach will provide a "quick fix" for citriculture's freeze problems. Reports that (2-chloroethyl) phosphonic acid (ethephon) increased the cold hardiness of dormant sweet cherry (*Prunus avicum* L.) with potential to develop into a commercial practice (Proebsting and Mills 1976) and that abscisic acid induced cold hardiness in cell suspension cultures of winter wheat (*Triticum aestivum* L.), winter rye (*Secale cereale* L.), and bromegrass (*Brano inermis* Leyss) without inductive cold temperatures (Chen and Gusta 1983) have stimulated continued interest in bioregulation of citrus frost hardiness.

In citrus, research has focused largely on growth regulators directed toward arresting growth, thereby artificially inducing the stage during which citrus expresses the greatest potential to cold-harden. Additional benefits of such treatments might include increases in protective osmotic relationships with significant cellular augmentation in cell walls, membranes, and cytoplasm. To date, growth cessation

TABLE 5.2 Cold-Hardiness Ratings of Citrus Types

Genus and species	Common name or number	Type seedlings when used as seed parent	Cold-hardiness rating[1]
Poncirus trifoliata (L.) Raf.	Trifoliate orange	Zygotic, nucellar	+++++
Fortunella sp.	Kumquat	Nucellar	++++
Citrus reticulata Blanco	Changsha	Nucellar	++++
C. reticulata	Owari	Nucellar	++++
C. reticulata × *C. reticulata*	54–1–2	Zygotic	++++
C. reticulata	Clementine	Zygotic	+++
[*C. sinensis* (L.) Osbeck × *C. reticulata*] × *C. reticulata*	Wilking	Zygotic	+++
C. paradisi Macf. × *C. reticulata*	Orlando	Nucellar	+++
C. reticulata hybrid	Murcott	Nucellar	+++
C. reticulata hybrid	Shekwasha	Zygotic, nucellar	+++
(*C. reticulata* × *C. sinensis*) × (*C. sinensis* × *C. reticulata*)	52–76–9	Zygotic	+++
C. reticulata × *C. reticulata*	54–3–1	Nucellar	+++
C. reticulata × (*C. paradisi* Macf. × *C. reticulata*	Bower	Zygotic	+++
C. reticulata × *C. reticulata*	48–9–6	Zygotic	+++
C. reticulata	Dancy	Nucellar	+++
C. reticulata × (*C. paradisi* × *C. reticulata*)	Robinson	Zygotic	+++
(*C. paradisi* × *C. reticulata*) × *C. reticulata*	Page	Zygotic, nucellar	+++
C. reticulata × (*C. paradisi* × *C. reticulata*)	Lee	Zygotic	+++
C. reticulata × (*C. paradisi* × *C. reticulata*)	Osceola	Nucellar	+++
C. reticulata × (*C. paradisi* × *C. reticulata*)	Nova	Nucellar	+++
C. reticulata × (*C. paradisi* × *C. reticulata*)	Fairchild	Nucellar	+++
(*C. sinensis* × *C. reticulata*) × *C. reticulata* × (*C. paradisi* × *C. reticulata*)	52–53–2	Nucellar	+++
C. reticulata × *C. sinensis*	Umatilla	Zygotic	++
C. reticulata × *C. sinensis* × *C. reticulata*	Kara	Nucellar	++
C. sinensis × *C. reticulata*	Kinnow	Nucellar	++
C. sinensis	Navel	Nucellar	++
C. sinensis	Jaffa	Nucllear	++
C. paradisi	Duncan	Nucellar	++
C. sinensis	Parson Brown	Nucellar	++
C. sinensis	Valencia	Nucellar	++
C. reticulata hybrid	Temple	Zygotic	+

Source: Adapted from Young and Hearn (1972).
[1] +++++ = superior. ++++ = excellent. +++ = good. ++ = fair. + = poor.

induced by growth regulators has not been associated with significant cold hardiness in citrus (Yelenosky 1979; Young 1971), except possibly in the case of maleic hydrazide [6-hydroxy-3-(2H)-pyridacone] (Burns 1970; Hendershott 1962a). However, maleic hydrazide has not received commercial consideration because of its serious adverse effects in prolonging arrested growth and in subsequent growth development. To what extent bioregulation can influence citrus cold hardiness and its practical application to citriculture remain to be determined (Holubowicz *et al.* 1982; Raese 1983).

VII. PRESENT AND FUTURE TRENDS

The potential of citrus to survive freezes is being optimized in various ways. Space technology has developed satellite systems capable of mapping wide areas for distribution patterns of minimum temperatures accurate to 1 °C (Martsolf 1982). This thermal imagery will improve freeze-warning systems, allowing growers to plan and react to dynamic freeze conditions for maximum effectiveness per unit of investment. Space technology will also be helpful in site selection and in monitoring annual grove conditions, including urban growth, insect and disease situations, surrounding water bodies, windbreaks, radiation losses, and suitability for customized protection. Frost protection will be made more accessible through computerized communication systems (Jackson and Ferguson 1983) that undoubtedly will stimulate replacement of costly and relatively outdated petroleum-fuel heating systems by innovative water-protection systems (Buchanan *et al.* 1982; Parsons *et al.* 1981, 1982a,b) and may even have impact on the use of soil banks or covers to protect young citrus trees (Yelenosky 1981).

Innovative approaches in classical breeding will ensure slow but steady progress in the development of more cold-hardy citrus trees with acceptable fruit. However, continued progress is dependent on sustaining present programs, which need the assistance of planned national and international germplasm repositories. Such repositories will also help to revitalize plant exploration and collecting and may lead to new developments in citrus cryobiology with supporting technology in cell and tissue culture, as well as cryopreservation of cells, asexual embryos, and meristems. There are no apparent signs that recombinant DNA and synthetic gene technology will have a significant impact on efforts to improve citrus frost hardiness for many years, but protoplast fusion (Vardi *et al.* 1975) represents a potential revolutionary change in citrus breeding for frost hardiness. Past emphasis on associated rather than causal factors in citrus frost hardiness probably will be gradually redirected, with attention shifting to cell and organelle fractions rather than whole plants (Steponkus *et al.* 1983). These changes may encourage acceptance of the general concept that multiple events through different pathways lead to citrus frost hardiness, rather than an isolated triggering mechanism for inducing the hardening process.

Cost-effective programs of frost protection may include ventures into the realm of INA agents such as INA bacteria, regardless of present legal opposition to introducing new life forms into the ambient environment. Cryoprotectants and other chemicals (including bactericides and yet-untried new growth regulators such as PP-333) (Williams and Edgerton 1983) will continue to be evaluated in citrus on a relatively small scale. Yet untried in citrus are the alkenylsuccinic acids and fluorinated compounds such as 1,5-difluro-2,4-dinitrobenzene, which protects young bean plants at −3°C for 8 hr (Kuiper 1967). Herbicides are used extensively in citriculture and need to be evaluated for their cold-hardening effects. Some herbicides lower the cold resistance of plants and may be classified as anti-cryoprotectants (St. John and Christiansen 1976). Regardless of the frost-protection method used, some modeling would greatly help the transfer of information into applied concepts such as has been done on citrus fruit freeze injury (Ishiuchi *et al.* 1977) and other woody plants (Kobayashi *et al.* 1983). Over all, constraints on the use of chemicals and other cultural practices to increase citrus frost hardiness probably will increase as urbanization competes with agriculture for prime land and available water.

The 1977 National Agricultural Lands Study in the United States (Fields 1977) indicates that Florida, which is estimated to produce one-fourth of the world's oranges and more than one-half of the world's grapefruit, is one of three states that risk losing all of their 'unique and prime" farmland in less than 20 years if 1967 to 1977 rates of farmland loss persist. Such changes in land use may force citrus production into less favorable or submarginal sites in freeze-prone areas, with concomitant increases in the risk of freeze injury since site location and tree health are important variables in citrus survival (McCown 1958), which depends on freeze (ice) avoidance, freeze (ice) tolerance, and regenerative capacity of injured trees. Nutritional deficiencies (Smith and Rasmussen 1958) and air pollution (Matsushima 1969) also need to be avoided. The effects on citrus frost hardiness of insidious, accumulating changes that develop over a number of years are most difficult to diagnose with a reasonable degree of certainty.

The extent to which major commercial citriculture will continue to exist in freeze-prone areas probably will reflect the balancing of concerns about economics and world food production and concerns about urbanization, technological advances, limiting natural resources, and "quality of life."

LITERATURE CITED

ALDEN, J. and R.K. HERMAN. 1971. Aspects of the cold hardiness in plants. Bot. Rev. 37:37–142.

ALEKSEEV, V.P. 1978. Citrus-ichangensis of the Rutaceae family and its hybrids. Subtrop. Kul't. 0.(5):73–75.

ALLEWELDT, G. 1969. The physiology of frost resistance. Agr. Meteorol. 6:97–110.

ANDERSON, J.A., D.W. BUCHANAN, and M.J. BURKE. 1983. Freeze tolerance vs. freeze avoidance in citrus leaves. Proc. Fla. State. Hort. Soc. 96:57–58.

ANDERSON, J.A., D.W. BUCHANAN, R.E. STALL, and C.B. HALL. 1982. Frost injury of tender plants increased by *Pseudomonas syringae* van Hall. J. Amer. Soc. Hort. Sci. 107:123–125.

ANDERSON, J.A., L.V. GUSTA, D.W. BUCHANAN, and M.J. BURKE. 1983. Freezing of water in citrus leaves. J. Amer. Soc. Hort. Sci. 108:397–400.

BAIN, F.M. 1949. Citrus and climate. Calif. Citrograph 34:382,412–414,426,448–449.

BARRETT, H.C. 1974. Colchicine-induced polyploidy in citrus. Bot. Gaz. 135:29–41.

BARRETT, H.C. 1977. Intergeneric hybridization of *Citrus* and other genera in citrus cultivar improvement. p. 586–589. In: W. Grierson (ed.), 1977 Proc. Intern. Soc. Citriculture, Vol. II, Orlando, Florida.

BARRETT, H.C. 1981. Breeding cold-hardy citrus scion cultivars. p. 61–81. In: K. Matsumoto (ed.), Proc. Intern. Soc. Citriculture, Vol. I, Tokyo, Japan.

BARRETT, H.C. and D.J. HUTCHISON. 1978. Spontaneous tetraploidy in apomictic seedlings of *Citrus*. Econ. Bot. 32:27–45.

BARRETT, H.C. and A.M. RHODES. 1976. A numerical taxonomic study of affinity relationships in cultivated *Citrus* and its close relatives. System. Bot. 1:105–136.

BECWAR, M.R. and M.J. BURKE. 1982. Winter hardiness limitations and physiography of woody timberline flora. p. 307–323. In: P.H. Li and A. Sakai (eds.), Plant cold hardiness and freezing stress: mechanisms and crop implications, Vol. 2. Academic Press, New York.

BIXBY, J.A. and G.N. BROWN. 1975. Ribosomal changes during induction of cold hardiness in black locust seedlings. Plant Physiol. 56:617–621.

BLONDEL, L. 1977. Resistance au froid conferee aux citru pour certains portegreffe. p. 97–101. In: O. Carpena (ed.), Proc. 1st Intern. Citrus Cong., Vol. 2, April 24–May 10, 1973. Murcia, Spain.

BRADLEY, J.T. 1975. Freeze probabilities in Florida. Tech. Bull. 777. Inst. Food and Agr. Sciences, Univ. of Florida, Gainesville.

BREWER, R.F. 1978. Soil applied irrigation water as a source of frost protection. Citrograph 63:283–284.

BUCHANAN, D.W., F.S. DAVIES, and D.S. HARRISON. 1982. High and low volume under-tree irrigation for citrus cold protection. Proc. Fla. State Hort. Soc. 95:23–26.

BURKE, M.J., L.V. GUSTA, H.A. QUAMME, C.J. WEISER, and P.H. LI. 1976. Freezing and injury in plants. Ann. Rev. Plant Physiol. 27:507–528.

BURNS, R.M. 1970. Testing foliar sprays for frost protection of young citrus. Proc. Fla. State Hort. Soc. 83:92–98.

CAMP, P.R. 1965. The formation of ice at water solid interfaces. Ann. New York Acad. Sci. 125:317–343.

CASTLE, W.S. 1983. Growth, yield and cold hardiness of seven-year-old 'Bearss' lemon trees on twenty-seven rootstocks. Proc. Fla. State Hort. Soc. 96:23–25.

CASTLE, W.S. and C.O. YOUTSEY. 1977. Root system characteristics of citrus nursery trees. Proc. Fla. State Hort. Soc. 90:39–44.

CHANDLER, W.H. 1913. The killing of plant tissues by low temperature. Res. Bull. 8. Agr. Expt. Sta., Univ. of Missouri, Columbia.

CHEN, P.M., M.J. BURKE, and P.H. LI. 1976. The frost hardiness of several solanum species in relation to the freezing of water, melting point depression, and tissue water content. Bot. Gaz. 137:313–317.

CHEN, T.H.H. and L.V. GUSTA. 1983. Abscisic acid-induced freezing resistance in cultured plant cells. Plant Physiol. 73:71–75.

COOPER, W.C. 1952. Influence of rootstock on injury and recovery of young citrus trees exposed to freezes of 1950–51 in the Rio Grande Valley. Proc. Rio Grande Valley Hort. Soc. 6:16–24.

COOPER, W.C. 1959. Cold hardiness in citrus as related to dormancy. Proc. Fla. State Hort. Soc. 72:61–66.

COOPER, W.C. 1963. Comparisons of freeze damage to citrus in Florida, Texas, Arizona, and California. Citrus Veg. Mag. 26:20,22,27.

COOPER, W.C. and A. PEYNADO. 1959. Winter temperatures of 3 citrus areas as related to dormancy and freeze injury of citrus trees. Proc. Amer. Soc. Hort. Sci. 74:333–347.

COOPER, W.C., A. PEYNADO, and J.R. FURR. 1962a. Effects of 1961–62 winter freezes on Valencia oranges in Florida, Texas, and California. Proc. Fla. State Hort. Soc. 75:82–88.

COOPER, W.C., P.C. REECE, and J.R. FURR. 1962b. Citrus breeding in Florida—Past, present and future. Proc. Fla. State Hort. Soc. 75:5–13.

COOPER, W.C., C.J. HEARN, G.K. RASMUSSEN, and R.H. YOUNG. 1963a. The Florida freeze. Calif. Citrograph 48:217–220.

COOPER, W.C., G.K. RASMUSSEN, A. PEYNADO, R. HILGEMAN, G. CAHOON, and K. OPITZ. 1963b. Injury and recovery of citrus trees as affected by the climate before and at time of freeze. Proc. Fla. State Hort. Soc. 76:97–104.

COOPER, W.C., R. FENTON, and W.H. HENRY. 1969. Recovery of Florida citrus trees from freeze injury. p. 565–569. In: H. D. Chapman (ed.), Proc. 1st Intern. Citrus Symp., Vol. 2, Univ. of California, Riverside.

DARBYSHIRE, B. and B.T. STEER. 1973. Dehydration of macromolecules. 1. Effect of dehydration-rehydration on indoleacetic and oxidase, ribonuclease, ribulose-diphosphate carboxylase, and ketose-1-phosphate aldolase. Austral. J. Biol. Sci. 26:591–604.

EAKS, I.L. 1960. Physiological studies of chilling injury in citrus fruits. Plant Physiol. 35:632–636.

FARHY, G.M. 1981. Simplified calculation of cell water content during freezing and thawing in nonideal solutions of cryoprotective agents and its possible application to the study of "solution effects" injury. Cryobiology 18:473–482.

FIELDS, S.F. 1977. Where have the farmlands gone? Bull. National Agr. Lands Study, USDA and Council on Envir. Quality, Washington, DC.

FRANKS, F., S.F. MATHIAS, P. GALFEE, S.D. WEBSTER, and D. BROWN. 1983. Ice nucleation and freezing in undercooled cells. Cryobiology 20:298–309.

FURR, J.R., R.T. BROWN, and E.O. OLSON. 1966. Relative cold tolerance of progenies of some citrus crosses. J. Rio Grande Valley Hort. Soc. 20:109–112.

GARDNER, F.E. and G.E. HORANIC. 1963. Cold tolerance and vigor of young citrus trees on various rootstocks. Proc. Fla. State Hort. Soc. 76:105–110.

GARNSEY, S.M. 1982. Increased freeze damage associated with exocortis infection in navel oranges on Carrizo citrange rootstock. Proc. Fla. State Hort. Soc. 95:3–6.

GEORGE, M.F., M.J. BURKE, H.M. PELLETT, and A.G. JOHNSON. 1974. Low temperature exotherms and woody plant distribution. HortScience 9:519–522.

GEORGE, M.F., M.R. BECWAR, and M.J. BURKE. 1982. Freezing avoidance by deep undercooling of tissue water in winter-hardy plants. Cryobiology 19:628–639.

GERBER, J.F. and F. HASHEMI. 1964. The freezing point of citrus leaves. Proc. Amer. Soc. Hort. Sci. 86:220–225.

GERBER, J.F. and J.D. MARTSOLF. 1965. Protecting citrus from cold damage. Fla. Agr. Expt. Sta. Circ. 287. Univ. of Florida, Gainesville.

GHAZALEH, M.Z.S. and C.H. HENDERSHOTT. 1967. Effect of drought and low temperature on leaf freezing points, and water-soluble protein and nucleic acid content of sweet orange plants. Proc. Amer. Soc. Hort. Sci. 90:93–102.

GRAHAM, P.R. and R. MULLIN. 1976. The determination of lethal freezing temperatures in buds and stems of deciduous azalea by a freezing curve method. J. Amer. Soc. Hort. Sci. 101:3–7.

GREGORY, W.C. and M.P. GREGORY. 1965. Induced mutations in quantitative characters: experimental basis for mutations to hardiness in *Citrus*. Proc. Fla. Soil & Crop Sci. 25:372–396.

GRIFFITH, M. and G.N. BROWN. 1982. Cell wall deposits in winter rye *Secale Cereale* L. 'Puma' during cold acclimation. Bot. Gaz. 143:486–490.

GUPTA, O.P., R.K. ARORA, and B.S. CHUNDAWAT. 1978. Variations in the extent of winter injury in different cultivars of sweet orange on various rootstocks. Hargana J. Hort. Sci. 7:112–115.

GUY, C.L. and J.V. CARTER. 1982. Effect of low temperature on the glutathione status of plant cells. p. 169–179. In: P. H. Li and A. Sakai (eds.), Plant cold hardiness and freezing stress: mechanisms and crop implications, Vol. 2. Academic Press, New York.

GUY, C.L., J.V. CARTER, G. YELENOSKY, and C.T. GUY. 1984. Changes in glutathione content during cold acclimation in *Cornus sericea* and *Citrus sinensis*. Cryobiology 21:443–453.

GUY, C.L., G. YELENOSKY, and H.C. SWEET. 1981. Distribution of ^{14}C photosynthetic assimilates in 'Valencia' orange seedlings at 10° and 25°C. J. Amer. Soc. Hort. Sci. 106:433–437.

HANSON, A.D. and N.A. SCOTT. 1980. Betaine synthesis from radioactive precursors in attached, water-stressed barley leaves. Plant Physiol. 66:342–348.

HATAKEYAMA, I. and I. KATO. 1965. Studies on the water relations of *Buxus* leaves. Planta 65:259–268.

HEARN, C.J. 1973. Development of scion cultivars of citrus in Florida. Proc. Fla. State Hort. Soc. 86:84–88.

HEARN, C.J. 1979. 'Sunburst' citrus hybrid. HortScience 14:761–762.

HEARN, C.J. 1981. The 'Sunburst' hybrid in Florida. p. 55–57. In: K. Matsumoto (ed.), Proc. Intern. Soc. Citriculture, Vol. I, Tokyo, Japan.

HEARN, C.J., W.C. COOPER, R.O. REGISTER, and R. YOUNG. 1963. Influence of variety and rootstock upon freeze injury to citrus trees in the 1962 Florida freeze. Proc. Fla. State Hort. Soc. 76:75–81.

HEBER, U. 1965. Freezing injury and uncoupling of phosphorylation from electron transport in chloroplasts. Plant Physiol. 42:1343–1350.

HEBER, U. 1968. Freezing injury in relation to loss of enzyme activities and protection against freezing. Cryobiology 5:188–201.

HEBER, U., L. TYANKOVA, and K.A. SANTARIUS. 1971. Stabilization and inactivation of biological membranes during freezing in the presence of amino acids. Biochem. Biophys. Acta. 24:578–592.

HENDERSHOTT, C.H. 1962a. The influence of maleic hydrazide on citrus trees and fruits. Proc. Amer. Soc. Hort. Sci. 80:241–246.

HENDERSHOTT, C.H. 1962b. The response of orange trees and fruit to freezing temperatures. Proc. Amer. Soc. Hort. Sci. 80:247–254.

HENDRIX, D.L. and W.S. PIERCE. 1983. Osmoregulation and membrane-mediated responses to altered water potential in plant cells. Cryobiology 20:466–486.

HIRANO, S.S., E.A. MAHER, A. KELMAN, and C.D. UPPER. 1978. Ice nucleation activity of fluorescent plant pathogenic pseudomonads. p. 717–725. In: Proc. 4th Intern. Conf. Pathogenic Bacteria, Angers, France.

HOLUBOWICZ, T., J. PIENIAZEK, and M.A. KHAMIS. 1982. Modification of frost resistance of fruit plants by applied growth regulators. p. 541–550. In: P. H. Li and A. Sakai (eds.), Plant cold hardiness and freezing stress: mechanisms and crop implications, Vol. 2. Academic Press, New York.

HORANIC, G.E. and G. YELENOSKY. 1969. Electrostatic heating of citrus trees. p. 539–544. In: H. D. Chapman (ed.), Proc. 1st Intern. Citrus Symp., Vol. III, Univ. of California, Riverside.

HORVATH, I., L. VIGH, A. BELEA, and T. FARKAS. 1980. Hardiness dependent accumulation of phospholipids in leaves of wheat cultivars. Physiol. Plant. 49:117–120.

HUNER, N.P.A., W.G. HOPKINS, B. ELFMAN, and D.B. HAYDEN. 1982. Influence of growth at cold-hardening temperatures on protein structure and function. p. 129–144. In: P. H. Li and A. Sakai (eds.), Plant cold hardiness and freezing stress: mechanisms and crop implications, Vol. 2. Academic Press, New York.

HUTCHISON, D.J. 1974. Swingle citrumelo—A primary rootstock hybrid. Proc. Fla. State Hort. Soc. 87:89–91.

IKEDA, I. 1982. Freeze injury and protection of citrus in Japan. p. 575–589. In: P. H. Li and A. Sakai (eds.), Plant cold hardiness and freezing stress: mechanisms and crop implications, Vol. 2. Academic Press, New York.

IKEDA, I., S. KOBAYASHI, and M. NAKATANI. 1980. Differences in cold resistance of various citrus varieties and cross-seedlings based on the data obtained from the freezes in 1977. Bull. Fruit Tree Res. Sta. (Akitsu Branch, Hiroshima, Japan) Series E No.3:49–65.

ISHIUCHI, D., N. OKUDAI, T. TAKAHARA, and T. SCHICHIJO. 1977. Studies on Fukuhara orange *Citrus sinensis* Osbeck. I. Effects of low temperature in winter on cold injury of fruit and effects of climatic conditions of fruit growing period on fruit quality. Bull. Fruit Tree Res. Sta. (Kuchinotsu, Nagasaki, Japan) Series D No.2:9–48.

IVANOV, S.M. 1939. Importance of temperature conditions in the hardening of citrous plants. Comptes Rendus (Doklady) de l'Academie des Sciences de l'URSS XXV:440–443.

IWASAKI, T., M. NISHIURA, and N. OKUDAI. 1966. New citrus varieties 'Okitsu-Wase' and 'Miho-Wase.' Bull. Hort. Res. Sta. (Kuchinotsu, Nagasaki, Japan) Series B No. 8:83–93.

JACKSON, J. and F. FERGUSON, JR. 1983. An experimental computer network for growers. Proc. Fla. State Hort. Soc. 96:5–7.

JACKSON, L.K. and J.F. GERBER. 1963. Cold tolerance and freezing points of citrus seedlings. Proc. Fla. State Hort. Soc. 76:70–74.

JOHANSSON, NIK-OLAF. 1970. Ice formation and frost hardiness in some agricultural plants. Natl. Swedish Inst. Plant Protection, Contribution 14:132–382.

JOHNSON, W.O. 1963. The meteorological aspects of the big freeze of December 1962. Proc. Fla. State Hort. Soc. 76:62–69.

KACPERSKA-PALACZ, A. 1978. Mechanism of cold acclimation in herbaceous plants. p. 139–152. In: P. H. Li and A. Sakai (eds.), Plant cold hardiness and freezing stress: mechanisms and crop implications, Vol. 1. Academic Press, New York.

KAKU, S. and M. IWAYA. 1978. Low temperature exotherm in xylem of evergreen and deciduous broad-leaved trees in Japan with reference to freezing resistance and distribution range. p. 227–239. In: P. H. Li and A. Sakai (eds.), Plant cold hardiness and freezing stress: mechanisms and crop implications, Vol. 1. Academic Press, New York.

KAPPEN, L., M. NOVIG, and M. MAIER. 1978. Seasonal relations between the content of amino acids and freezing tolerance of leaves of *Halimione portulacoides* under different salt stress. Biochem. Physiol. Pflanzen. 172:297–304.

KATO, T. and S. KUBOTA. 1982. Effects of low temperature in autumn on the uptake, assimilation and partitioning of nitrogen in citrus trees. J. Japan. Soc. Hort. Sci. 51:1–8.

KAWASE, K., K. YOSHINAGA, M. UCHIDA, and K. HIROSE. 1982. Studies on the frost hardiness of citrus trees. 1. Differences in the freezing resistance between citrus species and their seasonal variation in grafted nursery plant and seedlings. Bull. Fruit Tree Res. Sta. D (Japan) 4:42–48.

KENEFICK, D.G. 1964. Cold acclimation as it relates to winter hardiness in plants. Agr. Sci. Rev. 2:21–31.

KITTO, S.L. and M.J. YOUNG. 1981. *In vitro* propagation of Carrizo citrange. HortScience 16:305–306.

KOBAYASHI, K.D., L.H. FUCHIGAMI, and C.J. WEISER. 1983. Modeling cold hardiness of Red-osier dogwood. J. Amer. Soc. Hort. Sci. 108:376–381.

KOCHBA, J. and P. SPIEGEL-ROY. 1977. Cell and tissue culture for breeding and developmental studies of citrus. HortScience 12:110–114.

KOKAYA, Ts. D. 1977. Citrus juko—A promising frost resistant crop in humid subtropical regions of Georgia (in Russian). Bull. 'Vsesoyuznogo Ordena Lenina i Ordena Druzhby Narodov Institute Rastenievodstva imeni N.I. Vavilova 68:25–28. [English translation-USDA]

KUIPER, P.J.C. 1967. Surface-active chemicals as regulators of plant growth membrane permeability and resistance to freezing. Meded. Landbounhogeschool Wageningen 67(3):1–23.

LARCHER, W. 1971. Die Kaltereisistenz Von Obstbaumen und Ziergeholzen Subtropischer Herkunft. Ecol. Planta. 6:1–14.

LARCHER, W. 1982. Typology of freezing phenomena among vascular plants and evolutionary trends in frost acclimation. p. 417–436. In: P. H. Li and A. Sakai (eds.), Plant cold hardiness and freezing stress: mechanisms and crop implication, Vol. 2. Academic Press, New York.

LARCHER, W., U. HEBER, and K.A. SANTARIUS. 1973. Chapt. III. Limiting temperatures for life functions. p. 195–231. In: H. Precht, J. Christopherson, and W. Larcher (eds.), Temperature and life. Springer-Verlag, New York.

LEVITT, J. 1968. Symposium. Environmental cryobiology. Cryobiology 5:147–225.

LEVITT, J. 1980a. Chilling, freezing, and high temperature stress. p. 3–344. In: T. T. Kozlowski (ed.), Responses of plants to environmental stresses. Vol. 1, Physiological-Ecology Monograph Series. 2nd ed. Academic Press, New York.

LEVITT, J. 1980b. Factors related to freezing tolerance. p. 163–227. In: T. T. Kozlowski (ed.), Responses of plants to environmental stresses. Vol. 1, Physiological-Ecology Monograph Series. 2nd ed. Academic Press, New York.

LEVY, Y. 1980. Field determination of free proline accumulation and water-stress in lemon trees. HortScience 15:302–303.

LEYSLE, F.F. 1948. Ecological and physiological characteristics of leaves of evergreens from the Soviet moist subtropics. Akademia Nauk. USSR Minsk. Instytut Biialogii, Eksperimentalnia Botanika, 6:147–199.

LI, P.H. and J.P. PALTA. 1978. Frost hardening and freezing stress in tuber-bearing solanum species. p. 49–71. In: P. H. Li and A. Sakai (eds.), Plant cold hardiness of

freezing stress: mechanisms and crop implication, Vol. I. Academic Press, New York.
LI, P.H. and C.J. WEISER. 1969. Metabolism of nucleic acids in one-year-old apple twig during cold hardening and de-hardening. Plant & Cell Physiol. 10:21-30.
LINDOW, S.E. 1982b. Population dynamics of epiphytic ice nucleation active bacteria of frost sensitive plants and frost control by means of antagonistic bacteria. p. 395–416. In: P. H. Li and A. Sakai (eds.), Plant cold hardiness and freezing stress: mechanisms and crop implications, Vol. 2. Academic Press, New York.
LINDOW, S.E. 1983. Methods of preventing frost injury caused by epiphytic ice-nucleation-active bacteria. Plant Dis. 67:327–333.
LINDOW, S.E., D.C. ARNY, and C.D. UPPER. 1978. Distribution of ice nucleation-active bacteria on plants in nature. Applied & Environ. Microbiol. 36:831–838.
LINDOW, S.E., D.C. ARNY, and C.D. UPPER. 1982. Bacterial ice nucleation: A factor in frost injury to plants. Plant Physiol. 70:1084–1089.
LUCAS, J.W. 1954. Subcooling and ice nucleation in lemons. Plant Physiol. 29:246-251.
LUYET, B.J. and P.M. GEHENIO. 1939. The physical states of protoplasm at low temperatures. Biodynamica 48:1-128.
LUYET, B.J. and P.M. GEHENIO. 1940. Life and death at low temperatures. Biodynamica, Normandy, Missouri.
MAGNESS, J.R. and H.P. TRAUB. 1941. Climatic adaptation of fruit and nut crops. p. 400-420. In: Yearbook of Agriculture. USDA, Washington, DC.
MAISURADZE, N.I. 1978. Polyploidy in orange breeding (in Russian). Genetika 15:2195-2203. [English translation]
MARCELLOS, H. and W.V. SINGLE. 1979. Supercooling and heterogeneous nucleation of freezing in tissues of twenty plants. Cryobiology 16:74-77.
MARTSOLF, J.D. 1982. Satellite thermal maps provide detailed views and comparisons of freezes. Proc. Fla. State Hort. Soc. 95:14-19.
MATSUSHIMA, J. 1969. Problems in sulfur dioxide gas injury to citrus trees in Japan. p. 733-740. In: Proc. 1st Intern. Citrus. Symp., Vol. 2, Univ. of California, Riverside.
MAYLAND, H.F. and J.W. CARY. 1970. Frost and chilling injury to growing plants. Adv. Agron. 22:203-234.
MAZUR, P. 1965. The role of cell membranes in the freezing of yeast and other single cells. Ann. N.Y. Acad. Sci. 125:658-676.
MAZUR, P. 1969. Freezing injury in plants. Annu. Rev. Plant Physiol. 20:419-448.
McCOWN, J.T. 1958. Field observations of Florida citrus following the 1957-58 freezes. Proc. Fla. State Hort. Soc. 71:152-157.
McLEESTER, R.C., C.J. WEISER, and T.C. HALL. 1969. Multiple freezing points as a test for viability of plant stems in the determination of frost hardiness. Plant Physiol. 44:37-44.
MERYMAN, H.T. 1956. Mechanics of freezing in living cells and tissues. Science 124:515-521.
MERYMAN, H.T. 1966. Review of biological freezing. p. 2-114. In: H. T. Meryman (ed.), Cryobiology. Academic Press, New York.
MOLISCH, H. 1897. Investigations into the freezing of plants. (1982 English translation.) Cryoletters 3:331-390.
MORENO, J. and J. L. GARCIA-MARTINEZ. 1983. Seasonal variation of nitrogenous compounds in the xylem sap of *Citrus*. Physiol. Plant. 59:669-675.
MURASHIGE, T. 1974. Plant propagation through tissue cultures. Annu. Rev. Plant Physiol. 25:135-166.
NIIJAR, G.S. and J.W. SITES. 1959. Some effects of day length and temperature on cold hardiness in citrus. Proc. Fla. State Hort. Soc. 72:106-109.

NORDBY, H.E. and G. YELENOSKY. 1982. Relationships of fatty acids to cold hardening of citrus seedlings. Plant Physiol. 70:132-135.

NORDBY, H.E. and G. YELENOSKY. 1984. Effects of cold hardening and acyl lipids of citrus tissue. Phytochemistry 23:41-45.

OLIEN, C.R. 1981. Analysis of midwinter freezing stress. p. 35-60. In: C. R. Olien and M. N. Smith (eds.), Analysis and improvement of plant cold hardiness. CRC Press, Boca Raton, Florida.

OLIEN, C.R. and M.N. SMITH. 1981. Protective systems that have evolved in plants. p. 61-87. In: C. R. Olien and M. N. Smith (eds.), Analysis and improvement of plant cold hardiness. CRC Press, Boca Raton, Florida.

PARKER, J. 1963. Cold resistance in woody plants. Bot. Rev. 29:123-201.

PARSONS, L.R., T.A. WHEATON and J.D. WHITNEY. 1982. Undertree irrigation for cold protection with low volume microsprinklers. HortScience 17:799-801.

PARSONS, L.R., T.A. WHEATON, D.P.H. TUCKER, and J.D. WHITNEY. 1983. Low volume microsprinkler irrigation for citrus cold protection. Proc. Fla. State Hort. Soc. 95:20–23.

PEYNADO, A. 1982. Cold hardiness of young 'Ruby Red' grapefruit trees as influenced by rootstock, trickle and flood irrigation, and chloride and boron in the irrigation water. J. Rio Grande Valley Hort. Soc. 35:149–157.

POMEROY, M.K. and J.K. RAISON. 1981. Maintenance of membrane fluidity during development of freezing tolerance of winter wheat seedlings. Plant Physiol. 68:382-385.

PROEBSTING, E.L. 1978. Adapting cold hardiness concepts to deciduous fruit culture. p. 267-279. In: P. H. Li and A. Sakai (eds.), Plant cold hardiness and freezing stress: mechanisms and crop implications, Vol. I. Academic Press, New York.

PROEBSTING, JR. E.L., and P.K. ANDREWS. 1982. Supercooling young developing fruit and floral buds in deciduous orchards. HortScience 17:67–68.

PROEBSTING, JR. E.L., and H.H. MILLS. 1976. Ethephon increases cold hardiness of sweet cherry. J. Amer. Soc. Hort. Sci. 101:31–33.

PURVIS, A.C. 1980. Influence of canopy depth on susceptibility of 'Marsh' grapefruit to chilling injury. HortScience 15:731–733.

PURVIS, A.C. and G. YELENOSKY. 1982. Sugar and proline accumulation of grapefruit flavedo and leaves during cold hardening of young trees. J. Amer. Soc. Hort. Sci. 107:222–226.

QUAMME, H.A. 1978. Mechanisms of supercooling in overwintering peach flower buds. J. Amer. Soc. Hort. Sci. 103:57–61.

QUAMME, H., C. STUSHNOFF, and C.J. WEISER. 1972. The relationship of exotherms to cold injury in apple stem tissues. J. Amer. Soc. Hort. Sci. 97:608–613.

QUAMME, H.A., R.E.C. LAYNE, and W.G. RONALD. 1982. Relationship of supercooling to cold hardiness and the northern distribution of several cultivated and native *Prunus* species and hybrids. Can. J. Plant Sci. 62:137–148.

RAESE, J.T. 1983. Conductivity tests to screen fall-applied growth regulators to induce cold hardiness in young 'Delicious' apple trees. J. Amer. Soc. Hort. Sci. 108:172–176.

RAJASHEKAR, C., H.M. PELLET, and M.J. BURKE. 1982a. Deep supercooling in roses. HortScience 17:609–611.

RAJASHEKAR, C., M.N. WESTWOOD, and M.J. BURKE. 1982b. Deep supercooling and cold hardiness in genus *Pyrus*. J. Amer. Soc. Hort. Sci. 107:968–972.

RAJASHEKAR, C.B., P.H. LI, and J.V. CARTER. 1983. Frost injury and heterogeneous ice nucleation in leaves of tuber-bearing *Solanum* species. Plant Physiol. 71:749–755.

REECE, P.C. and F.E. GARDNER. 1959. Robinson, Osceola, and Lee—Early mat-

uring tangerine hybrids. Proc. Fla. State Hort. Soc. 72:49–51.
REECE, P.C., F.E. GARDNER, and C.J. HEARN. 1963. Page orange—A promising variety. Proc. Fla. State Hort. Soc. 76:53–54.
REECE, P.C., C.J. HEARN and F.E. GARDNER. 1964. Nova tangelo—An early ripening hybrid. Proc. Fla. State Hort. Soc. 77:109–110.
REED, H.S. and E.T. BARTHOLOMEW. 1930. The effects of dessicating winds on citrus trees. Agr. Expt. Sta. Bull. 484. Univ. of California, Berkeley.
RIEMENSCHNEIDER, C.H. 1983. World supply and demand for citrus and citrus products. Citrograph 68:243–245.
ROUSE, R.E. and W.J. WILTBANK. 1972. Effects of temperatures on citrus cold hardiness at different geographical locations in Florida. Proc. Fla. State Hort. Soc. 85:75–80.
ST. JOHN, J.B. and M.N. CHRISTIANSEN. 1976. Inhibition of linolenic acid synthesis and modification of chilling resistance in cotton seedlings. Plant Physiol. 57:257–259.
SAKAI, A. 1962. Studies on the frost hardiness of woody plants. 1. The causal relation between sugar content and frost hardiness. Contribution Series B, No. 11. Inst. Low Temperature Sci., Hokkaido Univ., Sapporo, Japan.
SAKAI, A. 1982. Freezing resistance of ornamental trees and shrubs. J. Amer. Soc. Hort. Sci. 107:572–581.
SAKAI, A. and S. YOSHIDA. 1968. The role of sugar and related compounds in variations of freezing resistance. Cryobiology 5:160–174.
SALAZAR, C.G. 1966. Comparative anatomy of citrus and cold-hardiness. J. Agr. Univ. Puerto Rico L:316–336.
SANTARIUS, K.A. and J. BAUER. 1983. Cryopreservation of spinach chloroplast membranes by low-molecular-weight carbohydrates. 1. Evidence for cryoprotection by a noncolligative-type mechanism. Cryobiology 20:83–89.
SCHNELL, R.C. and G. VALI. 1972. Atmospheric ice nuclei from decomposing vegetation. Nature 236:163–165.
SCHNELL, R.C. and G. VALI. 1973. World-wide source of leaf-derived freezing nuclei. Nature 246:212–213.
SCHNELL, R.C. and G. VALI. 1976. Biogenic ice nuclei. Part. 1. Terrestrial and marine sources. J. Atmos. Sci. 33:1554–1564.
SIMINOVITCH, D. and Y. CLOUTIER. 1983. Drought and freezing tolerance and adaptation in plants: Some evidence of near equivalences. Cryobiology 20:487–503.
SIMINOVITCH, D., B. RHEAUME, K. POMEROY, and M. LEPAGE. 1968. Phospholipid, protein, and nucleic acid increases in protoplasm and membrane structures associated with development of extreme freezing resistance in black locust tree cells. Cryobiology 5:202–225.
SIMINOVITCH, D., J. SINGH, and I.A. DE LA ROCHE. 1975. Studies on membranes in plant cells resistant to extreme freezing. I. Augmentation of phospholipids and membrane substance without changes in unsaturation of fatty acids during hardening of black locust bark. Cryobiology 12:144–153.
SINGLE, W.V. and H. MARCALLOS. 1981. Ice formation and freezing injury in actively growing cereals. p. 17–33. In: C. R. Olien and M. N. Smith (eds.), Analysis and improvement of plant cold hardiness. CRC Press, Boca Raton, Florida.
SMITH, P.F. and G.K. RASMUSSEN. 1958. Relation of fertilization to winter injury of citrus trees. Proc. Fla. State Hort. Soc. 71:170–175.
STEPONKUS, P.L. and F.O. LANPHEAR. 1967. Light stimulation of cold acclimation: production of a translocatable promoter. Plant Physiol. 42:1673–1679.
STEPONKUS, P.L., M.F. DOWGERT, and W.J. GORDON-KAMM.

1983. Destabilization of the plasma membrane of isolated plant protoplasts during a freeze-thaw cycle: the influence of cold acclimation. Cryobiology 20:448–465.

STOUT, D.G. 1981. Dehydration strain avoidance and tolerance in plant cold hardiness. J. Theor. Biol. 88:513–521.

SYVERTSEN, J.P. and M.L. SMITH, JR. 1983. Environmental stress and seasonal changes in proline concentration of citrus tree tissues and juice. J. Amer. Soc. Hort. Sci. 108:861–866.

TURRELL, F.M. 1973. The science and technology of frost protection. p. 338–446. In: W. Reuther (ed.), The citrus industry, Vol. III. Div. of Agr. Sci., Univ. of California, Berkeley.

UPHOF, J.C. Th. 1938. Wissenchaftliche Beobachtungen und Versuche an Argrumen IX. Der Einflub von frost. Gartenbauwissenschaft 11:391–412.

VARDI, A., P. SPIEGEL-ROY, and G. GALUN. 1975. Citrus cell culture: isolation of protoplasts, plating densities, effect of mutagens and regeneration of embryoids. Plant Sci. Lett. 4:231–236.

VASIL'YEV, I.M. 1956. Wintering of plants (in Russian). Translated by Royer and Royer, Inc.; edited by J. Levitt. Amer. Inst. Biol. Sci., Washington, DC.

WALLIN, J.R., D.V. LORNAN, and C.A.C. GARDNER. 1979. *Erwinia sterwartii*: An ice nucleus for frost damage to major seedlings. Plant Dis. Rptr. 63:751–752.

WEBBER, H.J. 1895. The two freezes of 1894–95 in Florida and what they teach. p. 159–174. In: Yearbook. USDA, Washington, DC.

WEBBER, H.J., W. REUTHER, and H.W. LAWTON. 1967. History and development of the citrus industry. p. 1–39. In: W. Reuther, H. J. Webber, and L. D. Batchelor (eds.), The citrus industry, Vol. I. Div. of Agr. Sci., Univ. of California, Berkeley.

WEBBER, H.J. and W.T. SWINGLE. 1904. New citrus creations of the Department of Agriculture. p. 749–826. In: Yearbook. USDA, Washington, DC.

WEISER, C.J. 1970. Cold resistance and injury in woody plants. Science 169:1269–1278.

WEISER, C.J., H.A. QUAMME, E.L. PROEBSTING, M.J. BURKE, and G. YELENOSKY. 1979. Plant freezing injury and resistance. p. 55–84. In: B. J. Barfield and J. F. Gerber (eds.), Modification of the aerial environment of crops. Amer. Soc. Agr. Engineers, St. Joseph, Michigan.

WILCOX, D.A. and F.S. DAVIES. 1981. Temperature-dependent and diurnal root conductivities in two citrus rootstocks. HortScience 16:303–305.

WILLEMOT, C. 1983. Rapid degradation of polar lipids in frost damaged winter wheat crown and root tissue. Phytochemistry 22:861–863.

WILLIAMS, M.W. and L.J. EDGERTON. 1983. Vegetative growth control of apple and pear trees with ICI PP333 (paclobutrazol), a chemical analog of bayleton. Acta Hort. 137:111–116.

WILLIAMS, R.J. 1981. Frost dessication: An osmotic model. p. 89–116. In: C. R. Olien and M. N. Smith (eds.), Analysis and improvement of plant cold hardiness. CRC Press, Boca Raton, Florida.

WUTSCHER, H.K. 1979. Citrus rootstocks. Hort. Rev. 1:237–269.

YANKOFSKY, S.A., Z. LEVIN, T. BERTOLD, and H. SANDLERMAN. 1981. Some basic characteristics of bacterial freezing nuclei. Amer. Meteor. Soc. 20:1013–1019.

YELENOSKY, G. 1971a. Effect of light on cold-hardening of citrus seedlings. HortScience 6:234–235.

YELENOSKY, G. 1971b. Relationship between the rate of ice growth in solutions, the molecular weight of the solutes and the osmotic pressure of solutions. Biodynamica 11:95–100.

YELENOSKY, G. 1975. Cold hardening in citrus stems. Plant Physiol. 56:540–543.

YELENOSKY, G. 1976a. Cold hardening young Valencia trees on Swingle citrumelo (C.P.B. 4475) and other rootstocks. Proc. Fla. State Hort. Soc. 89:9–10.

YELENOSKY, G. 1976b. Ice tolerance of cold hardened 'Valencia' orange wood. Cryobiology 13:243–247.

YELENOSKY, G. 1977. The potential of citrus to survive freezes. p. 199–203. In: W. Grierson (ed.), Proc. Intern. Soc. Citriculture, Vol. I, Orlando, Florida.

YELENOSKY, G. 1978a. Cold hardening 'Valencia' orange trees to tolerate −6.7°C without injury. Amer. Soc. Hort. Sci. 103:449–452.

YELENOSKY, G. 1978b. Freeze survival of citrus trees in Florida. p. 297–311. In: P. H. Li and A. Sakai (eds.), Plant cold hardiness and freezing stress: mechanisms and crop implications, Vol. 1. Academic Press, New York.

YELENOSKY, G. 1979a. Accumulation of free proline in citrus leaves during cold hardening of young trees in controlled temperature regimes. Plant Physiol. 64:425–427.

YELENOSKY, G. 1979b. Bark-splitting from freeze injury of young citrus trees on different rootstocks. Proc. Fla. State Hort. Soc. 92:28–31.

YELENOSKY, G. 1979c. Water-stress-induced cold hardening of young citrus trees. J. Amer. Soc. Hort. Sci. 104:270–273.

YELENOSKY, G. 1979d. Sensitivity of budded young orange trees and grapefruit seedlings to ancymidol in soil mix. HortScience 14:600–602.

YELENOSKY, G. 1981. A new insulator wrap to protect young citrus trees during freezes. HortScience 16:44–45.

YELENOSKY, G. 1982a. Chilling injury in leaves of citrus plants at 1.7°C. HortScience 17:385–387.

YELENOSKY, G. 1982b. Indicators of citrus cold-hardening in the field. Proc. Fla. State Hort. Soc. 95:7–10.

YELENOSKY, G. 1983. Ice nucleating active (INA) agents in freezing of young citrus trees. J. Amer. Soc. Hort. Sci. 108:1030–1034.

YELENOSKY, G. and W. GILBERT. 1974. Levels of hydroxyproline in citrus leaves. HortScience 9:375–376.

YELENOSKY, G. and C.L. GUY. 1977. Carbohydrate accumulation in citrus. Bot. Gaz. 138:13–17.

YELENOSKY, G. and C.L. GUY. 1982a. Protein scans of cold-hardened and freeze-injured 'Valencia' orange leaves. Cryobiology 19:646–650.

YELENOSKY, G. and C.L. GUY. 1982b. Seasonal variations in physiological factors implicated in cold hardiness of citrus trees. p. 561–573. In: P. H. Li and A. Sakai (eds.), Plant cold hardiness and freezing stress: mechanisms and crop implications, Vol. 2. Academic Press, New York.

YELENOSKY, G. and C.J. HEARN. 1967. Cold damage to young mandarin hybrid trees and different rootstocks on flatwood soil. Proc. Fla. State Hort. Soc. 80:53–56.

YELENOSKY, G. and G. HORANIC. 1969. Subcooling in wood of citrus seedlings. Cryobiology 5:281–283.

YELENOSKY, G. and R. YOUNG. 1977. Cold hardiness of orange and grapefruit trees on different rootstocks during the 1977 Florida freeze. Proc. Fla. State Hort. Soc. 90:49–53.

YELENOSKY, G., C.J. HEARN, and W.C. COOPER. 1968. Relative growth of trifoliate orange selections. Proc. Amer. Soc. Hort. Sci. 93:205–209.

YELENOSKY, G., G. HORANIC, and F. GALENA. 1970. Indicator of freezing in citrus seedlings. HortScience 5:270.

YELENOSKY, G., R.T. BROWN, and C.J. HEARN. 1973. Tolerance of trifoliate orange selection and hybrids to freezes and flooding. Proc. Fla. State Hort. Soc. 86:99–104.

YELENOSKY, G., H. BARRETT, and R. YOUNG. 1978. Cold hardiness of young hybrid trees of *Eremocitrus glauca* (Lindl.) Swing. HortScience 13:257–258.

YELENOSKY, G., R. YOUNG, C.J. HEARN, H.C. BARRETT, and D.J. HUTCHISON. 1981. Cold hardiness of citrus trees during the 1981 freeze in Florida. Proc. Fla. State Hort. Soc. 94:46–51.

YOSHIDA, T. 1981. Segregation of cold tolerance in citrus hybrid seedlings by artificial freezing. Bull. Fruit Tree Res. Sta. (Akitsu Branch, Hiroshima, Japan) Series B, no. 8:1–11.

YOSHIMURA, F. 1967. Studies on the cold injury of citrus trees. Mem. Facu. Agr. Kochi Univ. 18:79–134.

YOUNG, R. 1961. Influence of day length, light intensity, and temperature on growth, dormancy, and cold hardiness of Redblush grapefruit trees. Proc. Amer. Soc. Hort. Sci. 78:174–180.

YOUNG, R. 1963a. Freeze injury to young seedlings of citrus cultivars and related species in the Lower Rio Grande Valley. J. Rio Grande Valley Hort. Soc. 17:37–42.

YOUNG, R.H. 1963b. Climate—Cold hardiness—Citrus. J. Rio Grande Valley Hort. Soc. 17:3–14.

YOUNG, R.H. 1966. Freezing points and lethal temperatures of citrus leaves. Proc. Amer. Soc. Hort. Sci. 88:272–279.

YOUNG, R.H. 1969a. Cold hardening in citrus seedlings as related to artificial hardening conditions. J. Amer. Soc. Hort. Sci. 94:612–614.

YOUNG, R.H. 1969b. Cold hardening in Redblush grapefruit as related to sugars and water soluble proteins. J. Amer. Soc. Hort. Sci. 94:252–254.

YOUNG, R. 1969c. Effect of freezing on the photosynthetic system in citrus. p. 553–558. In: H. D. Chapman (ed.), Proc. 1st Intern. Citrus Symp., Vol. 2, Univ. of California, Riverside.

YOUNG, R. 1970. Induction of dormancy and cold hardiness in citrus. HortScience 5:411–413.

YOUNG, R. 1971. Effect of growth regulators on citrus seedling cold hardiness. J. Amer. Soc. Hort. Sci. 96:708–710.

YOUNG, R. 1981. Relationships between winter hardening temperatures and spring bud break in *Citrus* and related species. p. 318–321. In: K. Matsumoto (ed.), Proc Intern. Soc. Citriculture, Vol. I, Tokyo, Japan.

YOUNG, R. and C.J. HEARN. 1972. Screening citrus hybrids for cold hardiness. HortScience 7:14–18.

YOUNG, R. and M. MANN. 1974. Freeze disruption of sour orange leaf cells. J. Amer. Soc. Hort. Sci. 99:403–407.

YOUNG, R.H. and E.O. OLSON. 1963a. Freeze injury to citrus trees on various rootstocks in the Lower Rio Grande Valley of Texas. Proc. Amer. Soc. Hort. Sci. 83:337–343.

YOUNG, R.H. and E.O. OLSON. 1963b. Freeze injury to citrus varieties in the Lower Rio Grande Valley of Texas. Proc. Amer. Soc. Hort. Sci. 83:333–336.

YOUNG, R.H. and A. PEYNADO. 1962. Growth and cold-hardiness of citrus and related species when exposed to different night temperatures. Proc. Amer. Soc. Hort. Sci. 81:238–243.

YOUNG, R. and A. PEYNADO. 1963. Freeze injury to citrus trees of different ages and in different locations in the Rio Grande Valley of Texas in 1962. J. Rio Grande Valley Hort. Soc. 17:24–29.

YOUNG, R. and A. PEYNADO. 1965. Changes in cold hardiness and certain physiological factors of Red Blush grapefruit seedlings as affected by exposure to artificial hardening temperatures. Proc. Amer. Soc. Hort. Sci. 86:244–252.

YOUNG, R. and A. PEYNADO. 1967a. Freezing and water-soaking in citrus leaves.

Proc. Amer. Soc. Hort. Sci. 91:157-162.

YOUNG, R.H. and A. PEYNADO. 1967b. Freeze injury to 3-year-old citrus hybrids and varieties following exposure to controlled freezing conditions. J. Rio Grande Valley Hort. Soc. 1:80-88.

YOUNG, R.H., A. PEYNADO, and W.C. COOPER. 1960. Effect of rootstock-scion combination and dormancy on cold hardiness of citrus. J. Rio Grande Valley Hort. Soc. 14:58-65.

YOUNG, R., G. YELENOSKY, and W.C. COOPER. 1977. Hardening and freezing conditions for screening citrus trees for cold hardiness. p. 145-150. In: Proc. 1st Intern. Citrus Congr., Vol. 3, April 24-May 10, 1973, Murcia, Spain.

ZIEGLER, L.W. and H.S. WOLFE. 1961. Propagation of citrus fruit, p. 64-83. In: Citrus growing in Florida. Univ. of Florida Press, Gainesville.

6

Dormancy Release in Deciduous Fruit Trees[1]

M. C. Saure
Diplom-Agraringenieur, 2151 Moisburg, Federal Republic of Germany

[1] I am very grateful to Mrs. G. Schacht for preparing the drawings, to Miss I. Löhden who helped me very much in collecting the literature, and to many others who supported me with further literature and good suggestions.

Horticultural Reviews, Volume 7

ISBN 0-87055-492-1

I. INTRODUCTION

Dormancy in deciduous fruit trees and other woody perennials of the temperate zones is a phase of development that occurs annually and enables plants to survive cold winters. It is similar—although not necessarily identical—to dormancy in bulbs, tubers, and seeds. Because of its relation to frost hardiness, horticulturists initially dealt mainly with the induction of dormancy, especially in those parts of the world where cold winters are common. The release from dormancy was studied first by those who were engaged in early forcing of ornamentals. It received little attention in deciduous fruits until early attempts were made to grow temperate fruits in the subtropics or even in the tropics, where cold winters do not exist. In those warm-winter regions, prolonged dormancy became an important obstacle to economic production of temperate crops.

The dynamics of dormancy induction and release, i.e., the role of the factors involved both within and outside the plants, and the mode of their action, are still not yet fully understood, in spite of the countless publications dealing with dormancy or some of its aspects. Many interesting hypotheses and theories have been suggested, only to be rejected. This subject has been reviewed by Doorenbos (1953), Samish (1954), Vegis (1961, 1964, 1965a,b), Romberger (1963), Leike (1965), Wareing (1969), Wareing and Saunders (1971), Saunders (1978), and Noodén and Weber (1978).

Without a comprehensive theory of dormancy, it will be extremely difficult to improve the management of dormancy release, be it for the promotion or the retardation of growth and development or for the calculation of the actual state of dormancy as an important prerequisite for the proper timing of various cultural practices. This review is intended to contribute to our understanding of dormancy and of the mechanisms involved in dormancy release, especially in deciduous fruit trees, and possibly in other woody perennials, by incorporating the more recent findings of research on problems of prolonged dormancy in warmer regions into the fund of knowledge from cooler parts of the world. It is hoped that this review will encourage further experimental work in this interesting field.

II. THE PHENOMENON OF DORMANCY

A. Phases of Dormancy

Dormancy in a general sense may be defined as "a state in which visible growth is temporarily suspended" (Samish 1954), i.e., "in which a tissue predisposed to elongate does not do so" (Doorenbos 1953). Black (1952) has pointed out that "this state of inactivity may be due to any cause and may under certain conditions precede or extend beyond the rest period." Lavee (1973) has emphasized that dormancy does not necessarily entail a cessation of biological development, as it is known that important differentiation processes may occur normally in dormant organs, permitting a slow but steady increase in bud weight (Chandler and Tufts 1934; Brown and Kotob 1957; Zeller 1961, 1973; Stadler and Strydom 1967; El-Mansy and Walker 1969; Young *et al.* 1974).

Dormancy is not a uniform state within the development of plants but is rather a phenomenon covering a spectrum of different physiological conditions. Obviously, dormancy is the trough in the undulating course of growth activity, and consequently most scientists differentiate among several stages of rest intensity, or phases of dormancy. Doorenbos (1953) has proposed the following terms to describe these phases:

- *Summer-dormancy*, when extension growth of buds is prevented by physiological processes inside the plant but outside the bud
- *Winter-dormancy*, when extension growth of buds is prevented by an inhibitive system inside the bud
- *Imposed dormancy*, when extension growth of buds is prevented by external causes, directly and reversibly imposed by the environmental conditions, mainly in late winter

This terminology has been adopted in principle by many other authors. However, some of them have introduced other terms, which often are used more or less synonymously. Summer-dormancy has also been named *correlated inhibition* or *predormancy*. Winter-dormancy is identical in most cases with *innate dormancy*, *true dormancy*, or *rest*. The state of imposed dormancy after termination of winter-dormancy has also been termed *post-dormancy* or *after-rest*.

Unfortunately, the usage of certain terms varies among authors.

Williams *et al.* (1979) have not applied the term winter-dormancy synonymously to rest, but jointly for correlative inhibition + rest + environmental dormancy. Seibel and Fuchigami (1978) have also differentiated between dormancy and rest, considering rest to be only the deepest stage of winter-dormancy. The term *quiescence*, which has been used by Hill and Campbell (1949) and Romberger (1963) as a synonym for imposed dormancy, has been used by Samish (1954), Vegis (1964), and Noodén and Weber (1978) to cover both imposed dormancy and correlative inhibition, although there is a great difference between arrested growth from a lack of external growth promotion and arrested growth from an active endogenous inhibition.

The terminology of Doorenbos (1953) has some disadvantages. First, the term winter-dormancy does not indicate that this stage is reached during midsummer, the rest intensity being at its maximum at about leaf fall, and that rest may be completed in a cool climate by midwinter or even before (Section II.B). Further, the term does not apply to subtropical and tropical regions where virtually no real winter exists, but where dormancy is yet very pronounced. According to the system of Doorenbos, the final stage of imposed dormancy after termination of winter-dormancy should be broken as soon as favorable environmental conditions prevail in spring, when buds are expected to break and grow freely. This is what actually happens in a cool climate. However, there are many observations that after warm and/or short winters budbreak may be delayed, although the inhibition imposed by environment has already disappeared. Apparently, there can still be a very marked rest influence even when the buds are already able to break, resulting in a sluggish shoot growth (Brown and Kotob 1957; Chandler 1960; Kawase 1966; Couvillon and Hendershott 1974; Mielke and Dennis 1978; Amling and Amling 1980). Therefore, physiologically, the final stage of dormancy in a temperate climate may be quite different from the final stage of dormancy in a warm climate.

Because of the difficulties with the terminology of Doorenbos, a modified system will be used in this review. This system, which separates the source of inhibition from the season of inhibition, is illustrated in Fig. 6.1.

In this system, *predormancy* is the stage during which the lateral buds are directly prevented from breaking by the growing shoot tip and/or by adjacent leaves. The later termination of growth in apical buds, while growing conditions are still favorable, is assumed to result from an indirect inhibition by a negative feedback mechanism between shoots and roots: Phases of active shoot growth—and equally of rapid fruit development—cause a reduction of root growth, which then

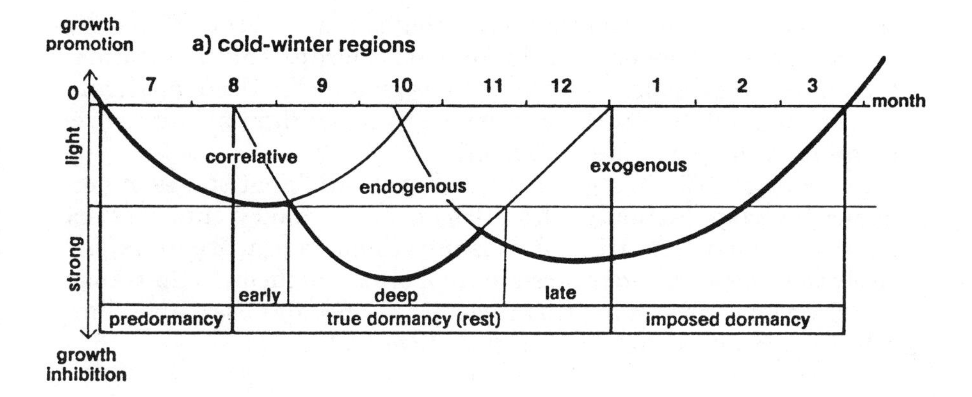

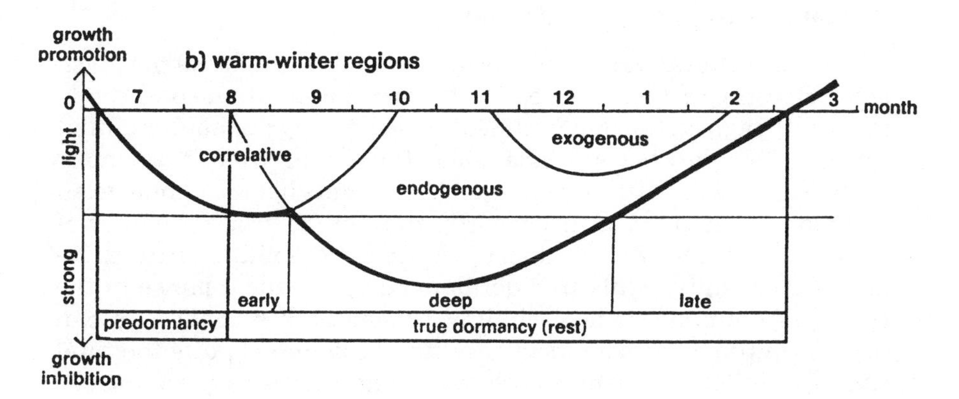

FIG. 6.1. The course of dormancy and its components as influenced by internal and external factors.

in turn slows down shoot growth (Saure 1971b, 1981). Accordingly, predormancy may be considered as the expression of a direct or indirect correlative inhibition. During the second stage, or *true dormancy (rest)*, the source of inhibition is located within the buds; true dormancy, therefore, may be considered as the expression of an endogenous inhibition. This true dormancy is the main subject of this review. As it is a characteristic of the individual bud, not all buds must be in the same stage of dormancy at a given time (Section III. B). After termination of true dormancy there may be—but need not be—inhibition of budbreak by adverse environmental conditions, especially low temperature. This is referred to as *imposed dormancy* and may be considered as the expression of an exogenous inhibition.

The resulting course of dormancy may vary considerably in different environments due to different proportions of endogenous and exogenous inhibition. Correlative inhibition may be little affected by environment, but endogenous inhibition is more extended in warmer than in cooler regions, for reasons to be reviewed in Section III, and exogenous inhibition is more extended in cooler than in warmer regions due to longer winters (Fig. 6.1).

Doorenbos (1953) has pointed out that the different stages of dormancy overlap. Actually, the division of dormancy into distinct phases is somewhat artificial. It is impossible in practice to fix precisely the inception and the termination of each of them. In the transition from one phase to another there are no specific observable characters to permit an unequivocal classification.

B. Identification of True Dormancy

The main characteristic of true dormancy is that plants may well enter this stage rather independently from the environmental conditions (although certain environmental factors may promote or inhibit this development) but are incapable of emerging from it autonomously. This peculiarity does not exist in all plants but is common to all deciduous fruit trees. Consequently, there is no true dormancy if plants can resume growth when environmental conditions are favorable. Accordingly, while true dormancy is the central phase in the course of dormancy where chilling or some other conditions can trigger continued development resulting in budbreak, it is the final stage of development where such triggering factors are permanently absent. Without renewed extension growth, the fluctuating curve of growth activity of the whole plant is damped and levels out rather soon, resulting in early senescence.

The identification of true dormancy is very important because a plant in the state of true dormancy responds quite differently to external influences than does a nondormant plant. It is well known that during true dormancy low temperatures may hasten and high temperatures may delay development (Sections III.A and IV.B), whereas the reverse is true in active plants. Similarly, Dostál (1942) has pointed out that many methods that are effective in forcing budbreak early in dormancy tend to delay budbreak when applied after termination of dormancy. Comparing results from identical treatments without considering the state of dormancy of the treated plants has contributed much to the confusion that exists about dormancy.

True dormancy is generally considered to have started when budbreak does not occur after either artificial defoliation and/or decapitation of the shoots although the external conditions are favorable for growth. The completion of true dormancy has been determined in many different ways. Some scientists have worked with isolated cultured buds, whereas many others have used excised shoots, either with or without terminal buds. Williams *et al.* (1979) suggested that it is valid to draw inferences about bud growth *in situ* from studies of detached shoots, although they observed that pruning the shoots may have a rest-breaking influence. However, Cerny (1963) got different responses from rooted cuttings and unrooted shoots, and Biggs (1966) found that the branch bioassay had a higher sensitivity than did greenhouse tests with seedlings in tests of the influence of chemical treatments on bud dormancy. Plancher (1983), working with black currant (*Ribes nigrum*), recently has warned against using budwood, one-node cuttings, and isolated buds for dormancy investigations, as the results may be modified by the method of sampling and by the material itself. Couvillon *et al.* (1975) found no significant differences in the rest period of rooted cuttings and mature trees. Therefore, rooted cuttings should be preferred for such experiments.

Besides different plant materials, numerous different treatments have been used to identify completion of true dormancy. Usually a series of samples is exposed consecutively to forcing temperatures following various temperature treatments in order to identify the present state of dormancy. However, some researchers have used constant temperatures, whereas others have worked with varying temperatures. Some have kept the material in permanent, others in intermittent, light. In addition to the light intensity, the source of irradiation has varied considerably. In some cases, in addition to forcing temperatures, different concentrations of gibberellic acid (GA) have been applied either to detached shoots (Hatch and Walker 1969) or to vigorous trees (Walser *et al.* 1981) in order to estimate the state of dormancy by the concentration of GA required to induce bud growth (Section VI.D.1).

In these experiments, various measures have been used to indicate the termination of dormancy:

- When buds showed green coloration (appearance of calyx through bud scales) within a 2-week period of forcing at 24°C (Couvillon and Hendershott 1974)
- When terminal buds had turned green or were swollen considerably within 2 weeks at greenhouse conditions (Hatch and Walker 1969)
- When 50% of the buds on the cuttings had reached a minimum of flowering stage 2 (green tip) within 21 days of forcing (Weinberger 1950) or 14 days at 21°C (Mielke and Dennis 1975)
- When 4 buds on a plant had broken at 21°–24°C (Norvell and Moore 1982)

However, Amling and Amling (1980) stated that the termination of, or release from, dormancy as determined by many researchers merely reflects a low enough intensity of true dormancy to permit budbreak within an arbitrary period of time under temperatures suitable for growth, but that a further reduction in dormancy can be accomplished by additional chilling. This agrees with the findings of Paiva and Robitaille (1978a) in apple and corresponds to the experience reported by many after warm and/or short winters (Section II. A). Accordingly, disagreements in calculating the duration of dormancy, mostly expressed as *chilling requirement* (Section III), may at least partially be due to a lack of standardization of evaluation methods on an international scale, as noted by Samish and Lavee (1962).

Obviously, true dormancy, like dormancy in the general sense, has a fluctuating intensity. The stage of lower intensity toward the end of true dormancy has been termed *after-rest* by Samish (1954) and *post-dormancy* by Vegis (1964). Vegis (1964) observed that in this phase dormant organs are able to start their growth within the limits of certain external conditions, which are characteristic for each species or cultivar. He pointed out that budbreak in this transitional stage may occur initially only in a narrow range of low temperature, which widens until growth activity has reached its maximum, and that temperatures exceeding this range may induce some kind of *secondary dormancy*. Since I have applied the term true dormancy for those cases where the inhibiting mechanism is assumed to be primarily located within the bud, and since the same mechanism is likely still to be acting to a lesser degree during this post-dormancy of Vegis, I prefer instead the term *late dormancy* within true dormancy as shown in Fig. 6.1.

The preceding discussion indicates that budbreak alone is not sufficient to determine the termination of true dormancy. Nor can the percentage of budbreak within a certain period be taken as a valid indicator of the degree to which dormancy is completed because many vegetative buds may fail to grow for reasons other than dormancy (Borkowska and Powell 1979) and because the influence of apical dominance may vary considerably among different species and cultivars.

A more valid indicator that plants have not only entered late dormancy but have fully completed true dormancy should be the speed of budbreak, parallel to the experience with seeds that the speed of germination is a better means of evaluating the stage of afterripening than the final percentage of germination (Kepczynski *et al.* 1977). Accordingly, Gurdian and Biggs (1964) proposed to take into account both the pattern of budbreak and the capacity for shoot growth. Couvillon and Hendershott (1974) observed that budbreak may occur within 72–84 hr after raising the temperature in peaches that had completed true dormancy but had been prevented from growing by imposed dormancy. Therefore, a rather short period of forcing appears sufficient to identify the completion of true dormancy definitely, and to reduce considerably the discrepancies in calculations of the chilling requirement.

C. Site of True Dormancy

True dormancy, as defined in Section II. A, is the expression of an endogenous inhibition having its source inside the buds. Therefore, the site of true dormancy quite naturally must be the bud. Additionally, the whole plant, or at least its aboveground parts, is in a resting state during true dormancy of the buds.

Several authors (Coville 1920; Denny and Stanton 1928; Black 1952; Doorenbos 1953; Jacobs *et al.* 1981) have come to the conclusion that dormancy is strictly limited to buds, i.e., it is not systemic. The work of Denny and Stanton (1928) with lilac (*Syringa vulgaris*) has provided especially strong evidence that this is so because they were able to force single buds to break and to grow normally without influencing neighboring buds. Within the buds, the scales seem to play an important role (Abbott 1969; Schneider 1968; de Maggio and Freeberg 1969; Tinklin and Schwabe 1970); their removal often leads to accelerated bud opening and leaf development. However, there may be no visible response to the removal of bud scales during deep dormancy (Leike 1967).

Some authors have postulated that other parts of the tree are also involved in true dormancy. This does not seem unreasonable, as other meristems, like the cambium, are considered to be temporarily in a state of dormancy, too (Reinders-Gouwentak 1965). Chandler (1960) concluded from his experiments with grafting nondormant apple shoots on either dormant or nondormant apple trees that some rest influence may move acropetally from the dormant tree to the grafted shoot through the graft union. This has been confirmed for pears by Westwood and Chestnut (1964), who found indications for a partial transfer of a rest influence from the host tree to a newly placed bud. Chilling experiments in apples (Nesterov 1967; Young and Werner 1983), but not in peach (Werner and Young 1984), suggest that a certain degree of dormancy in the rootstock influences the vigor of nondormant buds.

The roots themselves, or the root primordia, probably never become dormant (Doorenbos 1953; Samish 1954). However, H. A. A. van der Lek (cited in Doorenbos 1953) noted that although the root primordia in cuttings of poplar (*Populus candidans*) were not dormant, their elongation was restricted while the buds were in true dormancy. Richardson (1958) observed that in seedlings of *Acer saccharinum* the completion of bud dormancy and the initiation of new root growth are closely connected, and that root growth is possible only when at least one bud is physiologically nondormant. Active buds have also been found necessary for the resumption of cambial activity (Reinders-Gouwentak 1965).

Rooting ability, i.e., the capacity of the shoots to regenerate roots, is also low during true dormancy of the buds (Fadl and Hartmann 1967; Howard 1965, 1968). However, Howard (1968) expressed doubt that it is really the buds that influence rooting percentage. Young and Westwood (1975) and Erez and Yablowitz (1981) were also unable to establish unequivocally a relation between bud activity and inhibition or stimulation of rooting.

D. Prolonged Dormancy

As reported in Section II. A and illustrated in Fig. 6.1, in cool regions the period of true dormancy is terminated rather soon. However, budbreak is prevented for some time by a period of imposed dormancy due to adverse environmental conditions, although considerable growth and development of the buds may occur during brief periods of favorable weather, permitting speedy growth after the cessation of those limitations.

In warm-winter regions, the period of true dormancy is extended; budbreak occurs as soon as true dormancy is completed, or sometimes even during late dormancy, without an intervening period of imposed dormancy. If the winters become too mild and/or too short or, as in the tropics, are nonexistent, true dormancy may be extended still further. Indeed, in deciduous fruit trees, dormancy becomes nearly irreversible in the tropics. Under these conditions, the symptoms of *prolonged dormancy* or *delayed foliation* may occur. As described by Chandler *et al.* (1937), Hill and Cottingham (1949), Black (1952), and Skinner (1964), these symptoms include the following:

- Delayed, protracted, and very weak leafing
- Formation of bare, unbranched shoots that become increasingly shorter due to shorter internodes and increasingly swollen with each growth flush

- Shortage of spurs capable of forming flower buds
- Quickly declining growth vigor and, thus, early senescence of the trees; sometimes vigorous new growth from the base of the tree
- Delayed and protracted flowering season
- Poor fruit development and irregular ripening

In this review, the term prolonged dormancy will be used rather than delayed foliation because it is less specific and also indicates that the phenomenon is an extension of true dormancy whose duration is inversely proportional to the duration and/or the severity of the winter at the respective location.

In warm regions, the irregularities in flowering and fruiting associated with prolonged dormancy are the result of abnormal flower bud development. Flower buds are more sensitive than other buds when dormancy is abnormally prolonged; for example, leaf buds are usually not damaged or shed (Black 1952), and in rabbiteye blueberry (*Vaccinium ashei*) insufficient chilling affects floral buds more than vegetative buds (Spiers 1976). Black (1952) stated that during prolonged rest, flower buds in peach, and possibly in other deciduous fruits, are frequently structurally underdeveloped, the pistils in many cases being dwarfed. Consequently, the flower primordia often abort and the flower buds abscise in different stages of development. Zeller (1973) found, in apple trees grown in the tropics, that many flower bud primordia died during the later stages of flower bud development, but that new spurs or lateral inflorescences could grow in such buds from basal axillary meristems, thus contributing to the irregularity of flowering.

However, Samish (1954) suggested that delayed budbreak and flower bud abscission may be distinct phenomena. This is supported by observations of Brown (1958) that the abortion of flower primordia may occur rather early in autumn—and therefore cannot be attributed to insufficient chilling or prolonged dormancy—and by observations of Zeller (1961) that abortion can be observed under certain conditions and at certain periods even in a cool climate. Moreover, such abnormalities do not occur even without any chilling if flower buds (e.g., of peach) are forced to open prematurely by defoliation in late summer (Lloyd and Couvillon 1974). Therefore, it is best to distinguish between symptoms of prolonged dormancy and symptoms of abnormal flower bud development, which both may occur simultaneously in warm-winter regions.

III. THE CHILLING REQUIREMENT

A. Effective Temperatures

It is generally accepted that deciduous fruit trees must be subjected to low temperatures for a certain period for dormancy release to occur and that this chilling requirement is genetically determined. The term *chilling requirement*, therefore, has often been used nearly synonymously with *rest intensity*, on the assumption that the higher the intensity of dormancy, the more chilling is required for its release. Sufficient chilling in a cool-winter climate causes a rather quick decrease in the intensity of true dormancy; however, this decrease is masked under these conditions by a subsequent period of imposed dormancy. Insufficient chilling is the cause of the slower reduction of true dormancy in warm-winter climates, where, therefore, true dormancy extends over a longer period. When there is no chilling, the intensity of true dormancy would not decrease at all, leading to an extreme case of prolonged dormancy.

Coville (1920) pointed out that low temperatures do not necessarily mean freezing; he found temperatures some degrees above freezing to be sufficient for breaking dormancy in blueberry. Chandler and Tufts (1934) noted that −1°-0°C was quite effective in promoting budbreak of peach. However, Chandler *et al.* (1937) later reported that temperatures of 0.5°-4.5°C were as effective or even better than freezing temperatures, whereas temperatures of 9°C were less effective.

In more detailed experiments, Erez and Lavee (1971) obtained maximum rest-breaking efficiency in peach lateral leaf buds at 6°C, and in terminal buds at 8°C. In leaf buds, 10°C was only half as effective as 6°C in breaking dormancy, and 18°C was without any effect. Temperatures between 3° and 6°C had the same effect as 6°C; in lateral leaf buds of weaker lateral shoots, 3°C was even slightly better than 6°C. Unfortunately, there were no observations below 3°C.

Later, Erez and Couvillon (1982) reported that the maximum dormancy-breaking effect of continuous low temperature, for both vegetative and floral peach buds, occurred at 8°C; some small effect was observed at 0° and 12°C, and no effect at 14°C. Richardson *et al.* (1974) considered 6°C to be the optimum dormancy-breaking temperature in some peach cultivars. They did not assume a chilling effect above 12.5°C and, unlike Chandler and Tufts (1934), below 1.4°C. Gilreath and Buchanan (1981b) identified 8°C as the most effective temperature in peach cultivars having a low chilling requirement, with decreasing chilling effects observed at temperatures up to 14°C and down to 0°C. Norvell and Moore (1982), working with highbush blueberry (*Vaccinium corymbosum*), confirmed the values determined by Richardson *et al.* (1974) but got some chilling effect even below 1.4°C. In apple, Thompson *et al.* (1975) found 2°C to be more effective than 6°C, and 6°C more effective than 10°C. Shaltout and Unrath (1983b) obtained some small chilling effect in apple even at −0.6°C; however, the maximum effect was at 7.2°C, and the upper limit of chilling effect was 16.5°C. Rageau and Mauget (1982) have shown that in peach and walnut even temperatures above 15°C could induce a decrease of dormancy. However, although initially the bud growth capacity was increased, it finally remained below that observed in buds held at lower temperatures, especially in buds on whole trees.

Because of the varying efficiency of different temperatures in inducing dormancy release, Erez and Lavee (1971) concluded that the earlier methods of calculating the chilling requirement by adding all the hours at 7°C and below, as proposed by Weinberger (1950), would be insufficient. They recommended that chilling be measured not by such *chilling hours* but by *weighted chilling hours*. This concept has been adopted by Richardson *et al.* (1974) and others and used in calculating *chill units*: The smaller the effect of a temperature in completing dormancy, the lower its chill-unit value, and, correspondingly, the more hours are needed at this temperature to get the same effect as with more efficient temperatures.

All the methods for calculating chill units proceed on the assumption that the efficiency of each temperature in stimulating dormancy release is constant throughout the whole period of true dormancy. This has been criticized by Kobayashi *et al.* (1982), who hold that the temperature effect on the development of dormancy varies with the growth stage. This might explain why Thompson *et al.* (1975) found chilling of young apple trees early in the dormant period to be less effective than later chilling, with less growth of the main shoot and lower final number of buds growing.

Coville (1920) stated that low temperatures not only prepare the plant for budbreak but also lower the threshold values for bud growth; he found that buds of blueberry had begun to swell even at temperatures just below 0°C after being kept for about 9 months at this temperature. This lower temperature requirement for budbreak in the course of dormancy release has been discussed by Vegis (1964), as described in Section II.B.

In obvious contrast to these observations that temperatures above freezing are most effective for dormancy release are some experiments with black currant in which severe cold (−15° to −16°C) caused reasonable budbreak when applied for short periods of 5-60 min (Thomas and Wilkinson 1964; Tinklin and Schwabe 1970). Similar results were obtained by Sparks *et al.* (1976) in pecan (*Carya illinoensis*) with subfreezing temperatures down to −13°C for one night. The authors do not consider this effect to be from an accelerated completion of the chilling hours required for dormancy release. It could well be a response to sublethal conditions since extreme low-temperature treatment killed many buds in black currants and bud scales in pecans (A. Erez, personal communication).

B. Bud Differences

It is generally accepted that not all buds on a tree have identical chilling requirements, but each may behave individually. Normally, flower buds have a lower chilling requirement than vegetative buds, and terminal vegetative buds have a lower chilling requirement than lateral ones (Samish and Lavee 1962). Even among lateral leaf buds there may be noticeable differences, depending on the section of the shoot to which they belong, on the position within the tree, on the vigor of this shoot, etc. (Section IV.A).

Chandler (1960) found that in apple the rest influence was weakest in the latent buds of older wood. This agrees with earlier findings of Simon (1906) who assumed that those less-developed buds possibly lack the inhibiting mechanism that causes true dormancy. In long shoots of apple, Chandler (1960) observed that the rest influence was stronger in the basal than in the apical part in late summer and autumn, but in spring after a winter with adequate chilling, it was weaker in the basal part. Jacobs *et al.* (1981) stated that, apart from the terminal bud, the regrowth of apples suffering from prolonged dormancy occurred mainly from the lateral buds on the lower halves of 1-year-old shoots. Crabbé (1984), working with single-node cuttings, reported that, in upright shoots of apple rootstock M7, the percentage of nonbursting buds in December was lowest in the basal part of the shoots and increased toward the shoot apex, and the time required for 20% budbreak was shortest for basal buds. However, in March, budbreak was faster at the top than at the base, even though the percentage of nonbursting buds was the same in all parts of the shoot. This agrees with observations of Arias and Crabbé (1975) in sweet cherry buds where the first phase of true dormancy is characterized by a basitonic gradient of bursting capability, whereas toward the end of dormancy there is a return to acrotony. However, Borkowska and Powell (1979), working with isolated apple buds, found no consistent differences in growth responses among buds from apical, middle, and basal shoot regions. Terminal buds also may exhibit marked differences in chilling requirement; for example, apple spurs became dormant later and emerged earlier than terminal buds of short, medium, or long shoots (Latimer and Robitaille 1981). In apricot flower buds, the intensity of dormancy and the amount of chilling required to break it did not necessarily depend on the stage of bud development (Brown and Abi-Fadel, 1953).

Under conditions of limited chilling, terminal buds may break long before the lateral buds, due to their lower chilling requirement. This results in an apical dominance stronger than typical in a cooler climate, where all buds may start at once, and causes the growth habit typical for prolonged dormancy with long, slender branches and only a few spurs, as described in Section II. D. It may explain the observations of Erez *et al.* (1979b) that terminal bud opening was enhanced by all treatments that inhibited lateral bud opening. Actually, lateral buds thus can be prevented from breaking even if their chilling requirement finally has also been fulfilled (Erez and Lavee 1974).

There are some indications that apical dominance acts also during true dormancy. Leike (1967) proposed to work only with terminal buds because he concluded from the literature that lateral buds are only partially dormant or are inhibited only correlatively. Paiva and Robitaille (1978a) observed development of the uppermost remaining bud after decapitating detached apple shoots at all sampling dates from autumn to spring, but when a dormant terminal bud was present the lowermost bud on the shoots developed. Notching above a bud of a decapitated shoot causes this bud to break additionally. Williams *et al.* (1979) concluded that wounding (e.g., cutting of a shoot) has a rest-breaking influence. However, Latimer and Robitaille (1981) stated that the rate of development of the uppermost bud after topping the shoot varied with the intensity of rest. In apples suffering from prolonged dormancy, bud removal from upper or lower halves of shoots did not affect the sprouting of the remaining buds, except that terminal bud removal enhanced the sprouting of upper lateral buds, and buds induced to sprout had no effect on sprouting of untreated buds (Jacobs *et al.* 1981).

Meng-Horn *et al.* (1975) suggested that there may also be correlative effects between buds and stem tissues, causing less outgrowth of single buds on an otherwise disbudded shoot piece when they were situated at its base than when situated at its top. These "shoot effects" vary with the time of year and are strongest in autumn (Crabbé 1981).

Hatch and Walker (1969), forcing bud growth of peach and apricot from October to January by GA treatments, concluded from their experiments that the mechanism of dormancy might be different in leaf and flower buds. This idea has been supported by the findings of Lloyd and Couvillon (1974) that, in peach, defoliation promoted budbreak of flower buds increasingly from summer to autumn; in contrast, vegetative buds responded increasingly less to defoliation during the same period, thus indicating the progress of true dormancy. However, the physiological factors responsible for these differences need to be investigated further.

C. Genetic Differences

Among different species, and among cultivars within species, there are large differences in chilling requirement. In most cases, the chilling requirement of cultivars is stated as the minimum number of hours of chilling below 7°C (45°F). This statistic is rather rudimentary as it does not take into account the differing efficacies of temperatures below and above 7°C (Section III.A) or the negative effect of higher temperatures (Section IV. B). Therefore, the values for chilling requirement reported here serve only as a rough estimate. However, there is as yet no generally accepted method of calculation that might be more reliable (Section VI).

The bulk of information on the specific chilling requirements of different deciduous fruit tree cultivars from all over the world has been collected by Ruck (1975). Since then, some additional information, especially about newly bred cultivars, has been provided by Tabuenca (1976, 1979b), Sherman *et al.* (1977), Nakasu *et al.* (1982), and Mũnoz *et al.* (1984). Table 6.1 gives an impression of the range and variability of the chilling requirement in different species. In areas or in years with limited winter chill, low-chilling cultivars will flower much earlier than high-chilling cultivars, whereas they will flower at about the same time in cool areas. This may cause pollination problems because of insufficient overlapping of the blossom periods.

The low-chilling-requirement characteristic is heritable and therefore may be utilized in breeding of deciduous fruits for better adapta-

TABLE 6.1. Chilling Requirements in Hours below 7°C of Some Deciduous Fruit Species

Species	Number of chill hours
Almonds	0–800
Peaches	100–1250
Japanese plums	100–800
Apples and pears	200–1400
European plums	800–1500
Cherries	800–1700

Source: Ruck (1975).

tion to warm regions. This has already been done to some extent in different parts of the world (Black 1952; Oppenheimer 1962; Sherman and Sharpe 1970; Bowen 1971; Sharpe and Sherman 1971; Sherman *et al.* 1977; Nakasu *et al.* 1981). However, little is known about the genetics of the chilling requirement.

In peach, Lesley (1945) reported on bud mutations having a shorter dormancy. From his breeding experiments, he concluded that chilling requirement depends on multiple genes, with transgressions in both directions. Later, he found that the winter-chilling requirement in the F_1 was, as a rule, distinctly nearer to the shorter-chilling parent (Lesley 1957). Lammerts (1945) argued that the lower-chilling requirement is due to the cumulative effect of a series of recessive factors and probably several semidominant ones. In apple, Oppenheimer (1962) found a wide variation from very early to very late budbreak in progenies. He later assumed that earliness in budbreak might be dominant (Oppenheimer and Slor 1968). Nesterov (1969) recorded a shorter dormancy in seedlings of the Siberian crabapple, which requires very little chilling, than in seedlings of large-fruited 'Antonowka,' a high-chilling apple cultivar. In hybrids between them, the duration of dormancy was intermediate but closer to the mother cultivar.

The chilling requirement of the seeds and buds of a cultivar is often similar. This has been shown in apple (Pasternac-Orawiec and Powell 1983a,b), pear (Westwood and Bjornstad 1968) and peach (Chang and Werner 1982). Therefore, the chilling mechanism for seeds and buds is assumed to be similar.

However, little is known about the causes of the differences in chilling requirement. Bowen and Derickson (1978) reported that low-chilling peach cultivars commonly remain active later in the fall than do high-chilling cultivars. Gilreath and Buchanan (1981b) observed that in low-chilling peach and nectarine cultivars the inhibition of dormancy release by intermittent higher temperatures (Section IV. B) is less than in high-chilling cultivars. They suggested that low-chilling cultivars should be more tolerant to higher temperatures than high-chilling cultivars, especially in the upper range of positive chill units. Earlier, Gurdian and Biggs (1964) reported that in low-chilling peach cultivars, increasing the chilling temperature from 7.2° to 12.8°C caused little discernible effect on the subsequent growth of any cultivar or, in most cases, on the final percentage of buds terminating dormancy. These findings seem to be supported by observations of Nesterov (1969) that the buds of crabapples, which have a short dormant period, may develop even if kept at 16.1°C during dormancy.

IV. MODIFICATION OF CHILLING REQUIREMENT BY ENVIRONMENT

A. Prechill Growth Promotion

As mentioned in Section III. B, the chilling requirement of leaf or flower buds may differ within a given tree. Chandler and Tufts (1934) pointed out that such differences are quite common for deciduous fruit trees after warm winters and that they result from differences in the duration of the preceding shoot growth. They noted that after insufficient chilling, the blossoms on long suckers of peach trees were just opening when the fruits on shoots less than 30 cm in length were already nearly ripe. In leaf buds, budbreak on a young 'Northern Spy' apple occurred about 7 weeks later on long shoots than on shorter shoots, and 11 weeks later than on older, weaker trees. Apparently, various external factors enhancing vigor and duration of growth at the same time may enlarge the chilling requirement. However, this does not seem to apply to prolonged succulence of the shoots by shade in summer (Chandler 1960).

Chandler and Tufts (1934) found that the differences in chilling requirement of long shoots appeared only when growth continued exceptionally late in summer. It was not the nitrogen status of the trees but the presence of late-growing watersprouts that caused the delayed budbreak. Accordingly, nitrogen supplied to young pot-grown 'Golden Delicious' apple trees even promoted early budbreak and flowering if the nitrogen was provided during autumn and winter after the extension growth of even the longest shoots had already ceased, so that the duration of the growing season was not extended (Terblanche and Strydom 1973). In a later experiment, this effect of autumn N nutrition was confirmed by Terblanche *et al.* (1979).

There is some evidence that rootstocks may influence the chilling requirement of the scion in some species, e.g., pear (Westwood and Chestnut 1964), peach and almond (Tabuenca 1979a), and apple (Couvillon *et al.* 1984). However, the nature of this influence has not been established so far.

From the fact that an extended growth period of shoots may delay budbreak the following spring if there is insufficient chilling, one could conclude that a shortened growth period of the shoots caused by environmental factors might reduce the chilling requirement. Indeed, this has been confirmed already by Müller-Thurgau (1885), who reported that withholding the water supply to the roots was a common practice to promote early inception of dormancy and to shorten its duration. He further referred to observations of Knight (1801) and

Bouché (1873) that fruit trees forced to early flowering in one winter tended to flower earlier and to develop faster after forcing them again the next winter, although the capacity of the buds to set fruit was reduced. This reduced chilling requirement has been attributed to some kind of weakened bud development in these trees. Chandler and Brown (1957) noted that a deficiency of nitrogen and water may cause buds to open slightly earlier after a warm-winter period provided that this stress situation is not shortened by early autumn rains, which then could yield the opposite effect. This experience has long been utilized in apple and peach production in the tropics, where the withdrawal of irrigation water and the exposure of the roots for some time may help to force budbreak (Section V.A).

As dry conditions may induce early leaf fall and prechill growth promotion tends to delay leaf fall, the role of leaves in dormancy needs to be considered, too. In the fertilizer experiments of Terblanche *et al.* (1979), leaf drop in autumn was delayed whereas budbreak in spring was considerably enhanced. However, Walser *et al.* (1981) suggested that leaves retard dormancy release because they found that the longer leaves remained on the tree in autumn the longer and deeper dormancy was. The understanding of the role of leaves has been further confused by contradicting results after mechanical and/or chemical defoliation. In some experiments, bud development was inhibited and budbreak was delayed (Fuchigami 1977; Podešva *et al.* 1980; Mannini and Ryugo 1983). In others, no noticeable effect of defoliation on budbreak was found (Broome and Zimmerman 1976; Mielke and Dennis 1978). Crabbé (1984), removing the youngest leaves of *Prunus avium* in mid-June or mid-July, got a strongly reduced dormancy of the buds located in the defoliation zone in October. In the tropics, artificial defoliation is an efficient method to induce floral budbreak in deciduous fruit trees (Section V. A). An intensification of dormancy by well-functioning leaves seems also to be supported by the findings of Lakon (1917) that the less chlorophyll in the leaves on a branch of mottled and white-leafed forms of *Acer negundo*, the later the buds became dormant, and the earlier they were to force.

B. Intermittent Warm Periods

In warm regions, high temperatures during winter are well known to prolong dormancy. This effect has been analyzed to some extent by Weinberger (1954, 1956, 1967a,b). He found that intermittent days with rather high temperatures (above 27° and 32°C, respectively) delayed budbreak more than did a moderate continuous temperature elevation (2° or 3°C) throughout the winter. The magnitude of this delay caused by warmth in November and still more in December was positively correlated to the mean maximum but not to the mean minimum temperatures in these months. Weinberger (1967b) concluded that the delay in budbreak was primarily the result of high temperatures in November and December and that, consequently, the dormancy-breaking effect of chilling must be at least partially reversible. However, in leaf buds the negative effect of a warm December could be completely overcome by subsequent cool weather.

Weinberger (1967a) found that the way warm winter temperatures influence budbreak is quite different from the way they influence flower bud abscission. The recovery of flower buds from abnormalities due to detrimental high temperatures in December was limited, and the abscission of flower buds was not correlated to higher maximum temperatures but to higher minimum temperatures in December and January. A mean minimum temperature of 4.5°C in these months appeared to be critical for those cultivars susceptible to flower bud drop, and a slightly higher mean value was critical for less susceptible cultivars. However, Brown (1958) found that the decay of flower buds was associated with warm periods in late September to early October, and in late December to late January, but with higher minimum temperatures only in December.

Bennett (1949) had observed that only a few hours of daily warm treatment at 22.8°C were sufficient to partially offset the effect of a daily 18-hr cold treatment. This was confirmed by Overcash and Campbell (1955), who found that the same total number of chilling hours at 4°C was less effective when chilling was interrupted by 8 hr above 21°C in a daily cycle. Erez *et al.* (1979a,b) found that 8 hr at 15°C—a temperature that by itself could not remove dormancy—even increased the efficiency of chilling when used in a daily cycle with 6°C; however, no such promotion could be effected by 18°C, and higher temperatures increasingly negated the effect of chilling. However, even the highest temperature tested (24°C) lost its negating effect when the cycle length was extended, a low-chilling cultivar needing less extension than a high-chilling cultivar. This supports the thesis of Erez and Lavee (1971) that the reversal of the chilling effect by high temperatures may be prevented after some days by some kind of fixation process.

The "nullifying" effect of higher temperatures has led to the calculation of negative chill units corresponding to the positive chill units mentioned in Section III.A. Chill units are discussed in greater detail in Section VI.

Kobayashi *et al.* (1982), who have criticized the assumption of a constant rate of dormancy development at each temperature throughout the whole course of dormancy, perceived a biochemical or physiological shift in the effect of high temperatures between the deepening and the decreasing phase of dormancy. They consider this changing effect of high temperatures during dormancy as one factor contributing to the inconsistencies reported for the chill-unit models of other authors.

When temperatures become exceptionally high, reaching 35°C and above, they may lose their inhibiting effect and, instead of delaying, may promote budbreak even after only short periods of action. Immersing leafless shrubs or excised branches in warm water for about 12 hr has been a recommended practice for forcing woody ornamentals (Doorenbos 1953). According to Bennett (1949), a few hours or days at 43°–54°C will often start dormant buds into growth. Chandler (1960) found that hot weather periods in August, with maximum air temperatures in the orchard up to 43°C, had a strong dormancy-breaking influence. In heating experiments, he induced budbreak in apple trees held at 44°–45°C for 6 hr on two successive or even on only one day in July, October, or November. Weaver (1963) and Shulman *et al.* (1982) induced sprouting in grapevine cuttings by short-term hot-water treatments at 52° or 50°C.

However, when comparing a warm-water treatment at 45°C for 3.5 hr with 8–12 weeks of chilling at 3°C, Chaudhry *et al.* (1970) found that although both methods caused budbreak in vegetative buds of pear (*Pyrus communis*), the bud activity was much more pronounced and the sprouting more vigorous in the cold-treated plants. Powell and Huang (1982) pointed out that in apple trees heating can replace the chilling requirement only if buds are in a light state of dormancy; buds in deep dormancy did not respond.

C. Irradiation

Most of the experiments conducted to elucidate the role of both low and high temperatures in dormancy release have not considered the role of light. Such experiments generally have been carried out in the light, in the dark, or in light conditions varying with changes in temperature. This lack of concern about light seems to be justified by the findings of Coville (1920) that in blueberry plants, exposure to different amounts of light had nothing to do with the stimulation of growth: Chilled plants started into growth promptly whether the chilling was done in full light, in partial light, or in complete darkness.

In contrast, field observations in areas where plants may suffer from prolonged dormancy suggest that shade or frequent fogs during winter can promote budbreak, especially in plants with a high chilling requirement. For example, Buchanan *et al.* (1977) reported that nectarine trees covered with a 55% shade Saran cloth from November until after blossom time terminated their dormancy 10 days earlier than unshaded plants.

Freeman and Martin (1981) also found that low light intensity always significantly promoted floral budbreak in peach. However, shade may sometimes inhibit budbreak. Weinberger (1954) observed an appreciably delayed budbreak after reducing light by 50% when single branches of peaches were tightly covered with canvas. This effect was visible only when the branches were covered throughout November or December, but not when they were covered throughout January, when true dormancy possibly was already terminated.

Chandler *et al.* (1937) and Bennett (1949) have attributed the budbreak-promoting effect of reduced light in winter to lower temperatures because of less radiation energy. This view may explain the negative findings of Weinberger (1954) just mentioned because the maximum temperature under the canvas on sunny afternoons was about 9°C higher than in unshaded controls, whereas the differences in minimum temperatures were negligible. The results of Buchanan *et al.* (1977) do not exclude a role for temperature within the influence of light since they determined the termination of dormancy by the calculation of chill units. However, in the experiments of Freeman and Martin (1981), a temperature effect is excluded because low light intensity promoted budbreak even though the chilling temperatures were identical at the different light regimes.

Vegis (1964) concluded from his extensive review of the literature that deep dormancy can be overcome only by chilling and not by the action of light. However, he pointed out that during the early and the late phase of true dormancy, lengthening photoperiods can extend the temperature range at which budbreak may occur, thus permitting budbreak at increasingly lower and higher temperatures in these phases. This view has been supported in part by Erez *et al.* (1966, 1968), who found that peach leaf buds require a light stimulus, in addition to chilling, shortly before bud opening and that the response to this stimulation is increased by a preconditioning period in the dark. The effect of light on budbreak was more pronounced in long-day conditions than in continuous illumination. However, budbreak in terminal buds depended markedly less on light than did leaf buds in these experiments, and flower buds were inhibited by irradiation especially after dark-preconditioning, which could confirm the role of shade and frequent fogs mentioned already.

D. Leaching

Westwood and Bjornstad (1978) concluded that leaching of inhibiting substances by rain is an important factor in reducing the chilling requirement of buds. Their findings were based on the difference in behavior between resting trees exposed to natural rainfall and trees under plastic covers, and on experiments in which branches of 'Bartlett' pear were soaked in water for 1 or 2 days. These findings have been partially supported by Freeman and Martin (1981). However, Erez *et al.* (1980) attributed the effect of these immersion experiments to a reduction in oxygen level (Section VII.C) rather than to leaching. The effect of rain has been considered by Erez and Couvillon (1983) to be caused by reduced bud temperatures rather than by leaching. Gilreath and Buchanan (1981a) found no differences in the mineral content of the buds of irrigated trees that could be attributed to leaching, thus casting some further doubt on the importance of leaching to budbreak phenomena.

V. MODIFICATION OF CHILLING REQUIREMENT BY CULTURAL PRACTICES

A. Reduction of Chilling Requirement

Since deciduous fruit trees are grown in subtropical and even tropical regions, many attempts have been made to overcome the problems caused by insufficient chilling. Basicly, two approaches are possible: (1) preventing plants from entering true dormancy; and (2) hastening budbreak after plants have already entered true dormancy.

Preventing true dormancy is feasible only in regions without distinct seasons, where plants do not need a dormant period in order to survive adverse environmental conditions. True dormancy can be prevented by defoliating trees after the flower buds have been initiated but before these buds have proceeded from predormancy to true dormancy. This defoliation method has proven successful in several tropical countries, mainly with apple and peach, which thus may be forced to biannual cropping (Overcash 1967; Saure 1971a, 1973; Janick 1974; Giesberger 1975; Erez 1982; Sherman and Lyrene 1984). The leaves may be removed either directly by mechanical or chemical means, or indirectly by extended drought, which causes an early leaf fall. This method can be supplemented by pruning the roots and/or by exposing them to the air by removing the surface soil (Javaraya 1943).

The effect of defoliation can be improved by topping and/or bending the branches. Both practices, like the intermittent termination of extension growth by temporarily withholding irrigation water, reduce apical dominance and may either promote the induction and further development of lateral flower buds or, in rather vigorous trees, promote the development of spurs capable of flowering the next year. The opposite effects result from factors that extend the growing period, as discussed in Section IV.A, and by summer pruning, even though no new shoot growth may occur the same summer (Chandler and Brown 1957).

Hastening dormancy release after plants have entered true dormancy can be achieved by several methods:

- Increasing chilling through evaporative cooling (Gilreath and Buchanan 1981a; Erez and Couvillon 1983)
- Heat treatment at temperatures of 35°- 50°C (Chandler 1960; Shulman *et al.* 1982), described in Section IV. B)
- Late autumn application of N and irrigation (Terblanche *et al.* 1973, 1979), described in Section IV. A)
- Treatments with several chemicals

Most of the chemical treatments are effective at concentrations very near the lethal point (Black 1952) and would delay rather than accelerate development if applied after termination of dormancy (Dostál 1942). A list of chemicals that promote budbreak in one case or the other has been provided by Doorenbos (1953). Among them, oils, especially mineral oils, and several dinitrocompounds, such as dinitrophenol, dinitro-cyclohexyl-phenol, and dinitro-ortho-cresol (DNOC), have long held a reputation for promoting budbreak. In deciduous fruits, mainly DNOC, or a mixture of DNOC and mineral oil, has been applied commercially for this purpose in many parts of the world.

However, Strydom and Honeyborne (1971) often observed no beneficial effect of these sprays even in areas with a relatively warm winter. They warned against damage from excessive concentrations as stone fruits, especially the young trees, are rather sensitive. Black (1936) found that the effect of oil sprays depends on the cultivar treated: Those highly resistant to delayed foliation showed little response, while cultivars with a high chilling requirement were easily influenced. The effect depends further on the stage of bud development, with flower buds reaching a critical stage giving best response somewhat earlier than leaf buds. There was no effect when spraying was done during deep dormancy, and E.R. de Ong in California (cited in Black 1936) even found a delaying effect of oil sprays on blossoming in prunes when applied in November.

Samish (1945) mentioned that it is difficult to find a "most effective" time to spray DNOC + mineral oil because of the varying effects of timing of sprays. Early spraying during late dormancy has a "forcing" action, causing earlier foliation and blossoming without eliminating the irregularity of these events. Late spraying during late dormancy has only a very slight forcing action but a "normalizing" effect tending to shorten the period of main bloom, to reduce the time between bloom and foliation, and to reduce the number of buds remaining dormant. The difference between the "forcing" and the "normalizing" effect may be partially due to a slight impairment of the terminal bud causing a reduced apical dominance (Paiva and Robitaille 1978b). Erez and Zur (1981) later stated that in 'Golden Delicious' apple trees this breaking of lateral leaf buds was mainly effected by the mineral oil component. Erez *et al.* (1971) noticed that the efficacy of DNOC + mineral oil is increased by increasing temperatures during the few days following treatment, and Erez (1979) concluded that this might be the cause for the somewhat unpredictable effect of these sprays. A further increase in the efficacy of DNOC may be achieved by N applications in late autumn (Terblanche *et al.* 1973, 1979) and by pruning late in the winter (Black 1952).

Many other chemicals are known to have some effect in "breaking" dormancy, too (Doorenbos 1953; Samish 1954). The effects of applied GA and cytokinins are discussed in Sections VII.D. 1 and 2. Some other compounds that seem promising for commercial use in deciduous fruit trees are discussed in the remainder of this section.

Thiourea (TU) was first shown to be effective during the late stage of true dormancy by Blommaert (1964). Erez *et al.* (1971) found that TU—like DNOC + mineral oil—had a more "forcing" action when sprayed early, about 6 weeks prior to bud swelling, and a more "normalizing" effect when applied later, about 4 weeks prior to bud swelling. This normalizing effect might result from slight damage to the terminal buds, as TU has been reported to break apical dominance (Samish 1954). Erez *et al.* (1971) further noted that TU affected peach leaf buds more than flower buds, although the figures presented by them are inconclusive in this respect. Thiourea may have a synergistic effect when applied together with DNOC, DNOC + mineral oil, KNO_3, or GA + DNOC. In experiments by Petri and Pasqual (1982), TU did not significantly increase the number of lateral flower buds in 'Golden Delicious' apple trees when it was applied with KNO_3, but it very substantially increased the efficacy of a DNOC + mineral oil spray. However, in red raspberry (*Rubus idaeus*) the number of fruit laterals per cane was significantly increased even when TU was applied alone (Snir 1983).

Potassium nitrate (KNO_3) is probably more effective in peach flower buds than in peach leaf buds, when applied alone, and it may increase the level of flowering in stone fruits when applied together with DNOC during late dormancy (Erez *et al.* 1971).

Cyanamide (H_2CN_2) has been applied successfully to different species of deciduous fruits in Japan (Morimoto and Kumashiro 1978) and is used commercially to promote budbreak in grapevine in the warm areas of Israel. There it promotes a uniform and more rapid bud opening (Shulman *et al.* 1982). In raspberry, Snir (1983) found that cyanamide increased yield and advanced harvest date when it was applied at a time during late dormancy when DNOC, KNO_3, and TU did not yet work. She therefore concluded that cyanamide may be involved in earlier stages of the dormancy-breaking process than the other chemicals tested.

B. Increase of Chilling Requirement

Little is known about the possibilities of increasing the chilling requirement. Most efforts to delay the blossom period in order to avoid damage from late spring frosts have been carried out by delaying bud development *after* the termination of true dormancy. An example of this is evaporative cooling by sprinkling, which causes some kind of imposed dormancy by delaying heat-unit accumulation. Also, the retarded bloom of sweet cherry (*Prunus avium*) following late summer sprays with ethephon results more from inhibition of elongation than from extended dormancy (Proebsting and Mills 1976). Likewise, Siberian C rootstock, which reportedly may delay bloom in peach, apparently does so only after rest is completed, i.e., without affecting the number of chilling hours (Young and Olcott-Reid 1979).

Weaver (1959) and Brian *et al.* (1959) have discussed the possibility of retarding budbreak by autumn applications of GA, which prolong the dormancy of buds (Section VI. D. 1). However, as this may cause deleterious effects in flower buds, and as the normally extended growing period in autumn due to these sprays may increase the danger of damage to the shoots by early winter frosts, this proposal apparently has not been followed up intensively. These problems also apply to early defoliation in autumn (Section IV. A), which may cause some of the effects that may result from GA sprays in autumn, i.e., reduced bud development and delayed bud opening.

VI. CALCULATION OF CHILLING REQUIREMENT

The ideal way to achieve dormancy release in warm regions would be to plant cultivars adapted to those conditions, i.e., having a chilling requirement low enough to be met regularly by the given chill supply. In this case, the chilling requirement of the respective cultivar, and the chill supply of that area, would have to be determined only once at the minimum in order to be sure of a good adaptation.

Several methods have been developed to estimate whether cultivars with a known chilling requirement will be adapted to areas where detailed meteorological data are not available; these methods are described by Crossa-Raynaud (1955), Weinberger (1956), da Mota (1957), and Bidabé (1967) as cited in Ruck (1975). All involve the determination of the number of hours at temperatures of 7°C and below required to break dormancy. They differ in the formulae used, in the assessment of the period to be selected for these calculations, and obviously in the results obtained in different regions, as discussed by Muñoz (1969), Ruck (1975), and del Real-Laborde and Gonzalez-Cepeda (1982).

However, most of the cultivars that best satisfy the requirements of both the market and growers still have a rather high chilling requirement. In this case, problems caused by insufficient chilling can be avoided only by certain cultural practices, outlined in Section V.A. These necessitate proper management based on a high standard of skill and equipment. For proper timing of these cultural practices, the termination of dormancy must be predicted and the actual state of dormancy, therefore, must be determined regularly.

To predict dormancy release from concurrent ambient temperatures, Richardson *et al.* (1974) have developed a mathematical model, the so-called "Utah chill-unit model," that is based in principle on the calculation of chill units (Table 6.2). This approach relates the environmental temperature to the completion of dormancy by calculating estimated hourly temperatures, then multiplying them by the appropriate chill units. This model has been critically assessed and discussed by Aron (1975), Ruck (1975), Gilreath and Buchanan (1981b), Norvell and Moore (1982), and del Real-Laborde and Gonzalez-Cepeda (1982). Richardson *et al.* (1975) concede that their model works well mainly in continental climates, but not in exceptional years.

Similar models, all based on the calculation of chill units, have also been developed by Gilreath and Buchanan (1981b) and Shaltout and Unrath (1983b) (Table 6.2). The problems involved in establishing such models have been further discussed by Batten (1983) and Gil-

TABLE 6.2. Chill Units Calculated by Several Authors

Chill unit values	Corresponding temperatures (°C)		
	A	B	C
0	<1.4	−1.0	−1.1
0.5	1.5–2.4	1.8	1.6
1.0	2.5–9.1	8.0	7.2
0.5	9.2–12.4	14.0	13.0
0	12.5–15.9	17.0	16.5
−0.5	16.0–18.0	19.5	19.0
−1.0	>18	21.5	20.7
−1.5	—	—	22.1
−2.0	—	—	23.3

[1] A—Richardson *et al.* (1974), in high-chilling peaches.
B—Gilreath and Buchanan (1981b), in low-chilling nectarine.
C—Shaltout and Unrath (1983b), in high-chilling apple.

reath (1983). A disadvantage of all of them is that they are based solely on the influence of temperature, which by itself is imperfectly understood, and that they ignore the action of other environmental factors known to modify the chilling requirement. Accordingly, they cannot predict dormancy release accurately under all conditions. In particular, they fail under abnormal conditions caused by sprinkling or heat treatments (Gilreath and Buchanan 1981a; Walser *et al.* 1981). Gilreath (1983) concedes that for an accurate determination of chilling requirements the collection of data and comparisons over several years and/or at different sites would be needed, as the models developed rely on relatively scant experimental data. She is considering changes in the method of calculating chill-unit accumulation for low-chilling cultivars, including a reassignment of the chill-unit requirements of the cultivars.

Bowen (1971) has pointed out that a source of error common to almost all dormancy studies with peach is the inability to determine more precisely when chilling becomes effective as buds are entering true dormancy. This disadvantage applies also to the rest-prediction models mentioned here. This is also part of the fundamental criticism by Kobayashi *et al.* (1982). They emphasize that the present methods are all merely empirical and do not consider the physiological mechanisms involved and that basing these models on available temperature data rather than on physiological processes results in an oversimplification. Accordingly, Batten (1983) has pointed out "that we still have much to learn about the ecological physiology of stone fruit dormancy."

VII. ENDOGENOUS CONTROL OF DORMANCY RELEASE

The ecological physiology of dormancy, i.e., the inner mechanism of dormancy development as influenced by external factors, has been the subject of many investigations. These studies have consisted either of observations of the changes occurring within the plant during the course of dormancy and efforts to realize which are the cause and which the result of dormancy release; or of attempts to bring about these changes artificially in order to elucidate the mechanisms involved. The limitations of this approach have been discussed by Saunders (1978).

A. Nutrients

In the initial phases of dormancy research, the availability of sugars was believed to be the crucial factor in dormancy, an insufficient supply of sugars being the cause of dormancy and the release from dormancy by chilling being brought about by the promotion of processes converting starch into sugar (Müller-Thurgau 1885; Coville 1920). However, Doorenbos (1953) concluded that the increase in sugar is not important in breaking dormancy. From more recent investigations it appears more likely that the increase in sugar is the result rather than the cause of dormancy release (Hemberg 1965; Chaudhry *et al.* 1970; Bachelard and Wightman 1973; Borkowska 1980a).

Chaudhry *et al.* (1970) have tested several theories that nitrogen, phosphorus, and nucleic acid metabolism play a role in dormancy release. They found from their analyses of vegetative buds of pear (*Pyrus communis*) that increases in soluble N, soluble P, ester-P formation, DNA, and RNA were not strongly related to dormancy release. Neither could a reduced production of high-energy phosphate compounds (ATP) be considered a limiting factor in this process. Since very little protein was synthesized in dormant buds and in heat-treated buds (Section IV. B), even though sufficient soluble N was present, they regarded inactivated protein synthesis in the buds, due to binding of DNA to a phosphoprotein complex, to be responsible for the blocking of renewed plant development and growth. However, they did not explain how a "critical variation of temperature" as a dormancy-breaking treatment causes the liberation of DNA from this complex.

Although nutrients do not seem to be causally involved in dormancy release, Black (1952) suggested that a favorable nutritional status of the tree may mitigate to some extent the harmful effects of prolonged dormancy, which for him includes symptoms of both delayed budbreak and floral damage. The delayed and protracted flowering of potted apple trees receiving no nitrogen nutrition in autumn (Terblanche *et al.* 1973, 1979) seems to point in the same direction.

B. Enzymes

The low rate of protein synthesis in pear buds, observed by Chaudhry *et al.* (1970), may also indicate a low activity of proteolytic enzymes. However, Bachelard and Wightman (1974) assumed that enhancement of metabolic activity and mobilization of food reserves in the buds was only the second step within the first phase of dormancy release. This is in line with findings of Zarska-Maciejewska and Lewak (1983) on dormancy release of seeds. They distinguished three phases of apple seed afterripening: (1) the phase of removal of the primary cause of dormancy; (2) the catabolic phase with the highest metabolic activity; and (3) the phase of initiation of germination. They concluded that the high activity of enzymes hydrolyzing reserve proteins in the second phase of stratification is not directly induced by temperature and thus is not involved in the removal of the primary cause of dormancy. Likewise, Dawidowicz-Grzegorzewska *et al.* (1982) found that cold treatment of apple seeds within the fruits did not provoke any changes in the free amino acid level nor in the structure of storage protein bodies.

The massive protein hydrolysis during the second stage of afterripening of apple seeds may be stimulated by an aminopeptidase, which was found by Zarska-Maciejewska and Lewak (1983) to have its highest activity at about 5°C. They assume that it may also be involved in the release of hormones from their inactive conjugates. Hormones, in conjunction with inhibitors, are considered to regulate the activation and/or synthesis of hydrolytic enzymes (Amen 1968). For example, Bachelard and Wightman (1974) observed that in buds, enhanced metabolic activity was preceded by an increase in the GA:inhibitor ratio (Section VII.D.1.). The initially low protein synthesis in pear buds reported by Chaudhry *et al.* (1970) thus would not have been the immediate cause of dormancy release but an intermediate step in this process, indicating a low hormone level.

Another group of enzymes, the lipases, may play an important role in the removal of the primary cause of dormancy. Their involvement, at least as a necessary intermediate step in dormancy release, has been postulated for apple seeds by Smolenska and Lewak (1974), Zarska-Maciejewska and Lewak (1976), and Zarska-Maciejewska *et al.* (1980). They found that the activity of acid lipase, which precedes the activity of alkaline lipase during the progress of dormancy, has its maximum at a temperature as low as 5°C. From the literature reviewed by Vegis (1964), it appears that in cells of dormant organs lipids and perhaps other hydrophobic colloids accumulate on the surface of protoplasm, that these disappear during the late phase of true dormancy and after treatment with dormancy-breaking agents, and that consequently the permeability of the biomembranes to water and solutes may be considerably enhanced. The possible connection between these changes and the reported acid lipase activity at low temperatures deserves to be investigated further with respect to dormancy release in buds.

C. Oxygen

Previous reports reviewed by Samish (1954) have shown that products of anaerobic respiration, such as acetaldehyde, may promote dormancy release. Warm-water baths and oil sprays, both of which are known to hasten dormancy release at least in the late stage of true dormancy (Sections IV. B, V. A), are likely to act by limiting oxygen. Pollock (1953) has stated that intact buds are quite capable of respiration under anaerobic conditions. The limitation of oxygen in buds is most severe during August and September and is largely removed in late winter and early spring. Bachelard and Wightman (1973) confirmed that the predominant anaerobic metabolism in buds during winter changed to aerobic metabolism, but possibly only after true dormancy had been completed.

Tissaoui and Côme (1973) were able to break dormancy of apple embryos by anaerobiosis without chilling. The oxygen deprivation worked even more rapidly than a cold treatment. This has been confirmed for buds of poplar cuttings in which the effect of oxygen withdrawal was also faster at higher than at lower temperatures (Soudain and Regnard 1982). At low temperatures, the absence of oxygen and the effect of chilling were additive. Erez *et al.* (1980) obtained similar effects in peach leaf buds, where the rapidity of dormancy release was proportional to the degree of reduction in oxygen and to the duration of exposure to low oxygen. However, their experiments were carried out after partial chilling when the buds presumably had already reached the late stage of true dormancy; therefore, the reduced oxygen could only have had an enhancing but not a triggering function in dormancy release.

Several reports have suggested that a reduced oxygen supply in seeds did not promote but rather prevented dormancy release. The idea that the growth-inhibiting function of the bud scales could be based similarly on the prevention of oxygen uptake has been found unlikely. Schneider (1968) found that the scales inhibit growth even if there is no barrier to air. Preventing oxygen penetration with lanolin did not replace the growth-inhibiting function of the bud scales (Tinklin and Schwabe 1970).

D. Phytohormones

Most authors presently regard the interaction of different groups of phytohormones to be the most important factor in the control of dormancy. Although some aspects of their action have been elucidated, the present situation is accurately described by Dörffling (1982): "In the last years there has been little progress in the attempts to find out the hormonal mechanisms involved in the induction and termination of dormancy in buds and seeds. . . . Possibly completely new ways of thinking will be required."

1. Gibberellins. In many investigations an increase in endogenous GA has been observed in buds after some chilling. Therefore, GA has been thought to be closely related to dormancy release (Eagles and Wareing 1963, 1964; Smith and Kefford 1964; Leike 1967; Tumanov *et al.* 1970; Ramsay and Martin 1970a; Bachelard and Wightman 1974; Podešva *et al.* 1980). Wareing and Saunders (1971) assumed that the primary effect of low temperature was to remove a block in GA biosynthesis; thus, the observed increase in GA activity would be the result rather than the cause of dormancy removal. However, whereas it is generally accepted that temperature, light, and other phytohormones affect GA levels, according to Hedden *et al.* (1978) virtually nothing is known about how the changes caused by these factors are brought about: whether by the control of specific steps in GA biosynthesis, by the release of "bound" GA from membranes, by increased membrane permeability, or by a combination of these.

Eagles and Wareing (1963) noted that while certain fractions of endogenous GAs in the buds increased from December to April, others decreased during this period. Hedden *et al.* (1978) pointed out that of the 52 GAs known in 1978 few are very active and many may be considered to be biosynthetic dead ends, biosynthetic intermediates, or deactivated metabolites, although most of the literature does not distinguish between already deactivated and active GAs. This complicates the interpretation of the changes in the GA content of buds during dormancy development in respect to its role in dormancy release. Another complication noted by Saunders (1978) is that many

GAs have chromatographic properties similar to abscisic acid (ABA) and will interact with ABA in most bioassays. These and some other technical problems limit the value of many bioassays.

Several authors have tried to identify the role of gibberellins in dormancy release by treating dormant shoots or plants in different ways with either GA_3 or GA_{4+7}. Applications of GA in summer and autumn generally delayed budbreak and flowering in spring, especially in peach, apricot, sweet cherry, black currant, and grapevine, but not so or to only a small degree in apple. While GAs extended the growing period in autumn, they postponed leaf abscission, the accumulation of starch and other carbohydrates and, thus, the onset of frost hardiness. Additionally, they caused effects similar to those reported to result from intermittent warm periods, especially during the early stage of true dormancy (i.e., the abscission of undeveloped floral buds and the occurrence of damaged and incomplete flowers or flower clusters). Studies on the effects of GAs have been reported by Brian *et al.* (1959), Weaver (1959), Bradley and Crane (1960), Brown *et al.* (1960), Modliboska (1960), Corgan and Widmoyer (1970), Bowen and Derickson (1978), and Walser *et al.* (1981).

However, when GA was applied in winter, its effect on dormancy release was positive in most cases. Vegis (1964) pointed out that winter-applied GA may bring about a widening of the temperature range in which buds may grow and may also prevent and reverse the narrowing of the temperature range for budbreak that is induced by high temperatures during the late phase of true dormancy. Paiva and Robitaille (1978b) found that GA treatments increased the dormancy-breaking effect of DNOC before and after deep dormancy. In seeds, GA promotes the biosynthesis of several hydrolytic enzymes (e.g., amylases, proteases, and ribonucleases) that are required for the germination process and the development of the endoplasmic reticulum, which acts as an intracellular transport system (Walton 1980).

Hatch and Walker (1969), and similarly Walser *et al.* (1981), believed that GA treatments could substitute for chilling and, therefore, assumed that the concentration of GA required to induce bud swelling would be a good measure of rest intensity. However, several other authors observed that GA was able to promote budbreak only after some prior chilling (Donoho and Walker 1957; Walker and Donoho 1959; Brown *et al.* 1960; Leike 1967; Paiva and Robitaille 1978b). Brown *et al.* (1960) pointed out that GA in dormant pear buds does not seem to break or reduce the rest influence but rather stimulates growth of buds that have at least partially emerged from dormancy. Erez *et al.* (1979a) confirmed that GA_3 did not increase the final level of peach budbreak but advanced bud opening. As the terminal buds have a lower chilling requirement relative to the laterals, GAs thus advance primarily terminal budbreak, resulting in increased apical dominance (Walker and Donoho 1959; Erez *et al.* 1979b). This does not occur if budbreak is caused by sufficient chilling, and therefore it is unlikely that GA really can substitute for chilling during deep dormancy.

Leike (1967) concluded that an increased GA concentration in the buds is not the cause but the result of dormancy release by chilling. The concept that GA may not be a primary cause of dormancy release but is required only in later steps of bud development agrees in principle with findings obtained in seeds. Amen (1968) differentiated in the germination response of a dormant system between "triggering" agents, whose continued presence is not essential, and "germination" agents, whose continued presence is required. He considered GA to belong to the latter group. Bradbeer (1968) argued that the effect of chilling at 5°C in seeds of hazelnut (*Corylus avellana*) could consist of an activation of the mechanism for GA synthesis, which then would occur at the germination temperature (20°C); in this view, GA is not needed for dormancy release but for germination.

2. Cytokinins. Cytokinins are a group of compounds with hormonal character having in common the ability to induce cell division in certain plant tissue cultures. In growing shoots, they are known to counteract the inhibition of lateral bud growth resulting from apical dominance (Wickson and Thimann 1958; Sachs and Thimann 1967; Williams and Stahly 1968; Kender and Carpenter 1972) and from the competition of fruits (Chvojka *et al.* 1961).

The results of the rather few investigations of endogenous changes in the cytokinin content of plants during the development of dormant buds support the suggestion of Lavee (1973) that cytokinin probably has some supplementary function in dormancy release but is not its cause. The involvement of cytokinins in dormancy release apparently starts only after some other, prior changes are induced (Luckwill and Whyte 1968; Hewett and Wareing 1973; Bachelard and Wightman 1974; Staden and Dimalla 1981; Wood 1983).

This concept appears to be supported by experiments with applied cytokinins. In general, cytokinin treatments were not, or were only partially, effective without some prior cold treatment and cannot fully substitute for chilling (Pieniazek 1964; Leike 1965, 1967; Williams and Stahly 1968; Weinberger 1969; Broome and Zimmerman 1976; Shaltout and Unrath 1983a). However, Broome and Zimmerman (1976) considered the sometimes limited or irregular effects observed (Biggs 1966; Erez *et al.* 1971) to be due possibly to the failure of applied cytokinin to penetrate to the site of action within the buds.

Arias and Crabbé (1975) observed that isolated buds of sweet cherry

(*Prunus avium*) responded better to injections of aqueous solutions of cytokinins during the early phase of true dormancy, whereas the response to GA was better in the late phase of true dormancy. A similar biphasic character shown in the complementary action of GA and cytokinin has also been shown in dormant apple embryos. Ryć and Lewak (1982) found that certain enzyme activities were influenced by cytokinin during the initial period of culture, whereas GA became effective only after some days of culture. Therefore, the varying sensitivity toward hormones may be another explanation for the sometimes unpredictable effects of applied cytokinins.

The main effect of cytokinin apparently is to hasten the development of buds that have been at least partially released from dormancy. Borkowska (1980b) inferred from experiments with single-bud cultures that while cytokinin enhanced the effect of low temperatures in promoting shoot elongation, it did not alter the duration of dormancy. Another effect of cytokinin on development of lateral buds, especially those inhibited by apical dominance, could be the promotion of xylem differentiation and of the formation of connections to the vascular system of the main stem, which may be incomplete while bud development is inhibited by auxin (Sorokin and Thimann 1964; Gregory and Veale 1957).

Letham and Palni (1983) concluded from a survey of the presently known cytokinins that the actual mechanism and rate of their biosynthesis, and its control, are far from being understood. As with GAs, their activity varies from almost inactive to highly active; some of them possibly are storage forms. Bachelard and Wightman (1974) postulated an initial increase in GA that would be followed by an activation of the roots, which are known to be the most important source of cytokinin in intact plants (for references see Torrey 1976); however, Wood (1983) observed that in the buds of pecan (*Carya illinoenis*) cytokinin reached its peak prior to the peak of GA, which occurred during bud swelling. Hewett and Wareing (1973) found that the cytokinin content of poplar buds increased during chilling in the absence of roots. Staden and Dimalla (1981) suggested that the accumulation of cytokinin in dormant shoot material stored at low temperatures for prolonged periods of time could be due to a conversion of storage or bound cytokinin to free bases in the bark, which then would be utilized in the buds. They assumed that the buds do not synthesize cytokinins themselves but seem to have the capacity to hydrolyze storage cytokinins.

3. Auxins. Auxins are often classified among the growth-promoting substances, but the work of Thimann (1937) showed that although auxins do promote growth at low concentrations, they inhibit growth at higher concentrations. He found the concentration needed for inhibition to be lowest in roots, higher in buds, and highest in stems. Saure (1971b, 1981) has pointed out that the auxins thus may perform the central function of a controller in the feedback processes regulating growth.

There is much evidence that auxins, besides being involved in the inhibition of lateral bud development, participate also in the induction of dormancy, which is not the subject of this review. However, apical dominance could play a role in the maintenance of dormancy even in winter (Section III. B). Treatments with exogenous auxins generally did not promote dormancy release but may even have blocked budbreak (Thomas *et al.* 1965; Biggs 1966; Tinklin and Schwabe 1970; Pieniazek *et al.* 1970; Erez *et al.* 1971; Cheng 1971; Paiva and Robitaille 1976). Only Bennett and Skoog (1938) observed partial promotion of bud growth after injecting indoleacetic acid (IAA) into shoots of dormant peach trees. Biggs (1966) found that IAA, while ineffective alone, could promote the enhancing effect of GA when applied in combination. However, no such supporting effect was observed by Pieniazek *et al.* (1970) when applying naphthylacetic acid together with either GA or cytokinin.

The results of investigations on the changes in the auxin content of buds during dormancy differ considerably because some authors have determined all extractable auxin and others have searched for only free, active auxin and because, to some extent, the analytic methods may have been imperfect (Bachelard and Wightman 1974). However, it appears that auxin activity in buds decreases while rest intensity increases (Kassem 1946; Thom 1951; Eggert 1953; Hendershott and Walker 1959). Eggert (1953) concluded that the role of auxin was to inhibit, not to stimulate, budbreak. A new increase in auxin activity starts only rather late (Bennett and Skoog 1938; Hendershott and Walker 1959; Blommaert 1959; Luckwill and Whyte 1968; Wood 1983). Blommaert (1959) further found that auxin activity increased earlier in buds from chilled trees than in those from trees kept in warm conditions. Lavee (1973) concluded that auxins could not be a primary growth regulator in the emergence from dormancy, as its increase is markedly preceded by that in GA and cytokinin levels; instead, he proposed that auxins are only involved in the process of bud opening and the resulting outgrowth in spring. Accordingly, Reinders-Gouwentak (1965) suggested that a dormancy-breaking agent other than auxin is necessary to reactivate the spring growth of buds. She further observed that the cambium can be activated by auxin only when it is no longer dormant.

4. Abscisic Acid. A ubiquitous growth-inhibiting substance in plants, abscisic acid (ABA) was considered to be a component of the "β-inhibitor" identified by Bennet-Clark and Kefford (1953) after Hemberg (1949) first had postulated a role for growth inhibitors in the control of dormancy. Now ABA is thought to be the β-inhibitor.

Initially, inhibitors like naringenin in peach and phloridzin in apple were identified especially in the bud scales (Dennis and Edgerton 1961; Hendershott and Walker 1959; Pieniazek 1962). They are able to prevent cell elongation and were considered to be associated with dormancy because their content in the plant increased to a maximum in late autumn and decreased from then to a minimum at about budbreak (Hendershott and Walker 1959; Sarapuu 1965; El-Mansy and Walker 1969). Blommaert (1959) observed that adequate cold may inactivate such inhibitors and he assumed that delayed inactivation may disturb the auxin–inhibitor balance, thus leading to prolonged dormancy. Phillips (1962) suggested that the balance of GA and endogenous naringenin controlled dormancy in buds and that an increase in GA_3 could promote budbreak by stimulating an enzymatic destruction of naringenin. However, others could not confirm a close correlation between naringenin content and true dormancy (Dennis and Edgerton 1961; Corgan 1965), and Pieniazek (1962) could not find an influence of cold on the quantity of phloridzin. She therefore concluded that phloridzin accompanies dormancy but is not its cause.

Later, a nonphenolic inhibitor was identified within the β-inhibitor, which was first called dormin (Eagles and Wareing 1963) and afterwards was found to be identical with abscisin II, an abscission-accelerating hormone that has been renamed abscisic acid (ABA). The modes of action of ABA and naringenin appear similar in that they inhibit cell elongation (El-Mansy and Walker 1969). There are many indications that ABA acts as an antagonist to GA and most probably also to cytokinin and that its main function is to protect plants from different kinds of stress (Thomas *et al.* 1965; de Maggio and Freeberg 1969; Addicott and Lyon 1969; Milborrow 1974; Walton 1980; Borkowska 1980b). However, Hall (1973) pointed out that the interaction of cytokinin and ABA is more complex than a simple plus/minus relationship and that there are indications that at increasing concentrations of cytokinin, antagonism may change to synergism.

Numerous reports indicate that the concentration of free ABA increases in autumn, reaches its peak in early winter, and declines again toward a minimum at about budbreak, after which a new ascent is initiated. This time course of changes in ABA concentration has been found in buds (Corgan and Peyton 1970; Ramsay and Martin 1970a,b; Wright 1975; Seeley and Powell 1981), both in the bud scales and in the inner parts of the buds (Mielke and Dennis 1975; Tinklin and Schwabe 1970; Freeman and Martin 1981).

Several authors have suggested that ABA, or the ratio between ABA and growth-promoting substances like GA, is causally involved in the induction and maintenance of dormancy (Eagles and Wareing 1963; Thomas *et al.* 1965; Kawase 1966; Khan 1975). The importance of ABA seems to be supported by the observations of Bowen and Derickson (1978). They found that a positive correlation exists between the ABA concentration in peach flower buds in December and the respective chilling requirement of several peach clones; and that a high ABA concentration in these flower buds during the dormant season may reflect a high tree vigor during the previous season. This could explain why an extended growing season due to vigorous growth may delay budbreak (Section IV. A).

However, this role for ABA has been questioned by several workers. Mielke and Dennis (1978), for example, reported that mechanical defoliation in autumn prevented the increase of ABA but the intensity of dormancy remained unchanged in the flower primordia of sour cherry (*Prunus cerasus*). Likewise, an application of GA_3 in autumn lowered the ABA concentration in peach flower buds but delayed flower bud opening in spring (Bowen and Derickson 1978). Furthermore, the positive effect of chilling on dormancy release is not clearly reflected by corresponding changes in the ABA level. Tinklin and Schwabe (1970) observed that the decline in growth inhibitors in lateral buds of black currant (*Ribes nigrum*) was delayed in a warm greenhouse. In the flower primordia of sour cherry, chilling initially caused a slight decrease in ABA, but after some weeks the amount of ABA was even higher in buds of chilled than in those of unchilled trees. The buds still opened slowly, when chilling was insufficient, even though the ABA concentration had already reached a minimum several weeks before (Mielke and Dennis 1978). In single-bud cultures, the delaying effect of ABA was independent of the length of the cold period (Borkowska 1980b).

Other environmental factors promoting budbreak (mist, reduced light intensity) also do not necessarily reduce the ABA concentration in buds (Freeman and Martin 1981). The removal of bud scales containing high amounts of ABA did not influence release from dormancy when only the top halves of the scales were removed, and it was ineffective when leaves were present (Tinklin and Schwabe 1970). The effect of removing bud scales could be replaced by applications of GA_3 (de Maggio and Freeberg 1969).

A decrease of inhibitory activity at the time dormancy was broken was preceded by a sharp increase of GA-like activity in apricot buds (Ramsay and Martin 1970a). Pieniazek and Rudnicki (1971) concluded from their investigations in dormant apple buds that the breaking of dormancy was not so much dependent on the reduced level of endogenous ABA in the buds as on an accumulation of endogenous stimulators, among them cytokinin. Wood (1983) argued that ABA may be associated with post-budbreak growth inhibition, but that free ABA *per se* does not regulate budbreak.

Changes in ABA content associated with intermittent growth flushes occur also in *Citrus* which, like other tropical and subtropical trees, is known not to have a phase of true dormancy (Young 1970). Accordingly, periods of growth cessation in these species cannot be broken by chilling or other influences that break dormancy in temperate fruits (Chandler 1957).

Similar results have been obtained for embryo dormancy. Ryć and Lewak (1982) could not confirm earlier reports that ABA is the main factor in the maintenance of embryo dormancy, which results neither from the presence of ABA in a relatively high concentration nor from the appropriate ratio between ABA and GA and cytokinins.

5. Ethylene. Ethylene, a hormone that first became known for its action in abscission and in the promotion of senescence, is now thought to participate in many other physiological processes, including the inhibition of bud growth (Burg 1968; Pratt and Goeschl 1969; Lieberman 1979), but its possible involvement in the control of dormancy has been scarcely investigated. Noodén and Weber (1978) conceded that "ethylene could play some role in termination of dormancy, but this remains to be determined." In other reviews, the role of ethylene in the control of dormancy has not even been mentioned. This may be partially because some of the effects of ethylene closely mimic those of ABA. Since ethylene treatments cause changes in ABA concentration, Milborrow (1974) concluded that "the action of ethylene on processes which are also affected by ABA is extremely difficult to unravel." However, Lieberman (1979) argued that ABA and ethylene inhibit growth by different mechanisms.

Several authors, working with seeds or embryos, have deduced a promoting effect of ethylene in dormancy release. In many seeds, the evolution of ethylene has been found to increase during germination, more so in nondormant than in dormant cultivars (Esashi and Leopold 1969). Addition of exogenous ethylene increased the speed of germination of excised apple embryos from seed stratified for 3–8 weeks, while the inhibition of germination was accompanied by an inhibition of ethylene formation (Kepczynski *et al.* 1977). Sińska and Gladon (1984) obtained increased embryo germination by addition of ethephon, an ethylene-releasing compound, to the seed-stratification medium. They observed that trapping endogenously produced ethylene from the atmosphere did not affect embryo germination during stratification, but completely inhibited it during culture of isolated embryos. These authors suggest that ethylene produced by embryonic axes may have a physiological role in the regulation of dormancy and that ethylene produced by the cotyledons may regulate other processes, like subsequent seedling growth. However, Wan (1980) could find no evidence supporting a role for ethylene in breaking dormancy in apple embryos.

There are only a few investigations on the presence of ethylene in buds. Blanpied (1972) found a high content of extractable ethylene in the dormant buds of 'Golden Delicious' apple; this decreased considerably toward budbreak and was higher in vegetative than in floral buds. It must be left open whether this difference was due to the bud type or to the more advanced development of the blossom buds because no measurements were made before the middle of March. Podešva *et al.* (1980) observed high ethylene production in dormant flower buds of sweet cherry (*Prunus avium*). In contrast, Zimmerman *et al.* (1977) found that activation of dormant buds of the tea crabapple (*Malus hupehensis*) by a spray of a cytokinin resulted in a significant increase in the ethylene evolution, and Nell *et al.* (1983) also observed increased ethylene evolution in azalea flower buds 3 days after GA treatments and after the termination of chilling treatments. Possibly, these results do not really contradict the findings of Blanpied (1972), as IAA and free ABA concentrations have also been found to decrease steadily during the course of dormancy but increase again after the termination of dormancy.

The application of exogenous ethylene is generally believed to promote budbreak. Abbott (1969) stated that ethylene is one of the best artificial dormancy-breaking agents. He proposed a model in which ethylene acts by promoting bud scale senescence in dormant buds, which permits a true leaf primordium with a viable lamina to develop at the apex of the bud axis; this leads in turn to the resumption of the flow of growth substances and of transpiration, thus breaking dormancy.

The assumption that ethylene has a strong bud dormancy-breaking effect is based mainly on experiments of Denny and Stanton (1928), and many others thereafter, who worked with ethylene chlorhydrin (2-chloroethanol). However, Pratt and Goeschl (1969) stressed that this very toxic chemical is not equivalent to ethylene and that the ethylene-like effects it produces may be due to the injuries it causes.

One of these effects is to reduce the level of inhibitors in the buds (Hemberg 1949).

Tinklin and Schwabe (1970) observed only a rather limited effect of ethylene vapor on budbreak in dormant shoots of black currant compared with the effects of continuous light or GA. Ethephon did not stimulate budbreak in tea crabapple (Zimmerman *et al.* 1977) or, in general, in apple trees in the tropics, whether they were potted, field grown, or isolated buds cultured *in vitro* (Edwards 1980). Paiva and Robitaille (1978a) confirmed that ethylene-releasing compounds were ineffective when applied to apple shoots during dormancy, and they found no evidence for a role of ethylene in budbreak stimulated by wounding or DNOC. Lin and Powell (1981) also concluded that ethylene probably was not involved in the dormancy-breaking mechanism, as ethylene production was not increased in apple buds that were induced to break by removal of the bud scales.

VIII. MECHANISMS AND MODELS

The calculation of positive and negative chill units for certain temperatures by several authors (Table 6.2) has highlighted the importance of low temperatures for dormancy release and the retarding effect of higher temperatures between about 16° and 35°C in this process. However, as most of the methods for calculating chill units are empirical and reveal nothing about the physiological processes involved, they do not contribute much to the understanding of these processes. Therefore, many questions remain unanswered:

- How does chilling affect dormancy release?
- Why do higher temperatures *delay* bud development during true dormancy although they are known to *promote* bud growth at other times?
- Why does the retarding effect of higher temperature disappear when the temperature increases to 35°C and more?
- Why does the temperature range permitting bud growth widen during the late phase of true dormancy?
- Which factor(s) causes the difference between low-chilling and high-chilling cultivars?
- How do applied GA, DNOC, mineral oils, and other chemicals, as well as supplementary light, heating, and anoxia, contribute to earlier budbreak during late dormancy?

The fact that low temperatures promote and higher temperatures inhibit dormancy release has led to the hypothesis that there is initially a single reversible reaction (favored by chilling and reversed by higher temperatures) followed by some kind of fixation, after which different reactions occur at either low or high temperatures (Erez and Couvillon 1982). This model agrees in principle with that proposed by Purvis and Gregory (1952) for the devernalization of winter rye by high temperatures.

However, there may be two distinct temperature reactions from the beginning, which may be independent of one another, but which may overlap: one controlling the promotion of dormancy release and one controlling the inhibition of dormancy release (Fig. 6.2). In this case, the chill units as calculated for different temperatures by several authors indicate the balance between promotion and inhibition of dormancy release at a certain temperature (Fig. 6.2, bottom).

According to this model of a dual mode of temperature action in dormancy release, as diagrammed in Fig. 6.2, a temperature reduction from, e.g., 18° to 10°C would cause, separately, less inhibition of dormancy release and more promotion of dormancy release. In addition, a chill unit value of 0 would indicate no net promotion or inhibition of dormancy release. This would be the case when there is either a balance between promotion and inhibition at temperatures of about 18°C in this model, or a deficit of both at either low or high temperatures.

Most authors proceed on the assumption that chill-unit values corresponding to each temperature are constant throughout the dormant period. This has been questioned by Kobayashi *et al.* (1982); they suggested, instead, a physiological shift in the effect of higher temperatures between the deepening and the decreasing phase of true dormancy. Such a shift would be expected as the result of a continuous dual-temperature action. In deep dormancy after little chilling, there is only a small promotion of dormancy release and thus a still-strong potential for inhibition of dormancy release. Thus, as shown in Fig. 6.3a, in deep dormancy the chill units would become negative even at rather low temperatures. By late dormancy, however, sufficient chilling has occurred to considerably promote dormancy release, thus causing a diminishing inhibition potential. Accordingly, the chill units calculated in this stage would become negative only at much higher temperatures (Fig. 6.3b). As a consequence, the temperature range for the promotion of budbreak would widen, and the inhibition of budbreak by higher temperatures would diminish and finally disappear, just as postulated by Vegis (1964).

If this model is correct, then the chill units calculated by several authors would represent mean values. These values do not reflect, indeed they conceal, the proposed increase in positive chill-unit values when moving from deep to late dormancy and the decrease in negative

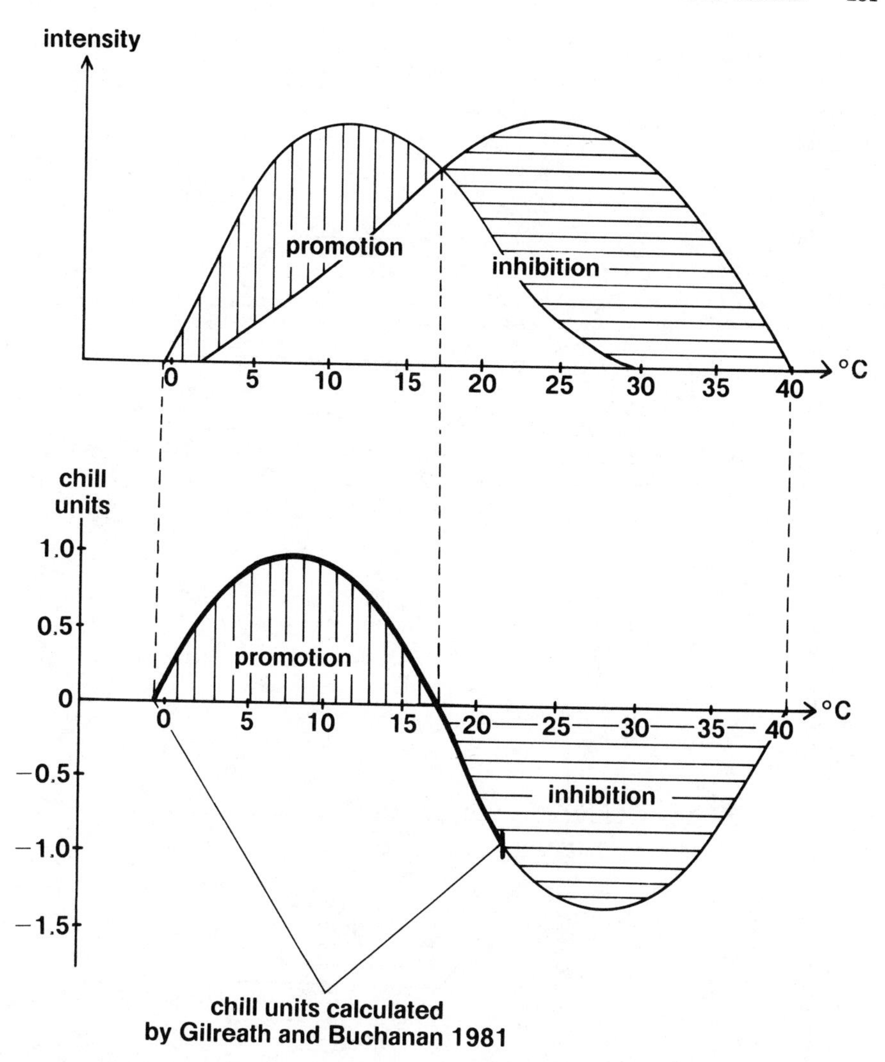

FIG. 6.2. Hypothetical model of a dual temperature action in dormancy release as derived from calculated chill units.

chill-unit values at higher temperatures or even their conversion into positive values within the process of dormancy release.

A model involving two distinct and independent temperature actions would explain the questions regarding temperature effects on dormancy release, raised at the beginning of this section. It would also suggest that the difference between low-chilling and high-chilling cultivars might be physiologically similar to the difference in high-chilling cultivars between late dormancy and deep dormancy, as low-

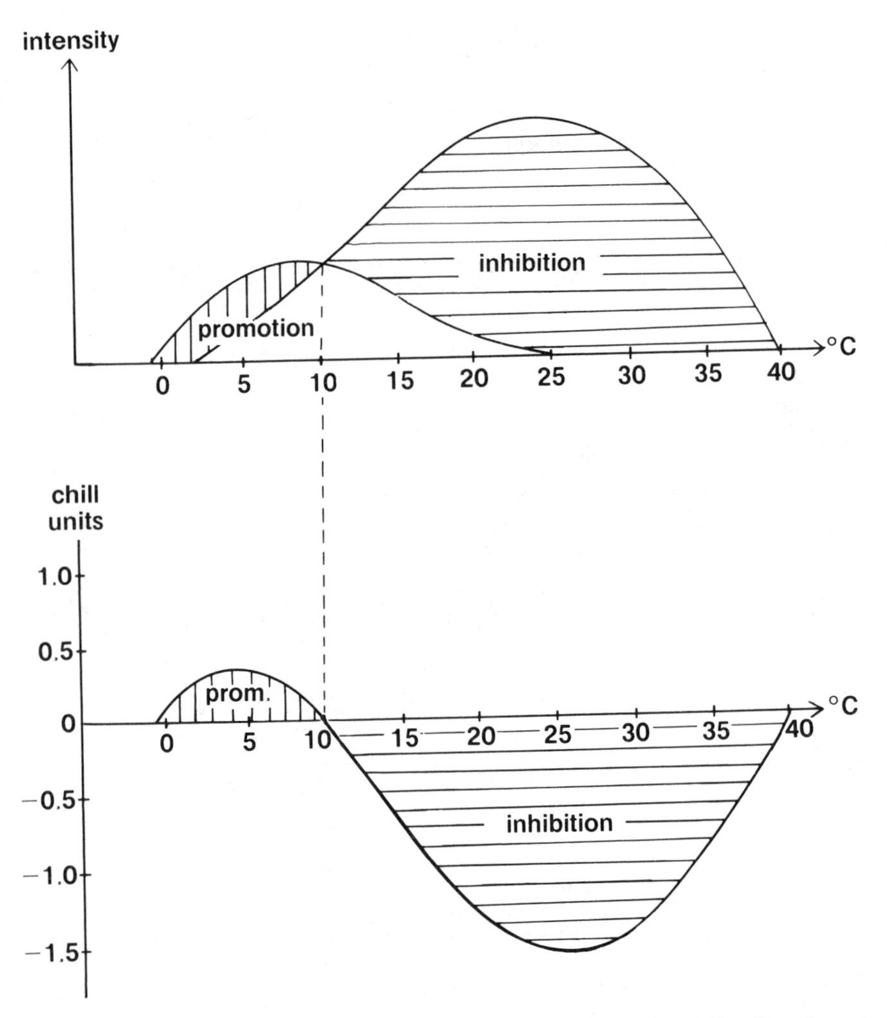

FIG. 6.3. The widening temperature range for the promotion of budbreak and the reduced range for an inhibiting effect of higher temperatures as deduced from a hypothetical model of dual temperature action in dormancy release.

chilling cultivars are less inhibited by higher temperatures (Section III. C). However, the proposed model needs to be verified experimentally by the following studies:

- Keeping dormant plant material at a continuous low temperature and exposing a series of samples consecutively for a limited, uniform period of time to forcing temperatures

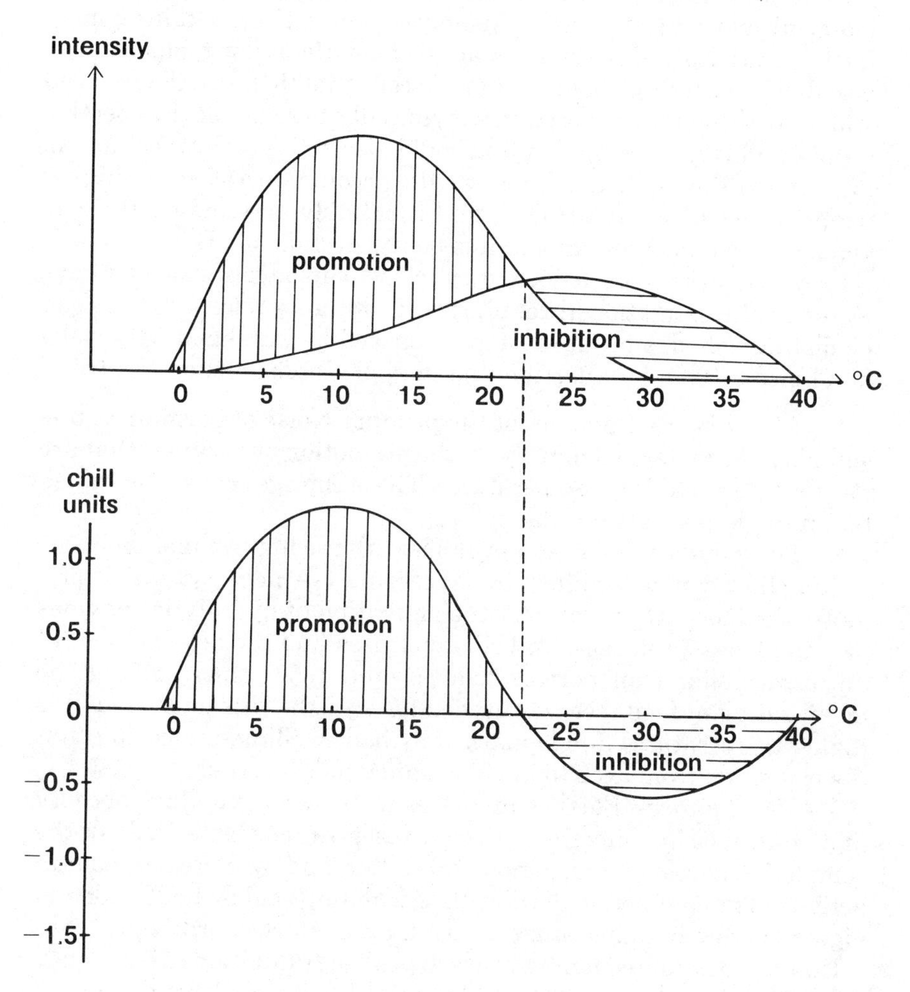

FIG. 6.3 (*Cont.*)

- Keeping dormant plant material at a continuous forcing temperature and exposing a series of samples consecutively for a limited, uniform period of time to chilling temperatures
- Exposing dormant plant material to a series of temperatures at the beginning and at later stages of dormancy

It is not yet clear by what mechanism the proposed dual action of chilling in the promotion of dormancy release is realized. Recent work on dormancy release in apple embryos suggests that the direct promotion of dormancy release by chilling in buds initially may involve an activation of lipolytic enzymes, which could cause changes in the biomembranes of the protoplasm (Section VII.B). Chilling may further enable other enzymes to act sufficiently at low temperatures, possibly by bringing about changes in cells that then stabilize or even stimulate the activity of certain enzymes like peroxidase (Kacperska-Palasz and Uliasz 1974). Chilling could also activate the mechanism for GA synthesis, which then would become operative at higher temperatures (Section VII.D.1), and it possibly may induce the conversion of bound cytokinins to free bases (Section VII.D.2).

Provided these steps in dormancy release can be confirmed for buds by further investigation, three phases of dormancy release in buds can be distinguished, similar to those proposed by Zarska-Maciejewska and Lewak (1983) for afterripening in apple seeds (Section VII.B):

1. *First phase*—Removal of the primary cause of dormancy, presumably characterized mainly by the promotion of enzymes that are most active at chilling temperatures. These changes cause the transition from deep to late dormancy.
2. *Second phase*—Catabolic phase with the highest metabolic activity, characterized mainly by increasing GA activity, which promotes the biosynthesis and/or the activation of hydrolytic enzymes like amylases, proteases, and ribonucleases and the development of an intracellular transport system (Walton 1980), thus causing the mobilization of food reserves needed for growth. This phase requires higher temperatures and is marked by bud swelling, indicating that the transition from deep to late dormancy has occurred.
3. *Third phase*—Further increases in GA and cytokinin activity and continued mobilization of food reserves, characterized by the complete removal of endogenous inhibition and by increasing auxin activity. The completion of this phase is manifested by budbreak and vigorous growth, indicating the termination of true dormancy.

How the postulated temperature-dependent inhibition of dormancy release inhibits or prevents budbreak is still obscure. Although most of the investigations reviewed in section VII.D.5 seem either to exclude an inhibiting role of ethylene in dormancy release, or even point to a promotion of dormancy release by ethylene, I believe further study of its possible inhibitory function is warranted. In most of the studies reported so far only the amount of endogenous ethylene production has been analyzed, without considering the fact that auxins, gibberellins, and especially cytokinins can promote ethylene formation in tissues, although at the same time they apparently protect the treated tissues from the inhibiting effects of ethylene (Burg and Burg 1968; Fuchs and Lieberman 1968; Pratt and Goeschl 1969; Lieberman 1979; Jackson and Campbell 1979). Furthermore, these investigations have been carried out with material that has passed at least the first phase of dormancy release. This applies especially to investigations on dormancy release with excised embryos. There are doubts whether true dormancy exists at all in seed embryos. Vegis (1964) has pointed out that the existence of seeds has not yet been proven where excised nonafterripened embryos are not able to germinate normally under any temperature, light, or other environmental conditions. In this case, a shift in the physiological function of ethylene may be assumed, parallel to the shift in the function of higher temperatures from inhibition during true dormancy toward promotion after termination of true dormancy, and parellel to the postulated change of ABA from antagonism to synergism at increasing cytokinin concentrations (Section VII.D.4). Therefore, again, the balance between cytokinins, and possibly other phytohormones, and ethylene seems to be important. Lewak *et al.* (1975) have found both cytokinins and GA at relatively high levels about the time when the positive response of apple seeds to ethylene is highest.

Other observations also suggest that ethylene may exert an inhibitory effect on dormancy release (Burg 1968; Pratt and Goeschl 1969; Lieberman 1979):

- Ethylene has been found to be involved in the inhibition of bud growth by apical dominance.
- It is well known for inducing senescence, which is the converse of what occurs during dormancy release;
- It inhibits cell elongation.
- It inhibits the basipetal IAA transport, presumably by causing a progressive, irreversible deterioration of the transport system.
- It affects cell dehydration.

Concerning the influence of temperature, the effect of ethylene has been found to decrease with decreasing temperature. Beyer (1973) found no inhibition of IAA transport by ethylene at 8°C or less. It is well known that during fruit storage at low temperature, quite high ethylene concentrations can be tolerated because the fruits will not react at these temperatures. Yu *et al.* (1980) observed that ethylene production reached the maximum at 30°C and declined at higher temperatures, so that only trace amounts were produced at 40°C. Heat-treated tissues have been shown either to be incapable of metabolizing ethylene or to have a severely reduced ethylene metabolism. An inhibition of ethylene production may also be caused by anaerobiosis and by dinitro compounds (Yu *et al.* 1980).

If a temperature-dependent inhibition of dormancy release by ethylene could be confirmed, the "fixation" of the chilling effect observed by Erez and Lavee (1971) could be explained by the suppression of the inhibiting effects of ethylene, like that of other inhibitors such as ABA, by increasing amounts of cytokinins, and possibly other phytohormones, in the course of dormancy release. This assumption is supported by the findings of Gilreath and Buchanan (1981b) that a minimum of chilling must be received before a temperature-dependent fixation process can occur.

The promotion of dormancy release by cultural practices such as heat treatments, anoxia, and applied DNOC, mineral oil, and other chemicals thus could be explained by a suppression of the inhibition caused by ethylene. These measures, however, would be effective only during the second or third phase of dormancy release, when the promotion of dormancy release has already reached a high level. They would be ineffective if promoting factors are still weak, even when inhibition is removed. Therefore, the efficacy of DNOC may be improved by growth-promoting cultural practices (Section V. A).

The application of GA directly supports the promotion of dormancy release by endogenous GA, beginning in the second phase of dormancy release. Supplementary light during late dormancy (Section IV. C), which may partially compensate for insufficient chilling in peach leaf buds, is considered to act by promoting GA biosynthesis, provided it has already started. Accordingly, it is quite natural that the more advanced blossom buds do not respond to additional irradiation, or GA treatments, or may even be inhibited by them, as a shortage of GA is not the bottleneck for their development.

Dark-preconditioning at low temperatures has been found to promote dormancy release in peach already in an early phase. One possible explanation is the promotion of enzymes like acid lipase by the absence of light (Zarska-Maciejewska and Lewak 1976). Another, possibly supplementary effect of chilling during dark-preconditioning is to prevent an increase in ethylene production in the dark at higher temperatures. For example, Chandler and Tufts (1934) noted that with chilling in the dark the petiole stubs of defoliated peach trees did not abscise for more than 1 month, whereas in the greenhouse they did so after 5 days.

All the cultural practices that have been found to influence dormancy differ from chilling in that only chilling exhibits the proposed dual action in dormancy release; the other treatments act in one way only, either to reduce the inhibition or to increase the promotion of dormancy release. Therefore, combining different treatments might provide a good means of enhancing dormancy release, as indicated by some experiments of Erez *et al.* (1971) and Paiva and Robitaille (1978b). However, it should be kept in mind that most of these practices differ from chilling also in that they can work only on the surface, while the action of chilling includes the whole aboveground part of the tree.

Despite the findings of Denny and Stanton (1928), who caused single dormant buds of lilac to grow freely without treating other parts of the plant, treatment of buds alone may be insufficient to promote dormancy release. Chandler (1960) and others have reported that a rest-promoting influence moves upward from unchilled trees to chilled budwood (Section II. B). In terms of the proposed model of dual temperature action, such an upward movement of a rest influence is quite unlikely. The reason for the limitation of growth after initial budbreak, as reported by them, may be a lowered transport capacity of the unchilled aboveground part of the plant even though the growth inhibition within the buds has been removed by chilling. In this case, auxin produced in the growing shoot tips could probably not induce the roots to grow enough and to provide sufficient newly synthesized GA and cytokinin after the reserves mobilized from storage forms have been used up. This explanation could be verified by comparing root growth in trees having received chilling only in the top with fully chilled trees.

Thus, chilling could be involved in the reactivation of the transport system to its full capacity. It should be noted that Denny and Stanton (1928) worked with lilac plants in which the function of the conductive tissues was not restricted. This would mean that even though the cause of true dormancy is located in the buds, it is not necessarily confined to the buds alone.

LITERATURE CITED

ABBOTT, D.L. 1969. The role of budscales in the morphogenesis and dormancy of the apple fruit bud. p. 65–82. In: L.C. Luckwill and C.V. Cutting (eds.), Physiology of tree crops. Academic Press, London and New York.

ADDICOTT, F.T., and J.L. LYONS. 1969. Physiology of abscisic acid and related substances. Annu. Rev. Plant Physiol. 20:139–164.

AMEN, R.D. 1968. A model of seed dormancy. Bot. Rev. 34:1–31.

AMLING, H.J. and K.A. AMLING. 1980. Onset, intensity, and dissipation of rest in several pecan cultivars. J. Amer. Soc. Hort. Sci. 105:536–540.

ARIAS, O. and J. CRABBÉ. 1975. Les gradients morphogénétiques du rameau d'un an des végétaux ligneux, en repos apparent. Physiol. Vég. 13:69–81.

ARON, R.H. 1975. Comments on a model for estimating the completion of rest for 'Redhaven' and 'Elberta' peach trees. HortScience 10:559–560.

BACHELARD, E.P. and F. WIGHTMAN. 1973. Biochemical and physiological studies on dormancy release in tree buds. I. Changes in degree of dormancy, respiratory capacity, and major cell constituents in overwintering vegetative buds of *Populus balsamifera*. Can. J. Bot. 51:2315–2326.

BACHELARD, E.P. and F. WIGHTMAN. 1974. Biochemical and physiological studies on dormancy release in tree buds. III. Changes in endogenous growth sub-

stances and a possible mechanism of dormancy release in overwintering vegetative buds of *Populus balsamifera*. Can. J. Bot. 52:1483–1489.
BATTEN, D.J. 1983. Criticisms. HortScience 18:13–14.
BENNET-CLARK, T.A. and N.P. KEFFORD. 1953. Chromatography of the growth substances in plant extracts. Nature 171:645–647.
BENNETT, J.P. 1949. Temperature and bud rest period. Calif. Agr. 3(11):9,12.
BENNETT, J.P. and F. SKOOG. 1938. Preliminary experiments on the relation of growth-promoting substances to the rest period in fruit trees. Plant Physiol. 13:219–225.
BEYER, E.M. 1973. Abscission: support for a role of ethylene modification of auxin transport. Plant Physiol. 52:1–5.
BIDABÉ, B. 1967. Action de la température sur l'évolution des bourgeons de pommier et comparaison de méthodes de contrôle de l'époque de floraison. Ann. Physiol. Vég. 9:65–86 (cited in Ruck 1975).
BIGGS, R.H. 1966. Screening chemicals for the capacity to modify bud dormancy of peaches. Proc. Fla. State Hort. Sci. 79:383–386.
BLACK, M.W. 1936. Some physiological effects of oil sprays upon deciduous fruit trees. J. Pomol. Hort. Sci. 14:175–202.
BLACK, M.W. 1952. The problem of prolonged rest in deciduous fruit trees. p. 1122–1131. In: Proc. 13th Intern. Hort. Congr., Vol. 2, London.
BLANPIED, G.D. 1972. A study of ethylene in apple, red raspberry, and cherry. Plant Physiol. 49:627–630.
BLOMMAERT, K.L.J. 1959. Winter temperature in relation to dormancy and the auxin and growth inhibitor content. S. African J. Agr. Sci. 2:507–514.
BLOMMAERT, K.L.J. 1964. New spray material controls delayed foliation of peaches. Decid. Fruit Grow. 14:165–166.
BORKOWSKA, B. 1980a. Effect of chilling period, benzylaminopurine (BA) and abscisic acid (ABA) on physiological and biochemical changes in non-growing and growing apple buds. Fruit Sci. Rpt. 7:101–115.
BORKOWSKA, B. 1980b. Releasing the single apple buds from dormancy under the influence of low temperature, BA and ABA. Fruit Sci. Rpt. 7:147–153.
BORKOWSKA, B. and L.E. POWELL. 1979. The dormancy status of apple buds as determined by an *in vitro* culture system. J. Amer. Soc. Hort. Sci. 104:796–799.
BOWEN, H.H. 1971. Breeding peaches for warm climates. HortScience 6:153–157.
BOWEN, H.H. and G.W. DERICKSON. 1978. Relationship of endogenous flower bud abscisic acid to peach chilling requirements, bloom dates, and applied gibberellic acid. HortScience 13:694–696.
BRADBEER, J.W. 1968. Studies in seed dormancy. IV. The role of endogenous inhibitors and gibberellin in the dormancy and germination of *Corylus avellana* L. seeds. Planta 78:266–276.
BRADLEY, M.V. and J.C. CRANE. 1960. Gibberellin-induced inhibition of bud development in some species of *Prunus*. Science 131:825–826.
BRIAN, P.W., J.H.P. PETTY, and P.T. RICHMOND. 1959. Extended dormancy of deciduous woody plants treated in autumn with gibberellic acid. Nature 184:69.
BROOME, O.C. and R.H. ZIMMERMAN. 1976. Breaking bud dormancy in tea crabapple (*Malus hupehensis* (Pamp.) Rehd.) with cytokinins. J. Amer. Soc. Hort. Sci. 101:28–30.
BROWN, D.S. 1958. The relation of temperature to the flower bud drop of peaches. Proc. Amer. Soc. Hort. Sci. 71:77–87.
BROWN, D.S. and J.F. ABI-FADEL. 1953. The stage of development of apricot flower buds in relation to their chilling requirement. Proc. Amer. Soc. Hort. Sci. 61:110–118.

BROWN, D.S. and F.A. KOTOB. 1957. Growth of flower buds of apricot, peach, and pear during the rest period. Proc. Amer. Soc. Hort. Sci. 69:158–164.

BROWN, D.S., W.H. GRIGGS, and B.T. IWAKIRI. 1960. The influence of gibberellin on resting pear buds. Proc. Amer. Soc. Hort. Sci. 76:52–58.

BUCHANAN, D.W., J.F. BARTHOLIC, and R.H. BIGGS. 1977. Manipulation of bloom and ripening dates of three Florida grown peach and nectarine cultivars through sprinkling and shade. J. Amer. Soc. Hort. Sci. 102:466–470.

BURG, S.P. 1968. Ethylene, plant senescence and abscission. Plant Physiol. 43:1503–1511.

BURG, S.P. and E.A. BURG. 1968. Ethylene formation in pea seedlings: its relation to the inhibition of bud growth caused by indole-3-acetic acid. Plant Physiol. 43:1069–1074.

CERNY, L. 1963. Vegetační klid a odpočinek pupenu jabloní. Scient. Stud. Pomol. Res. Inst. Fruit Growing Holovousy 2:121–141.

CHANDLER, W.H. 1957. Deciduous orchards. 3rd ed. Lea & Febiger, Philadelphia.

CHANDLER, W.H. 1960. Some studies of rest in apple trees. Proc. Amer. Soc. Hort. Sci. 76:1–10.

CHANDLER, W.H. and D.S. BROWN. 1957. Deciduous orchards in California winters. Calif. Agr. Ext. Serv. Cir. 179.

CHANDLER, W.H., D.S. BROWN, M.H. KIMBALL, G.L. PHILP, W.P. TUFTS, and G.P. WELDON. 1937. Chilling requirements for opening of buds on deciduous orchard trees and some other plants in California. Bull. 611. Calif. Agr. Expt. Sta., Berkeley.

CHANDLER, W.H. and W.P. TUFTS. 1934. Influence of the rest period on opening of buds of fruit trees in spring and on development of flower buds of peach trees. Proc. Amer. Soc. Hort. Sci. 30:180–186.

CHANG, S. and D.J. WERNER. 1982. Relationship of seed germination and respiration with varietal chilling requirement in peach. HortScience 17:146 (Abstr.).

CHAUDHRY, W.M., T.C. BROYER, and L.C.T. YOUNG. 1970. Chemical changes associated with the breaking of the rest period in vegetative buds of *Pyrus communis*. Physiol. Plant. 23:1157–1169.

CHENG, C.-Y. 1971. Experimente zur Regulation der Knospenruhe. Thesis, Univ. of Giessen, West Germany.

CHVOJKA, L., K. VEREŠ, and J. KOZEL. 1961. The effect of kinins on the growth of apple-tree buds and on incorporation of radio-active phosphate (in Russian). Biol. Plant. (Prague) 3:140–147.

CORGAN, J.N. 1965. Seasonal change in naringenin concentration in peach flower buds. Proc. Amer. Soc. Hort. Sci. 86:129–132.

CORGAN, J.N. and F.B. PEYTON. 1971. Abscisic acid levels in dormant peach flower buds. J. Amer. Soc. Hort. Sci. 95:770–774.

CORGAN, J.N. and F.B. WIDMOYER. 1971. The effects of gibberellic acid on flower differentiation, date of bloom, and flower hardiness of peach. J. Amer. Soc. Hort. Sci. 96:54–57.

COUVILLON, G.A. and C.H. HENDERSHOTT. 1974. A characterization of the "after-rest" period of flower buds of two peach cultivars of different chilling requirement. J. Amer. Soc. Hort. Sci. 99:23–26.

COUVILLON, G.A., C.M. MOORE, and P. BUSH. 1975. Obtaining small peach plants containing all bud types for "rest" and dormancy studies. HortScience 10:78–79.

COUVILLON, G.A., N. FINARDI, M. MAGNONI, and C. FREIRE. 1984. Rootstock influences the chilling requirement of 'Rome Beauty' apple in Brazil. HortScience 19:255–256.

COVILLE, F.V. 1920. The influence of cold in stimulating the growth of plants. J. Agr. Res. 20:151–160.
CRABBÉ, J.J. 1981. The interference of bud dormancy in the morphogenesis of trees and shrubs. Acta Hort. 120:167–172.
CRABBÉ, J.J. 1984. Correlative effects modifying the course of bud dormancy in woody plants. Z. Pflanzenphysiol. 113:465–469.
CROSSA-RAYNAUD, P. 1955. Effet de hivers doux sur le comportement des arbres á feuilles caduques. Ann. Serv. Bot. Agron. Tunisie 28:1–22 (cited in Ruck 1975).
DA MOTA, F.S. 1957. Os invesnos de pelotas, RS, em relacão ás exigências das árvores frutiferas de fôlhas caducas. Rio Grande do Sul. Bol. Tec. Inst. Agr. Sul. 18:38 (cited in Ruck 1975.)
DAWIDOWICZ-GRZEGORZEWSKA, A., N. WEISMAN, C. THÉVENOT, and D. CÔME. 1982. Proteolysis and dormancy release in apple-embryos. Z. Pflanzenphysiol. 107:115–121.
DEL REAL-LABORDE, J.I. and I.A. González-Cepeda 1982. The behavior of the chilling accumulation methods under mild winter conditions. P. 1141:In: Abstr.-21stIntern. Hort. Congr., Vol. 1, Hamburg.
DE MAGGIO, A.E. and J.A. FREEBERG. 1969. Dormancy regulation: hormonal interaction in maple (*Acer platanoides*). Can. J. Bot. 47:1165–1169.
DENNIS, F.G. and L.J. EDGERTON 1961. The relationship between an inhibitor and rest in peach flower buds. Proc. Amer. Soc. Hort. Sci. 77:107–116.
DENNY, F.E. and E.N. STANTON. 1928. Localization of response of woody tissues to chemical treatments that break the rest period. Amer. J. Bot. 15:337–344.
DONOHO, C.W. and D.R. WALKER. 1957. Effect of gibberellic acid on breaking of rest period in Elberta peach. Science 126:1178–1179.
DOORENBOS, J. 1953. Review of the literature on dormancy in buds of woody plants. Meded. Landbouwhogeschool Wageningen 53:1–24.
DÖRFFLING, K. 1982. Das Hormonsystem der Pflanzen. G. Thieme, Stuttgart-New York.
DOSTÁL, R. 1942. Über das Frühtreiben der Fliederzweige (*Syringa vulgaris*) und Kartoffelknollen (*Solanum tuberosum*) und die hormonale Deutung dafür. Gartenbauwissenschaft 16:195–206.
DUARTE, O. and R. FRANCIOSI. 1974. Temperate zone fruit production in Peru, a special situation (Abstr.). p. 519. In: Proc. 19th Intern. Hort. Congr., Vol. 1B, Warsaw.
EAGLES, C.F. and P.F. WAREING. 1963. Dormancy regulators in woody plants. Experimental induction of dormancy in *Betula pubescens*. Nature 199:874–875.
EAGLES, C.F. and P.F. WAREING. 1964. The role of growth substances in the regulation of bud dormancy. Physiol. Plant. 17:697–709.
EDWARDS, G.R. 1980. Some physiological limitations to the culture of apple in the tropics. Paper presented at Symp. Current Problems of Fruits and Vegetables in the Tropics, Los Baños, California.
EGGERT, F.P. 1953. The auxin content of spur buds of apple as related to the rest period. Proc. Amer. Soc. Hort. Sci. 62:191–200.
EL-MANSY, H.J. and D.R. WALKER. 1969. Seasonal fluctuations of flavanones in 'Elberta' peach flower buds during and after the termination of rest. J. Amer. Soc. Hort. Sci. 94:298–301.
EREZ, A. 1979. The effect of temperature on the activity of oil + dinitro-o-cresol sprays to break the rest of apple buds. HortScience 14:141–142.
EREZ, A. 1982. Die Winterruhe beim Apfel und ihre Unterbrechung auf natürliche und künstliche Weise. Erwerbsobstbau 24:116–118.
EREZ, A. and G.A. COUVILLON. 1982. The effect of climatic conditions on break-

ing the rest in peach buds: a reassessment of chilling requirement. p. 1142. In: Abstr. 21st Intern. Hort. Congr., Vol. 1, Hamburg.

EREZ, A. and G.A. COUVILLON. 1983. Evaporative cooling to improve rest breaking of nectarine buds by counteracting high daytime temperatures. HortScience 18:480–481.

EREZ, A. and S. LAVEE. 1971. The effect of climatic conditions on dormancy development of peach buds. I. Temperature. J. Amer. Soc. Hort. Sci. 96:711–714.

EREZ, A. and S. LAVEE. 1974. Breaking the dormancy of deciduous fruit trees in subtropical climates. p. 69–78. In: Proc. 19th Intern. Hort. Congr., Vol. 3, Warsaw.

EREZ, A. and Z. YABLOWITZ. 1981. Rooting of peach hardwood cuttings for the meadow orchard. Scientia Hort. 15:137–144.

EREZ, A. and A. ZUR. 1981. Breaking the rest of apple buds with narrow-distillation-range oil and dinitro-o-cresol. Scientia Hort. 14:47–54.

EREZ, A., R.M. SAMISH, and S. LAVEE. 1966. The role of light in leaf and flower bud break of the peach (*Prunus persica*). Physiol. Plant. 19:650–659.

EREZ, A., S. LAVEE, and R.M. SAMISH. 1968. The effect of limitation in light during the rest period on leaf bud break of the peach (*Prunus persica*). Physiol. Plant. 21:759–764.

EREZ, A., S. LAVEE, and R.M. SAMISH. 1971. Improved methods for breaking rest in the peach and other deciduous fruit species. J. Amer. Soc. Hort. Sci. 96:519–522.

EREZ, A., G.A. COUVILLON, and C.H. HENDERSHOTT. 1979a. Quantitative chilling enhancement and negation in peach buds by high temperatures in a daily cycle. J. Amer. Soc. Hort. Sci. 104:536–540.

EREZ, A., G.A. COUVILLON, and C.H. HENDERSHOTT. 1979b. The effect of cycle length on chilling negation by high temperatures in dormant peach leaf buds. J. Amer. Soc. Hort. Sci. 104:573–576.

EREZ, A., G.A. COUVILLON, and S.J. KAYS. 1980. The effect of oxygen concentration on the release of peach leaf buds from rest. HortScience 15:39–41.

ESASHI, Y. and A.C. LEOPOLD. 1969. Dormancy in subterranean clover seeds by ethylene. Plant Physiol. 44:1470–1472.

FADL, M.S. and H.T. HARTMANN. 1967. Relationship between seasonal changes in endogenous promoters and inhibitors in pear buds and cutting bases and the rooting of pear hardwood cuttings. Proc. Amer. Soc. Hort. Sci. 91:96–112.

FREEMAN, M.W. and G.C. MARTIN. 1981. Peach floral bud break and abscisic acid content as affected by mist, light, and temperature treatments during rest. J. Amer. Soc. Hort. Sci. 106:333–336.

FUCHIGAMI, L.H. 1977. Ethephon-induced defoliation and delay of spring growth in *Cornus stolonifera* Michx. J. Amer. Soc. Hort. Sci. 102:452–454.

FUCHS, R. and M. LIEBERMAN. 1968. Effects of kinetin, IAA, and gibberellin on ethylene production, and their interaction in growth of seedlings. Plant Physiol. 43:2029–2036.

GIESBERGER, G. 1975. Growing deciduous fruit trees in the tropics—a new approach to an old problem. Acta Hort. 49:109–111.

GILREATH, P.R. 1983. Author's reply. HortScience 18:14–15.

GILREATH, P.R. and D.W. BUCHANAN. 1981a. Floral and vegetative bud development of 'Sungold' and 'Sunlite' nectarine as influenced by evaporative cooling by overhead sprinkling during rest. J. Amer. Soc. Hort. Sci. 106:321–324.

GILREATH, P.R. and D.W. BUCHANAN. 1981b. Rest prediction model for low-chilling 'Sungold' nectarine. J. Amer. Soc. Hort. Sci. 106:426–429.

GREGORY, F.G. and J.A. VEALE. 1957. A reassessment of the problem of apical dominance. Symp. Soc. Expt. Biol. (London) 11:1–20.

GURDIAN, R.J. and R.H. BIGGS. 1964. Effect of low temperature on terminating bud dormancy of 'Okinawa,' 'Flordawon,' 'Flordahome,' and Nemaguard peaches. Proc. Fla. State Hort. Sci. 77:370–379.

HALL, R.H. 1973. Cytokinins as a probe of development processes. Annu. Rev. Plant Physiol. 24:415–444.

HATCH, A.H. and D.R. WALKER. 1969. Rest intensity of dormant peach and apricot leaf buds as influenced by temperature, cold hardiness, and respiration. J. Amer. Soc. Hort. Sci. 94:304–307.

HEDDEN, P., J. MACMILLAN, and B.O. PHINNEY. 1978. The metabolism of the gibberellins. Annu. Rev. Plant Physiol. 29:149–192.

HEMBERG, T. 1949. Growth-inhibiting substances in terminal buds of *Fraxinus*. Physiol. Plant. 2:37–44.

HEMBERG, T. 1965. The significance of inhibitors and other chemical factors of plant origin in the induction and breaking of rest periods. p. 669–698. In: W. Ruhland (ed.), Encyclopedia of Plant Physiology, Vol. 15/2. Springer, Berlin-Heidelberg-New York.

HENDERSHOTT, C.H. and D.R. WALKER. 1959. Seasonal fluctuation in quantity of growth substances in resting peach flower buds. Proc. Amer. Soc. Hort. Sci. 74:121–129.

HEWETT, E.W. and P.F. WAREING. 1973. Cytokinins in *Populus* × *robusta*: changes during chilling and bud burst. Physiol. Plant. 28:393–399.

HILL, A.G.G. and G.K.G. CAMPBELL. 1949. Prolonged dormancy of deciduous fruit-trees. Empire J. Expt. Agr. 17:259–264.

HOWARD, B.H. 1965. Increase during winter in capacity for root generation in detached shoots of fruit tree rootstocks. Nature 208:912–913.

HOWARD, B.H. 1968. Effects of bud removal and wounding on rooting in hardwood cuttings. Nature 220:262–264.

JACKSON, M.B. and D.J. CAMPBELL. 1979. Effects of benzyladenine and gibberellic acid on the response of tomato plants to anaerobic root environments and to ethylene. New Phytol. 82:331–340.

JACOBS, G., P.J. WATERMEYER, and D.K. STRYDOM. 1981. Aspects of winter rest of apple trees. Crop. Prod. 10:103–104.

JANICK, J. 1974. The apple in Java. HortScience 9:13–15.

JAVARAYA, H.C. 1943. Bi-annual cropping of apple in Bangalore. Ind. J. Hort. 1:31–34.

KACPERSKA-PALACZ, A. and M. ULIASZ. 1974. Cold induced changes in peroxidase activities in the winter rape leaves. Physiol. Vég. 12:561–570.

KASSEM, M.M. 1941. The seasonal variation of hormones in pear buds in relation to dormancy. Ph.D. Thesis, Univ. of Calif., Berkeley (cited in Eggert 1953).

KAWASE, M. 1966. Growth-inhibiting substances and bud dormancy in woody plants. Proc. Amer. Soc. Hort. Sci. 89:752–757.

KENDER, W.J. and S. CARPENTER. 1972. Stimulation of lateral bud growth of apple trees by 6-benzylamino purine. J. Amer. Soc. Hort. Sci. 97:377–380.

KEPCZYNSKI, J., R.M. RUDNICKI, and A.A. KHAN. 1977. Ethylene requirement for germination of partly after-ripened apple embryo. Physiol. Plant. 40:292–295.

KHAN, A.A. 1975. Primary, preventive and permissive roles of hormones in plant systems. Bot. Rev. 41:391–420.

KNIGHT, T.A. 1801. Account of some experiments on the ascent of the sap in the trees. Philos. Trans. Roy. Soc. London B 9:333–353. (cited in Pollock 1953.)

KOBAYASHI, K.D., L.H. FUCHIGAMI, and M.J. ENGLISH. 1982. Modeling temperature requirements for rest development in *Cornus serica*. J. Amer. Soc. Hort. Sci. 107:914–918.

LAKON, G. 1917. Über die Festigkeit der Ruhe panaschierter Holzgewächse. Ber. Deut. Bot. Ges. 35:643–652.

LAMMERTS, W.E. 1945. The breeding of ornamental edible peaches for mild climates. Amer. J. Bot. 32:53–61.

LATIMER, J.G. and H.A. ROBITAILLE. 1981. Source of variability in apple shoot selection and handling for bud rest determinations. J. Amer. Soc. Hort. Sci. 106:794–798.

LAVEE, S. 1973. Dormancy and bud break in warm climates; considerations of growth regulator involvement. Acta Hort. 34:225–234.

LEIKE, H. 1965. Neuere Ergebnisse über die Ruheperiode (Dormancy) der Gehölzknospen. Wiss. Z. Univ. Rostock, Math.-naturwiss. R. 14:475–492.

LEIKE, H. 1967. Wirkung von Gibberellinsäure und Kinetin auf ruhende Knospen verschiedener Gehölze. Flora (Jena) A 158:351–362.

LESLEY, J.W. 1945. Peach breeding in relation to winter chilling requirement. Proc. Amer. Soc. Hort. Sci. 44:243–250.

LESLEY, J.W. 1957. A genetic study of inbreeding and of crossing inbred lines of peaches. Proc. Amer. Soc. Hort. Sci. 70:93–103.

LETHAM, D.S. and L.M.S. PALNI. 1983. The biosynthesis and metabolism of cytokinins. Annu. Rev. Plant Physiol. 34:163–197.

LEWAK, S., A. RYCHTER, and B. ZARSKA-MACIEJEWSKA. 1975. Metabolic aspects of embryonal dormancy in apple seeds. Physiol. Vég. 13:13–22.

LIEBERMAN, M. 1979. Biosynthesis and action of ethylene. Annu. Rev. Plant Physiol. 30:533–591.

LIN, C. and L.E. POWELL. 1981. The effect of bud scales in dormancy of apple buds. HortScience 16:441 (Abstr.).

LLOYD, D.A. and G.A. COUVILLON. 1974. Effects of date of defoliation on flower and leaf bud development in the peach. J. Amer. Soc. Hort. Sci. 99:514–517.

LUCKWILL, L.C. and P. WHYTE. 1968. Hormones in the xylem sap of apple trees. p. 87–101. In: Plant growth regulators. Monograph 31. Soc. Chem. Industry.

MANNINI, F. and K. RYUGO. 1983. Effect of 2-chloroethylphosphonic acid (ethephon) on the endogenous levels of gibberellin-like substances and abscisic acid in buds and developing shoots of three grape varieties. J. Enol. Vitic. 33:164–167.

MENG-HORN, C., P. CHAMPAGNAT, P. BARNOLA, and S. LAVARENNE. 1975. L' axe caulinaire, facteur de préséances entre bourgeons, sur le rameau de l'année du *Rhamnus frangula* L. Physiol. Vég. 13:335–348.

MIELKE, E.A. and F.G. DENNIS. 1975. Hormonal control of flower bud dormancy in sour cherry (*Prunus cerasus* L.). II. Levels of abscisic acid and its water soluble complex. J. Amer. Soc. Hort. Sci. 100:287–290.

MIELKE, E.A. and F.G. DENNIS. 1978. Hormonal control of flower bud dormancy in sour cherry (*Prunus cerasus* L.). III. Effects of leaves, defoliation and temperature on levels of abscisic acid in flower primordia. J. Amer. Soc. Hort. Sci. 103:446–449.

MILBORROW, B.V. 1974. The chemistry and physiology of abscisic acid. Annu. Rev. Plant Physiol. 25:259–307.

MODLIBOWSKA, I. 1960. Breaking the rest period in black currants with gibberellic acid and low temperature. Ann. Appl. Biol. 48:811–816.

MORIMOTO, F.R. and K. KUMASHIRO. 1978. Studies on the rest-breaking of buds of deciduous fruit trees by chemical treatment. J. Fac. Agr. Shinushu Univ. 15:1–17 (cited in Snir 1983.)

MÜLLER-THURGAU, H. 1885. Beitrag zur Erklärung der Ruheperioden der Pflanzen. Landw. Jahrb. 14:851–907.

MUÑOZ, S.M.G. 1969. Evaluacíon de fórmulas para el cálculo de foras-frio en algunas zonas fruticolas de México. Proc. Trop. Reg. Amer. Soc. Hort. Sci. 13:345–366.

MUÑOZ, C., J. VALENCUELA, A. IBACACHE, and W.B. SHERMAN. 1984. Preliminary evaluation of low-chilling peaches and nectarines in warm-winter areas of Chile. Fruit Var. J. 38:40–43.
NAKASU, B.H., M. DO CARMO BASSOLS, and A.J. FELICIANO. 1981. Temperate fruit breeding in Brazil. Fruit Var. J. 35:114–122.
NAKASU, B.H., M.C.B. RASEIRA, and A.J. FELICIANO. 1982. Peach cultivars for fresh market and processing in southern Brazil. p. 1314f. In: Abstr. 21st Intern. Hort. Congr., Vol. 1, Hamburg.
NELL, T.A., W.H. BODNARUK, J.N. JOINER, and T.J. SHEEHAN. 1983. Ethylene evolution and flowering of cold- and GA-treated 'Redwing' azaleas. HortScience 18:454–455.
NESTEROV, J.S. 1967. The vegetative propagation of fruit crops in relation to the dormant period (in Russian). Vestn. Sel.-hoz. Nauki 12:79–85. [Hort. Abstr. (1968) 38:2576.]
NESTEROV, J.S. 1969. Times of occurrence of the dormant period in fruit crops (in Russian). Vestn.Sel'skohoz.Nauk 9:58–64. [Plant Breeding Abstr. (1970) 40:3666.]
NOODÉN, L.D. and J.A. WEBER. 1978. Environmental and hormonal control of dormancy in terminal buds of plants. p. 221–226. In: M.E. Clutter (ed.), Dormancy and developmental arrest. Academic Press, New York.
NORVELL, D.J. and J.N. MOORE. 1982. An evaluation of chilling models for estimating rest requirements of highbush blueberries (*Vaccinium corymbosum* L.). J. Amer. Soc. Hort. Sci. 107:54–56.
OPPENHEIMER, C. 1962. Breeding of apples for a subtropical climate. p. 18–24. In: Proc. 16th Intern. Hort. Congr., Vol. 3, Brussels.
OPPENHEIMER, C. and E. SLOR. 1968. Analysis of two F_2 and nine back cross populations. Theor. Appl. Genet. 38:97–102.
OVERCASH, J.P. 1967. Pear and apple variety testing in Kenya, East Africa. Fruit Var. J. 21:38–39.
OVERCASH, J.P. and J.A. CAMPBELL. 1955. Daily warm periods and total chilling-hour requirements to break the rest in peach twigs. Proc. Amer. Soc. Hort. Sci. 66:87–92.
PAIVA, E. and H.A. ROBITAILLE. 1976. Mechanism of budbreak in insufficiently chilled apple shoots. HortScience 11:319 (Abstr.).
PAIVA, E. and H.A. ROBITAILLE. 1978a. Breaking bud rest on detached apple shoots: effects of wounding and ethylene. J. Amer. Soc. Hort. Sci. 103:101–104.
PAIVA, E. and H.A. ROBITAILLE. 1978b. Breaking bud rest on detached apple shoots: interaction of gibberellic acid with some rest-breaking chemicals. HortScience 13:57–58.
PASTERNAK-ORAWIEC, G. and L.E. POWELL. 1983a. Changes in abscisic acid and gibberellin during the stratification of low and high chilling apple seeds. HortScience 18:560 (Abstr.).
PASTERNAK-ORAWIEC, G. and L.E. POWELL. 1983b. Chilling requirements for seeds of late-blooming apples. HortScience 18:561 (Abstr.).
PETRI, J.L. and M. PASQUAL. 1982. Quebra da dormência em macieira. Estado de Santa Catarina, Florianopolis, Bol. Tec. 18.
PHILLIPS, I.D.J. 1962. Some interactions of gibberellic acid with naringenin (5,7,4′-trihydroxy flavonene) in the control of dormancy and growth of plants. J. Expt. Bot. 13:213–226.
PHILLIPS, I.D.S. and P.F. WAREING. 1958. Studies in dormancy of sycamore. I. Seasonal changes in the growth-substance content of the shoot. J. Expt. Bot. 9:350–364.
PIENIAZEK, J. 1962. The content of endogenous growth substances in the chilled

and unchilled buds of apple variety 'Antonovka' during winter rest period. p. 395–404. In: Proc. 16th Intern. Hort. Congr., Vol. 5, Brussels.

PIENIAZEK, J. 1964. Kinetin induced breaking of dormancy in 8-month-old apple seedlings of 'Antonovka' variety. Acta Agrobot. 16:157–169.

PIENIAZEK, J. and R. RUDNICKI. 1971. The role of abscisic acid (ABA) in the dormancy of apple buds. Bull. Acad. Polon. Sciences, Sci. Biol. 19:201–204.

PIENIAZEK, J., M. SANIEWSKI, and L.S. JANKIEWICZ. 1970. The effect of growth regulators on cambial activity in apple shoots. In: Physiologische Probleme im Obstbau. Tagungsber.DAL Berlin 99:61–71.

PILET, P.E. 1971. Abscisic acid action in basipetal auxin transport. Physiol. Plant. 25:28–31.

PLANCHER, B. 1983. Kältebedürfnis bei bewurzelten Abrissen, abgeschnittenen Trieben und einnodigen Triebstücken von *Ribes nigrum* L. Gartenbauwissenschaft 48:248–255.

PODESVA, J., V. KLEJZAROVÁ, and J. BECKA. 1980. Contribution to the physiology of dormancy in flower buds of the cherry cultivar 'Napoleonova chrupka' (in Czech). Acta Univ. Agr. Brno, Fac. Agr. 28:65–80. [Landw. Lit. Tschech. (1980) 3/4:758 (Abstr.)]

POLLOCK, B.M. 1953. The respiration of Acer buds in relation to the inception and termination of the winter rest. Physiol. Plant. 6:47–64.

POWELL, L.E. and H. HUANG. 1982. The effect of heat on bud break in apple cultivars having different chilling requirements. p. 1140. In: Abstr. 21st Intern. Hort. Congr., Vol. 1, Hamburg.

PRATT, H.K. and J.D. GOESCHL. 1969. Physiological roles of ethylene in plants. Annu. Rev. Plant Physiol. 20:541–584.

PROEBSTING, E.L. and H.H. MILLS. 1976. Ethephon increases cold hardiness of sweet cherry. J. Amer. Soc. Hort. Sci. 101:31–33.

PURVIS, O.N. and F.G. GREGORY. 1952. Studies in vernalisation. XII. The reversibility by high temperature of the vernalised condition in Petkus winter rye. Ann. Bot. N.S. 16:1–21.

RAGEAU, R. and J.-C. MAUGET. 1982. Influence of temperatures above 15°C on dormancy breaking in peach and walnut leaf buds. p. 1187. In: Abstr. 21st Intern. Hort. Congr., Vol. 1, Hamburg.

RAMSAY, J. and G.C. MARTIN. 1970a. Seasonal changes in growth promoters and inhibitors in buds of apricot. J. Amer. Soc. Hort. Sci. 95:569–574.

RAMSAY, J. and G.C. MARTIN. 1970b. Isolation and identification of a growth inhibitor in spur buds of apricot. J. Amer. Soc. Hort. Sci. 95:574–579.

REINDERS-GOUWENTAK, C.A. 1965. Physiology of the cambium and other secondary meristems of the shoot. p. 1077–1105. In: W. Ruhland (ed.), Encyclopedia of Plant Physiology, Vol. 15/1. Springer, Berlin-Heidelberg-New York.

RICHARDSON, E.A., S.D. SEELEY and D.R. WALKER. 1974. A model for estimating the completion of rest for 'Redhaven' and 'Elberta' peach trees. HortScience 9:331–332.

RICHARDSON, E.A., S.D. SEELEY and D.R. WALKER. 1975. Author's reply. HortScience 10:561–562.

RICHARDSON, S.D. 1958. Bud dormancy and root development in *Acer saccharinum*. p. 409–425. In K.V. Thimann (ed.), The physiology of forest trees. Ronald Press, New York.

ROMBERGER, J.A. 1963. Meristems, growth, and development in woody plants. Forest Serv. Tech. Bull. 1293. USDA.

RUCK, H.C. 1975. Deciduous fruit tree cultivars for tropical and subtropical regions. Commonwealth Agr. Bureaux, East Malling, United Kingdom.

RYĆ, M. and S. LEWAK. 1982. Hormone interactions in the formation of the photosynthetic apparatus in dormant and stratified embryos. Z. Pflanzenphysiol. 107:15–24.

SACHS, T. and K.V. THIMANN. 1967. The role of auxins and cytokinins in the release of buds from dominance. Amer. J. Bot. 54:136–144.

SAMISH, R.M. 1945. The use of dinitrocresol–mineral oil sprays for the control of prolonged rest in apple orchards. J. Pomol. Hort. Sci. 21:164–179.

SAMISH, R.M. 1954. Dormancy in woody plants. Annu. Rev. Plant Physiol. 5:183–203.

SAMISH, R.M. and S. LAVEE. 1962. The chilling requirement of fruit trees. p. 372–388. In: Proc. 16th Intern. Hort. Congr., Vol. 5, Brussels.

SARAPUU, L.P. 1965. The physiological effect of phloridzin as a β-inhibitor during growth and dormancy in the apple tree (in Russian). Fiziol Rast. 12:134–145. [Hort. Abstr. (1965) 35:7203.]

SAUNDERS, P. 1978. Phytohormones and bud dormancy. p. 423–445. In: Letham, D.S., P.B. Goodwin, and T.J.V. Higgins (eds.), Phytohormones and related compounds—a comprehensive treatise. Elsevier, North Holland.

SAURE, M. 1971a. Beobachtungen über den Apfelanbau im tropischen Indonesien als Beitrag zur Frage der Blütenbildung, des Kältebedürfnisses und des Alterungsprozesses unserer Apfelbäume. Gartenbauwissenschaft 36:71–86.

SAURE, M. 1971b. Grundlagen der Regelung von Wachstum und Blütenbildung bei Apfelbäumen. Gartenbauwissenschaft 36:405–416.

SAURE, M. 1973. Successful apple growing in tropical Indonesia. Fruit Var. J. 27:44–45.

SAURE, M. 1981. Der Obstbaumschnitt als Eingriff in pflanzliche Regelungsvorgänge —Beitrag zum Verständnis der Schnittwirkungen. Erwerbsobstbau 23:79–83, 113–120.

SCHNEIDER, E.F. 1968. The rest period of Rhododendron flower buds. I. Effect of the bud scales on the onset and duration of rest. J. Expt. Bot. 19:817–824.

SEELEY, S.D. and L.E. POWELL. 1981. Seasonal changes of free and hydrolyzable abscisic acid in vegetative apple buds. J. Amer. Soc. Hort. Sci. 106:405–409.

SEIBEL, J.R. and L.H. FUCHIGAMI. 1978. Ethylene production as an indicator of seasonal development in red-osier dogwood. J. Amer. Soc. Hort. Sci. 103:739–741.

SHALTOUT, A.D. and C.R. UNRATH. 1983a. Effect of some growth regulators and nutritional compounds as substitutes for chilling of 'Delicious' apple leaf and flower buds. J. Amer. Soc. Hort. Sci. 108:898–901.

SHALTOUT, A.D. and C.R. UNRATH. 1983b. Rest completion prediction model for 'Starkrimson Delicious' apples. J. Amer. Soc. Hort. Sci. 108:957–961.

SHARPE, R.H. and W.B. SHERMAN. 1971. Peaches for warm climates. Fruit Var. Hort. Dig. 25:37–41.

SHERMAN, W.B. and P.M. LYRENE. 1984. Biannual peaches in the tropics. Fruit Var. J. 38:37–39.

SHERMAN, W.B. and R.H. SHARPE. 1970. Breeding plums in Florida. Fruit Var. J. 24:3–4.

SHERMAN, W.B., J. SOULE, and C.P. ANDREWS. 1977. Distribution of Florida peaches and nectarines in the tropics and subtropics. Fruit Var. J. 31:75–78.

SHULMAN, Y., G. NIR, and S. LAVEE. 1982. The effect of cyanamide and its Ca salts in termination of dormancy in grapevine buds (*Vitis vinifera*). p. 1186. In: Abstr. 21st Intern. Hort. Congr., Vol. 1, Hamburg.

SIMON, S. 1906. Untersuchungen über das Verhalten einiger Wachstumsfunktionen sowie der Atmungstätigkeit der Laubhölzer während der Ruheperiode. Jahrb. Wiss. Bot. 43:1–48.

SINSKA, I. and R.J. GLADON. 1984. Ethylene and the removal of embryonal apple seed dormancy. HortScience 19:73–75.

SKINNER, E.J. 1964. Delayed foliation. Decid. Fruit Grower 14:195–197.

SMITH, H. and N.P. KEFFORD. 1964. The chemical regulation of the dormancy phases of bud development. Amer. J. Bot. 51:1002–1012.

SMOLÉNSKA, G. and S. LEWAK. 1974. The role of lipases in the germination of dormant apple embryos. Planta 116:361–370.

SNIR, I. 1983. Chemical dormancy breaking of red raspberry. HortScience 18:710–713.

SOROKIN, H.P. and K.V. THIMANN. 1964. The histological basis for inhibition of axillary buds in *Pisum sativum* and the effects of auxins and kinetin in xylem development. Protoplasma 59:326–350.

SOUDAIN, P. and J.-L. REGNARD. 1982. Breaking dormancy of hardwood poplar cuttings by anoxia. p. 1144. In: Abstr. 21st Intern. Hort. Congr., Vol. 1, Hamburg.

SPARKS, D., J.A. PAYNE, and B.D. HORTON. 1976. Effect of subfreezing temperature on bud break of pecan. HortScience 11:415–416.

SPIERS, J.M. 1976. Chilling regimes affect bud break in 'Tifblue' rabbiteye blueberry. J. Amer. Soc. Hort. Sci. 101:84–86.

STADLER, J.D. and D.K. STRYDOM. 1967. Flower bud development of two peach cultivars in relation to their winter chilling requirements. S. African J. Agr. Sci. 10:831–840.

STRYDOM, D.K. and G.E. HONEYBORNE. 1971. Delayed foliation of pome- and stonefruit. Decid. Fruit Grower 21:126–129.

TABUENCA, M.C. 1976. Necesidades del frio invernal de variedades de nectarina. Anal. Est. Exp. Aula Dei 13:256–260. [Hort. Abstr. (1978) 48:6307.]

TABUENCA, M.C. 1979a. Influencia del patrón en el periodo de responso invernal de variedades de melocotonero y del almendra. Anal. Est. Exp. Aula Dei 14:469–476. [Hort. Abstr. (1981) 51:7674.]

TABUENCA, M.C. 1979b. Duracion del periodo de responso a distintas temperaturas y evalución del las necesidades de frio en variedades de albaricoquerco y almendro. Anal. Est. Exp. Aula Dei 14:519–531. [Hort. Abstr. (1981) 51:7656.]

TERBLANCHE, J.H. and D.K. STRYDOM. 1973. Effects of autumnal nitrogen nutrition, urea sprays and a winter rest-breaking spray on budbreak and blossoming of young 'Golden Delicious' trees grown in sand culture. Decid. Fruit Grower 23:8–14.

TERBLANCHE, P.J.C., I. HESEBECK, and D.K. STRYDOM. 1979. Effects of autumn nitrogen nutrition and a winter rest-breaking spray on the growth, development and chemical composition of young 'Golden Delicious' apple trees grown in sand culture. Scientia Hort. 10:37–48.

THIMANN, K.V. 1937. On the nature of inhibition caused by auxin. Amer. J. Bot. 24:407–412.

THOM, L.C. 1951. A study of the respiration of Hardy pear buds in relation to the rest period. Ph.D. Thesis, Univ. of California, Berkeley (cited in Hemberg 1965).

THOMAS, G.G. and E.H. WILKINSON. 1964. Winter dormancy in the black currant. Hort. Res. 4:78–88.

THOMAS, T.H., P.F. WAREING, and P.M. ROBINSON. 1965. Action of the sycamore 'dormin' as a gibberellin antagonist. Nature 205:1270–1272.

THOMPSON, W.K., D.L. JONES, and D.G. NICHOLS. 1975. Effects of dormancy factors on the growth of vegetative buds of young apple trees. Austral. J. Agr. Res. 26:989–996.

TINKLIN, I.G. and W.W. SCHWABE. 1970. Lateral bud dormancy in the blackcurrant *Ribes nigrum* (L.). Ann. Bot. 34:691–707.

TISSAOUI, T. and D. CÔME. 1973. Levée de dormance de l' embryon de pommier (*Pyrus malus* L.). Planta 111:315–322.

TORREY, J.G. 1976. Root hormones and plant growth. Annu. Rev. Plant Physiol. 27:435–459.

TUMANOV, I.I., G.V. KUZINA, and L.D. KARNIKOVA. 1970. Effect of gibberellins on the period of dormancy and frost resistance of plants (in Russian). Fiziol. Rast. 17:885–895. [Bibliogr. Dt.Wetterdienst 26, 1971.]

VAN DER LEK, H.A.A. 1934. Over den invloed der knoppen op de wortelforming van stekken. Meded. Landbouwhogeschool Wageningen 38(2) (cited in Doorenbos 1953.)

VAN STADEN, J. and G.G. DIMALLA. 1981. The production and utilization of cytokinins in rootless, dormant almond shoots maintained at low temperature. Z.Pflanzenphysiol. 103:121–129.

VEGIS, A. 1961. Samenkeimung und vegetative Entwicklung der Knospen. p. 168–298. In: W. Ruhland (ed.), Encyclopedia of Plant Physiology, Vol. 16. Springer, Berlin-Heidelberg-New York.

VEGIS, A. 1964. Dormancy in higher plants. Annu. Rev. Plant Physiol. 15:185–224.

VEGIS, A. 1965a. Ruhezustände bei höheren Pflanzen; Induktion, Verlauf und Beendigung: Übersicht, Terminologie, allgemeine Probleme. p. 499–533. In: W. Ruhland (ed.), Encyclopedia of Plant Physiology, Vol. 15/2. Springer, Berlin-Heidelberg-New York.

VEGIS, A. 1965b. Bedeutung von Aussenfaktoren bei Ruhezuständen bei höheren Pflanzen. p. 534–668. In: W. Ruhland (ed.), Encyclopedia of Plant Physiology, Vol. 15/2. Springer, Berlin-Heidelberg-New York.

WALKER, D.R. and C.W. DONOHO. 1959. Further studies of the effect of gibberellic acid on breaking the rest of young peach and apple trees. Proc. Amer. Soc. Hort. Sci. 74:87–92.

WALSER, R.H., D.R. WALKER, and S.D. SEELEY. 1981. Effect of temperature, fall defoliation, and gibberellic acid on the rest period of peach leaf buds. J. Amer. Soc. Hort. Sci. 106:91–94.

WALTON, D.C. 1980. Biochemistry and physiology of abscisic acid. Annu. Rev. Plant Physiol. 31:453–489.

WAN, C.-K. 1980. Fruit induced dormancy on apple (*Malus domestica* Borkh.) seed. Ph.D. Thesis, Michigan State University.

WAREING, P.F. 1969. The control of bud dormancy in seed plants. Symp. Soc. Expt. Biol. 23:241–262.

WAREING, P.F. and P.R. SAUNDERS. 1971. Hormones and dormancy. Annu. Rev. Plant Physiol. 22:261–288.

WEAVER, R.J. 1959. Prolonging dormancy in *Vitis vinifera* with gibberellin. Nature 183:1189–1199.

WEAVER, R.J. 1963. Use of kinins in breaking rest in buds of *Vitis vinifera*. Nature 198:207–208.

WEINBERGER, J.H. 1950. Chilling requirements of peach varieties. Proc. Amer. Soc. Hort. Sci. 56:122–128.

WEINBERGER, J.H. 1954. Effect of high temperature during the breaking of the rest of Sullivan Elberta peach buds. Proc. Amer. Soc. Hort. Sci. 63:157–162.

WEINBERGER, J.H. 1956. Prolonged dormancy troubles in peaches in the Southeast in relation to winter temperatures. Proc. Amer. Soc. Hort. Sci. 67:107–112.

WEINBERGER, J.H. 1967a. Studies on flower bud drop in peaches. Proc. Amer. Soc. Hort. Sci. 91:78–83.

WEINBERGER, J.H. 1967b. Some temperature relations in natural breaking of the rest of peach flower buds in the San Joaquin Valley, California. Proc. Amer. Soc. Hort. Sci. 91:84–89.

WEINBERGER, J.H. 1969. The stimulation of dormant peach buds by a cytokinin. HortScience 4:125–126.

WERNER, D.J. and E. YOUNG. 1984. Effect of shoot and root chilling during rest on resumption of growth in peach. HortScience 19:202 (Abstr.).

WESTWOOD, M.N. and H.O. BJORNSTAD. 1968. Chilling requirements of dormant seeds of 14 pear species as related to their climatic adaptation. Proc. Amer. Soc. Hort. Sci. 92:141–149.

WESTWOOD, M.N. and H.O. BJORNSTAD. 1978. Winter rainfall reduces rest period in apple and pear. J. Amer. Soc. Hort. Sci. 103:142–144.

WESTWOOD, M.N. and N.E. CHESTNUT. 1964. Rest period chilling requirement of Bartlett pear as related to *Pyrus calleryana* and *P. communis* rootstocks. Proc. Amer. Soc. Hort. Sci. 84:82–87.

WICKSON, M. and K.V. THIMANN. 1958. The antagonism of auxin and kinetin in apical dominance. Physiol. Plant. 11:62–74.

WILLIAMS, M.W. and E.A. STAHLY. 1968. Effect of cytokinins on apple shoot development from axillary buds. HortScience 3:68–69.

WILLIAMS, R.R., G.R. EDWARDS, and B.G. COOMBE. 1979. Determination of the pattern of winter dormancy in lateral buds of apples. Ann. Bot. 44:575–581.

WOOD, B.W. 1983. Changes in indoleacetic acid, abscisic acid, giberellins, and cytokinins during bud break in pecan. J. Amer. Soc. Hort. Sci. 108:333–338.

WRIGHT, S.T.C. 1975. Seasonal changes in the levels of free and bound abscisic acid in blackcurrant (*Ribes nigrum*) buds and beech (*Fagus sylvatica*) buds. J. Expt. Bot. 26:161–174.

YOUNG, E. and B. OLCOTT-REID. 1979. Siberian C rootstock delays bloom of peach. J. Amer. Soc. Hort. Sci. 104:178–181.

YOUNG, E. and D.J. WERNER. 1983. Root chilling effects on post-dormancy growth in apple. HortScience 18:586 (Abstr.).

YOUNG, L.C.T., J.T. WINNEBERGER, and J.P. BENNETT. 1974. Growth in resting buds. J. Amer. Soc. Hort. Sci. 99:146–149.

YOUNG, M.J. and M.N. WESTWOOD. 1975. Influence of wounding and chilling on rooting of pear cuttings. HortScience 10:399–400.

YOUNG, R. 1970. Induction of dormancy and cold hardiness of Citrus. HortScience 5:411–413.

YU, Y.B., D.O. ADAMS and S.F. YANG. 1980. Inhibition of ethylene production by 2,4-dinitrophenol and high temperature. Plant Physiol. 66:286–290.

ZARSKA-MACIEJEWSKA, B. and S. LEWAK. 1976. The role of lipases in the removal of dormancy in apple seeds. Planta 132:177–181.

ZARSKA-MACIEJEWSKA, B. and S. LEWAK. 1983. The role of proteolytic enzymes in the release from dormancy of apple seeds. Z. Pflanzenphysiol. 110:409–417.

ZARSKA-MACIEJEWSKA, B., E. WITKOWSKA, and S. LEWAK. 1980. Low temperature, gibberellin and acid lipase activity in removal of apple seed dormancy. Physiol. Plant. 48:532–535.

ZELLER, O. 1961. Entwicklungsgeschichte der Blütenknospen und Fruchtanlagen an einjährigen Langtrieben von Apfelbüschen. I. Entwicklungsverlauf und Entwicklungsmorphologie der Blüten am einjährigen Langtrieb. Z. Pflanzenzüchtung 44:175–214.

ZELLER, O. 1973. Blührhythmik von Apfel und Birne im tropischen Hochland von Ceylon. Gartenbauwissenschaft 38:327–342.

ZIMMERMAN, R.H., M. LIEBERMAN, and O.C. BROOME. 1977. Inhibitory effect of a rhizobitoxine analog on bud growth after release from dormancy. Plant Physiol. 59:158–160.

7

Physiological Control of Water Status in Temperate and Subtropical Fruit Trees

Hamlyn G. Jones
East Malling Research Station, Maidstone, Kent, ME19 6BJ, United Kingdom
Alan N. Lakso
New York State Agricultural Experiment Station, Geneva, New York 14456
J. P. Syvertsen
Citrus Research and Education Center, University of Florida, Lake Alfred, Florida 33850

Horticultural Reviews, Volume 7

ISBN 0-87055-492-1

I. INTRODUCTION

Water is one of the more important factors determining fruit size and total yield of fruit crops. Because of the high value of many fruit crops, it is often economically profitable to irrigate to alleviate water deficits, even in humid climates such as southern England. In more arid environments irrigation can be essential.

Plant water status can be controlled by alterations in frequency and timing of irrigation. In more humid climates where irrigation of horticultural crops is only of marginal economic benefit, it is particularly important to understand the natural processes involved in the control of plant water status. This information can be valuable in the formulation of effective techniques for optimizing plant water status by means of breeding or by management practices, such as fertilization, pruning, orchard systems, or chemical control. In addition, an understanding of control mechanisms is also useful in determining the most appropriate method of irrigation, for example, by flooding, controlled trickle, microsprinkler, or over-tree mist.

Before proceeding it is necessary to consider how water status is important. It is well known that drought tends to lower net productivity and fruit production and affects fruit quality (see Landsberg and Jones 1981; Kriedemann and Barrs 1981). Although often detrimental, mild water deficits sometimes are beneficial; they may, for example, substitute for a cold requirement or promote flowering (especially in tropical fruit species). Water deficits can also shift the balance from excessive vegetative growth toward reproductive growth (see Chalmers *et al.* 1981, and, for a review, Grierson *et al.* 1982) and appear to increase cold hardiness (Wildung *et al.* 1973; Yelenosky 1979).

Water deficits can affect a wide range of physiological and developmental processes involved in fruit production, including growth (both cell division and cell expansion can be inhibited), photosynthesis (stomatal aperture, and photosynthetic and respiratory enzymes are all affected), and a wide range of other biochemical and developmental processes. These phenomena have been well reviewed in recent years (Slatyer 1967; Kozlowski 1968, 1972, 1974; Hsiao 1973; Paleg and Aspinall 1981) and will not be discussed in detail here.

In this review we restrict our discussion to temperate and subtropical fruit trees.

II. MEASURES OF WATER STATUS

It is necessary to consider the various measures of plant water status that are available, and their relevance to the physiology and yield of fruit trees. The most popular measure of water status is total water potential (Ψ), which is a thermodynamically based quantity defined in terms of the chemical potential of water (μ_w), which in turn is the amount by which the Gibbs free energy in a system changes as water is added or removed, with temperature, pressure, and other constituents remaining constant (see Slatyer 1967; Nobel 1974):

$$\Psi = (\mu_w - \mu_w{}^0) / \overline{V}_w \qquad (1)$$

where $\mu_w{}^0$ is the chemical potential at a reference state consisting of pure free water at the same temperature, elevation, and atmospheric pressure, and $\overline{V}_w$ is the partial molal volume (used to convert units of energy per mole into the more convenient units of pressure in MPa). It follows from this definition that the prime relevance of Ψ to plants is in determining the *direction* of water flow (from a higher to a lower Ψ), as in the transpiration stream. The evidence that the chemical potential is directly concerned in the regulation of cell physiological processes is scant (Hsiao 1973), and as he pointed out, it is hard to see how the small changes in water activity that correspond to typical changes in Ψ could effectively be sensed.

The total water potential may be partitioned into several components:

$$\Psi = \Psi_p + \Psi_s + \Psi_g + \Psi_m \qquad (2)$$

The components of most relevance to plants are the pressure potential (Ψ_p), osmotic potential (Ψ_s), and matric potential (Ψ_m). The gravitational potential (Ψ_g) increases by only 0.01 MPa for each 1-m gain in height so would be less than 0.05 MPa for most fruit trees and is usually ignored.

Within cells the pressure potential, or turgor, measures the difference between internal and external pressure. Excess pressure inside the cells provides the driving force for cell expansion and for processes such as stomatal movement. Turgor is also involved in regulation of many biochemical processes, both by feedback control acting through altered substrate concentrations in the cell as growth is affected and by means of a turgor sensor in the plasma membrane (Bisson and Gutknecht 1980). Turgor may also affect membrane hydraulic conductivity (Zimmermann and Steudle 1974). There are, in addition, many ways in which turgor changes are indirectly involved in the control of plant functioning, for example, through effects on the metabolism of endogenous plant growth regulators. In particular, there is evidence that the synthesis of ABA may be stimulated by the loss of turgor (Pierce and Raschke 1980), and it is possible that turgor-induced reductions in growth of rapidly expanding organs could inhibit auxin or gibberellin production. Another indirect effect could be on cytokinin distribution; since the transpiration stream contains significant levels of cytokinins (Itai *et al.* 1968), cytokinin distribution within the plant would be expected to be influenced by stomatal opening and hence by turgor.

The cell osmotic potential depends on the concentration of solutes in the cell but, as with total potential, there is little evidence for its directly affecting metabolic processes. It is possible, however, that solute concentration could affect enzyme activity via alterations in protein conformation (see, for example, Hsiao 1973). The fact that large changes in cell solute concentration can occur in fruit trees during osmotic adjustment (Section V) provides good evidence that Ψ_s is not directly involved in the control of most plant responses to water deficits, though the role of Ψ_s in turgor maintenance is obviously critical.

In older work, water status was frequently measured in terms of water content rather than in terms of its energy status. Relative water content (RWC) has been a widely used (see Slatyer 1967), and still is, a useful measure of water status. RWC is, for example, usually more closely related to turgor than to Ψ. In fact, experiments with permeating and nonpermeating osmotica (Jones 1973; Kaiser 1982) confirm that the closely related parameters of cell volume, RWC, and Ψ_p are more relevant to the control of photosynthesis than is Ψ_s or Ψ.

Unfortunately, direct measurement of Ψ_p is difficult (Zimmermann 1978), though indirect sensors have been used (Barrs 1968; Heathcote *et al.* 1979). Therefore, estimates of turgor are usually obtained from the difference between Ψ and Ψ_s, both of which also are difficult to measure accurately. Because total leaf water potential is readily determined using a pressure chamber, many workers rely on this as the sole measure of plant water status even though total Ψ is probably only directly relevant to water flow and is not necessarily the best indicator of physiological stress. However, particularly in those plants, such as citrus, where active adjustment of Ψ_s is not important in mature leaves, Ψ is a useful indicator of physiological stress.

Even when Ψ is in equilibrium throughout a plant, the potential components usually differ from cell to cell and from tissue to tissue. In a transpiring plant there are also gradients of Ψ from tissue to tissue, leading to further difficulties in measuring and defining plant water status.

An additional problem relates to the specific tissues where water status has its effects. Traditionally, measurements were made on leaves, but many processes involved in the control of productivity occur in other tissues such as meristems, flowers, fruits, or roots. In addition, processes in leaves may be controlled by hormone transport from the roots, which itself is sensitive to root water status (Itai *et al.* 1968). A further problem with using leaf Ψ as an indicator of physiological stress is that it fluctuates rapidly with environmental demand, whereas many stress responses (especially the development of organs) integrate water status over periods of days or weeks (Schulze and Hall 1981). Therefore, a method of measurement that integrates stress over time is necessary. For this reason the relatively stable predawn Ψ, which can estimate soil Ψ, is often used. Unfortunately, the use of soil Ψ to estimate Ψ in other parts of the plant is likely to lead to as much error as does the use of leaf Ψ. The accumulation of free proline in leaves has also been proposed as a useful integrated measure of stress (Levy 1980b; Syvertsen and Smith 1984), but proline also accumulates in leaves in response to cold and salinity (Aspinall and Paleg 1981).

Jones (1983a) has described a technique for estimating Ψ at the surface of the roots and has argued that it gives a good general measure of plant water stress. The technique is based on simple measurements of leaf conductance and leaf water potential on control and stressed plants, using well-watered controls to estimate the hydraulic flow resistance in the plant. This can then be used to estimate the average Ψ at the root surface of stressed plants.

In any study of the effects of drought or water stress, it is clearly necessary to consider not only what the relevant measure of water status is but also where in the plant it should be measured and over what time period it should be integrated. In particular, the disadvantages of the commonly used leaf water potential as a measure of stress should be noted when designing future experiments. Unfortunately, however, in this review we have had to rely heavily on measurements of leaf water potential as the main measure of stress because of the lack of more relevant alternative measurements in the literature.

III. CONTROL OF WATER POTENTIAL—PRINCIPLES

The physical processes involved in the control of water status in any plant tissue are, at least in principle, fairly well understood, though the complexity of the soil-plant-atmosphere system means that prediction in any particular situation can be only approximate (see reviews by Slatyer 1967; Monteith 1973; Milburn 1979; Jones 1983b). In this section and the next, we describe factors affecting total Ψ; osmotic adjustment and control of turgor are discussed in Section V.

The basic processes involved in the control of water status in any tissue can be most readily understood from a consideration of water flow in the soil-plant system. Electrical analogues for the complex flow path in a real plant are shown in Fig. 7.1, but it has been found that a simple catenary model (Fig. 7.1b) is often adequate to describe steady state flow (Jones 1983b). Using this model, steady state leaf water potential (Ψ_l) is given by:

$$E_l = (\Psi_l - \Psi_{soil}) / R_{sp} \quad (3)$$

where Ψ_{soil} is the bulk soil water potential, E_l is the transpiration rate, and R_{sp} is the total hydraulic resistance in the soil-plant pathway. This can be written as:

$$\Psi_l = \Psi_{soil} - E_l R_{sp} \quad (4)$$

It is clear, therefore, that Ψ_l is primarily determined by three processes, which will be considered separately below: (1) flow rate through the plant, (2) hydraulic resistance, and (3) soil water potential.

A. Flow through the Plant

Hydraulic flow is primarily determined by transpiration rate, which at any instant may be regarded as a constant current generator in an electrical circuit (Fig. 7.1). Although equation 3 indicates that R_{sp} and Ψ_{soil} can have direct effects on transpirational flow, in practice their main effect is to influence flow indirectly via effects on tissue Ψ and hence stomatal aperture. The transpiration rate (E_l) is determined by a combination of environmental factors (radiation, temperature, humidity, and wind speed) and crop factors (e.g., canopy structure, leaf area, and stomatal conductance). A particularly useful general equation for calculating total evaporation (E) has been derived by Monteith (1965). This can be expressed in the form:

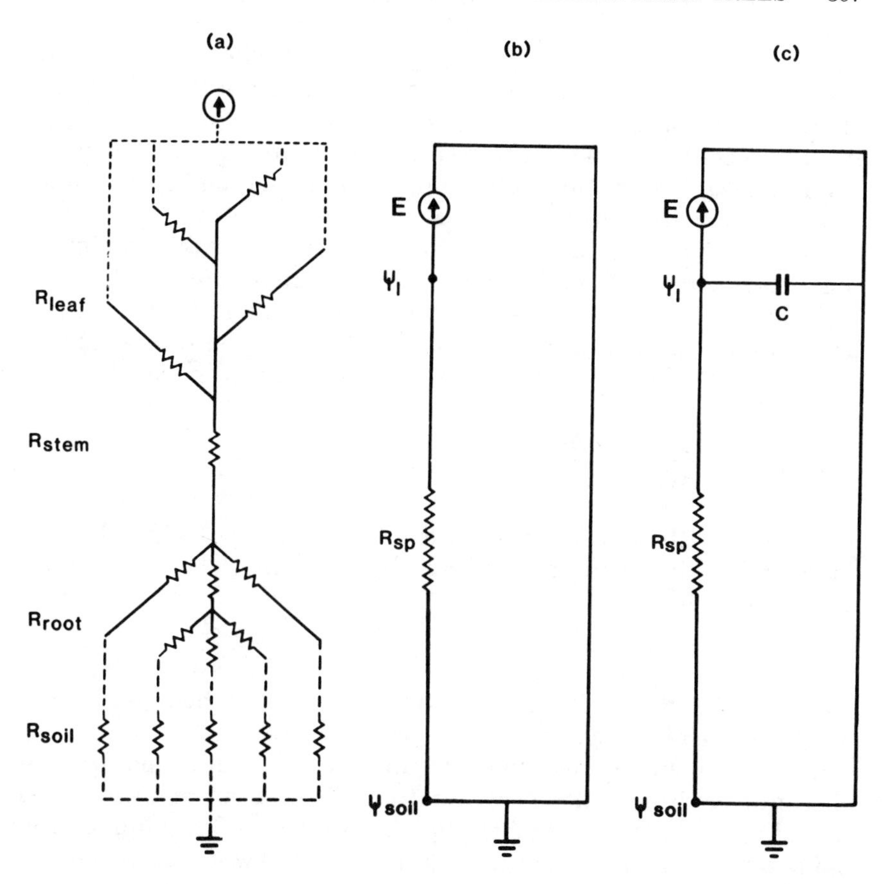

FIG. 7.1. (a) Electrical analogue of the hydraulic flow pathways in a transpiring plant. (b) Simplified model reduced to a single resistor for the whole soil–plant pathway with flow driven by a constant current generator *E*. (c) Same as (b), but including a lumped tissue capacitance.

$$E = \frac{\Delta Q_n + \rho_a c_p g_{aH} \delta e}{\lambda [\Delta + (\gamma g_{aH}/g_W)]} \tag{5}$$

where Δ is the slope of the saturation vapor pressure curve (Pa/K), ρ_a is the air density (kg/m^3), c_p is the specific heat of air (J kg^{-1}K^{-1}), γ is the psychrometric constant (Pa/K), λ is the latent heat of vaporization of water (J/kg), δe is the water vapor pressure deficit of the ambient air (Pa), Q_n is the net radiation absorbed by the leaves (W/m^2), g_{aH} is the conductance to heat transfer through the boundary layer (m/s), and g_W is the total conductance to water vapor loss. This last quantity equals $g_{aW}\, g_{lW}/(g_{aW} + g_{lW})$, where g_{aW} is the boundary layer conductance and g_{lW} is the leaf conductance (consisting of the sum of the stomatal and cuticular paths in parallel). In the case of orchards, g_a includes a contribution from both single leaves and the crop boundary layer, while g_l is a physiological conductance including appropriately weighted contributions of all leaves (see Thom 1975).

Equation 5 does not allow for changes in the longwave radiation balance of leaves as stomata close and leaf temperatures rise; therefore, Jones (1976) has proposed the replacement of Q_n by the net isothermal radiation (Q_{ni}), which is the net radiation that would be received by the canopy if it were at air temperature. This leads to:

$$E = \frac{\Delta Q_{ni} + \rho_a\, c_p\, g_{HR}\, \delta e}{\lambda\,[\Delta + \gamma(g_{HR}/g_W)]} \tag{6}$$

where g_{HR} is the parallel sum of conductances due to heat loss by longwave radiation (g_R) and sensible heat transfer. The conductance g_R is given by:

$$g_R = 4\sigma\epsilon T^3/\ \rho_a\, c_p \tag{7}$$

where σ is the Stefan-Boltzmann constant, T (K) is the temperature, and ϵ is the emissivity.

Equation 6 enables one to determine the effect on evaporation of changing environment (due to climatic or crop management changes) and of altered plant characteristics (as a result of breeding or crop management changes). The implications for leaf water potential can then be obtained from equation 3. A detailed analysis of equations 5 and 6 may be found in Jones (1983b).

When E is changing rapidly, the flow rate through the plant can lag behind E because water stored within the plant tissues can enter the transpiration stream. This can be particularly important in fruit trees because of their massive framework. In such cases, incorporation of capacitances into the resistance models for flow (Fig. 7.1c) allows effective prediction of Ψ_l (Powell and Thorpe 1977; Jones 1978; Jarvis *et al.* 1981).

B. Hydraulic Resistance

It follows from equation 3 that a low hydraulic resistance between any tissue and the source of water in the soil tends to minimize the depression of Ψ at midday or at high transpiration rates. The overall value of R_{sp} depends on the number and size of conducting vessels, on the length of the pathway, on the number and distribution of roots, and on the type and water content of the soil (Slatyer 1967; Jones

1983b). In addition, there can be both flow-rate and Ψ-dependent effects on R_{sp}. Although R_{sp} is often partitioned (e.g., into soil, root, and stem components), this can only be approximate, as is illustrated by the observation that increasing the number of roots may reduce both soil and root resistances. Increasing root density can enhance soil-root contact and thereby inhibit transient soil water depletion at the soil–root interface by reducing the transpirational demand on individual roots (McCoy *et al.* 1984).

The actual Ψ in any tissue could, in principle, be derived from the model in Fig. 7.1a, if flows and resistances were known in each component resistor. Similarly, Ψ of tissues off the main flow pathway tend to equilibrate with the nearest point on the main flow, though the rate of equilibration depends both on the hydraulic capacity of the tissue in question and on the hydraulic resistance in the connecting pathway. The fruits of apple and orange trees, for example, are hydraulically well isolated from the main transpiration stream and have large hydraulic capacitances (Landsberg and Jones 1981); fruits can maintain a much more stable Ψ than the leaves, remaining as much as 0.5 MPa higher than leaves at midday (Goode *et al.* 1978b; Syvertsen and Albrigo 1980). In cherry, although some damping of diurnal fluctuations of Ψ occurs in the fruit, Ψ differences between leaves and fruit tend to be much smaller (Tvergyak and Richardson 1979) than in apple and orange.

C. Soil Water Potential

The bulk soil water potential sets the upper limit for plant Ψ and provides the baseline below which the transpiration-dependent depression of Ψ operates. Even predawn values of Ψ_l may not rise as high as Ψ_{soil} because of the substantial internal resistances to water movement within plants. In addition, Ψ_{soil} has a direct effect on the soil component of the hydraulic resistance (see Slatyer 1967). It is also important to remember that Ψ_{soil} itself depends on previous evaporation and the rate at which soil water is depleted. Because of the importance of other controls of water potential, Ψ_l is often not closely related to Ψ_{soil} in fruit trees (Kaufmann and Elfving 1972; Jones 1983a).

IV. CONTROL OF WATER POTENTIAL—APPLICATION TO FRUIT TREES

In general, diurnal variations in leaf water potential in fruit trees (Fig. 7.2) are similar to those in other species and are explicable in terms of the mechanisms outlined in Section III. There are marked diurnal changes in Ψ_l, with minimum values of between −1.0 and −2.5 MPa usually occurring in the early afternoon at the time of highest transpiration rates (Klepper 1968; Goode and Higgs 1973; Landsberg *et al.* 1975; Kriedemann and Barrs 1981; Chalmers *et al.* 1983). Although these values are similar to those found in many cereal crops, they tend to be lower than those for many dicotyledonous annuals. Surprisingly perhaps, there is little difference between the minimum

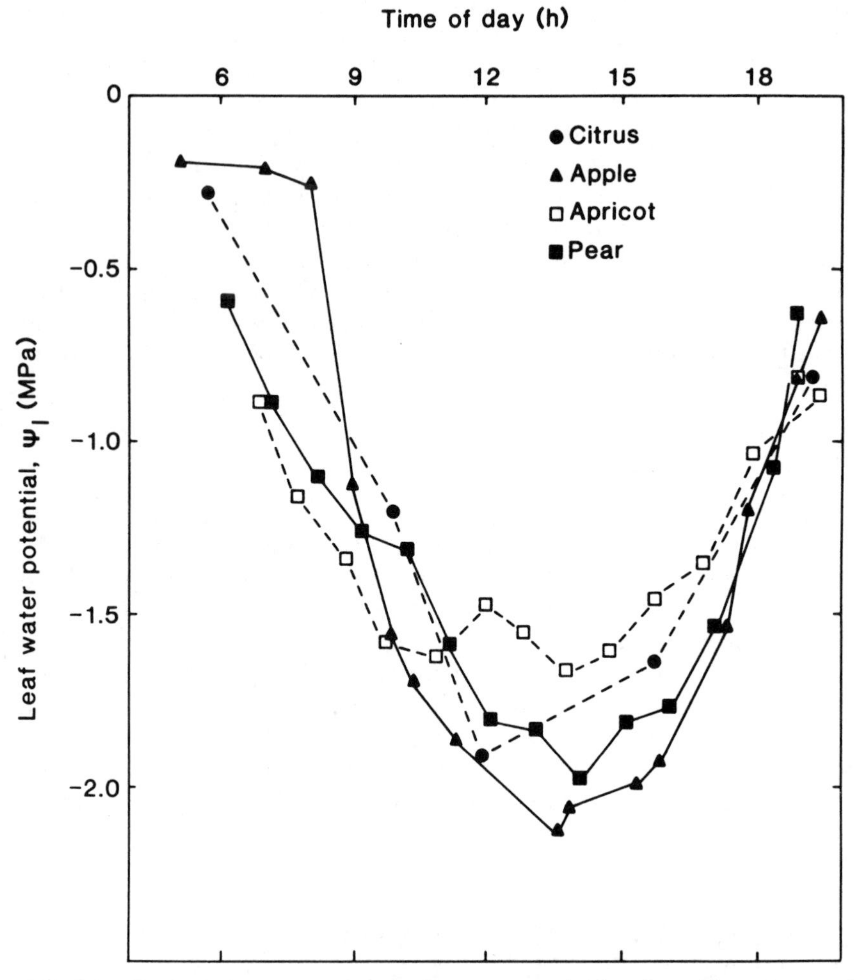

FIG. 7.2. Typical diurnal trends of leaf water potential for citrus (*Syvertsen and Albrigo 1980*), apricot (*Klepper 1968*), apple (*Goode and Higgs 1973*), and pear (*Klepper 1968*).

Ψ_l achieved in well-watered trees in humid and arid environments (Levy and Syvertsen 1981). This indicates an effective physiological control of Ψ_l, largely by means of control of transpiration rate under conditions of high evaporative demand (Schulze *et al.* 1974; Jones 1983a). With severe soil water deficits, Ψ_l can, of course, fall much lower—e.g., to at least −6.6 MPa in orange (Fereres *et al.* 1979)—and still recover on rewatering.

The different mechanisms involved in the control of Ψ in fruit trees will be discussed individually, together with some evidence for their importance. It should be recognized, however, that the complex interactions and feedbacks existing between the different processes often make it difficult to determine their relative importance.

A. Factors Affecting Transpiration

It is useful to distinguish between the water transpired by the leaves and all other water evaporated by a crop, whether from the soil or from free water on the surface of the leaf. Only the transpired water is involved in the flux-dependent lowering of leaf Ψ described by equation 3. This can be shown by the fact that treatments (such as misting) that reduce the proportion of water lost by transpiration from a crop act directly to raise tissue Ψ. The total evaporation can, however, be important in determining long-term changes in soil water content.

The direct effects of the various factors highlighted by equations 5 and 6 are important in the control of transpiration and will be discussed separately.

1. Radiation Balance. The net radiation (Q_n) absorbed by the leaves is a major factor determining tissue temperature, stomatal aperture, and crop water use. The effects of radiation vary both as a function of radiation available (varying with region and time of year) and with the fraction actually intercepted. Radiative properties of orchard systems have been reviewed by Jackson (1980) so will not be discussed in detail. In addition to any direct effects on stomata, it is apparent from equation 5 that a decrease in radiation absorption tends to decrease transpiration and hence to raise Ψ_l (see Jarvis 1981).

Orchard crops are characterized by areas of high leaf area density (transmitting little radiation) within the tree outline and by areas without leaves (see Jackson 1980). The reflection coefficients of whole orchards and of individual trees are low (around 0.17) (Kriedemann and Barrs 1981; Landsberg and Jones 1981; Chalmers *et al.* 1983). The total interception of shortwave radiation by the leaves is very dependent on tree spacing and on age, with interception being as low as 11%

in young apple orchards or (for deciduous crops) early in the season and as high as 80% in mature orchards with large trees (Jackson 1980). Maximum interception may reach 85% in peach orchards (Chalmers *et al.* 1983) and 90% in mature citrus (Kalma and Stanhill 1972; Jahn 1979). Reductions in radiation interception (e.g., by wide spacing of plants) can certainly reduce total evaporative loss, particularly with dry soil, and hence slow the development of soil and plant water deficits. Indeed, wide spacing is common practice in nonirrigated orchards in rainfall-deficient areas. Atkinson (1978), for example, showed that total water use for dwarf trees grown with bare soil was directly dependent on density of planting (proportional to ground cover in this case). As outlined by Jackson (1980), radiation interception can be increased by increasing tree height and reducing alley width.

Because of the generally high leaf density in fruit tree canopies (Jahn 1979; Jackson 1980), alterations in leaf angle are rarely likely to have a major effect on total radiation absorption, only influencing radiation distribution within the canopy. Similarly, although leaf pubescence and waxiness may vary markedly within one species (see Gates 1980; Baker and Procopiou 1980) and thus affect individual leaf reflection coefficients, it is unlikely that these differences would significantly affect the canopy reflection coefficient (Baldocchi *et al.* 1983). It is possible, however, that the radiation distribution within the canopy could be altered in such a way as to reduce total evaporation (Baldocchi *et al.* 1983). Increasing soil reflectivity (see Jackson 1980) may also tend to increase radiation absorbed by the leaves, and hence crop water loss.

Although the radiative properties and distribution of individual leaves probably have only a small effect on total crop transpiration, they can have important effects on the transpiration and Ψ of those individual leaves. For example, water potentials of shaded leaves tend to be significantly higher than those of leaves fully exposed to the sun (Olsson and Milthorpe 1983; Jones and Cumming 1984), and sun-exposed leaves tend to have higher temperatures and different anatomy and physiology (Syvertsen and Smith 1984; Syvertsen 1984). Leaves of apples, peaches, and citrus that are in exterior canopy positions are often somewhat folded along the midvein and borne in a variety of orientations, whereas the interior canopy leaves are often more horizontal (Syvertsen and Smith 1984). The severity of midday stress can sometimes be ameliorated by arranging the canopy foliage in a narrow vertical hedgerow, rather than in a horizontal T-form (Fig. 7.3).

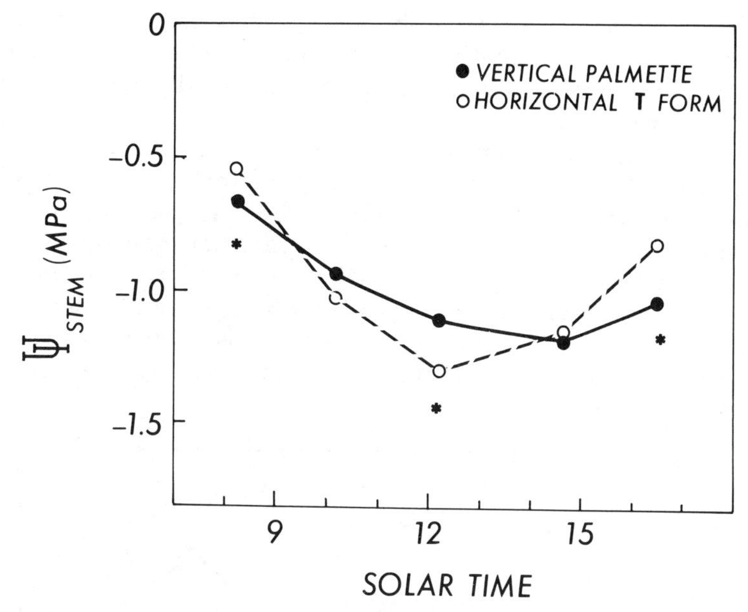

FIG. 7.3. Diurnal trend of stem water potentials (estimated by enclosing leaves in foil bags) of vertical and horizontal canopy forms of apple trees. Midday exposed leaf water potentials were about 0.6 MPa lower than stem potentials; * indicates statistical significance at the 5% level.
From A.N. Lakso, unpublished data.

2. Ambient Humidity. Decreases in the humidity deficit of ambient air (δe) have the direct effect of reducing evaporation (equations 5 and 6). This may be partially offset by increases in stomatal aperture (Section IV. A. 4. a). There is, however, little scope for modification of δe by changing plant or orchard characteristics, although the use of over-tree misting can be effective. For example, Lomas and Mandel (1973) reported a 27% increase in air humidity when misting within an avocado orchard, but the effects are often much smaller (Kohl and Wright 1974). Other effects of misting on Ψ are discussed in Section VI. A.

3. Boundary Layer Conductance. Orchards tend to be aerodynamically extremely rough and nonhomogeneous surfaces. As a result, the boundary layer conductance of orchards tends to be much greater than for field crops at any wind speed. Much of the crop boundary layer conductance is attributable to that of the individual leaves (see Landsberg and Jones 1981). Precise prediction of g_a is difficult, however, because transfer processes in this type of system are very complex and incompletely understood. Actual values may depend on many factors, including wind orientation relative to the rows, canopy forms, and tree spacing.

The generally high boundary layer conductance of forests or orchards, 100–300 mm/s (Jarvis 1981), implies that they are closely coupled to environment. This results in the transpiration of trees being much more sensitive to changes in total leaf conductance than is transpiration in many other crops (Fig. 7.4). Similarly, the individual leaf boundary layers (which depend on wind speed and on leaf size, shape, and hairiness) can also be important. It is interesting to note, however, that leaf sizes and shapes of many fruit crops are fairly

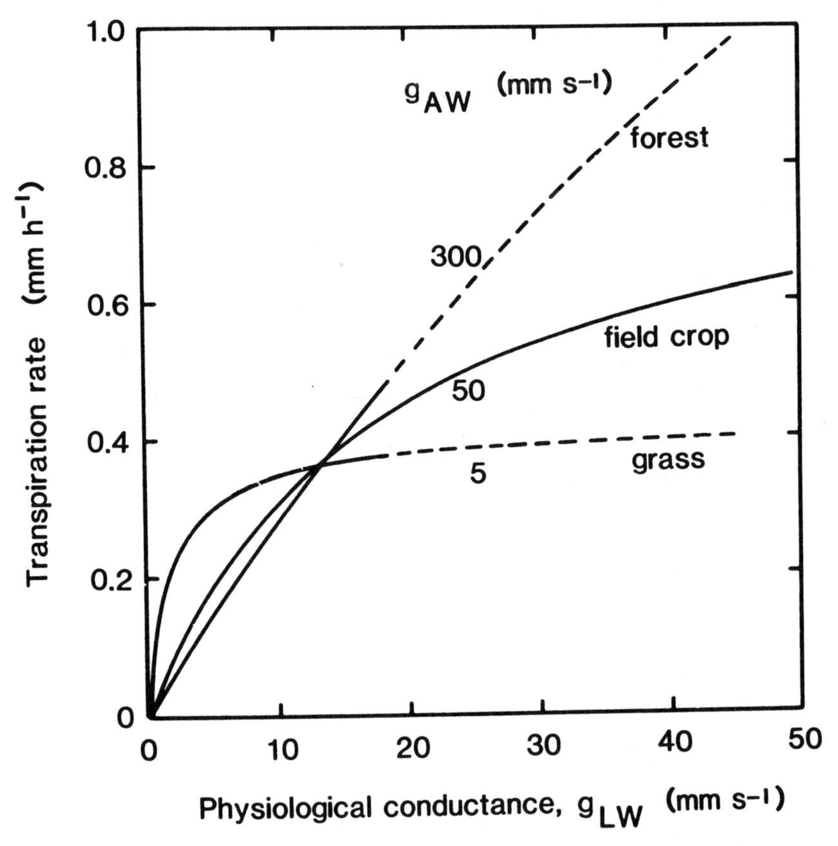

FIG. 7.4. Effect of changing physiological and boundary layer conductances on transpiration from different crops. Note the greater sensitivity to physiological conductance in tall crops such as forests or orchards.
After Jarvis (1981).

similar (varying by at most one order of magnitude in linear dimensions among species and by much less within species). This may be because most tree fruit crops originated in environments where a relatively narrow range of leaf sizes were favored by selection.

4. Physiological Conductance. As discussed already, the stomata provide the dominant short-term physiological control over evaporation, whereas leaf area index is equally, if not more, important for long-term control. The cuticular conductance as well as factors affecting the leaf boundary layer can also contribute. The canopy physiological conductance can be approximated from individual values of g_l (measured by porometry) using the relationship:

$$\text{canopy } g_l = \Sigma\,(g_{li}\,L_i) \qquad (8)$$

where L_i is the leaf area index in a particular stratum of canopy and g_{li} is the corresponding mean leaf conductance for that stratum (Jones 1983b). It follows from equation 8 that leaf area index and stomatal aperture are both important in determining overall canopy physiological conductance and hence transpiration rate and Ψ.

a. Stomatal Component. Typical values of g_l for leaves of well-watered fruit trees vary widely, depending on many factors. For apple, g_l is typically between about 3 and 8 mm/s (Landsberg *et al.* 1975; Lakso 1979; Warrit *et al.* 1980; Jones *et al.* 1983), though values as high as 14–18 mm/s have been reported (Fanjul and Jones 1982; Jackson *et al.* 1983). For various citrus species, g_l is commonly only about 1–3 mm/s (van Bavel *et al.* 1967; Elfving *et al.* 1972; Levy 1980a; Syvertsen 1982; Sinclair and Allen 1982), though values as high as 10 mm/s (Hall *et al.* 1976) have been reported. Values of g_l for other fruit trees are equally variable. In apricot, it ranges from 2.7 to 11 mm/s (Schulze *et al.* 1972; De Jong 1983); in peach, from 1.5 to 7.0 mm/s (Natali 1982; Tan and Buttery 1982; De Jong 1983); in avocado, from 1 to 10 mm/s (Sterne *et al.* 1977; Farré 1979); and plum and cherry both have maximum values of g_l of about 7 mm/s (De Jong 1983).

Stomata of all these species are sensitive to environment, tending to close in dry air (see especially Schulze *et al.* 1972; West and Gaff 1976; Hall *et al.* 1976). This humidity response (Fig. 7.5) can be extremely rapid. In apple, for example, much of the closing response to a decrease in ambient humidity occurred within 20 s (Fanjul and Jones 1982), though opening responses were slower. A humidity response is probably also involved in the gradual closure of stomata during the day that has been observed in apple (Landsberg *et al.* 1975; Lakso 1985), pear (Kriedemann and Canterford 1971), and citrus (Sinclair and Allen

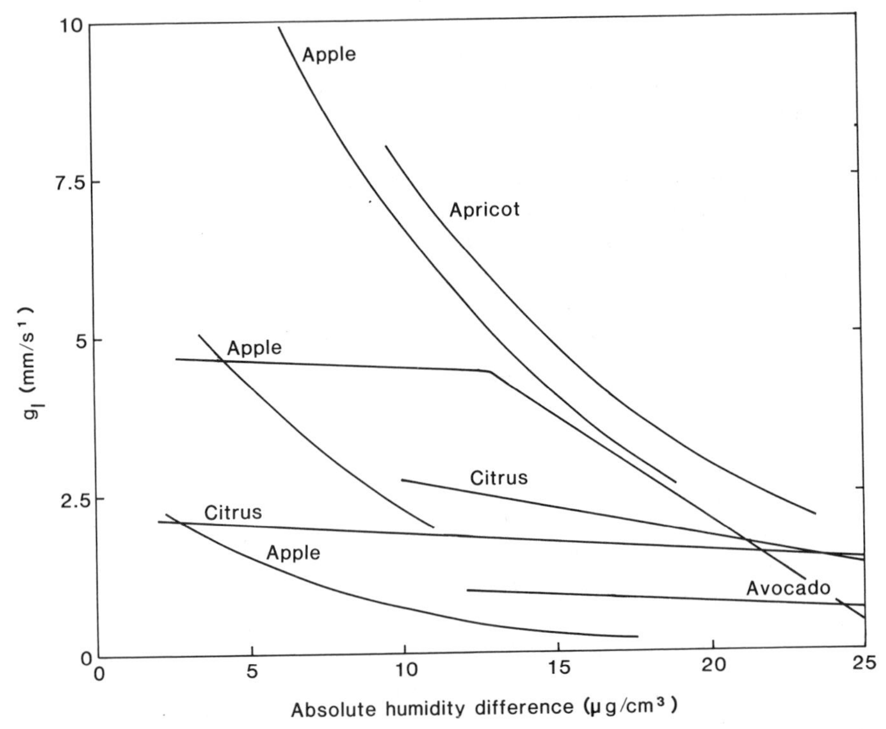

FIG. 7.5. Typical relationships between leaf conductance to water vapor (g_l) and the absolute humidity difference between leaf and air for well-watered apricot (*Schulze et al. 1972*), apple (*Lakso 1983; West and Gaff 1976; Fanjul and Jones 1982*), citrus (*Hall et al. 1974; Levy and Syvertsen 1981*), and avocado (*Sterne et al. 1977*) trees.

1982; Cohen and Cohen 1983). The fact that a similar diurnal trend has also been found for apple trees in constant environment chambers (Monselise and Lenz 1980b; Lakso 1983) suggests that internal factors such as carbon balance may play an important role as well as humidity. Stomata also respond to temperature and irradiance (Hall *et al.* 1976; Warrit *et al.* 1980) independently from humidity, though the temperature response is quite small in both apple and citrus (Hall *et al.* 1974).

Stomatal responses to humidity and temperature generally act to minimize the effect of changing environment on E, and hence Ψ_l, via both feedback and feedforward control (see Farquhar 1978; Jones 1983b). This is, therefore, an important mechanism acting to maintain favorable tissue Ψ even in severely desiccating environments. The degree of stomatal control can vary among species (Davies and Kozlowski 1974); for example, citrus stomata may reopen more slowly than stomata in several temperate tree species. In apricot trees growing in the Negev desert, stomatal responses to environment were so effective in controlling transpiration that only small changes in total daily transpiration were observed over a wide range of humidity deficits (Schulze *et al.* 1974). Good stomatal control of leaf water potential over a range of evaporative demands has also been reported for citrus (Levy 1980a). This may provide an important mechanism enabling citrus to acclimate to both humid and arid environments without much difference in water usage (Levy and Syvertsen 1981).

Stomata are also sensitive to plant water status, tending to close as Ψ declines (Jarvis 1980; Ludlow 1980). The closure in response to bulk leaf water status provides another important feedback control of plant water potential, tending to stabilize leaf Ψ, because the resulting decrease in E reduces the water potential difference between the soil and the leaves (equation 3). It is likely, however, that the stomata are really responding to turgor rather than to Ψ, because the relationship between g_l and Ψ changes dramatically during the season (Lakso 1979) with leaf age (Lakso *et al.* 1984; Syvertsen 1982) and with stress preconditioning (Ludlow 1980) as a result of osmotic adaptation.

A good example of the effectiveness of stomatal control of leaf water potential over a range of soil water contents is provided by data for apple (Jones *et al.* 1983; Jones 1983a). In these studies, leaf water potentials of well-irrigated and droughted trees were maintained within 0.25 MPa of each other in spite of large differences in Ψ_{soil}. In one year, it was even observed that stomatal closure in the droughted trees reduced transpiration to such an extent that their water potentials remained significantly *above* those of well-irrigated controls for several weeks. Similarly, moderate water stress can decrease citrus transpiration and thereby actually increase Ψ_l above that of well-watered trees (Levy 1983), and previously stressed and reirrigated citrus trees can have g_l above that of nonstressed controls (Cohen and Cohen 1983). Midday stomatal closure can also result in an increase in Ψ_l in spite of high evaporative demand (Levy and Syvertsen 1981). Such stomatal "overcompensation" for soil water deficits has also been observed in potted trees (H.G. Jones, unpublished data).

Normal diurnal fluctuations in leaf Ψ do not usually cause stomatal closure when soil moisture is adequate, so that over short periods g_l is often negatively correlated with Ψ at higher water potentials. The mechanism whereby declining Ψ (or Ψ_p) causes stomatal closure is incompletely understood, but there is good evidence that responses are mediated by abscisic acid produced within the leaf mesophyll and translocated to the stomata (see review by Addicott 1983).

Although it is well known that the dimensions and frequency of stomata on leaf surfaces of fruit trees vary with environmental conditions (particularly water status), season, and leaf age (Slack 1974; Syvertsen and Albrigo 1980), these changes are not generally well correlated with leaf conductance (Jones 1977). This is because stomatal morphology is mainly involved in determining only the maximum value of g_l. A factor that may greatly influence stomatal opening and thus transpiration is the variable response of stomatal opening to internal leaf CO_2. Changes in the photosynthetic rate induced by changes in the source–sink balance, aging processes, nutrient status, etc., will likewise affect transpiration (Monselise and Lenz 1980b; Lakso 1983). The mechanism controlling the response of stomata to CO_2 is not understood, but it has been suggested that abscisic acid levels may be important (Raschke 1979; Wilson 1981).

b. Leaf Area. The leaf area index is the other major factor determining canopy conductance and hence E. It should of course be noted that canopy g_l does not necessarily increase in proportion to leaf area index. This is because the proportion of shaded leaves increases with increasing leaf area and these shaded leaves have lower stomatal conductances (Syvertsen and Albrigo 1980; Warrit *et al.* 1980; Jones and Cumming 1984).

With a low leaf area index, crop water loss can be approximately proportional to foliage area, as has been reported for almond in Egypt (El-Sharkawi and El-Monayeri 1976) and for apple (Atkinson 1978), though this may be partly due to altered radiation interception (Section IV. A. 1). Seasonal changes in evaporation from citrus orchards have correlated well with changes in overall physiological conductance (van Bavel *et al.* 1967; Stanhill 1972), with leaf area changes being a major component of the changes in canopy g_l. For deciduous crops, seasonal changes in leaf area are even more important.

Leaf area changes are particularly important as a natural drought adaptation mechanism (see Kriedemann and Barrs 1981; Jones 1983b) because they (1) reduce water loss, thus conserving soil water and maintaining a favorable Ψ, and (2) decrease the evaporating surface per unit cross-section of the hydraulic pathway within the soil–plant system, thus decreasing the depression of Ψ_l below Ψ_{soil}.

Decreases in leaf area can result from a reduced rate of leaf production, early cessation of shoot or leaf expansion growth, or enhanced leaf abscission. All these processes operate in fruit trees (Goode and Higgs, 1973; Jones and Higgs 1979a; Kriedemann and Barrs 1981; Syvertsen 1982; Lakso 1983). Leaf abscission is dominant in pot experiments (H.G. Jones, unpublished data) and also in field experiments when severe stress occurs, both in citrus (Fereres *et al.* 1979) and other genera. The actual leaf fall in response to water deficits may, in some cases, occur only after rewatering at the end of the drought period (Suranyi 1977; Jones 1982). Leaf area may also be adjusted artificially by pruning (Section VI. B. 3).

c. Cuticular Conductance. Cuticular water loss usually makes only a very small contribution to the total canopy g_l. Although cuticular thickness varies among species, being thicker, for example, in lemon and clementine leaves than in mandarin orange leaves (Baker *et al.* 1975), these differences are only likely to be significant in extreme drought when stomata are almost completely closed. This is because cuticular conductances can be as low as 0.006–0.025 mm/s in citrus (Schönherr and Schmidt 1979); this approximates the values in apple fruits (Jones and Higgs 1982) but is about two orders of magnitude lower than the conductance of leaf surfaces with open stomata.

Evaporation from the bark of stems and twigs, which may constitute a significant proportion of crop surface area, does not appear to have been studied in detail, though Forshey *et al.* (1983) reported that branches and shoots composed only about 12% of the total surface area for 8-year-old apple trees in midsummer. This is unlikely, however, to be a major source of water loss as surface conductance of these parts may be as low as that of fruit (H.G. Jones, unpublished data for Bramley apple).

B. Hydraulic Flow Resistances

The flow resistances in the soil–plant pathway are crucial in determining tissue Ψ (equation 4). The magnitudes of the component resistances in citrus and apple, respectively, have been reviewed by Kriedemann and Barrs (1981) and Landsberg and Jones (1981). The general conclusions of these authors are that the main resistance resides in the root system (half or more of the total) and that the soil resistance only becomes significant at low soil water contents.

Genotypic differences in flow resistances have been observed in both citrus (Syvertsen 1981) and apple (Olien and Lakso 1984) rootstocks. In the case of citrus, both sour orange and Cleopatra mandarin had higher resistances than did rough lemon or carrizo citrange (more vigorous) rootstocks, while in apple, M.9 and M.26 appeared to have higher root resistances than other, more invigorating, rootstocks.

Although the hydraulic resistance is often observed to be flow-rate dependent, the majority of recent reports for trees (including Lands-

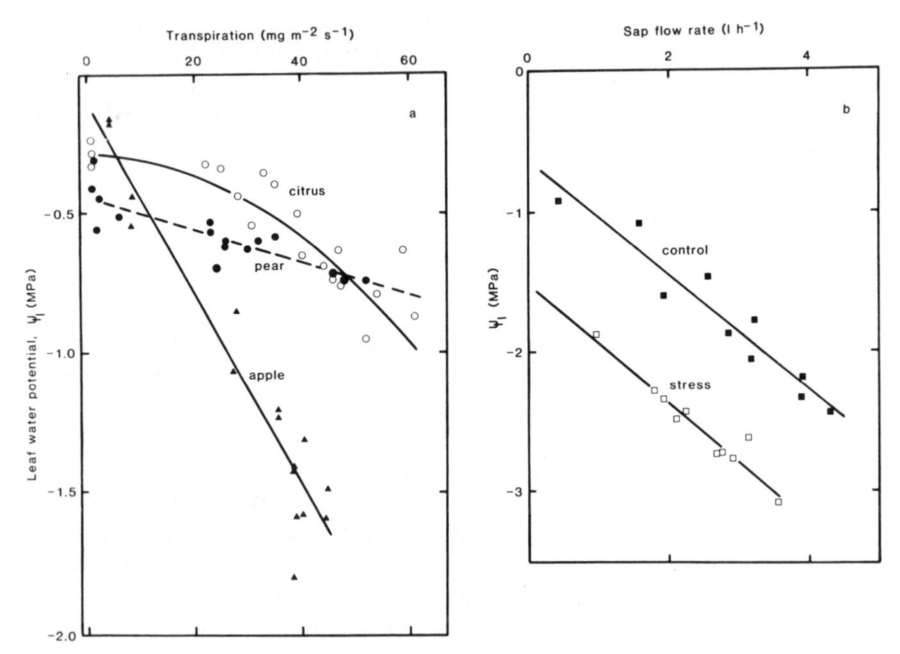

FIG. 7.6. Relationships between leaf water potential and transpiration rate in different fruit trees. A straight line indicates that use of a constant resistance (R_{sp}) in equation 3 is valid. (a) Growth-chamber data for citrus and pear (*Camacho-B et al. 1974*) and field data for apple (*Landsberg et al. 1975*). (b) Field data for stressed and well-watered orange (*Cohen et al. 1983a*).

berg *et al.* 1975; Cohen *et al.* 1983a; Jones 1983a) indicate that use of a constant diurnal value of R in equations 3 and 4 is valid in predicting leaf Ψ in most cases (Fig. 7.6).

The total length and density of roots in the soil are other factors in determining the plant hydraulic resistance. These aspects have been reviewed by Atkinson (1980), who concluded that fruit trees, including a range of *Prunus* and *Malus* species, tended to have extremely low root densities (0.8–70 cm of root under each cm^2 of soil surface) compared with those reported for herbs, which averaged one to two orders of magnitude greater. The sparsity of fruit tree roots implies that the soil component of the hydraulic resistance can become significant at higher soil Ψ than for herbaceous species due to localized drying around the few roots (Landsberg and Jones 1981; Cohen *et al.* 1983a). In spite of this, the great extent of many fruit tree roots enables fruit trees to reach a greater soil volume and thus maintain adequate leaf Ψ and transpiration for longer periods than many shallow-rooting species.

It is well known that many pests and diseases can affect plant hydraulic resistances (Ayres 1978). For example, Young and Garnsey (1977) and Cohen *et al.* (1983b) reported that citrus blight resulted from restricted water movement in trunk xylem caused by plugging materials, while *Phytophthora* root rot was shown to increase root resistance in avocado (Sterne *et al.* 1978). Similarly, feeding on roots by the larvae of the fruit tree root weevil (*Leptopius squalidus*) can greatly reduce water uptake by sweet cherry and apple (Baxter and West 1977). These and other infections, such as the vascular wilts, all increase flow resistances, thus tending to lower leaf Ψ.

Vesicular-arbuscular mycorrhizal infections, on the other hand, tend to increase water uptake by plant roots (Safir *et al.* 1972; Levy *et al.* 1983). The decreased hydraulic resistance of the root systems of infected plants may result both from enhanced root growth because of improved nutrition, especially of phosphorous, (Graham and Syvertsen 1984) and from an enlarged surface area (provided by the fungus) for water absorption from the soil. The decreased root resistance of mycorrhizal plants tends to raise leaf Ψ and thus increase g_l and E. Levy *et al.* (1983), however, reported that when rough lemon seedlings were subjected to soil-drying cycles, mycorrhizal infection had the opposite effect. Root resistance was increased by soil drying, with the effect being greatest for mycorrhizal plants. This effect was possibly a result of more rapid soil water depletion by the infected plants.

Root hydraulic resistance is also sensitive to soil conditions such as temperature, flooding (anaerobiosis), and ethanol (Andersen and Proebsting 1984). As in other species, root hydraulic resistance in citrus and pears has been found to increase with a decrease in root temperature from 33° to 16°C (Ramos and Kaufmann 1979; Wilcox and Davies 1981; Syvertsen *et al.* 1983; Wilcox *et al.* 1983; Andersen and Proebsting 1984). It is interesting to note, however, that growing seedlings at soil temperatures of 16°C can decrease root resistance compared with seedlings grown in warmer soil (Syvertsen *et al.* 1983). Flooding has also been found generally to increase root hydraulic resistance, especially at higher temperatures (Maotani *et al.* 1976; Syvertsen *et al.* 1983). It has also been reported (Ramos and Kaufmann 1979; Syvertsen 1981) that water stress can increase root hydraulic resistance in lemon, but this effect is not always observed (Hoare and Barrs 1974; Cohen *et al.* 1983a). Davies and Lakso (1979) found that the root resistance of apple trees grown outdoors was as little as one-third of that of greenhouse-grown trees. This resulted in the outdoor-grown trees being much less sensitive to subsequent water stress.

The generally low hydraulic resistances in the stems and branches of fruit trees (at least in citrus and apple) result in rather small gradients of Ψ (less than 0.1 MPa/m) throughout much of the tree canopy. Differences in the leaf Ψ of different leaves can often be attributed to differences in transpiration rates from those leaves (Jones and Cumming 1984).

There is now increasing evidence that cavitations or embolisms can occur within the xylem of transpiring plants (see Milburn 1979 for a review). These gaps in continuity of water in the xylem would be expected to increase the hydraulic resistance, the effect being largest for cavitations within the largest xylem elements. Studies using an acoustic click detector (West and Gaff 1976) and based on the decreases in stem density observed using a gamma-probe (Brough 1983) show that cavitations can occur in apple trees. The effects on stem hydraulic resistance, however, have not yet been quantified.

C. Other Genetic Differences

Much of the genetic variations in leaf water potential cannot be attributed unambiguously to any one of the mechanisms described so far. For example, observations that Ψ_l is often lower in dwarfing than in invigorating rootstocks can in some cases be attributed to the greater fruit load, and hence greater transpiration rate, of the dwarfing stocks (Giulivo and Bergamini 1982). In another case, the lower Ψ_l was apparently due to higher root resistances in the dwarfing rootstocks (Olien and Lakso 1984) than in invigorating stocks. Similarly, differences in Ψ_l between peach trees on different rootstocks have been reported (Young and Houser 1980; Natali *et al.* 1983). In this case, Ψ_l was not directly related to tree size since trees on the intermediate-sized St Julien GF 655/2 rootstock used the most water and suffered the greatest stress.

It is even more difficult to interpret differences in Ψ_l among species. In one study (Klepper 1968), Ψ_l of pear trees was found to vary more diurnally than did that of apricot, falling to a midday minimum of −2.0 MPa compared with −1.7 MPa for apricot. The relatively more stable Ψ_l of apricot may reflect better stomatal control over transpiration in apricot (Section IV. A. 4. a). In another study (Smart and Barrs 1973), minimum values of Ψ_l on hot, dry days ranged between −1.3 and −1.4 MPa for grapes, between −1.8 and −2.3 MPa for citrus, between −2.4 and −2.8 MPa for prunes, and averaged −3.1 MPa for peaches. The lower values in each range were obtained prior to irrigation, whereas the upper values were obtained under well-watered conditions. For all four species, multiple regressions of leaf water potential against radiation, temperature, and vapor pressure deficit indicated that radiation was the dominant factor controlling Ψ_l, presumably through its effect on transpiration rate. Soil Ψ was, however, not included in the analysis.

Species differences in the control of leaf water potential may also reflect different mechanisms of turgor maintenance (Section V). A species such as apple that can osmotically adjust (Section V. A) can maintain turgor above critical levels with high transpiration rates and low leaf water potentials. Citrus, which does not appear to show significant osmotic adjustment, would then depend on controlling transpiration rates so that leaf water potentials and turgor would be maintained. Thus, tight stomatal control over transpiration rate and leaf water potential would not appear to be as critical for osmotically adjusting species. It would be of interest to see if a correlation exists between the ability to adjust osmotically and the degree of stomatal control of transpiration.

V. CONTROL OF TURGOR

As pointed out in Section II, turgor affects many processes in the fruit tree, especially those related to expansive growth or stomatal opening, and is probably the most important water potential component. Maintenance of turgor by means of osmotic adjustment provides a major physiological method for minimizing the detrimental effects of water stress.

The turgor potential (Ψ_p) in any cell can be obtained from equation 2 as the difference between the total potential (Ψ) and the osmotic potential (Ψ_s), which is itself simply related to the cell sap concentration by the Van't Hoff relationship:

$$\Psi_s = -RTn_s/V = -RTc_s \tag{9}$$

where R is the universal gas constant, T is the absolute temperature (K), V is the symplastic water volume, n_s is the number of moles of solute, and c_s is the solute concentration. It follows that any change of cell volume as a cell loses or gains water will be reflected in a change in Ψ_s. A loss of water, for example, causes a decrease in Ψ_s, partially offsetting the decline in Ψ that caused the loss. The magnitude of this *passive* compensatory osmotic adjustment by dehydration depends on the elasticity of the cell walls: that is, the adjustment is small for cells with very rigid cell walls because the volume change is small for a given drop in Ψ.

The rigidity of cell walls is described by the bulk elastic modulus (ϵ_B), which is defined as:

$$\epsilon_B = V\,d\Psi_p/dV \tag{10}$$

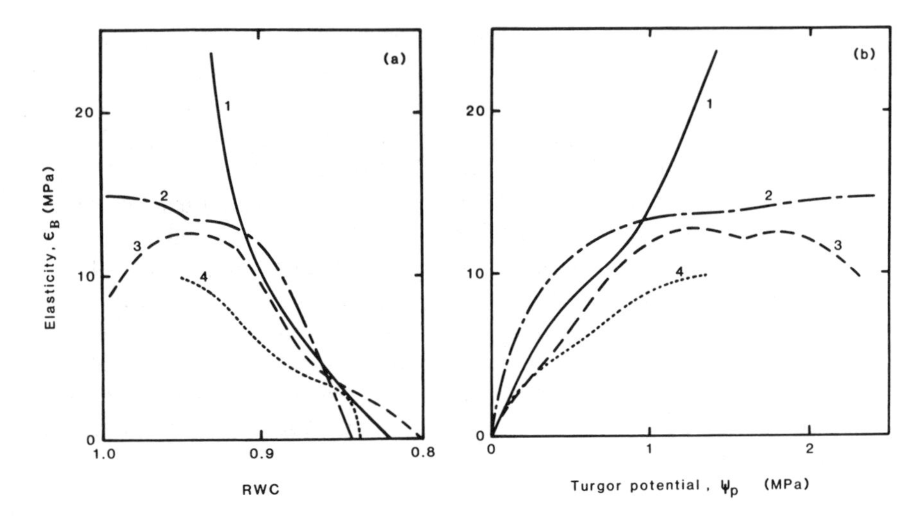

FIG. 7.7 Dependence of bulk elastic modulus for apple leaves on (a) relative water content (RWC) and (b) turgor pressure. Curves 1, 2, and 3 for leaves dried in a pressure chamber, and curve 4 for a leaf dried on the bench. All curves except 2 are for irrigated trees.
Calculated from data of Jones and Higgs (1979).

A higher value of ϵ_B corresponds to more rigid cells. The bulk elastic modulus can be conveniently obtained from pressure–volume curves and is generally a strong function of turgor, decreasing to zero at zero turgor (Fig. 7.7), regardless of the initial leaf water status.

Another way in which Ψ_s may change without active changes in the number of solute molecules is by altered partitioning of total tissue water between the osmotic volume (the symplast) and the apoplast (which includes water in cell walls and in xylem vessels). Changes in the proportion of apoplastic water can therefore contribute to apparent osmotic adjustment over the long periods necessary for the synthesis of thicker cell walls or new types of cells.

In addition to these passive changes in cell sap concentration, in many plants turgor maintenance is based on *active* osmotic adjustment where the *numbers* of solute molecules increase in response to stress.

When considering the responses of different plants, it is useful to distinguish between passive and active adjustment. In order to study the active component, it is necessary to eliminate passive changes. One way is to convert observed osmotic potentials to the corresponding values at 100% RWC, using equation 9 and a knowledge of the relationship between RWC and Ψ (Jones and Higgs 1979b). An alternative is to rehydrate the leaf to 100% RWC before determining Ψ_s. Unfortunately, rehydration may give inaccurate answers because some change in solute concentration may occur during the several hours required for rehydration. The main types of turgor maintenance are illustrated schematically in Fig. 7.8, and the evidence for their occurrence in fruit trees is considered next.

A. Osmotic Adjustment

The general decrease of the relative water content of plant tissues at midday (e.g., Klepper 1968; Goode and Higgs 1973; Davies and Lakso 1979; Fanjul and Rosher 1984), and the consequent passive concentration of solutes in the cells, occurs in all fruit trees. In addition, there is likely to be an active diurnal photosynthate accumulation in fruit tree leaves since translocation rates, at least in apple, are quite slow and probably do not equal photosynthetic rates in the light period (Priest-

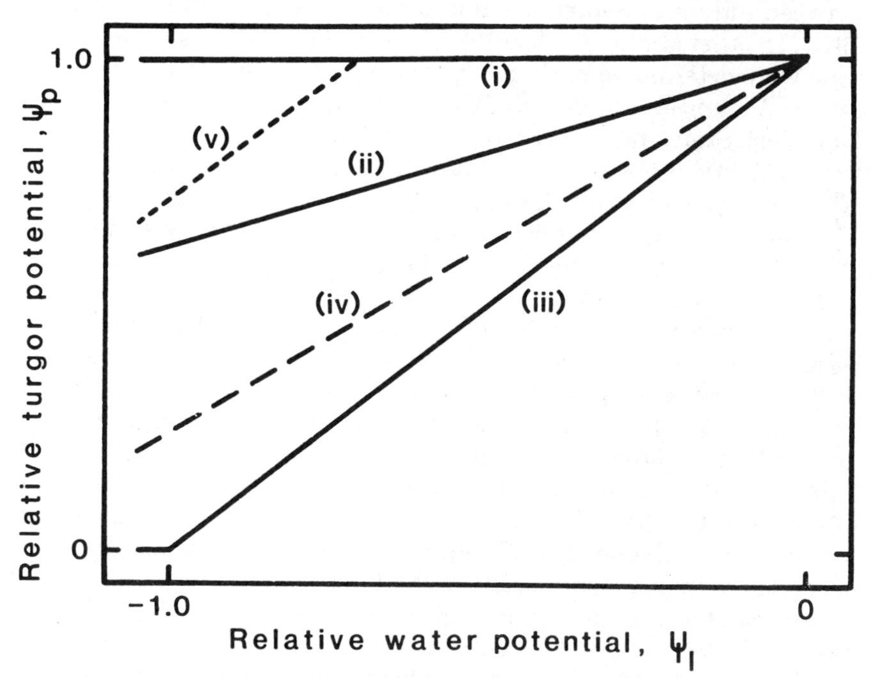

FIG. 7.8. Schematic representation of the relationships between turgor and total potential (both normalized to a scale of 0 to 1). (i) Full turgor maintenance (active osmotic adjustment); (ii) partial turgor maintenance; (iii) no turgor maintenance and no passive change in cell solutes (extremely rigid cell walls); (iv) partial turgor maintenance (passive adjustment only); (v) full turgor maintenance over a limited range of Ψ.
From Jones (1983b).

ley 1973). The net effect is that turgor potentials tend to decrease much less than total potentials during a daily cycle. For example, the data of Goode and Higgs (1973), for apple, indicated a total diurnal osmotic adjustment of between 0.5 and 1.0 MPa, of which perhaps 0.2 to 0.4 MPa was attributable to active photosynthate accumulation and the rest was passive. This corresponds to case (ii) in Fig. 7.8. Davies and Lakso (1979), also for apple, reported a total diurnal osmotic adjustment of 1.65 MPa, of which only a small proportion (0.37 MPa) could be attributed to active adjustment. In contrast, experiments with peach seedlings did not detect any active diurnal osmotic adjustment although the seedlings were growing in very low light (Young *et al.* 1982).

Osmotic adjustment can also occur in response to slowly developed water stress and during the course of the growing season, though there are few reports of studies on field-grown trees. For apple in southeast England, a decline in Ψ_s of 0.5 MPa (adjusted to a water potential of zero and therefore indicating active osmotic adjustment) between July and September has been reported (Goode and Higgs 1973). In other studies (Lakso 1983; Lakso *et al.* 1984), osmotic potentials were determined early in the morning to estimate the value at 100% RWC and eliminate normal diurnal changes. In these studies, in New York State, an apparently active adjustment over the season of 2.0–2.5 MPa was observed in mature field trees. In another study from New York, however, no change of Ψ_s (at 100% RWC) was detectable between July and September in potted apple trees (Davies and Lakso 1979). The reason why potted trees do not osmotically adjust as much as field trees is not clear.

Although a total osmotic adjustment of about 3 MPa has been observed in response to severe water stress in citrus trees (Fereres *et al.* 1979), a very large proportion of this was attributable to passive changes, with only about 0.3 MPa being related to changing Ψ_s at 100% RWC. In other work, with Valencia oranges (Syvertsen and Albrigo 1980), leaf osmotic potential was found to vary over the course of a season by about 1.0 MPa, but again very little of this could be attributed to active changes. The general pattern of response in citrus species is therefore close to case (iv) in Fig. 7.8.

Unfortunately, relatively little is known about the capacity for osmotic adjustment in other fruit tree species. The leaf water potential required to close stomata in field-grown peach trees has been found to decrease during a season (Chalmers *et al.* 1983), as it has also in apple (Lakso 1979). This indicated that some adjustments may have occurred, but osmotic potentials were apparently not measured in these studies. Comparisons of osmotic changes in several fruit crops grown in one environment are still required, with particular emphasis on determining the active osmotic adjustment using comparable techniques.

Osmotic adjustment tends to occur most readily when water stress develops slowly. The rapid desiccation usually found in stress experiments with potted trees tends to limit osmotic adjustment. Where stress has been allowed to develop slowly, however, by rewatering pots daily to a given weight, changes of Ψ_s of up to 0.3 MPa at 100% RWC, and of 0.5 MPa at zero turgor, have been observed in mature apple leaves (Fanjul and Rosher 1984). With these potted trees, osmotic potentials recovered to control values within about 2–3 days after rewatering. The large soil volume explored by fruit tree roots tends to slow the rate of stress development compared with annual crops and may be a factor allowing the time required for osmotic adaptation to stress in field-grown fruit trees. In relation to the diagrammatic responses of osmotic adjustment shown in Fig. 7.8, fruit tree responses are commonly between partial active adjustment (ii) and passive adjustment alone (iv), but extreme cases of full adjustment (i) or zero adjustment (iii) may occur.

Different organs on one plant may or may not show osmotic adjustment. A seasonal study of the water relations of mature apple leaves and shoot tips (meristem plus folded leaves) on the same shoots showed that the mature leaves adjusted osmotically, while the shoot tips did not (Lakso 1983; Lakso *et al.* 1984). This work also showed that the shoot tips had significantly less negative osmotic potentials than the mature leaves, as has also been reported for citrus (Fig. 7.9). These results may partially explain the different responses of stomata and of leaf area expansion to water stress.

Even within the canopy of a fruit tree, marked differences occur in the osmotic potentials of mature leaves with different exposures. For example, exposed leaves may experience midday water potentials 0.3–0.7 MPa lower than those of heavily shaded leaves (Olsson and Milthorpe 1983; Jones and Cumming 1984). Exposed leaves had correspondingly lower osmotic potentials than shaded leaves. The result is that leaf turgor tended to be equal for the two types of leaf (A.N. Lakso, unpublished data).

The contribution of different solutes to active osmotic adjustment has been little studied in fruit trees. Fanjul and Rosher (1984) reported a 50% increase in total sugars (largely glucose) during osmotic adjustment in pot-grown apple trees. Effective osmotic adjustment requires that the solutes involved be "compatible," that is, they can increase within the cytoplasm without toxic effects (Wyn Jones and Gorham 1983). Compatible solutes in different higher plant species

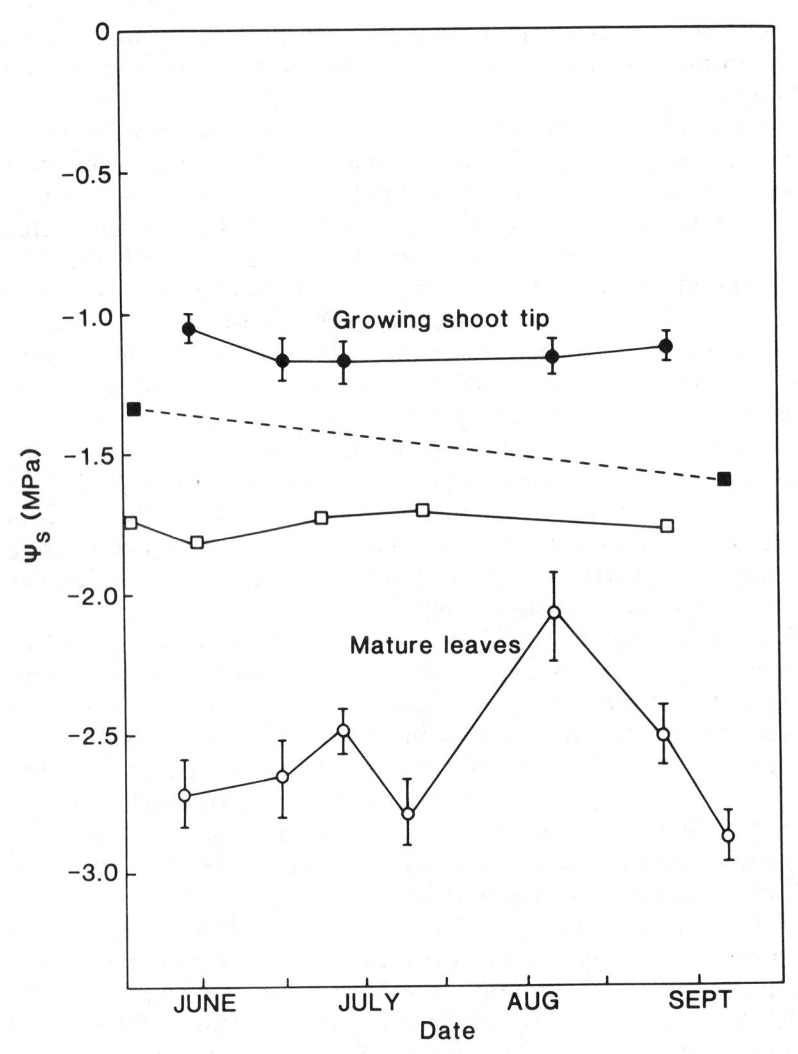

FIG. 7.9. Seasonal trends of osmotic potentials of mature apple leaves (○) and growing shoot tips (●) (*Lakso et al. 1984*), and of mature (□) and immature (■) citrus leaves (*Syvertsen et al. 1981; Syvertsen and Albrigo 1980*).

include various betaines, proline, sucrose, and sorbitol. Of particular interest here is the fact that all rosaceous fruit trees—including apple, apricot, cherry, peach, and pear—produce large quantities of sorbitol. In photosynthetic tissues such as leaves, sorbitol can approach 60–80% of the total soluble carbohydrate (Bieleski 1982). Sorbitol may be involved in osmotic adjustment, though direct evidence is lacking.

For citrus there is some evidence that both water stress and low temperatures can increase leaf proline (Levy 1980b; Syvertsen and Smith 1983). An interesting finding for sugarcane, which may be relevant in fruits, is that ABA can greatly increase invertase activity (Gayler and Glaziou 1969). Since ABA is synthesized in response to a loss of turgor, ABA-enhanced invertase activity may provide a mechanism of osmotic adjustment by conversion of sucrose to glucose and fructose.

The most obvious organ showing osmotic adjustment is the fruit itself. Most ripe fruits contain 10–20% soluble sugars. Since osmotic potential is a colligative property (dependent on the number, not the size, of solute molecules), it requires less carbon investment to adjust osmotically with small molecules such as sorbitol than with sucrose. The relative balance of sugars in ripe fruits varies greatly, with sucrose accounting for 1, 4, 14, 26, 33, 51, and 67% of the total sugars in sweet cherry, tart cherry, pear, apple, plum, orange, and peach, respectively (Attaway and Carter 1971; Wrolstad and Shallenberger 1981). The proportion of sorbitol ranges from about 5% of the soluble carbohydrate in loquat, apple, apricot, cherry, and peach to about 50% in plum and pear (Bieleski 1982). The osmotic potentials of ripe fruit sap have been reported to be between −1.3 and −4.0 MPa (Thomas *et al.* 1973; Cook and Papendick 1978).

The osmotic potential of the xylem sap is often neglected, as it is usually above −0.15 MPa (e.g., Jones 1983b; Andersen and Proebsting 1984). However, solute concentrations do vary significantly during the season in apple, and probably also in other temperate fruits. In general, the concentrations of sugars and inorganic ions in the xylem sap seem to vary in parallel, with the total soluble sugars in apple decreasing from about 20 $\mu g/cm^3$ in early May to less than 5 $\mu g/cm^3$ in June and during the rest of the season (J. Baker and H.G. Jones, unpublished data). The concentration of calcium has been found to decline from 150 $\mu g/cm^3$ to 50 $\mu g/cm^3$ over the same period (Jones and Samuelson 1983), while other inorganic ions are known to vary in parallel.

If osmotic adjustment is to be a successful mechanism for drought resistance, it must be integrated with other physiological processes. For example, Lakso (1983) has pointed out that osmotic adjustment in mature apple leaves is combined with nonadjustment of the growing shoot tip. This provides a mechanism that retards leaf area development during stress while maintaining the photosynthetic activity of the mature leaves. An additional key to this mechanism is the slow photosynthetic aging of apple leaves (Porpiglia and Barden 1980), a phenomenon that maintains photosynthesis even when new leaves are not produced. Citrus leaves remain alive for up to 3 years. This type of broad perspective on the water relations of fruit trees in the context of growth and fruiting habits and carbon physiology will be needed more as we attempt to integrate our knowledge of specific water status control mechanisms into our understanding of overall tree physiology.

B. Role of Wall Elasticity and Bound Water

As pointed out already, the elasticity of the cell walls surrounding the protoplasts is an important factor in turgor control. A simple analogy is the comparison between an inflated balloon and an inflated polyethylene bag with the same initial pressure. As air is released from the rigid polyethylene bag, the pressure (corresponding to turgor) decreases rapidly and reaches zero with a very small volume change. The elastic balloon, however, maintains positive pressures over a wide range of volumes. Similarly in plant cells, the elasticity of the cell walls affects the relationship between water potential, turgor, and RWC.

Unfortunately, little is known of the elasticities of fruit tree organs, though they can readily be calculated from pressure–volume curves. Figure 7.7 shows the dependence of bulk leaf ϵ_B on turgor pressure and RWC, as calculated from data for apple leaves (Jones and Higgs 1979b). There is, however, little evidence that changes in ϵ_B provide a major control of water status in fruit trees (Fanjul and Rosher 1984). In any case, it has been pointed out that, theoretically, very large changes in ϵ_B are required to affect turgor maintenance significantly (Turner 1979).

Perhaps the only organ where a low ϵ_B (highly elastic cells) may be important in the control of turgor in fruit trees is the fruit. It could be argued that the slow water exchange between the fruit and the rest of the tree (e.g., Goode *et al.* 1978b), coupled with the high capacitance of the fruit (see Landsberg and Jones 1981), results in maintenance of turgor in the fruit at midday. This can occur even though Ψ_p and Ψ may fall dramatically elsewhere in the tree. The fact that leaf and fruit water potentials follow each other much more closely in small fruits such as cherries (Tvergyak and Richardson 1979) than in large fruits suggests a less important role for this mechanism in cherries.

Fereres *et al.* (1979) found that much of the alteration in osmotic properties of orange tree leaves after severe stress could be related to a change in bound water. The proportion of bound water increased from about 3 to 31% with severe stress, though there was also a decrease in elasticity. In this study, bound water was more important than differences in elasticity between young and old leaves or between stressed and control leaves because changes in the relation between Ψ and RWC occurred below the point of zero turgor.

VI. INTERACTION OF CROP MANAGEMENT AND PHYSIOLOGICAL CONTROL OF WATER STATUS

In this section we consider some of the more important interactions between physiological control of water status and crop management practices that have not already been discussed.

A. Irrigation

The benefits of irrigation where fruit are grown in semiarid climates are undeniable, but many thousands of hectares of tree fruits are grown in semihumid climates without irrigation (Hilgeman and Reuther 1967; Uriu and Magness 1967; Shalhevet *et al.* 1979; Kriedemann and Barrs 1981; Elfving 1982). It is economically profitable, however, to irrigate virtually all subtropical fruit crops because of the low water-holding capacity of subtropical soils and uneven distribution of rainfall in subtropical regions. In Florida, for example, more than 66% of citrus is irrigated and virtually all new orchards have some form of irrigation (Harrison and Koo 1976). Although irrigation will not be discussed in detail here, certain aspects are relevant to the subject of this review.

1. Effects on Root Growth. Plant root systems adapt to their existing water regime. In particular, root production and extension tend to be less under conditions of ample water supply than when water supply is limited. In general, with irrigation, roots proliferate mainly within the wetted soil (Levin *et al.* 1979; Delver 1980; Jones *et al.* 1983). This effect is particularly marked with trickle irrigation (Levin *et al.* 1979; Goode *et al.* 1978b; Delver 1980; Elfving 1982), such that the *total* amount of root may even be much reduced by irrigation. In areas of high rainfall, however, roots are not confined to that portion of the soil wetted by irrigation since they tend to proliferate when water is freely available. As has been pointed out by Landsberg (1980), this can have a large effect on the root–soil hydraulic resistance (particularly important if irrigation ceases for any reason) and hence on the levels of water stress in the leaves. Where water is continuously supplied, it is possible for Ψ_l to be maintained at a reasonable level with a much smaller investment of carbohydrate in the roots than is possible in drier situations. However, where large amounts of water are applied by trickle irrigation, it is possible for root growth to be promoted over a large volume of soil (Giulivo and Uriu 1980).

2. Mist Irrigation. Application of irrigation water as a mist can lead to leaf water potentials as much as 0.8–1.0 MPa (Goode *et al.* 1978b; Brough 1983) higher than those in soil-irrigated plants. Although part of the effect of mist irrigation on Ψ_l can result from increased ambient humidity in the orchard (and hence decreased evaporation), the main effects result from the wetting of the leaves. This in turn acts in two ways: (1) by lowering leaf temperature (see, e.g., Stang *et al.* 1978) and hence decreasing the leaf-to-air vapor pressure difference, and (2) by allowing much of the water evaporated to come from the leaf surface rather than from within the leaf, thus decreasing the flow within the plant and hence the leaf water deficits (equation 4).

Goode *et al.* (1979) reported that the effect of mist persisted even after the leaves had dried, with Ψ_l of misted leaves still being 0.5 MPa higher than that of control leaves between periods of misting. The maintenance of a high Ψ_l after leaves had dried out presumably resulted from the capacitance of the tissues.

B. Other Management Practices

1. Soil Management. Any soil management practice that conserves or increases available water in the soil is clearly of value in maintaining plant Ψ over the long term. Treatments such as deep fertilization and deep ploughing that can improve root penetration and proliferation in the soil (see, e.g., Atkinson 1980) increase the total available soil water and also decrease the soil–root component of the hydraulic resistance. These effects can be beneficial to plant water status.

Any ground cover in an orchard (e.g., grass alleyways and weeds) can contribute as much as 20% of the total water use (Atkinson and White 1976; Delver 1980). In addition to this direct effect on water use, grass competition can limit tree root growth (Atkinson and White 1976), particularly below the grass alleys, with obvious implications for tree hydraulic resistances.

Water conservation practices, such as the use of soil mulches (both straw and polyethylene), can also be valuable for improving crop water balance (Måge 1982). In addition, the whole range of soil factors such as temperature (Gur *et al.* 1972), mineral nutrition (Natr 1972) and toxicity (Horton and Edwards 1976), salinity, and waterlogging

(Rowe and Beardsell 1973; Maotani *et al.* 1976; Bradford and Yang 1981; Syvertsen *et al.* 1983) affect plant water status indirectly via their effects on growth, hydraulic conductances, or even stomatal conductance. A discussion of these effects is, however, outside the scope of this review.

2. Crop Load. Crop load has a major effect on the water relations of fruit trees. A heavily cropping tree tends to put less carbohydrate into vegetative growth, including root growth (Maggs 1963; Hansen 1971b; Lenz and Doring 1975; Atkinson 1980; Lenz 1985), thus reducing water uptake capacity. In addition, there is a general tendency for stomatal conductance to be higher on spurs with than without fruit and on heavily cropping trees (Maggs 1963; Avery 1970, 1975; Hansen 1971a; Chalmers *et al.* 1975; Lenz and Doring 1975; Monselise and Lenz 1980a; Jones and Cumming 1984; Lenz 1985).

The seasonal water use of cropping apple and peach trees has been found to be 15–100% more than that of noncropping trees; the differences are even greater if expressed on a leaf area basis (Hansen 1971b; Chalmers *et al.* 1983; Lenz 1985). Lenz and Doring (1975), however, using rooted citrus cuttings, found no increase in water use with moderate crop load and a marked decrease at high crop load. This occurred because very large reductions in leaf area counterbalanced the increased water loss rates per leaf. Whether as great an effect on leaf area would occur in larger field citrus trees needs to be determined.

Despite the variability in responses observed in different experiments, the general principle that cropping induces greater water use per unit leaf area can be accepted for temperate fruits at least. Thinning fruits to improve marketable size is an established practice that may help reduce water use, although this needs to be examined.

3. Pruning and Training. The crop load effects described in the previous section can be achieved by practices such as thinning, pruning, spreading, tying-down, and use of precocious rootstocks. In addition, these practices directly affect both total leaf area and radiation interception, both of which are important determinants of crop transpiration rate (Section IV.A).

Summer pruning can be used to control orchard water use in apple (Taylor and Ferree 1981), and severe pruning (dehorning) of peach and pear has been found to reduce transpiration to such an extent that the pruned plants could survive a period without water that killed unpruned trees (Proebsting and Middleton 1980). In citrus, where shoot growth occurs in distinct flushes, shoot and root growth occur alternately. The effect of hedging and topping citrus at different times of year on subsequent vegetative and reproductive growth is yet to be determined. Techniques of pruning, training, and orchard design as they affect fruit tree water status deserve more attention.

4. Antitranspirants. There has been considerable interest over the past 20 years in the use of antitranspirants (see reviews by Gale and Hagan 1966; Solarova *et al.* 1981). A wide range of materials have been tried, which fall into three main categories: (1) compounds, such as abscisic acid, phenyl mercuric acetate, and decenylsuccinic acid, that close stomata; (2) film-forming compounds, such as silicone emulsions or plastic films; and (3) reflecting materials, such as kaolinite, that reduce transpiration by decreasing absorption of incident radiation.

The main objective of much of the research has been to improve the efficiency of crop water use (i.e., the ratio of photosynthesis or of yield to water use), on the premise that decreased stomatal conductance would normally be expected to increase water use efficiency (Jones 1983b). Indeed, there are reports that water stress, which decreases total transpiration, can improve water use efficiency in citrus (Bielorai 1982; Moreshet *et al.* 1983). In practice, the results of applying antitranspirants have been variable (see, for example, Weller and Ferree 1978), partly because the effects of these materials are often not very longlasting. Other problems have included the phytotoxicity of some compounds (e.g., phenyl mercuric acetate) or their expense (abscisic acid).

One widely used application of film antitranspirants has been for conservation of water where photosynthetic productivity is not a concern (extreme stress, newly dug nursery stock, new grafts, or transplants) (Solarova *et al.* 1981). In other cases, compounds such as abscisic acid have been found to have beneficial effects on leaf water potential (Goode *et al.* 1978a) resulting from reduced transpiration. Since abscisic acid is a naturally occurring growth regulator, the possibility exists that it may be possible to breed genotypes producing large amounts of abscisic acid in response to stress (see Addicott 1983).

5. Growth Regulators. Growth retardants, such as Alar and, of course, abscisic acid, may reduce water use of fruit trees by decreasing leaf area (Ludders and Fischer-Bölükbasi 1979). Other growth retardants such as paclobutrazol (PP333 from ICI) also decrease water use and may even find a role in the improvement of water use efficiency. Atkinson *et al.* (1983) provided evidence for a direct effect of paclobutrazol on stomatal conductance in addition to a large reduction in shoot growth.

An improved understanding of the mode of action of new types of growth regulators and of the ways in which they can be used to manipulate the water status of fruit crops is a priority for future research.

VII. CONCLUSIONS

The main principles involved in the physiological control of plant water status are reasonably well understood. Much work is still needed to understand the interaction of these principles in those fruit trees (especially pears, stone fruits, and subtropicals) that have been studied less than apples and citrus.

An additional consideration is that tree fruit crops are much more genetically wild-type than most agricultural crops due to relatively little systematic breeding of these heterozygous, long-generation plants. As a result, understanding how these species evolved and responded to the environment in their natural habitats should give useful insights into how they respond in the agricultural habitat. For example, as discussed earlier, fruiting inhibits growth of both leaf area and roots. For a young deciduous tree competing for light and soil moisture with shrubs in a Mediterranean climate, or for an understory citrus seedling in subtropical or tropical forests, a long juvenility period to avoid fruiting-induced stresses could be very advantageous.

Since fruit trees are manipulated far more than most crops, there is a great potential to utilize knowledge about the physiological principles of control of water status to practical advantage. However, further work to evaluate the complex interactions of genotype, environment, and management on tree water status is necessary if fruit tree management and productivity are to be optimized in the future.

LITERATURE CITED

ADDICOTT, F.T. (ed.). 1983. Abscisic acid. Praeger, New York.

ANDERSEN, P.C. and W.M. PROEBSTING. 1984. Water and ion fluxes of abscisic acid-treated root systems of pear, *Pyrus communis*. Physiol. Plant. 60:143–148.

ASPINALL, D. and L.G. PALEG. 1981. Proline accumulation; physiological aspects. p. 206–242. In: L.G. Paleg and D. Aspinall (eds.), The physiology and biochemistry of drought resistance in plants. Academic Press, New York.

ATKINSON, D. 1978. Use of soil resources in high density planting systems. Acta Hort. 65:79–89.

ATKINSON, D. 1980. The distribution and effectiveness of the roots of tree crops. Hort. Rev. 2:424–490.

ATKINSON, D. and G.C. WHITE. 1976. Soil management with herbicides: the response of soils and plants. p. 873-884. In: Proc. 1976 British Crop Protection Conf.—Weeds, Vol. 3.

ATKINSON, D., C.M. CRISP, T.E. ASAMOAH, and J.S. CHAUHAN. 1983. Effects of plant growth regulators on root growth. p. 30-31. In: Annu. Rpt. East Malling Res. Sta. for 1982.

ATTAWAY, J.A. and R.D. CARTER. 1971. Some new analytical indicators of processed orange juice quality. Proc. Fla. State Hort. Soc. 84:200-205.

AVERY, D.J. 1970. The effects of fruiting on the growth of apple trees on four rootstock varieties. New Phytol. 69:19-30.

AVERY, D.J. 1975. Reduction in growth increments by crop competition. p. 103-106. In: H.C. Pereira (ed.), Climate and the orchard. Res. Rev. 5. Commonwealth Agricultural Bureaux, Farnham Royal, Slough, United Kingdom.

AYRES, P.G. 1978. Water relations of diseased plants. p. 1-60. In: T.T. Kozlowski (ed.), Water deficits and plant growth, Vol. V. Academic Press, New York.

BAKER, E.A. and J. PROCOPIOU. 1980. Effect of soil moisture status on leaf surface wax yield of some drought-resistant species. J. Hort. Sci. 55:85-87.

BAKER, E.A., J. PROCOPIOU, and G.M. HUNT. 1975. The cuticles of *Citrus* species. Composition of leaf and fruit waxes. J. Sci. Food Agr. 26:1093-1101.

BALDOCCHI, D.D., S.B. VERMA, N.J. ROSENBERG, B.L. BLAD, A. GARAY, and J.E. SPECHT. 1983. Leaf pubescence effects on the mass and energy exchange between soybean canopies and the atmosphere. Agron. J. 75:537-543.

BARRS, H.D. 1968. Determination of water deficits in plant tissues. p. 236-368. In: T.T. Kozlowski (ed.), Water deficits and plant growth, Vol. I. Academic Press, New York.

BAXTER, P. and D. WEST. 1977. The flow of water into fruit trees. I. Resistances to water flow through roots and stems. Ann. Appl. Biol. 87:95-101.

BIELESKI, R.L. 1982. Sugar alcohols. p. 158-192. In: F.A. Loewus and W. Tanner (eds.), Plant carbohydrates. I: Intracellular carbohydrates. Springer-Verlag, Berlin, Heidelberg, New York.

BIELORAI, H. 1982. The effect of partial wetting of the root zone on yield and water use efficiency in a drip- and sprinkler-irrigated mature grapefruit grove. Irrig. Sci. 3:89-100.

BISSON, M.A. and J. GUTKNECHT. 1980. Osmotic regulation in algae. p. 131-146. In: R.M. Spanswick, W.K. Lucas, and J. Dainty (eds.), Plant membrane transport; current conceptual issues. Elsevier, Amsterdam.

BRADFORD, K.J. and S.F. YANG. 1981. Physiological responses of plants to waterlogging. HortScience 16:25-30.

BROUGH, D.W. 1983. The role of cavitation in the water relations of irrigated and non-irrigated apple trees. Ph.D. Thesis, University of Edinburgh.

CAMACHO-B, S.E., A.E. HALL, and M.R. KAUFMANN. 1974. Efficiency and regulation of water transport in some woody and herbaceous species. Plant Physiol. 54:169-172.

CHALMERS, D.J., R.L. CANTERFORD, P.H. JERIE, T.R. JONES, and T.D. UGALDE. 1975. Photosynthesis in relation to growth and distribution of fruit in peach trees. Austral. J. Plant Physiol. 2:635-645.

CHALMERS, D.J., P.D. MITCHELL, and L. VAN HEEK. 1981. Control of peach tree growth and productivity by regulated water supply, tree density, and summer pruning. J. Amer. Soc. Hort. Sci. 106:307-312.

CHALMERS, D.J., K.A. OLSSON, and T.R. JONES. 1983. Water relations of peach trees and orchards. p. 197-232. In: T.T. Kozlowski (ed.), Water deficits and plant growth, Vol. VII. Academic Press, New York.

COHEN, S. and Y. COHEN. 1983. Field studies of leaf conductance response to environmental variables in citrus. J. Appl. Ecol. 20:561-570.

COHEN, Y., M. FUCHS, and S. COHEN. 1983a. Resistance to water uptake in a mature citrus tree. J. Expt. Bot. 34:451-460.

COHEN, M., R. PELOSI, and R. BRLANSKY. 1983b. Nature and location of xylem blockage structures in trees with citrus blight. Phytopathology 73:1125-1130.

COOK, R.J. and R.I. PAPENDICK. 1978. Role of water potential in microbial growth and development of plant disease, with special reference to postharvest pathology. HortScience 13:559–564.

DAVIES, F.S. and A.N. LAKSO. 1979. Diurnal and seasonal changes in leaf water potential components and elastic properties in response to water stress in apple trees. Physiol. Plant. 46:109-114.

DAVIES, W.J. and T.T. KOZLOWSKI. 1974. Stomatal responses of five woody angiosperms to light intensity and humidity. Can. J. Bot. 52:1525-1534.

DE JONG, T.M. 1983. CO_2 assimilation characteristics of five *Prunus* tree fruit species. J. Amer. Soc. Hort. Sci. 108:303-307.

DELVER, P. 1980. Drip irrigation and root development in a humid climate and problems of irregular dripping. p. 89–101. In: Seminaires sur l'irrigation localisée. III. Influence de l'irrigation localisée sur la morphologie et la physiologie des racines. L'institute pour l'irrigation du CNR et l'institute d'hydraulic agricole de l'université de Naples, Sorrento.

ELFVING, D.C. 1982. Crop response to trickle irrigation. Hort. Rev. 4:1–48.

ELFVING, D.C., M.R. KAUFMANN, and A.E. HALL. 1972. Interpreting leaf water potential measurements with a model of the soil–plant–atmosphere continuum. Physiol. Plant. 27:161–168.

EL-SHARKAWI, H.M. and M. EL-MONAYERI. 1976. Response of olive and almond orchards to partial irrigation under dry-farming practices in semi-arid regions. III. Plant–soil water relations in almond during the growing season. Plant & Soil 44:113–128.

FANJUL, L. and H.G. JONES. 1982. Rapid stomatal responses to humidity. Planta 154:135–138.

FANJUL, L. and P.H. ROSHER. 1984. Effects of water stress on internal water relations of apple leaves. Physiol. Plant. 62:321–328.

FARQUHAR, G.D. 1978. Feedforward responses of stomata to humidity. Austral. J. Plant Physiol. 5:787–800.

FARRÉ, J.M. 1979. Water use and productivity of fruit trees: effects of soil management in irrigation. Ph.D. Thesis, Wye College, University of London.

FERERES, E., G. CRUZ-ROMERO, G.J. HOFFMANN, and S.L. RAWLINS. 1979. Recovery of orange trees following severe water stress. J. Appl. Ecol. 16:833–842.

FORSHEY, C.G., R.W. WEIRES, B.H. STANLEY, and R.C. SEEM. 1983. Dry weight partitioning of 'McIntosh' apple trees. J. Amer. Soc. Hort. Sci. 108:149–154.

GALE, J. and R.M. HAGAN. 1966. Plant antitranspirants. Annu. Rev. Plant Physiol. 17:269–282.

GATES, D.M. 1980. Biophysical ecology. Springer-Verlag, New York.

GAYLER, K.R. and K.T. GLAZIOU. 1969. Plant enzyme synthesis: hormonal regulation of invertase and peroxidase synthesis in sugar cane. Planta 84:185–194.

GIULIVO, C. and A. BERGAMINI. 1982. Effect of rootstock–scion combination on water balance of apple tree, cv. Golden Delicious. Abstr. 1264. In: Proc. 21st Intern. Hort. Congr., Hamburg.

GIULIVO, C. and K. URIU. 1980. Effect of drip irrigation on root distribution of mature french prune trees. p. 65–71. In: Seminaires sur l'irrigation localisée. III.

Influence de l'irrigation localisée sur la morphologie et la physiologie des racines. L'institute pour l'irrigation du CNR et l'institute d'hydraulic agricole de l'université de Naples, Sorrento.

GOODE, J.E. and K.H. HIGGS. 1973. Water, osmotic and pressure potential relationships in apple leaves. J. Hort. Sci. 48:203–215.

GOODE, J.E., K.H. HIGGS, and K.J. HYRYCZ. 1978a. Abscisic acid applied to orchard trees of Golden Delicious apple to control water stress. J. Hort. Sci. 53:99–103.

GOODE, J.E., K.H. HIGGS, and K.J. HYRYCZ. 1978b. Nitrogen and water effects on the nutrition, growth, crop yield and fruit quality of orchard-grown Cox's Orange Pippin apple trees. J. Hort. Sci. 53:295–306.

GOODE, J.E., K.H. HIGGS, and K.J. HYRYCZ. 1979. Effects of water stress control in apple trees by misting. J. Hort. Sci. 54:1–11.

GRAHAM, J.H. and J.P. SYVERTSEN. 1984. Influence of VAM on hydraulic conductivity of two citrus rootstocks. New Phytol. 97:277–284.

GRIERSON, W., J. SOULE, and K. KAWADA. 1982. Beneficial aspects of physiological stress. Hort. Rev. 4:247–271.

GUR, A., B. BRAVDO, and Y. MIZRAHI. 1972. Physiological responses of apple trees to supraoptimal root temperature. Physiol. Plant. 27:130–138.

HALL, A.E., S.E. CAMACHO-B, and M.R. KAUFMANN. 1974. Regulation of water loss by citrus leaves. Physiol. Plant. 33:62–65.

HALL, A.E., E.-D. SCHULZE, and O.L. LANGE. 1976. Current perspectives of steady state stomatal responses to environment. p. 169–188. In: O.L. Lange, L. Kappen, and E.-D. Schulze (eds.), Ecological studies, Vol. 19. Springer-Verlag, Berlin, Heidelberg, New York.

HANSEN, P. 1971a. The effect of fruiting upon transpiration rate and stomatal opening in apple leaves. Physiol. Plant. 25:181–183.

HANSEN, P. 1971b. The effect of cropping on the distribution of growth in apple trees. Tidsskrift Planteavl 75:119–127.

HARRISON, D.S. and R.C.J. KOO. 1976. Irrigation methods and equipment for production of citrus in Florida. WRC-10. Inst. of Food and Agr. Sciences, Univ. of Florida.

HEATHCOTE, D.G., J.R. ETHERINGTON, and F.I. WOODWARD. 1979. An instrument for nondestructive measurement of the pressure potential (turgor) of leaf cells. J. Expt. Bot. 30:811–816.

HILGEMAN, R.H. and W. REUTHER. 1967. Evergreen tree fruits. p. 704–718. In: R.M. Hagan, H.R. Haise, and T.W. Edminster (eds.), Irrigation of agricultural lands. No. 11 Agron. Ser. Amer. Soc. Agron., Madison, Wisconsin.

HOARE, E.R. and H.D. BARRS. 1974. Water relations and photosynthesis amongst horticultural species as affected by simulated soil water stress. p. 321–334. In: Proc. 19th Intern. Hort. Cong., Vol. 3, Warsaw.

HORTON, B.D. and J.H. EDWARDS. 1976. Diffusive resistance rates and stomatal aperture of peach seedlings as affected by aluminium concentration. HortScience 11:591–593.

HSIAO, T.C. 1973. Plant responses to water stress. Annu. Rev. Plant Physiol. 24:519–570.

ITAI, C., A. RICHMOND, and Y. VAADIA. 1968. The role of root cytokinins during water and salinity stress. Israel J. Bot. 17:187–195.

JACKSON, J.E. 1980. Light interception and utilization by orchard systems. Hort. Rev. 2:208–267.

JACKSON, J.E., J.W. PALMER, G.C. WHITE, and E.E. CANNADINE. 1983. Modelling light interception and utilisation. p. 34–35. In: Annu. Rpt. East Malling Res. Sta. for 1982.

JAHN, O.L. 1979. Penetration of photosynthetically active radiation as a measurement of canopy density of citrus trees. J. Amer. Soc. Hort. Sci. 104:557–560.

JARVIS, P.G. 1980. Stomatal response to water stress in conifers. p. 105–112. In: N.C. Turner and P.J. Kramer (eds.), Adaptation of plants to water and high temperature stress. John Wiley & Sons, New York.

JARVIS, P.G. 1981. Stomatal conductance, gaseous exchange and transpiration. p. 175–204. In: J. Grace, E.D. Ford, and P.G. Jarvis (eds.), Plants and their atmospheric environment. Blackwells, Oxford.

JARVIS, P.G., W.R.N. EDWARDS, and H. TALBOT. 1981. Models of plant and crop water use. p. 151–194. In: D.A. Rose and D.A. Charles-Edwards (eds.), Mathematics and plant physiology. Academic Press, London.

JONES, H.G. 1973. Photosynthesis by thin leaf slices in solution. II. Osmotic stress and its effects on photosynthesis. Austral. J. Biol. Sci. 26:25–33.

JONES, H.G. 1976. Crop characteristics and the ratio between assimilation and transpiration. J. Appl. Ecol. 13:605–622.

JONES, H.G. 1977. Transpiration in barley lines with differing stomatal frequencies. J. Expt. Bot. 28:162–168.

JONES, H.G. 1978. Modelling diurnal trends of leaf water potential in transpiring wheat. J. Appl. Ecol. 15:613–626.

JONES, H.G. 1982. Effects of water stress on flowering. p. 151. In: Annu. Rpt. East Malling Res. Sta. for 1981.

JONES, H.G. 1983a. Estimation of an effective soil water potential at the root surface of transpiring plants. Plant, Cell & Environ. 6:671–674.

JONES, H.G. 1983b. Plants and microclimate. Cambridge University Press, Cambridge.

JONES, H.G. and I.G. CUMMING. 1984. Variation of leaf conductance and leaf water potential in apple orchards. J. Hort. Sci. 59:329–336.

JONES, H.G. and K.H. HIGGS. 1979a. Control of water loss by apple leaves. p. 165–166. In. Annu. Rpt. East Malling Res. Sta. for 1978.

JONES, H.G. and K.H. HIGGS. 1979b. Water potential—water content relationships in apple leaves. J. Expt. Bot. 30:965–970.

JONES, H.G. and K.H. HIGGS. 1982. Surface conductance and water balance of developing apple (*Malus pumila* Mill) fruits. J. Expt. Bot. 33:67–77.

JONES, H.G. and T.J. SAMUELSON. 1983. Calcium uptake by developing apple fruits. II. The role of spur leaves. J. Hort. Sci. 58:183–190.

JONES, H.G., M.T. LUTON, K.H. HIGGS, and P.J.C. HAMER. 1983. Experimental control of water status in an apple orchard. J. Hort. Sci. 58:301–316.

KAISER, W.M. 1982. Correlation between changes in photosynthetic activity and changes in total protoplast volumes in leaf tissue from hygro-, meso-, and xerophytes under osmotic stress. Planta 154:538–545.

KALMA, J.D. and G. STANHILL. 1972. The climate of an orange orchard: physical characteristics and microclimate relationships. Agr. Meteorol. 10:185–201.

KAUFMANN, M.R. and D.C. ELFVING. 1972. Evaluation of tensiometers for estimating plant water stress in citrus. HortScience 7:513–514.

KLEPPER, B. 1968. Diurnal pattern of water potential in woody plants. Plant Physiol. 43:1931–1934.

KOHL, R.A. and J.L. WRIGHT. 1974. Air temperature and vapour pressure changes caused by sprinkler irrigation. Agron. J. 66:85–88.

KOZLOWSKI, T.T. (ed.). 1968. Water deficits and plant growth. Vol. I. Academic Press, New York.

KOZLOWSKI, T.T. (ed.). 1972. Water deficits and plant growth. Vol. III. Academic Press, New York.

KOZLOWSKI, T.T. (ed.). 1974. Water deficits and plant growth. Vol. IV. Academic Press, New York.

KRIEDEMANN, P.E. and H.D. BARRS. 1981. Citrus orchards. p. 325-417. In: T.T. Kozlowski (ed.), Water deficits and plant growth, Vol. VI. Academic Press, New York.

KRIEDEMANN, P.E. and R.L. CANTERFORD. 1971. The photosynthetic activity of pear leaves (*Pyrus communis* L.). Austral. J. Biol. Sci. 24:197-205.

LAKSO, A.N. 1979. Seasonal changes in stomatal response to leaf water potential in apple. J. Amer. Soc. Hort. Sci. 104:58-60.

LAKSO, A.N. 1983. Morphological and physiological adaptations for maintaining photosynthesis under water stress in apple trees. p. 85-93. In: R. Marcelle, H. Clijsters, and M. Van Poucke (eds.), Stress effects on photosynthesis. Nijhoff/Junk, The Hague.

LAKSO, A.N. 1985. Photosynthesis in fruit trees in relation to environmental factors. In: A.N. Lakso and F. Lenz (eds.), Regulation of photosynthesis in fruit trees. N.Y. State Agr. Expt. Sta. Spec. Bull. (in press).

LAKSO, A.N., A.S. GEYER, and S.G. CARPENTER. 1984. Seasonal osmotic relations in apple leaves of different ages. J. Amer. Soc. Hort. Sci. 109:544-547.

LANDSBERG, J.J. 1980. Optimal root systems for different conditions: a theoretical assessment in relation to trickle irrigation. p. 19-29. In: Seminaires sur l'irrigation localisée. III. Influence de l'irrigation localisée sur la morphologie et la physiologie des racines. L'institute pour l'irrigation du CNR et l'institute d'hydraulic agricole de l'université de Naples, Sorrento.

LANDSBERG, J.J. and H.G. JONES. 1981. Apple orchards. p. 419-469. In: T.T. Kozlowski (ed.), Water deficits and plant growth, Vol. VI. Academic Press, New York.

LANDSBERG, J.J., C.L. BEADLE, P.V. BISCOE, D.R. BUTLER, B. DAVIDSON, L.D. INCOLL, G.B. JAMES, P.G. JARVIS, P.J. MARTIN, R.E. NEILSON, D.B.B. POWELL, M. SLACK, M.R. THORPE, N.C. TURNER, B. WARRIT, and W.R. WATTS. 1975. Diurnal energy, water and CO_2 exchanges in an apple (*Malus pumila*) orchard. J. Appl. Ecol. 12:659-684.

LENZ, F. 1985. Fruit effects on transpiration and dry matter production in apples. In: A. Lakso and F. Lenz (eds.), Regulation of photosynthesis in fruit trees. N.Y. State Agr. Expt. Sta. Spec. Bull. (in press).

LENZ, F. and H.W. DORING. 1975. Fruit effects on growth and water consumption in citrus. Gartenbauwissenschaft 6:257-260.

LEVIN, I., R. ASSAF, and B. BRAVDO. 1979. Soil moisture and root distribution in an apple orchard irrigated by tricklers. Plant & Soil 52:31-40.

LEVY, Y. 1980a. Effect of evaporative demand on water relations of *Citrus limon*. Ann. Bot. 46:695-700.

LEVY, Y. 1980b. Field determinations of free proline accumulation and water stress in lemon trees. HortScience 15:302-303.

LEVY, Y. 1983. Acclimation of citrus to water stress. Sci. Hort. 20:267-273.

LEVY, Y. and J.P. SYVERTSEN. 1981. Water relation of citrus in climates with different evaporative demands. Proc. Intern. Soc. Citriculture 2:501-503.

LEVY, Y., J.P. SYVERTSEN, and S. NEMEC. 1983. Effect of drought stress and vesicular-arbuscular mycorrhiza on citrus transpiration and hydraulic conductivity of roots. New Phytol. 93:61-66.

LOMAS, J. and M. MANDEL. 1973. The quantitative effects of two methods of sprinkler irrigation on the microclimate of a mature avocado plantation. Agr. Meteorol. 12:34–48.

LUDDERS, P. and T. FISCHER-BÖLÜKBASI. 1979. Einfluss von Alar und Tiba auf den Wasserverbrauch unterschiedlich fruchtender Apfelbäume. Gartenbauwissenschaft 44:171–177.

LUDLOW, M.M. 1980. Adaptive significance of stomatal responses to water stress. p. 123-138. In: N.C. Turner and P.J. Kramer (eds.), Adaptation of plants to water and high temperature stress. John Wiley & Sons, New York.

MÅGE, F. 1982. Black plastic mulching compared with other orchard soil management methods. Scientia Hort. 13:131-136.

MAGGS, D.H. 1963. The reduction in growth of apple trees brought about by fruiting. J. Hort. Sci. 38:119-128.

MAOTANI, T., Y. MACHIDA, K. YAMATSU, and T. YAMAZAKI. 1976. Studies on leaf water stress in fruit trees. III. Effect of soil factors on leaf water potential of citrus trees. J. Japan. Soc. Hort. Sci. 44:367-374.

McCOY, E.L., L. BOERSMA, M.L. UNGS, and S. AKRATANAKCUL. 1984. Toward understanding soil water uptake by plant roots. Soil Sci. 137:69–77.

MILBURN, J.A. 1979. Water flow in plants. Longmans, London.

MONSELISE, S.P. and F. LENZ. 1980a. Effects of fruit load on stomatal resistance, specific leaf weight and water content of apple leaves. Gartenbauwissenschaft 45:188-191.

MONSELISE, S.P. and F. LENZ. 1980b. Effect of fruit load on photosynthetic rates of budded apple trees. Gartenbauwissenschaft 45:220-224.

MONTEITH, J.L. 1965. Evaporation and environment. Symp. Soc. Expt. Biol. 19:205-234.

MONTEITH, J.L. 1973. Principles of environmental physics. Edward Arnold, London.

MORESHET, S., Y. COHEN, and M. FUCHS. 1983. Response of mature 'Shamouti' orange trees to irrigation of different soil volumes at similar levels of available water. Irrig. Sci. 3:233–236.

NATALI, S. 1982. Relationship between soil moisture, leaf water potential and transpiration. Riv. Ortoflorofrutt. It. 66:261-276.

NATALI, S., C. XILOYANNIS, J. LITTARDI, and A. BARBIERI. 1983. Growth, yield and seasonal and diurnal water relations of peach trees grafted on different rootstocks. Riv. Ortoflorofrutt. It. 67:329-341.

NATR, L. 1972. Influence of mineral nutrients on photosynthesis of higher plants. Photosynthetica 6:80-99.

NOBEL, P.S. 1974. Introduction to biophysical plant physiology. Freeman, San Francisco.

OLIEN, W.C. and A.N. LAKSO. 1984. A comparison of the dwarfing character and water relations of five apple rootstocks. Acta Hort. 146:151–158.

OLSSON, K.A. and F.L. MILTHORPE. 1983. Diurnal and spatial variation in leaf water potential and leaf conductance of irrigated peach trees. Austral. J. Plant Physiol. 10:291-298.

PALEG, L.G. and D. ASPINALL (eds.). 1981. Biochemistry and physiology of drought resistance in plants. Academic Press, London.

PIERCE, M. and K. RASCHKE. 1980. Correlation between loss of turgor and accumulation of abscisic acid in detached leaves. Planta 148:174-182.

PORPIGLIA, P.J. and J.A. BARDEN. 1980. Seasonal trends in net photosynthetic potential, dark respiration, and specific leaf weight of apple leaves as affected by canopy position. J. Amer. Soc. Hort. Sci. 105:920-923.

POWELL, D.B.B. and M.R. THORPE. 1977. Dynamic aspects of plant-water relations. p. 259-279. In: J.J. Landsberg and C.V. Cutting (eds.), Environmental effects on crop physiology. Academic Press, London.

PRIESTLEY, C.A. 1973. The use of apple leaf discs in studies of carbon translocation. p. 121-128. In: Proc. 3rd Symp. Accumulation and Translocation of Nutrients and Regulators in Plant Organism, Vol. 3, Warsaw.

PROEBSTING, E.L. and J.E. MIDDLETON. 1980. The behaviour of peach and pear trees under extreme drought stress. J. Amer. Soc. Hort. Sci. 105:380-385.

RAMOS, C. and M.R. KAUFMANN. 1979. Hydraulic resistance of rough lemon roots. Physiol. Plant. 45:311-314.

RASCHKE, K. 1979. Movements of stomata. p. 383-441. In: Encyclopedia of Plant Physiology, N.S., Vol. 7.

ROWE, R.N. and D.V. BEARDSELL. 1973. Waterlogging of fruit trees. Hort. Abstr. 43:533-548.

SAFIR, G.P., J.J. BOYER, and J.W. GERDEMANN. 1972. Nutrient status and mycorrhizal enhancement of water transport in soybean. Plant Physiol. 49:700-703.

SCHÖNHERR, J. and H.W. SCHMIDT. 1979. Water permeability of plant cuticles: dependence of permeability coefficients of cuticular transpiration on vapour pressure saturation deficit. Planta 144:391-400.

SCHULZE, E-D. and A.E. HALL. 1981. Short-term and long-term effects of drought on steady state and time-integrated plant processes. p. 217-235. In: C.B. Johnson (ed.), Physiological processes limiting plant productivity. Butterworths, London.

SCHULZE, E-D., O.L. LANGE, U. BUSCHBOM, L. KAPPEN, and M. EVENARI. 1972. Stomatal responses to changes in humidity in plants growing in the desert. Planta 108:259-270.

SCHULZE, E-D., O.L. LANGE, M. EVENARI, L. KAPPEN, and U. BUSCHBOM. 1974. The role of air humidity and leaf temperature in controlling stomatal resistance of *Prunus armeniaca* L. under desert conditions. I. A simulation of the daily course of stomatal resistance. Oecologia 17:159-170.

SHALHEVET, J., A. MANTELL, H. BIELORAI, and D. SHIMSHI (eds.). 1979. Irrigation of field and orchard crops under semi-arid conditions. Rev. ed. Intern. Irrig. Info. Cntr. Bet Dagan, Israel.

SINCLAIR, T.R. and L.H. ALLEN. 1982. Carbon dioxide and water vapour exchange of leaves on field-grown citrus trees. J. Expt. Bot. 33:1166-1175.

SLACK, E.M. 1974. Studies of stomatal distribution on the leaves of four apple varieties. J. Hort. Sci. 49:95-103.

SLATYER, R.O. 1967. Plant water relationships. Academic Press, London.

SMART, R.E. and H.D. BARRS. 1973. The effect of environment and irrigation interval on leaf water potential of four horticultural species. Agr. Meteorol. 12:337-346.

SOLAROVA, J., J. POSPISILOVA, and B. SLAVIK. 1981. Gas exchange regulation by changing of epidermal conductance with antitranspirants. Photosynthetica 15:365-400.

STANG, E.J., D.C. FERREE, F.R. HALL, and R.A. SPOTTS. 1978. Overtree misting for bloom delay in 'Golden Delicious' apple. J. Amer. Soc. Hort. Sci. 103:82-87.

STANHILL, G. 1972. Recent developments in water relations studies: some examples from Israel citriculture. p. 367-379. In: Proc. 18th Intern. Hort. Cong., Vol. 4, Israel.

STERNE, R.E., M.R. KAUFMANN, and G.A. ZENTMYER. 1977. Environmental effects on transpiration and leaf water potential in avocado. Physiol. Plant. 41:1-6.

STERNE, R.E., M.R. KAUFMANN, and G.A. ZENTMYER. 1978. Effect of phytophthora rot on water relations of avocado: interpretation with a water transport model. Phytopathology 68:595-602.

SURANYI, D. 1977. The effect of rootstocks and artificial infection on the transpiration of apricot trees. Kertgazdasag 9:15–26 [in Hungarian].

SYVERTSEN, J.P. 1981. Hydraulic conductivity of four commercial citrus rootstocks. J. Amer. Soc. Hort. Sci. 106:378–381.

SYVERTSEN, J.P. 1982. Minimum leaf water potential and stomatal closure in citrus leaves of different ages. Ann. Bot. 49:827–834.

SYVERTSEN, J.P. 1984. Light acclimation in citrus leaves. II. CO_2 assimilation and light, water and nitrogen use efficiency. J. Amer. Soc. Hort. Sci. 109:812–817.

SYVERTSEN, J.P. and L.G. ALBRIGO. 1980. Seasonal and diurnal citrus leaf and fruit water relations. Bot. Gaz. 144:440–446.

SYVERTSEN, J.P. and M.L. SMITH, JR. 1983. Environmental stress and seasonal changes in proline concentration of citrus tree tissues and juice. J. Amer. Soc. Hort. Sci. 108:861–866.

SYVERTSEN, J.P. and M.L. SMITH, JR. 1984. Light acclimation in citrus leaves. I. Changes in physical characteristics, chlorophyll and nitrogen content. J. Amer. Soc. Hort. Sci. 109:807–812.

SYVERTSEN, J.P., M.L. SMITH, and J.C. ALLEN. 1981. Growth rate and water relations of citrus leaf flushes. Ann. Bot. 47:97–105.

SYVERTSEN, J.P., R.M. ZABLOTOWICZ, and M.L. SMITH. 1983. Soil temperature and flooding effects on two species of citrus. I. Plant growth and hydraulic conductivity. Plant & Soil 72:3–11.

TAN, C.S. and B.R. BUTTERY. 1982. Response of stomatal conductance, transpiration, photosynthesis, and leaf water potential in peach seedlings to different watering regimes. HortScience 17:222–223.

TAYLOR, B.H. and D.C. FERREE. 1981. The influence of summer pruning on photosynthesis, transpiration, leaf abscission, and dry weight accumulation of young apple trees. J. Amer. Soc. Hort. Sci. 106:389–393.

THOM, A.S. 1975. Momentum, mass and heat exchange of plant communities. p. 57–109. In: J.L. Monteith (ed.), Vegetation and the atmosphere, Vol I. Academic Press, London.

THOMAS, M., S.L. RANSON, and J.A. RICHARDSON. 1973. Plant physiology. 5th ed. Longmans, London.

TURNER, N.C. 1979. Drought resistance and adaptation to water deficits in crop plants. p. 343–372. In: H. Mussell and R.C. Staples (eds.), Stress physiology in crop plants. Wiley-Interscience, New York.

TVERGYAK, P.G. and D.G. RICHARDSON. 1979. Diurnal changes of leaf and fruit water potentials of sweet cherries during the harvest period. HortScience 14:520–521.

URIU, K. and J.R. MAGNESS. 1967. Deciduous tree fruits and nuts. p. 686–703. In: R.M. Hagan, H.R. Haise, and T.W. Edminster (eds.), Irrigation of agricultural lands. Bull. 11. Agron. Ser. Amer. Soc. Agron., Madison, Wisconsin.

VAN BAVEL, C.H.M., J.E. NEWMAN, and R.H. HILGEMAN. 1967. Climate and estimated water use by an orange orchard. Agr. Meteorol. 4:27–37.

WARRIT, B., J.J. LANDSBERG, and M.R. THORPE. 1980. Responses of apple stomata to environmental factors. Plant, Cell & Environ. 3:12–22.

WELLER, S.C. and D.C. FERREE. 1978. Effect of a pinolene-base antitranspirant on fruit growth, net photosynthesis, transpiration, and shoot growth of 'Golden Delicious' apple trees. J. Amer. Soc. Hort. Sci. 103:17–19.

WEST, D.W. and D.F. GAFF. 1976. The effect of leaf water potential, leaf temperature and light intensity on leaf diffusion resistance and the transpiration of leaves of *Malus sylvestris*. Physiol. Plant. 38:98–104.

WILCOX, D.A. and F.S. DAVIES. 1981. Temperature-dependent and diurnal root conductivities in two citrus rootstocks. HortScience 16:303–305.

WILCOX, D.A., F.S. DAVIES, and D.W. BUCHANAN. 1983. Root temperature, water relations, and cold hardiness in two citrus rootstocks. J. Amer. Soc. Hort. Sci. 108:318–321.

WILDUNG, D.K., C.J. WEISER, and H.M. PELLETT. 1973. Temperature and moisture effects on hardening of apple roots. HortScience 8:53–55.

WILSON, J.A. 1981. Stomatal responses to applied ABA and CO_2 in epidermis detached from well-watered and water-stressed plants of *Commelina communis* L. J. Expt. Bot. 32:261–269.

WROLSTAD, R.E. and R.S. SHALLENBERGER. 1981. Free sugars and sorbitol in fruits—a compilation from the literature. J. Assoc. Offic. Anal. Chem. 64:91–103.

WYN JONES, R.G. and J. GORHAM. 1983. Osmoregulation. p. 35–58. In: O.L. Lange, P.S. Nobel, C.B. Osmond, and H. Ziegler (eds.), Physiological plant ecology III. Springer-Verlag, Berlin, Heidelberg, New York.

YELENOSKY, G. 1979. Water-stress-induced cold hardening of young citrus trees. J. Amer. Soc. Hort. Sci. 104:270–273.

YOUNG, E. and J. HOUSER. 1980. Influence of Siberian C rootstock on peach bloom delay, water potential and pollen meiosis. J. Amer. Soc. Hort. Sci. 105:242–245.

YOUNG, E., J.M. HAND, and S.C. WIEST. 1982. Osmotic adjustment and stomatal conductance in peach seedlings under severe water stress. HortScience 17:791–793.

YOUNG, R.H. and S.M. GARNSEY. 1977. Water uptake patterns in blighted citrus trees. J. Amer. Soc. Hort. Sci. 102:751–756.

ZIMMERMANN, U. 1978. Physics of turgor- and osmo-regulation. Annu. Rev. Plant Physiol. 29:121–148.

ZIMMERMANN, U. and E. STEUDLE. 1974. The pressure dependence of the hydraulic conductivity, the membrane resistance and membrane potential during turgor pressure regulation in *Valonia utricularis*. J. Membr. Biol. 16:331–352.

8

CO_2 Enrichment of Protected Crops[1]

Michael A. Porter[2] *and Bernard Grodzinski*
Department of Horticultural Science, Ontario Agricultural College, University of Guelph, Guelph, Ontario, Canada N1G 2W1

[1] We thank Lorna Woodrow and John T.A. Proctor for their advice and assistance during the preparation of this review.
[2] Present address: Oak Ridge National Laboratory, Biology Division, P.O. Box Y, Oak Ridge, Tennessee 37881.

Horticultural Reviews, Volume 7

ISBN 0-87055-492-1

I. INTRODUCTION

> The art of exploiting microclimate is not new; many traditional techniques of husbandry are means of adjustment to, or manipulation of, microclimate even though the solutions adopted may be empirical rather than calculated. The wind sheltered gardens of the Scillies, the shade trees of tea estates, horticultural mulches and glass culture are obvious examples. (Caborn 1973)

Many techniques have been used to shelter valuable horticultural crops and enhance yield. The open field environment may be improved by proper site selection, orientation of the crop with respect to prevailing winds, the use of living or artificial windbreaks, and the installation of horizontal shading structures. Temporary structures such as cloches or plastic tunnels may be erected to prolong the growing season, particularly in the spring when days are lengthening and nights are cool. Greenhouses, which represent a more sophisticated level of crop shelter, not only protect the crop and extend the season but also allow the creation of an entirely new growing season with potential control over the intensity, duration, and quality of light, humidity, temperature, water, nutrient supply, and gas exchange.

It is generally recognized that the canopy microclimate influences the growth and productivity of individual plants within it. The microclimate is dependent on the interaction of the canopy with the atmosphere above, and any shelter imposed between the crop and the atmosphere will alter the microclimate inside the canopy. Shelters can alter light interception by the crop, reduce or increase heat accumulation inside the canopy, and modify evapotranspirational water loss. Possible effects on air exchange, specifically CO_2 exchange, between the leaf and the canopy microclimate, between the crop canopy and the shelter environment, and between the shelter and the external atmosphere have been studied less rigorously. An appreciation of the characteristics of the shelter, canopy, and leaf boundary layers is most essential to a discussion of CO_2 enrichment practices, since CO_2 exchange may be limited at these interfaces. However, historically the practice of CO_2 enrichment by horticulturalists is easier to document in terms of our current understanding of photosynthetic carbon fixation.

Over 3000 pages in six volumes were published in 1984 by the organizers of the VI International Congress of Photosynthesis, held the previous summer in Brussels. Over three centuries earlier, in 1648, a citizen of that city, Jean-Baptiste van Helmont, published a modest but classic paper that contributed to the future understanding of photosynthesis and plant productivity. He planted a small (5-lb) willow tree in a pot of soil and added only water to the soil for the next 5 years. At the end of this period, the tree had gained 164 lb 3 oz, while the soil and the pot had lost only 2 oz. Van Helmont established that plants do not obtain their mass from the soil and laid the path for the fundamental observation that CO_2 is a major substrate for plant growth. In 1804 DeSaussure presented evidence that "CO_2 fertilization" could enhance the growth and yield of pea plants. By the beginning of the twentieth century, many reports indicated that plant and crop yields could be improved by growth in approximately 1000 μl/liter CO_2 (Brown and Escombe 1902; Cummings and Jones 1918). Blackman (1905) included CO_2 as one of the limiting factors to plant growth, and it was realized that an accelerated rate of photosynthesis could result from CO_2 enrichment. Pollution problems associated with the combustion of wood and fossil fuel frustrated the early workers; however, today with improved sources of CO_2, the benefits of CO_2 enrichment in greenhouse crop production are widely recognized (Enoch *et al.* 1973; Hanan *et al.* 1978; Hand 1982; Kimball 1983a,b; Kramer 1981; Krizek 1979, 1984; Rogers *et al.* 1981; Wittwer 1983; Wittwer and Robb 1964; Zelitch 1971).

In the following pages we attempt to assess how the CO_2 environment of a crop is modified by sheltering practices, how resultant CO_2 depletion may limit plant growth, and why CO_2 enrichment can play a vital role in overcoming the restrictions on crop growth and yield that can arise in a protective structure. This review is intended to be explanatory as well as descriptive. In it we examine the effects of different sheltering structures on the environment inside the crop canopy, CO_2 exchange rates, photosynthetic activity, growth patterns, and plant productivity; and we summarize the benefits that can accrue from efficient CO_2 enrichment in these situations. A subsequent review, planned for a later volume of this series, will present a more detailed examination of the biochemical and physiological bases of plant growth and development during acclimation to elevated CO_2 levels.

II. PROTECTED ENVIRONMENTS

A. The Crop Canopy

In order to understand why CO_2 enrichment practices can benefit plant growth, an appreciation of the environment created by a crop canopy is essential. The crop microclimate plays a major role in controlling the metabolic activity and CO_2 exchange of individual plants in a stand. As the plant stand matures, either in the field or inside a greenhouse, the protection and the restriction provided by the vegetative canopy increases in significance.

Gradients and fluctuations of CO_2 concentration within crop canopies have been documented in the field (Biscoe *et al.* 1975; Brown and Rosenberg 1972; Desjardins *et al.* 1978; Inoue 1965; Lemon and Wright 1969; Monteith 1973; Ohtaki 1980). These changes reflect the magnitudes of photosynthetic, photorespiratory, and respiratory activities and the degree of turbulent air exchange between the crop and the atmosphere, which replenishes the CO_2 (Ohtaki 1980; Verma and Rosenberg 1976, 1981). In the protected environment of a shelter or an enclosure, the factors that can contribute to potential CO_2 depletion are greatly magnified. Under such conditions of intensive crop cultivation, localized or general CO_2 depletion may present a significant limitation to photosynthesis and crop production (Hand 1982; Wittwer 1981, 1983; Zelitch 1971). Enriching the atmosphere inside a greenhouse or other enclosure with CO_2 to maintain either ambient (330 μl/liter) or higher concentrations serves to overcome the unfavorable CO_2 gradient between the leaf and its environment when CO_2 would otherwise be limiting.

1. Determining CO_2 Gradients. A crop canopy obtains CO_2 for assimilation from the atmosphere and from the soil (Oke 1978; Redmann 1974a,b,c; Ripley 1974). During the day, the canopy is a net sink for CO_2 and both the soil and the atmosphere function as sources. At night, the crop becomes a CO_2 source and the atmosphere serves as the sink for crop- and soil-derived CO_2. The CO_2 concentration 16m above field crops in the American Midwest can fluctuate between 300 and 350 μl/liter over a 24-hr period, illustrating the capacity of a canopy as a CO_2 sink or source (Verma and Rosenberg 1976, 1981). It is widely recognized that the variations in CO_2 concentration between an open canopy and the atmosphere are subtle; the major physical barrier to CO_2 movement into the leaves is the stomata (Zelitch 1971). However numerous workers have described CO_2 concentration profiles in dense canopies, thus demonstrating that measurable gradients do exist

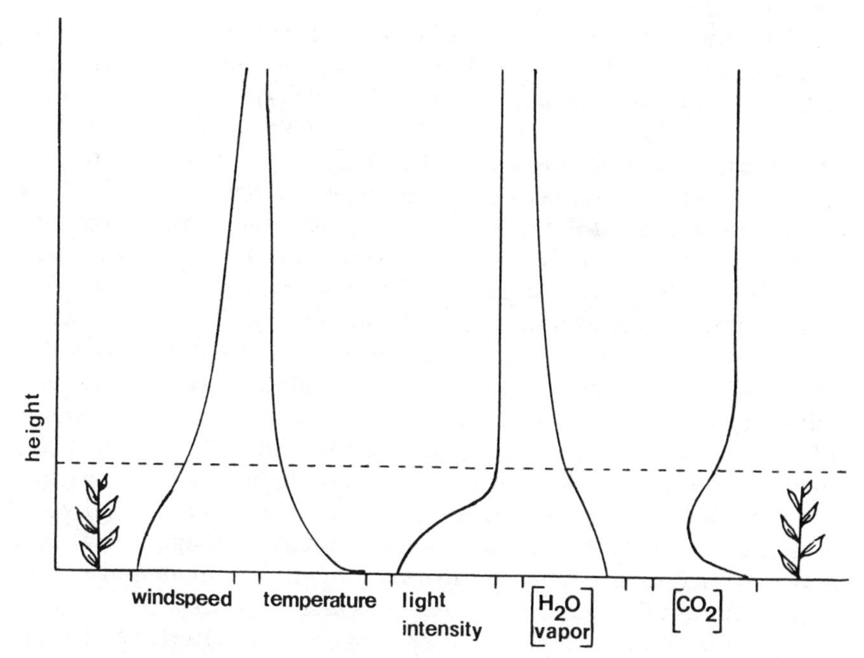

FIG. 8.1. Representative vertical profiles of wind speed, temperature, light intensity, water vapor concentration, and CO_2 concentration in a crop canopy. The values for each parameter increase from left to right. The (———) line indicates the top of the canopy.
Adapted from Allen (1975) and Lemon et al. (1969).

(Allen 1979; Aoki *et al.* 1975; Biscoe *et al.* 1975; Chapman *et al.* 1954; Desjardins *et al.* 1978; Inoue 1965; Ohtaki 1980; Spittlehouse and Ripley 1974). Microclimate profiles have been used to illustrate gradients in CO_2, temperature, water vapor, momentum, and light in a canopy (Allen 1975; Brown and Rosenberg 1971; Inoue 1965; Lemon *et al.* 1969). Figure 8.1 represents typical gradient patterns that might be encountered in a vegetative canopy. Although CO_2 concentration profiles within and above a canopy do not represent photosynthetic or respiratory rates, they do demonstrate that metabolic activity is intense and that environmental conditions can be directly modified by the crop.

The diurnal exchange of CO_2 among the soil, canopy, and atmosphere is one component of the mass and energy transfer that occurs in the form of light, water vapor, momentum, O_2,CO_2, and nutrients (Oke 1978; Ripley and Saugier 1975; Rosenberg 1974). Calculations of vertical CO_2 flux (g $CO_2/m^2/hr$) in and out of a canopy are often based on the CO_2 concentration differences between the canopy interior and the atmosphere and on the assumption that the coefficient of CO_2 transfer can be approximated by analogy with those obtained for water vapor, heat, and momentum (Lemon and Wright 1969; Oke 1978; Rosenberg 1974; Verma and Rosenberg 1976, 1981). However, this assumption is often not valid (Rosenberg 1974) because the pathway of CO_2 exchange from plant to atmosphere involves cellular and mesophyll resistances, as well as stomatal, leaf boundary layer, and crop boundary layer resistances (Monteith 1973; Oke 1978). The complexity of the exchange pathway and the difficulty in accurately measuring CO_2 concentration and exchange in canopies have resulted in very few studies which directly address CO_2 flux. It is also extremely difficult to determine the relative importance of CO_2 flux compared with that of other parameters in the field (e.g., water stress) as a limitation to crop yield (Gifford 1979a; Sionit *et al.* 1980). Carbon dioxide flux in a grass stand (Ripley and Saugier 1975) was found to drop both diurnally and seasonally in response to suboptimal moisture supply. Soybeans grown in open-topped field chambers demonstrated an altered starch-sugar content in water-stressed leaves, which appeared to be negated by CO_2 enrichment to 800 μl/liter (Rogers *et al.* 1984). Water use efficiency has been observed to increase with CO_2 concentration in many crops (Kimball and Idso 1983; Wong 1979), and recently Krizek (1984) observed that the combined effects of CO_2 supply and water supply on crop growth are difficult to predict since they are not well understood at the field level.

Carbon dioxide flux may also be determined using the principle of eddy correlation and vertical wind speed to obtain an estimate of CO_2 exchange (Rosenberg 1974). This approach emphasizes the importance of turbulent air movement in maintaining a continual supply of CO_2 to the leaves and reducing boundary layer resistance to CO_2 uptake (Monteith 1973). The CO_2 concentration in the eddies moving up from the canopy into the atmosphere has been recorded at 6–10 μl/liter lower than that of the atmosphere above the crop (Ripley 1974). The wind is important in modifying vertical canopy gas exchange as increasing wind speed results in more rapid eddy diffusion and speeds the dispersal of gases. Momentum is also transferred to the canopy by the wind (the greatest drop in wind speed corresponds with the layer of densest foliage), resulting in bending and distortion of the canopy (Lemon and Wright 1969; Monteith 1973). This distortion changes the architecture of the canopy, the height and resistance of the crop boundary layer, and the depth of wind penetration into the canopy. Changes in resistance and flux at the crop boundary layer are translated to the leaf and ultimately influence gas exchange and photosynthesis (Farquhar and Sharkey 1982; Zelitch 1971). Any shelter or practice that restricts this vertical air exchange can alter CO_2 flux and CO_2 concentrations within the crop.

2. CO_2 Gradients in the Field. Plants play an active role in the moderation of their own environment (Oke 1978) through photosynthetic activity, transpiration, respiration, and the absorption and dissipation of energy. The CO_2 levels around a plant growing in a natural stand vary dramatically depending upon the extent to which the plant is protected by neighboring plants and/or sheltering structures. A mature forest canopy, for example, forms a natural protection for numerous species of shrubs, herbaceous plants, mosses, and fungi growing on the forest floor. The presence of the canopy increases humidity and moderates extremes of daytime heating and nighttime cooling while restricting the growth of species adapted to high irradiance. The mixing and diffusion of gases may be restricted in these environments resulting in accumulations of CO_2 underneath the tree canopy where light levels are reduced (Bjorkman *et al.* 1972; Sparling and Alt 1966). This effect becomes even more pronounced on foggy, overcast days (Wilson 1948) and at night when temperature inversions may restrict vertical air exchange further (Woodwell and Dykeman 1966). Indications of restricted gas movement and mixing within and above dense forest canopies can often be seen with the naked eye. The redwood forests of the west coast of the United States are frequently shrouded in fog throughout the day though exposed to both sun and wind. Similarly, the bluish haze that hangs over the forests of the Great Smoky Mountains of the southeast United States results from the accumulation of gaseous isoprenes, particularly when temperature inversions and local climatic conditions restrict air movement.

The CO_2 budget of a vegetative stand may be substantially affected by soil respiratory activity. The soil and the atmosphere were each found to contribute one-half of the CO_2 available to a prairie grass stand (Ripley 1974) during periods in the season when the soil temperature and moisture levels favored microbial and root respiration. Soil CO_2 concentrations under the grass stand ranged from 2,000 to 6,000 μl/liter through the soil profile, creating a gradient for diffusion up into the canopy (Redmann 1974a,b,c). Fluxes of CO_2 from cultivated land after the incorporation of organic matter (MacDonald *et al.* 1973) and from muck soils high in organic matter (Hopen and Ries 1962b; Lamborg *et al.* 1983) can contribute to the crop CO_2 requirement. Practices that increase soil temperature and moisture retention (e.g., organic or artificial mulches) would favor the production of soil-derived CO_2. An organic mulch (e.g., straw, woodchips) is a further substrate source for the soil microbial populations. The role of the soil as a CO_2 source in the field raises a question as to whether the move to inorganic growing media (e.g., rockwool, gravel) and hydroponics in greenhouses may be effectively removing a substantial CO_2 source from the crop as well as drastically altering the normal CO_2 environment of the roots. Even in a sterilized conventional soil mix the CO_2 content may be very low.

The CO_2 concentration at a height of 1 m in an immature cornfield can be 50μl/liter lower than the concentration in the air above the canopy during periods of low wind speed (Chapman *et al.* 1954). In a vertical CO_2 profile, the height at which the concentration is lowest generally corresponds with the height of maximum light interception and is just below the height at which the maximum drop in wind speed occurs (Fig. 8.1; Lemon and Wright 1969). At night, CO_2 levels inside the same canopy may reach 400 μl/liter particularly if there is little wind movement. Inoue (1965) calculated that a similar though smaller decline in CO_2 concentration occurred at a height of 20 cm in a wheat canopy. In a barley stand, a daytime gradient of 9 μl/liter CO_2 existed from the top to the base of the canopy (Biscoe *et al.* 1975); in a native prairie grass stand, a gradient of 15 μl/liter was observed through a height of 30 cm (Spittlehouse and Ripley 1974). As wind speed increased, the gradients were diminished and the resistance of the boundary layers was reduced.

3. CO_2 Flux in Sheltered Field Crops. A number of studies by N.J. Rosenberg and his colleagues have demonstrated that the CO_2 concentration profiles in field-crop canopies are affected by windbreaks erected to modify the microclimate of the crop. Microclimate modification in the field is practiced routinely through site selection and optimal use of the local topography. Slopes may be chosen to maximize air drainage and the interception of radiation and to avoid "frost pockets" in depressions and valleys. These measures often result in improved air mixing and exchange through the canopy. Site modification—for example, ridge and furrow grading to enhance soil warming or proper orientation of row crops with respect to prevailing winds—is also a common means of modifying the microclimate. Site selection and windbreaks have also been used to shelter greenhouse structures from the extreme stresses of the local climate. The orientation and structure of the greenhouse influence how the environment inside the greenhouse must be controlled (Hanan *et al.* 1978).

Living windbreaks such as tree or shrub hedges are often established to reduce the length of an open sweep of wind across a flat landscape or to form barriers for a specific plot of land (Oke 1978; Rosenberg 1974). The intensity of turbulent air exchange in a sugar beet crop was reduced by planting a double row of corn every 15 m as windbreaks (Brown and Rosenberg 1971, 1972). In the sheltered plots the average daytime temperature was 1.8°C higher, the water vapor pressure was 400 Pa greater, and the CO_2 concentration was 1 μl/liter lower than in the unsheltered plots. The wind speed 25 cm above the sheltered canopy was 40% of that over the open canopy; however, it appeared that, despite this reduction in wind speed, vertical turbulence was still sufficient for adequate air exchange within the sheltered crop. In a similar study with soybeans protected by a 2-m slat fence (50% wind porosity), vapor pressure, air temperature, and CO_2 gradients were all intensified on the leeward side of the fence, and a 20% decrease in evapotranspirational water loss was observed over a 6-day period (Miller *et al.* 1973). Estimates of photosynthetic CO_2 flux rates were not markedly altered by the presence of the shelter in either study, but some benefit was obtained from the increased water use efficiency.

Crop microclimates can also be modified through the use of horizontal barriers in the form of shading structures such as thatches, wooden slats, white, black, or colored cloth, and netting. The primary purpose of these structures (Fig. 8.2) may be to reduce light intensity (Willey 1975), as is the case with ginseng (Proctor 1980). Ginseng cultivation in Korea has historically been carried out under tightly thatched shade houses, which provide protection from the elements and assure interception of the proper level of solar radiation (Fig. 8.2A). The environment inside these houses is cooler and more humid than that of the surrounding countryside and more favorable for the growth of this valuable crop. Very little is known about the response of semi-shade species, such as ginseng, to CO_2 gradients in either artificial or natural habitats. Konjak (*Amorphophallus konjak* K. Koch, Araceae), a semishade C_3 species, is cultivated in Japan and China for its edible corm. It is normally found in the warm forests of Indochina, where large CO_2 gradients can exist because of extensive soil respiration and relatively still air conditions under the canopy (Aoki *et al.* 1975). Imai and Coleman (1983) report that the corm yield obtained at 700 μl/liter CO_2 is double that obtained under ambient conditions. In modern ginseng production systems, the desire for mechanization has led to the use of continuous horizontal shade structures, such as wooden slats or, more recently, black polypropylene shade cloth suspended by poles at a convenient working height (Fig. 8.2B).

FIG. 8.2. Protected environments for commercial production of ginseng. A—Traditional impermeable thatched straw or reed shades in Korea. B—Modern black woven polypropylene shade cloth covers in Lytton, British Columbia, Canada.
Courtesy of J. T. A. Proctor, University of Guelph, Ontario.

When a semiporous barrier is introduced between the crop canopy and the atmosphere, the degree of vertical turbulent gas exchange may be reduced, thus limiting CO_2 flux into the crop canopy. This effect of a shade cloth tent on vertical air exchange over a soybean crop was observed by Allen (1975). The shade cloth reduced light intensity, reduced the wind speed at the top of the soybean canopy, and increased the water vapor concentration, creating a "humidity blanket" beneath the cloth. The CO_2 gradient was steeper in the shaded canopy (20 μl/liter below the atmospheric level) than in the unshaded canopy (5 μl/liter below the atmospheric level) during peak periods of photosynthetic activity. The difference was partially attributed to greater water stress in the unshaded beans, which resulted in lower net photosynthetic activity. Based on eddy diffusivity measurements, the shade cloth did reduce vertical air mixing at cloth height to 16% of that occurring at the same height over an unshaded field. The horizontal porous barrier acts as a second canopy layer with its own boundary layer characteristics underneath which the crop itself responds. It was not evident in this study whether the lowered CO_2 levels in the shaded crop actually reduced photosynthesis below the potential given that water stress was reduced in the protected environment. Parallel effects may occur in greenhouses where shade cloths are suspended horizontally above the crop canopy.

B. CO_2 Enrichment in the Field

Because of the responsiveness of many field crops, such as wheat (Gifford 1979a,b; Marc and Gifford 1983; Sionit *et al.* 1980, 1981a,b), cotton (Krizek 1979, 1984), and soybeans (Finn and Brun 1982; Rogers *et al.* 1984; Williams *et al.* 1981), to experimental CO_2 enrichment, the feasibility of directly raising CO_2 levels in the field has received considerable attention in the recent literature.

In a field trial where CO_2 was released into an unsheltered cotton field (leaf area index, 2.34) the crop utilized 7–33% of the supplemental CO_2, depending on the incident solar radiation (Harper *et al.* 1973). The CO_2 concentration inside the canopy remained at least 100 μl/liter higher than that in a control plot during CO_2 release from the supply lines. This persistance of CO_2 in the canopy was unexpected because a simulation study by Allen *et al.* (1971) predicted much more rapid dispersal from the canopy. Carbon dioxide release into a crop beneath a shade cloth barrier has also been tested (Allen 1975). However, the authors of both studies stressed that during field application of CO_2 under the test conditions large amounts of CO_2 were lost. The crop response (e.g., dry matter accumulation and yield) to CO_2 enrichment in the field was not assessed in these studies, as the procedure did not extend over a long enough period of time.

Recent studies using open-topped field chambers have demonstrated a clear crop response to CO_2 enrichment in the field. Utilizing this technique, Rogers *et al.* (1984) were able to show that elevated CO_2 levels (up to 910 μl/liter) resulted in reduced transpirational water losses and delayed the development of severe water stress in soybeans.

C. Cloches and Tunnels

Effective use of topography, windbreaks, shading, etc., provide important protection for many horticultural crops (Caborn 1973). Although these practices greatly moderate the microclimate of the crop canopy, they do not markedly change the growing season. Cloches were introduced in France during the fourteenth century to reduce stress on tender plants during the "Little Ice Age." By 1912 the continuous cloche (tunnel) had been patented by L. H. Chase and was in wide use in Europe. Modern technology has modified the continuous glass cloche into a continuous plastic tunnel (Proulx 1984), which is now used worldwide. Growing plants inside polyethylene sheet or synthetic fiber cloches and tunnels is a technique employed in horticultural production mainly to accelerate growth early in the season and obtain earlier yields. In some cases, the extended period of growth in the cloche relative to the unprotected situation allows the production of crops in regions where they would not otherwise mature before autumn frosts.

In Great Britain, strawberries are brought to market early by protecting young plants in cloches ranging in size from those large enough to walk into to those standing less than a meter in height and spanning only a single row (Swabey 1977). The warm environment of the cloche in spring ensures proper fruit set and development. In Japan, young melon transplants are protected from pest infestation by covering individual plants with a rice paper and wire cloche. This shelter is temporary and is removed when the plant outgrows it. Many small-scale producers around the world protect individual plants in a similar manner because the cloches reduce transpiration losses during the day and protect tender transplants from nighttime radiative chilling. On a larger commercial scale, growers in California and Israel protect many tender horticultural crops (e.g., tomato, cucumber, eggplant, zucchini) from mechanical damage from wind (e.g., sandblasting) and from water loss by use of plastic cloches and tunnels. Plastic tunnels have also proved to be valuable aids to crop production in arid and semiarid climates where water supply and extreme temperature fluctuations limit crop growth (Wittwer 1981).

The use of polyethylene cloches has also been investigated in some conifer-growing areas, primarily Scotland, as a means of obtaining large transplant material in shorter periods of time than is possible with open seedbeds (Biggin 1983; Canham and McCavish 1981; Thompson and Biggin 1980). Larger seedlings with lower shoot-to-root ratios were obtained in the cloches than under control bird-netting. The effect was dependent on the edges of the tunnels being buried, implying that the internal atmosphere of the cloche is critical to the success of the technique.

Temperature fluctuations (Guttormsen 1972a,b; Shadbolt *et al.* 1962; Tan *et al.* 1984; Wiest *et al.* 1976), light quality (Guttormsen 1972a,b), and water and nutrient cycling (Biggin 1983) inside cloches have been examined by several workers. However, changes in the CO_2 concentration inside cloches have received less attention as a component of the environment (Enoch *et al.* 1970).

In closed greenhouses without CO_2 enrichment, the CO_2 concentration fluctuates over 24 hr as a result of photosynthetic activity and both plant and soil organism respiration. A similar phenomenon occurs in a cloche in which ventilation is restricted. During the day, the CO_2 level may rapidly reach a minimum of about 100–150 μl/liter as a result of photosynthetic activity and the large amount of plant tissue per unit volume (Enoch *et al.* 1970). At this point the plant is under CO_2 limitation (stress) and no net dry matter accumulation can occur. At night, the CO_2 concentration rises as a result of plant and soil respiration but is pulled back down within an hour of daybreak (Enoch *et al.* 1970). As shown in Fig. 8.3, the diurnal range of CO_2 concentrations encountered in cloches and greenhouses is more extreme than that encountered in an open field. Slitted tunnels provide some ventilation and are usually employed to moderate the heat buildup that can occur inside the tunnels. However, the effect of this practice on CO_2 availability during the day and escape at night has not been well documented.

D. Greenhouses and Growth Chambers

The growth of crops outdoors is dictated by the length of the growing season and the ability of the crop to withstand temporary climatic changes during the growing season. The shelters discussed in previous sections, including tightly sealed plastic cloches, moderate the canopy microclimate, but generally their use merely extends existing

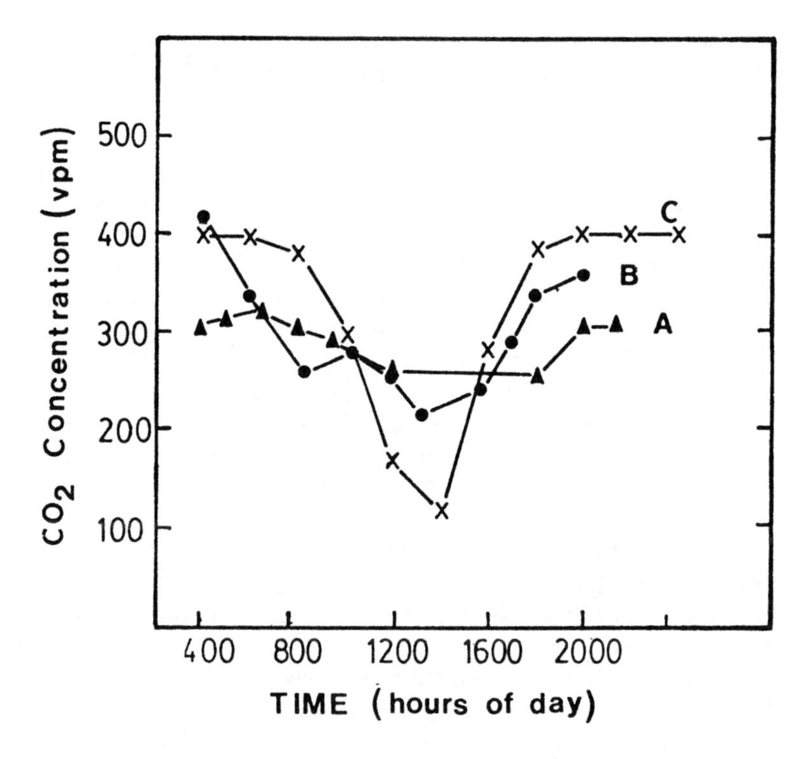

FIG. 8.3. Typical diurnal fluctuations in CO_2 concentrations. A—In a corn canopy in the field (*Chapman et al. 1954*). B—In a glass greenhouse (*Wittwer and Robb 1964*). C—In a tightly sealed plastic house. (*Enoch et al. 1970*).

growing seasons by a week to a month. By comparison, the use of modern climate-controlled greenhouses and growth chambers, which also provide a "rigid" boundary layer between the local climate and the canopy microclimate, offer horticulturists the opportunity of multiple cropping during the year and the potential for extending cultivation into regions where it is normally excluded (White 1979). For the present discussion, the major distinction made between a modern computerized greenhouse and a growth chamber is the link between the greenhouse and the external environment. The primary source of light for CO_2 fixation in the greenhouse is natural solar radiation, whereas in climate-controlled growth chambers, the lighting is obtained entirely from electric lamp sources.

Many articles and reviews have been written about the value of CO_2 enrichment in greenhouses (Enoch *et al.* 1973; Hanan *et al.* 1978; Hand 1982; Kimball 1982, 1983a,b; Kramer 1981; Krizek 1979; Lemon 1983; Wittwer and Robb 1964). Wittwer and Robb (1964) demonstrated that CO_2 levels could decline below 200 μl/liter CO_2 on a bright day (Fig. 8.3B). The diurnal pattern of CO_2 exchange in the greenhouse is similar to that described for other sealed structures such as cloches. In a greenhouse with an actively growing crop, the CO_2 concentration rarely drops as low as the CO_2 compensation point (e.g., 50– 70 μl/liter for a C_3 crop plant) because of the "leakiness" of most greenhouses, which allows air exchange with the external environment through doors, seams, and gaps (Hand 1982; Kimball 1982, 1983a,b; Kimball and Mitchell 1979). The demand for more energy-efficient greenhouses has led to development of new materials and construction methods that have increased the "tightness" of these structures (White 1979). Double-glazing, reduced or lapped seams, thermal curtains, internal air recycling, and more effective roof panel insulation reduce heat and moisture losses but also reduce CO_2 exchange between the environment and the greenhouse. During cold periods, gas exchange may be further limited when ice seals form in the small gaps in the structure. Plastic greenhouses provide less opportunity for air exchange with the environment than do glass greenhouses because of fewer openings and seams in their construction.

The CO_2 concentrations and fluxes occurring inside greenhouse-crop canopies are dependent upon the rate of air exchange between the greenhouse and the external air as well as on crop activity. In the field, measurable gradients of temperature, humidity, and CO_2 in canopies are evident even under conditions of normal air movement. Inside a greenhouse, these gradients are magnified because air movement is severely reduced and the degree of depletion is great. Humidity and temperature profiles have been studied in greenhouses particularly in relation to ventilation requirements. Horizontal and vertical temperature gradients develop in closed greenhouses according to the placement of heaters and the degree of air mixing (Hanan *et al.* 1978). A wind must be generated artificially inside a greenhouse to prevent hot and cold spots from forming and to reduce humidity to a level that does not favor fungal and bacterial growth. Carbon dioxide gradients encountered in the field are also exaggerated in the greenhouse to the extent that various degrees of CO_2 depletion occur in the crop canopy. Carbon dioxide enrichment combined with adequate air movement to reduce boundary layer resistances can help overcome this limitation to growth. Because of the depletion that may occur in a closed greenhouse, the actual crop response obtained with CO_2 enrichment may be greater than that predicted from controlled-environment studies in which the enriched condition is compared with a constant ambient concentration of approximately 320 μl/liter.

The tightly sealed, energy-efficient greenhouse also requires frequent ventilation to reduce the heat load on the plants during periods of high incident solar radiation. The resultant fluctuations in CO_2 concentration and other microclimate parameters confound the interpretion of much of the CO_2 enrichment literature (Kimball 1982, 1983a,b). A modern phytotron (Kramer *et al.* 1970), in which light, temperature, humidity, airflow, and CO_2 levels are not left to chance, represents one of the more desirable structures for experimental work on CO_2 enrichment. However, even in this seemingly "tight" structure, variations in CO_2 levels can be observed, particularly as a result of nearby human or industrial activity (Pallas 1979). Carbon dioxide levels can also rise in the dark and fall in the light in spite of apparently rapid air exchange rates between the growth chamber and a fresh air supply (Bugbee 1984; Tibbitts and Krizek 1978). These observations only serve to underscore the ability of a crop to alter its CO_2 microclimate (Oke 1978; Rosenberg 1974) and the value of CO_2 enrichment in sheltered environments.

III. BENEFITS OF CO_2 ENRICHMENT

A. Expectation of Increased Yield

Although the beneficial effects of CO_2 enrichment on crop yield, advanced harvest date, and crop quality have been demonstrated by numerous workers (Lemon 1983; Rogers *et al.* 1982, 1984; Strain and Armentano 1982), it has been difficult to compare the results of different studies. Even within a species, cultivars can have different photosynthetic efficiencies (Augustine *et al.* 1979; Nilwik *et al.* 1982). In a trial of 16 lettuce cultivars, yield increases under high CO_2 varied from 21 to 56% (Johnston 1972). The parameters that have been used to measure growth and yield vary tremendously. Comparison of yield in greenhouse studies has been confounded by the use of various units, such as g dry weight/m^2, fresh fruit weight/plant, and number of flowers/unit floor area. Tomato studies provide a good example of the complexity of the data; growth has been evaluated both before and during harvest, and changes in flowering date, flower and fruit number, and fruit size have been documented by various workers but rarely in the same report. Because of these difficulties, Kimball (1982, 1983a,b) conducted an analysis of previously published CO_2 enrichment reports in order to standardize the results and allow meaningful comparisons of the crops and their responses. The study is comprehensive and distinguishes between species, cultivar, and photosynthetic type (C_3 vs. C_4). The segregation of the crops by end-use and developmental stage aids the analysis of yield from different experiments. Leafy crops (e.g., swiss chard, lettuce) are distinguished from other vegetatively growing crops (e.g., potato, radish) and from flower and fruit crops. Although Kimball omitted certain crops (e.g., poinsettia) from the analysis because of uncertainty as to how to express enhanced growth, such crops definitely benefit from enrichment (Wada and Kristofferson 1974, Tsujita, personal communication). Data on yield of marketable produce are compiled under "mature crop yields" and data on growth before harvest are compiled under "immature crop yields."

The data for each experiment were standardized against their own unenriched control by calculating the ratio (harvest yield) of the CO_2-enriched yield to the control yield (Kimball 1983a). Since several studies were available for each crop, a mean harvest yield was computed to provide an average response. Values of the mean yield ratio and the number of reports analyzed (N) for selected crops from the study are summarized in Table 8.1. For brevity we omitted confidence intervals but refer the reader to Kimball (1982, 1983a,b).

Several features of this analysis stand out. Most important from a commercial viewpoint is that enrichment to 1000–1200 μl/liter CO_2 (three to four times current ambient levels) has a marked positive effect on the growth and yield of most crop plants tested. Setting aside for the present the additional problems of assessing indeterminate growth and harvest index, several significant points are evident in the data of Table 8.1. The average mean yield ratio of about 1.25 (i.e., a 25% increase due to CO_2 enrichment) of all mature crops was significantly lower than the yield ratio for immature crops (about 1.5). These data tend to support the observation that young tissue benefits more from enrichment than older tissue or that different developmental stages of a tissue may vary in sensitivity to CO_2 enrichment. As discussed later, the effect of CO_2 on growth may not be due solely to enhanced photosynthesis. In this context, it is interesting to note the rather large difference in the mean yield responses of C_3 and C_4 plants.

Many scientists have recently become concerned with the way plants will respond to the predicted rise in atmospheric CO_2 levels (Keeling *et al.* 1976; Lemon 1983; Revelle 1982; Rosenberg 1981, 1982; Rycroft 1982; Wittwer 1983). Although C_4 crops such as maize and sorghum are not classical greenhouse crops, many of the major weeds occurring in C_3 crops are C_4 plants and vice versa (Patterson and Flint 1980, 1982; Patterson *et al.* 1983), and an understanding of the potential relative growth enhancement by CO_2 enrichment is desirable. The intuitive view is that C_4 plants will benefit less from enrichment than C_3 plants (Mauney *et al.* 1978), even though the water use efficiency of

TABLE 8.1 Mean Yield Response of Crops to CO_2 Enrichment

Crop	Immature crops		Mature crops	
	Mean ratio[a]	N[b]	Mean ratio	N
VEGETATIVE				
Cabbage (*Brassica oleracea*)	1.28	10	1.05	2
Clover (*Trifolium* sp.)	—	—	1.12	2
Endive (*Cichorium endivia*)	—	—	1.15	1
Fescue (*Fescue* sp.)	1.51	2	—	—
Lettuce (*Lactuca sativa*)	1.68	7	1.35	54
Swiss chard (*Beta vulgaris* var. cicla)	—	—	1.67	17
All Leafy Crops (average)	1.44	19	1.40	76
ROOT AND TUBER CROPS				
Potato (*Solanum tuberosum*)	—	—	1.44	18
Radish (*Raphanus sativus*)	1.79	6	1.28	5
Sugar Beet (*Beta saccarifera*)	1.75	7	—	—
All Root Crops	1.77	13	1.40	23
FLOWER CROPS (number of blooms)				
African violets (*Saintpaulia* sp.)	2.36	3	—	
Carnation (*Dianthus caryophyllus*)	—	—	1.07	33
Chrysanthemum (*Chrysanthemum × morifolium*)	—	—	1.06	66
Cyclamen (*Cyclamen persicum*)	—	—	1.35	3
Nasturtium (*Tropaeolum majus*)	—	—	1.86	3
Geraniums (*Pelargonium × hortorum*)	1.19	3	—	
Rose (*Rosa* spp.)	—	—	1.27	24
Snapdragon (*Antirrhinum majus*)	—	—	1.03	1
All Flower Crops (average)	1.28	6	1.12	130

FRUIT CROPS				
Cucumber (*Cucumis sativus*)	1.46	26	1.43	31
Eggplant (*Solanum melongena*)	—		1.88	3
Grape (*Vitis vinifera*)	2.49	1	1.88	—
Muskmelon (*Cucumis melo*)	—		1.13	7
Okra (*Hibiscus esculentus*)	2.74	3	—	
Pepper (*Capsicum annuum*)	2.41	1	1.60	4
Strawberry (*Fragaria* × *ananassa*)	—		1.17	13
Tomato (*Lycopersicon esculentum*)	1.52	24	1.17	131
All Fruit Crops (average)	1.56	55	1.22	189
WOODY CROPS				
Apple (*Malus pumila*)	1.32	4	—	—
Birch (*Betula* sp.)	1.06	4	—	—
Cottonwood (*Populus deltoides*)	1.69	2	—	—
Fir (*Abies* sp.)	1.18	2	—	—
Gum (*Nyssa* sp.)	1.10	3	—	—
Leea sp.	2.58	1	—	—
Maple (*Acer saccharinum*)	1.74	2	—	—
Pine (*Pinus* sp.)	1.36	26	—	—
Spruce (*Picea pugens*)	1.58	17	—	—
Sycamore (*Platanus occidentalis* L.)	1.21	2	—	—
All Woody Crops (average)	1.40	63	—	—
ALL C_3 PLANTS	1.51	211	1.25	542
ALL C_4 PLANTS	1.11	14	2.01	5
ALL SPECIES	1.48	225	1.26	547

Source: Modified from Kimball (1983 a,b).
[a] Ratio of CO_2 = enriched yield to nonenriched yield.
[b] Number of observations.

C_4 as well as C_3 plants could be enhanced by CO_2-induced stomatal closure (Gifford 1979a,b; Kimball and Idso 1983; Krizek 1984; Rosenberg 1981; Wong 1979).

The presence of CO_2-concentrating mechanisms in C_4 plants implies that photosynthetic rates may be relatively efficient at atmospheric CO_2 concentrations (Ehleringer and Bjorkman 1977; Mauney *et al.* 1978). Although there is some debate about this oversimplified view of C_3/C_4 dogma, it is clear from the data in Table 8.1 that vegetative growth of C_4 plants is not greatly stimulated (1.17 yield ratio) by enrichment, whereas vegetative growth of C_3 plants is markedly enhanced (1.57 yield ratio). In contrast, however, the yield of mature C_4 plants (admittedly based on relatively few reports) doubles with CO_2 enrichment (2.01 yield ratio), whereas C_3 crop yields only increase by one-quarter (1.25 yield ratio). Clearly, the generalizations relating vegetative growth to reproductive growth are not yet fully supported. Though young C_3 plants may appear to have a significant growth advantage at elevated levels of CO_2, C_4 plants respond particularly well to CO_2 at high irradiance (Bjorkman 1971; Ehleringer and Bjorkman 1977), and the increased seed production by a mature C_4 plant may favor the long-term dominance of C_4 plants in mixed open-field populations.

However, increases in global CO_2 concentrations might be associated with changes in other environmental parameters such as light intensity, temperature, and water supply (Gifford 1979a,b; Rosenberg 1982; Sionit *et al.* 1980, 1981a,b), which could interact in creating a local climate that favors C_3- or C_4-type metabolism (Pearcy and Ehleringer 1984). These possibilities complicate predictions about the future relative advantages of C_4 plants over C_3 plants if atmospheric CO_2 levels increase substantially. In fact, many authors (Elmore 1980; Gifford and Evans 1979; Kramer 1981; Lemon 1983; Zelitch 1971) have analyzed the data on crop photosynthetic rates and yield and warn against equating in absolute terms increased photosynthetic efficiency with increased yield.

B. Vegetative Growth

Based on the observations summarized in Table 8.1, there seems to be little doubt that CO_2 enrichment enhances plant growth, especially vegetative growth. Insofar as plant productivity can be related to leaf photosynthesis (i.e., CO_2 fixation), the effect of CO_2 enrichment seems predictable. As Lemon (1983) succinctly observes, everyone knows that a "squirt of CO_2" will enhance the photosynthetic rate. Blackman (1905) and Gaastra (1966) showed many years ago that CO_2 is a limiting factor and that at full sunlight the photosynthetic carbon fixation rate (measured on a short-term basis of minutes or hours) is not saturated at ambient (300 ± 30μl/liter) CO_2 levels (Fig. 8.4). Furthermore, the CO_2 level around a leaf required to saturate photosynthetic carbon fixation rises as the temperature and light levels increase. However, short-term measurements of maximal leaf carbon fixation rates are by themselves very misleading. Although leaf carbon fixation may be two- to threefold greater at 1000 μl/liter CO_2 than at 300 μl/liter CO_2, this increased carbon fixation by a source leaf is not necessarily matched by an equivalent increase in whole plant dry matter production or yield over an extended growing period (Elmore 1980; Gifford and Evans 1979; Kramer 1981; Lemon 1983; Zelitch 1971).

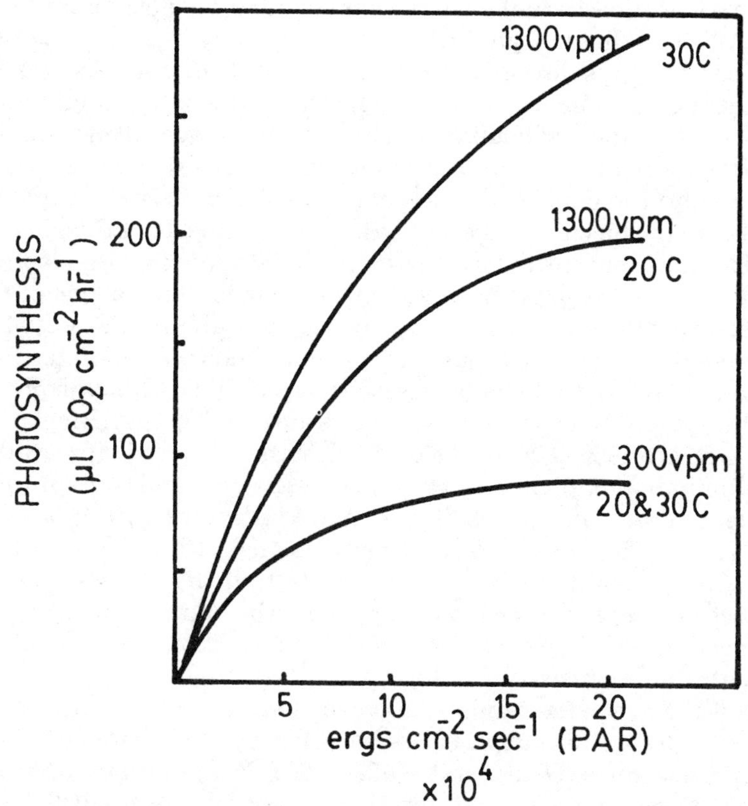

FIG. 8.4. The effect of high CO_2 and increased temperature on the photosynthetic light response curve of cucumber, a typical C_3 plant. vpm = μl/liter. After Gaastra (1966).

Leaf photosynthetic rates are *also* influenced by temperature, humidity, light intensity, leaf age (Akita and Moss 1973; Aoki and Yabuki 1977; Brun and Cooper 1967; Hand and Soffe 1971; Hardman and Brun 1974; Hicklenton and Jolliffe 1978, 1980a; Lee and Whittingham 1974; Liu 1983; Madore and Grodzinski 1984a,b), duration of exposure to high CO_2 (Aoki and Yabuki 1977; Porter and Grodzinski 1984), leaf nitrogen status (Nevins and Loomis 1970; Porter and Grodzinski 1984; Wong 1979) and sink demand (Gifford and Evans 1981; Ho 1977; Hofstra and Hesketh 1975; Huber *et al.* 1984). Many physiological and biochemical factors are involved in the response of leaves to high CO_2, both in the short term and the long term, which will be addressed in more detail in a second review in this series. It suffices here to point out that there are several possible explanations of why exposure to high CO_2 alters measured rates of net CO_2 fixation.

In the short term, assuming that high CO_2 does not induce severe stomatal closure (Farquhar and Sharkey 1982; Goudriaan and van Laar 1978), high atmospheric CO_2 during the day increases the CO_2 gradient from outside the leaf to inside the leaf. The resistances to CO_2 flux into the leaf and to high carbon fixation rates, which might arise due to canopy and leaf boundary layers and/or as a result of stomatal resistance, are in effect minimized by artificially raising the level of CO_2 around the crop. With respect to mesophyll resistance, loosely equated to the biochemical events associated with CO_2 diffusion and CO_2 fixation, current theory favors the view that ribulose bisphosphate carboxylase-oxygenase, the key carboxylation enzyme of the reductive pentose phosphate pathway (i.e., the Calvin cycle), would be stimulated by a greater supply of CO_2 and photorespiration (i.e., oxygenase activity) would be suppressed (Ogren 1984). Although photosynthesis is enhanced at elevated levels of substrate CO_2 (Blackman 1905; Gaastra 1966) it is not at all clear that the enhanced rates are due to a diminution in photorespiration alone (Bravdo and Canvin 1979; Hicklenton and Jolliffe 1980b; Lee and Whittingham 1974; Madore and Grodzinski 1984a,b; Porter and Grodzinski 1984; Zelitch 1971). Nor is it always possible to demonstrate that the increased plant productivity or yield associated with long-term maintenance of a crop at high CO_2 is due to a "faster" rate of CO_2 fixation since high CO_2 concentrations can even be inhibitory.

It is well known, for example, that elevated levels of CO_2 can induce stomatal closure in many species, which tends to diminish carbon assimilation rates (Delieu and Walker 1981; Farquhar and Sharkey 1982; Goudriaan and van Laar 1978; Jones and Mansfield 1970; Raschke 1975; Zelitch 1971). Furthermore, Woo and Wong (1983) showed that inhibition of CO_2 assimilation by supraoptimal CO_2 levels occurs in cotton, cowpea, and sunflower leaves grown at ambient CO_2 levels. The inhibition of photosynthesis by high CO_2 (800 μl/liter) is most pronounced in plants grown at low light and nitrate (nutrient) levels, an effect that cannot be attributed solely to changes in stomatal aperture. Ultimately, photosynthetic acclimation to higher CO_2 levels is linked to the total fluxes, movement, and metabolism of carbon and nitrogen in a plant that itself is growing more rapidly (Huber *et al.* 1984; Porter and Grodzinski 1984).

An increased carbon supply would be expected to be balanced by an increased demand for nitrogen and other nutrients (Finn and Brun 1982; Masterson and Sherwood 1978; Nevins and Loomis 1970; Sionit *et al.* 1981a; Skoye and Toop 1973; Strain and Armentano 1980; Williams *et al.* 1981). Traditionally, greenhouse growers practicing "CO_2 fertilization" also maintain a relatively high nutrient regime (Hand 1982; Wittwer 1981). It cannot be determined from the large number of studies we have surveyed whether or not plants growing under high CO_2 remobilize fixed nitrogen from older tissue to younger, growing regions of the plant at faster rates than do plants at ambient levels of CO_2. In many of the studies on CO_2 enrichment, crowding and mutual shading of the lower leaves in the crop canopy may affect the photosynthetic efficiency of these tissues (Hand 1982; Hand and Postlethwaite 1971). However, in several controlled studies where mutual shading effects do not appear to be a problem, it appears that the protein content and activity of certain key photosynthetic enzymes are lower in the expanded mature leaves of plants grown at high CO_2 (1200 μl/liter) (Porter and Grodzinski 1984; Vu *et al.* 1983; Wong 1979). In fact, when assayed in ambient CO_2 concentrations, the mature leaves of plants acclimated to CO_2 showed reduced photosynthetic rates (Aoki and Yabuki 1977; Frydrych 1976; Ho 1977; Liu 1983; Wong 1979). In bean plants grown under high CO_2, the activities of carbonic anhydrase and ribulose bisphosphate carboxylase (the major soluble leaf protein) were suppressed but only in mature fully expanded leaves (Porter and Grodzinski 1984); the activities in young leaves were not reduced. Although a decline in carbon fixation potential of older expanded leaves appears to act as a "drag" on total production, these observations should be viewed cautiously as they may be reflecting greater growth and turnover rates within the plant. Several studies show that photosynthetic activity of young tissue is more sensitive to high CO_2, and on a total leaf area basis net *plant* carbon fixation rate (i.e., photosynthesis) is enhanced by increased CO_2 (Aoki and Yabuki 1977; Hicklenton and Jolliffe 1980a; Liu 1983). Enoch *et al.* (1984)

calculate that over 40% of the carbon in tomato plants is derived from the artificially added CO_2 when supplied at a rate of 1100±0.100 μl/liter.

As a plant ages and young leaves mature faster, the emergence and growth of active young leaves under high CO_2 may be balanced by the loss in photosynthetic activity of older exporting leaves. As discussed already, the efficiency of CO_2 enrichment clearly depends on the developmental age of the crop (Hand and Postlethwaite 1971; Hardman and Brun 1971; Newton 1965) and on the rate of turnover of both photosynthetically active leaves and growing sinks (Gifford and Evans 1979; Kramer 1981; Zelitch 1971). The rate of assimilate export is a function of both carbon fixation and leaf reserves, so that over an extended time the translocation rate is determined by the changing source–sink relationships (Enoch *et al.* 1973; Ho 1977; Hofstra and Hesketh 1975; Huber *et al.* 1984). In recent studies with cucumbers grown under high CO_2, proportionally more carbon was mobilized as transport sugars (e.g., stachyose, verbascose, and sucrose) than as amino acids, indicating that CO_2 enrichment alters the types of transport assimilates as well as the amounts of newly fixed carbon retained in the leaves as starch. (Madore and Grodzinski 1984b). Manipulating the levels of the key substrate of photosynthesis (i.e., CO_2) will not increase yield in the absence of a balanced program for partitioning the carbon for improved plant growth and development.

The effect of high CO_2 on both the morphology and phenology of many flowering plants may alter yield potential. At a basic level, cell division in cultured cells of undifferentiated sycamore (*Platanus occidentalis* L.) was dependent on the presence of CO_2 in the growing media, a requirement that could not be replaced by organic carbon sources (Gathercole *et al.* 1976). Elevated CO_2 concentrations also influence vegetative development in several other noticeable ways. Skytt-Andersen (1976) observed that 1400 μl/liter CO_2 suppressed apical dominance in peas, which indicates a potential influence on axillary growth. Carbon dioxide concentrations up to 1500 μl/liter result in more branching in young trees (Zimmerman *et al.* 1971), soybeans (Cooper and Brun 1967; Rogers *et al.* 1981), tomato (Hand and Postlethwaite 1971; Hand and Soffe 1971; Hicklenton and Jolliffe 1978), cucumber (Krizek *et al.* 1974; Madore and Grodzinski 1984b), and chrysanthemums (Eng 1982; Hughes and Cockshull 1971; Mortensen and Moe 1983a,b); more tiller development in wheat (Fischer and Aguilar 1976); and more leaf and flower initiation in gerbera (Liu 1983). Soybeans grown in 910 μl/liter CO_2 initiated more main stem leaves, branches, and total leaves than did plants grown at ambient levels of CO_2 (Rogers *et al.* 1981, 1982). The initiation of leaves was accelerated by growth in high CO_2 as indicated by a decline in the plastochron ratio from 2.10 to 1.75 days in the enriched atmospheres (Rogers *et al.* 1981). Individual leaves reached their maximum area 1 day sooner when grown in elevated CO_2, but there was no final difference in the individual leaf areas (Rogers *et al.* 1981). *Phaseolus* (O'Leary and Knecht 1981) and soybean (Thomas and Harvey 1983) leaves grown in high CO_2 had more stomata than those grown in ambient conditions, but Madsen (1973) showed the opposite effect of high CO_2 on tomato leaves. Leaves of tomato (Madsen 1973), soybean, sweetgum (*Liquidamber styraciflua* L.), and loblolly pine (*Pinus taeda* L.) (Thomas and Harvey 1983) were thicker after growth in high CO_2. Microscopic examination of soybean and sweetgum revealed an extra layer of palisade cells in the leaves of high CO_2-grown plants (Thomas and Harvey 1983), but this was not observed in tomato (Madsen 1973). Young soybean mesophyll cells differentiated into palisade and spongy parenchyma more rapidly in high CO_2 (772 or 910 μl/liter), and there was an apparently higher rate of cell division in both cell types (Rogers *et al.* 1981). Examination of leaf thickness throughout the day, at various enrichment levels, revealed that leaves in low CO_2 thicken slightly as the day progresses but that leaves grown in high CO_2 (645 or 910 μl/liter) are thickest at dawn and dusk, not during the late afternoon (Rogers *et al.* 1982). It is not clear whether these observations are part of a general pattern; however, they indicate that CO_2 at any concentration may have physiological effects on plant growth separate from its role as a photosynthetic substrate.

1. Leaf Crops. Growth of leafy vegetative crops such as lettuce represents the simplest situation for yield analysis (Hand 1982; Kimball 1982, 1983a,b). Lettuce yields increase 30–50% when plants are grown in 600 μl/liter CO_2 (Johnston 1972; Knecht and O'Leary 1983; Wittwer and Robb 1964), but higher concentrations of CO_2 may provide little additional benefit (Enoch *et al.* 1970). For example, Krizek *et al.* (1974) showed that lettuce ('Grand Rapids') grown in 2000 μl/liter CO_2 had a greater leaf area than controls but did not have a greater dry weight under identical light and temperature conditions. The data in Table 8.1 for all leaf crops indicate that CO_2 enrichment enhances vegetative growth (1.44 mean yield ratio) to the same extent as mature yield (1.40 mean yield ratio). Since the whole top of leaf-crop plants is harvested, this observation is not surprising. In fact, the definition of maturity in the case of leaf crops is misleading since it reflects market maturity rather than physiological maturity. Commercial lettuce is not allowed to bolt or flower, much less set seed.

The commercial benefits of CO_2 enrichment for crops harvested during vegetative stages of growth can be very important. In the nursery industry, more rapid development of young trees shortens the holding time of the crop and reduces costs. Under high CO_2, conifer seedlings grew 30–60% larger depending on the species (Yeatman 1970), and young crabapple trees demonstrated significant growth increases in 2000 μl/liter and greater CO_2 levels (Krizek *et al.* 1971; Zimmerman *et al.* 1970). Treatment for 4 weeks in 4000 μl/liter led to growth differences that were still evident 2–3 months after the treatment period ended. Tobacco seedling growth was increased 5–25%, depending on nutrient status, by exposure to 1000 μl/liter CO_2 (Thomas *et al.* 1975). Based on increases in the numbers of cuttings, stem length, leaf area, fresh weight, and dry weight, daytime CO_2 enrichment was beneficial in the production of high-quality chrysanthemum stock plants (Eng 1982; Eng *et al.* 1983). Light and CO_2 are recognized as critical in the production of carnation stock plants (Goldsberry 1966; Holley and Altstadt 1966).

The basis of expressing vegetative growth, which has been reported both as a growth rate and as an absolute measure of growth gain, may disguise the actual effect of CO_2 enrichment. In response to CO_2 enrichment, young crabapple trees show the greatest relative growth rate proportional to their initial size, and older plants show the greatest absolute plant growth (Zimmerman *et al.* 1970). As plants grow larger, relative growth enhancement declines, which may explain some of the reports of declining growth rates with prolonged exposure to high CO_2 (Aoki and Yabuki 1977; Thomas *et al.* 1975). Often, increased growth rates are the basis for predictions that more crops can be grown in a given production season. Wittwer and Robb (1964) reported that with high CO_2 three lettuce crops could be grown in a season rather than two, but Dullforce (1966) reported only a 4-day advancement of the harvest of marketable heads. In contrast, soybeans, cotton, and sunflower show actual delays in reproductive development when grown in 620 μl/liter CO_2 (Hesketh and Hellmers 1973).

2. Root Crops. Normally, CO_2 enrichment of the shoot results in an increase in root growth. The root is a major sink of new photoassimilates, and CO_2 enrichment results in substantial increases in the yield of those vegetative crops in which root tissue represents the marketable organ (Collins 1976; Imai and Coleman 1983; Wyse 1980). The mean yield ratio for mature root crops is 1.40 (Table 8.1), equal to that of leafy vegetative crops, while the immature mean yield ratio of 1.77 is superior to that of leafy crops.

However, experiments in which root tissue was directly exposed to high CO_2 resulted in marked effects on development (Geisler 1963; Phillips *et al.* 1976; Stolwijk and Thimann 1957). Misting with CO_2-enriched water improved root development in difficult-to-root woody ornamentals (French and Lin 1984; Lin and Molnar 1980, 1982). Very high levels (45%) of CO_2 applied directly to potato root tissue for a single 12-hr period doubled the number of tubers initiated (Arteca *et al.* 1979). The interactions between high soil CO_2 (Hopen and Ries 1962b; Lamborg *et al.* 1983) and root development are not understood, but the observations support the view that high CO_2 alters the rate of development of the plant beyond a simple substrate stimulation of photosynthesis. The time of CO_2 application relative to growth stage may greatly alter the initiation of new organs and alter the time of transition from vegetative to reproductive growth (Rogers *et al.* 1981).

C. Reproductive Development

1. Flowers. The development of consumer demand for fresh flowers during the winter season has stimulated improvement of flower-crop management for the greenhouse. Work has focussed on high-value floriculture crops such as roses (Biran *et al.* 1973; Hand and Cockshull 1975; Mattson and Widmer 1971a,b), carnations (Goldsberry 1966; Holley and Altstadt 1966; Koths and Adzima 1966), and chrysanthemums (Eng 1982; Eng *et al.* 1983; Hughes and Cockshull 1971, 1972; Mortensen and Moe 1983a,b; Skoye and Toop 1973).

High CO_2 increases the flower production of roses, apparently by reducing the proportion of blind shoots to blossoms (Enoch *et al.* 1973; Lindstrom 1965; Zeislin *et al.* 1972). Overall plant size is increased and greater numbers of side shoots and buds have been observed (Mattson and Widmer 1971), although this may not be a general phenomenon in roses (Zeislin *et al.* 1972). An increase in flower number can also be obtained with increased light levels (Enoch *et al.* 1973; Hurd 1968), emphasizing that response to CO_2 enrichment is affected by other environmental parameters (e.g., light, temperature, nutrient supply) which may be limiting to plant productivity.

The assessment of developmental rates under CO_2 enrichment is complicated by the effects that high CO_2 may have on photoperiodic responses. Hesketh and Hellmers (1973) showed that floral development in sorghum, soybean, and sunflower was delayed during growth at high CO_2 levels. Sorghum and soybean are known to be photoperiod sensitive, and experiments with *Pharbitus nil* and *Xanthium pennsylvanicum* (Hicklenton and Jolliffe 1980b; Ireland and Schwabe 1982; Purohit and Tregunna 1974) have shown alteration in day-length requirements in high CO_2. Marc and Gifford (1983) reported that high CO_2 can alter the rate of flower differentiation in sorghum and attributed the response to a direct morphogenetic effect. Alterations in the number of flowers and the timing of anthesis can significantly affect fruit development and final yield. Whether common greenhouse crops show similar responses has not been determined, but practical experience has not revealed any problems associated with floral initiation under high CO_2 (ca. 1200 μl/liter).

2. Fruit Development. The most extensively studied horticultural crops in which fruit production has been studied under CO_2 enrichment are tomato and cucumber. While most experiments report only total fruit yield, some yield analysis data have been accumulated.

An increased number of fruit in tomato apparently contributes more to CO_2-induced yield increases than gains in individual fruit weight (Hand and Postlethwaite 1971; Hicklenton and Jolliffe 1978; Wittwer and Robb 1964). Tomatoes grown in high CO_2 produce more trusses per plant, more open flowers per truss, and more fruit per truss (Calvert and Slack 1975; Hand and Soffe 1971; Knecht and O'Leary 1974; Hicklenton and Jolliffe 1978). The effects of CO_2 enrichment are more pronounced during the early part of the harvest period, in part because anthesis, and therefore initial yields, are advanced (Hand and Postlethwaite 1971; Krizek *et al.* 1974). As a result, the enhancement of tomato fruit production by CO_2 enrichment may appear to be greater during the early part of the season than during later periods.

The effect of CO_2 enrichment on yield also depends on the reproductive strategy of the plant. As discussed by Mauney *et al.* (1978), plants with terminal inflorescences, such as sorghum and sunflower, have more limited yield enhancement potentials than do plants that flower from many nodes, such as cotton and soybeans. We are not aware of any CO_2-enrichment studies that explicitly compare the yield of determinate and indeterminate flowering types within a single species. Both determinate and indeterminate tomato cultivars are commercially available, with indeterminate types favored for greenhouse use.

Total fruit yield of cucumbers grown at 3000 μl/liter was 26% higher than that of control plants grown at 300 μl/liter (Enoch *et al.* 1976). The yield enhancement during the first month of harvest was 100% higher under the CO_2-enriched condition, which may be attributed to earlier flowering, increased number of flowers, and a shortened harvest period. The increase in fruit number may be further attributed to the increased branching of enriched plants (Enoch *et al.* 1976; Krizek *et al.* 1974; Peet *et al.* 1983) and the higher proportion of female flowers found on the side shoots relative to the main shoot (Enoch *et al.* 1976; Madore and Grodzinski 1984). The increased fruit yield of sweet pepper grown under high CO_2 also appears to be the result of greater fruit number (Enoch *et al.* 1970, 1976).

Accelerated growth during the relatively immature stages of tomato and cucumber development may prove to be the most significant effect of CO_2 enrichment on yield in these crops. Greater axillary bud development under high CO_2 leads to more shoot, flower, and fruit development, which may result in a larger sink demand for photosynthate. In many crops, increased sink demand has been shown to enhance photosynthetic rates in source leaves (Gifford and Evans 1979). Thus, the basis for the beneficial responses to CO_2 enrichment appears to have several components. First, as discussed in the section on vegetative growth, adding CO_2 reduces the possibility of CO_2 depletion in the canopy and overcomes canopy and leaf boundary layer resistances. An enhanced supply of CO_2 outside the leaf increases the internal leaf CO_2 concentration, which favors photosynthesis. The maintenance of photosynthetic activity and the enhanced rate of photosynthesis favor greater photoassimilate movement and the development of healthy growing sinks. The maintenance of elevated levels of CO_2 during fruit filling ensures an adequate supply of CO_2 for photosynthesis to meet increased sink demand.

D. Quality of Produce

Growers expect a very high return for their capital investment in CO_2-enrichment facilities. Successful marketing of floriculture crops such as chrysanthemum, rose, and carnation and of out-of-season vegetables such as lettuce, tomato, and cucumber is dependent on the production of a high-quality product. The appearance, texture, aroma, and keeping quality of higher priced crops are very important commercial considerations.

Roses (Mattson and Widmer 1971b), chrysanthemums (Hughes and Cockshull 1971, 1972; Lindstrom 1968; Mortensen and Moe 1983a,b), and gerbera (Liu 1983) have larger blooms when grown in high CO_2. Cut flowers grown under high CO_2 have thicker, longer, and stronger stems (Eng 1982; Hughes and Cockshull 1971; Koths and Adzima 1966; Liu 1983; Mattson and Widmer 1971b). Liu (1983) showed that enrichment of gerbera had relatively little effect on total flower yield but significantly increased the proportion of blooms in the premier grades, improving the overall crop value. Roses grown in high CO_2 demonstrated a lower incidence of "bluing" than flowers grown at ambient CO_2 concentrations (Biran *et al.* 1973; Enoch *et al.* 1973), though the reason for the pigment improvement is unclear. Data published on the relative keeping quality of flowers grown in high CO_2

are inconclusive; however, Mattson and Widmer (1971a) observed one floribunda rose cultivar with a half-day longer vase life after growth in 1000 μl/liter CO_2.

Fruit grading is highly dependent on size and shape. In tomato, enrichment increases the size of fruit marginally (Hand and Postlethwaite 1971; Kretchman and Howlett 1970) but produces more uniformly shaped fruit (Kretchman and Howlett 1966), which leads to a higher proportion of top-grade produce (Hand 1982; Kretchman and Howlett 1966, 1970; Wittwer and Robb 1964). The harvest date, however, influences the degree to which CO_2 enrichment improves fruit size. Numerous reports of higher yield with no effect on fruit quality have appeared (Calvert and Slack 1975; Hand and Postlethwaite 1971), and in one case, growth in high CO_2 increased the incidence of fruit cracking (Kretchman and Howlett 1966). Taste tests have indicated that the acceptability of tomato fruit (Kimball and Mitchell 1981) and radish roots (Knecht 1975) grown in high CO_2 is similar to the acceptability of these crops grown at ambient CO_2 levels. Vitamin A content of tomato fruit was increased under growth in high CO_2 (Kimball and Mitchell 1981).

These various observations suggest that CO_2-induced improvements in product quality may be as important as increases in yield. As Wittwer (1983) points out, the potential economic benefit of CO_2 enrichment is undisputed; however, it is essential that crop production be matched with local market demand and the availability of cheap CO_2.

IV. ENRICHMENT PROCEDURES

Three major groups practicing CO_2 enrichment of whole plants and crops can be identified: commercial greenhouse growers, greenhouse researchers, and research scientists (e.g., physiologists, pedologists, ecologists, agrometeorologists) whose interest may not be directed to immediate practical applications (Bach 1980; Keeling *et al.* 1976; Lemon 1983; Rogers *et al.* 1981, 1982; Wittwer 1983; Zelitch 1971). Requirements for CO_2 vary according to the scale of the operation, the quantity of CO_2 handled, and the degree of control over CO_2 concentration required.

A. Sources of CO_2

Anyone undertaking CO_2 enrichment of plants requires a source of CO_2 that is free of contaminants, readily available, and effectively dispensed to the crop. For commercial growers, in particular, the source of CO_2 and its application must also be cost efficient.

1. Biologically Derived CO_2. Carbon dioxide is produced naturally by bacteria and fungi as they degrade organic matter (Lamborg *et al.* 1983). The CO_2 concentration of soil is high (1000–30,000 μl/liter) relative to that of the air. Soils high in organic matter (e.g., muck) produce more CO_2 than mineral soils (Hopen and Ries 1962b), and inside a plastic tunnel or under mulch, the contribution of CO_2 from the soil is probably greater than in the open. Accumulation of CO_2 above ambient levels in canopies at night (Fig. 8.3) demonstrates the importance of these respiratory processes. However, CO_2 derived from the organic matter of the soil can only supply part of the CO_2 requirement of crops, since evolution rates do not compare with carbon fixation rates on sunny days.

Increased yield of greenhouse cucumbers has been attributed to the use of straw mulch from which CO_2 is evolved. The choice of organic mulch clearly affects the rate of breakdown to CO_2. For example, Hicklenton and Jolliffe (1978) found very little nighttime rise in greenhouse CO_2 using wood chips, which represent suberized and lignified matter with a low nitrogen content. The lack of control over the rate of release of CO_2 from organic mulches and soils, as well as the increased possibility of accidental infestation of the greenhouse crop, raises serious questions about the use of decomposing organic matter as a CO_2 source. Soil sterilization would only further limit the effectiveness of this method, as the recolonization of the soil by microbes would be too slow to produce significant CO_2, particularly during the crucial early periods of crop development.

2. Combustion. A common method of generating CO_2 is by controlled combustion (Hand 1982; Wittwer and Robb 1964). Many growers still favor this method as the raw materials (e.g., natural gas, propane, kerosene, or paraffin) are available, reasonably inexpensive, and produce heat as a useful by-product. Furthermore, the equipment for generating CO_2 by efficient combustion of fossil fuels has been commercially available for many years (Hanan *et al.* 1978; Nelson 1981).

However, major problems can arise with combustion. The water vapor generated during combustion can complicate humidity control inside a greenhouse, particularly during winter months when excess water vapor condenses on the cooler outer walls. During warmer seasons the associated heat production may necessitate earlier venting and loss of CO_2. In addition, potentially toxic pollutants can be generated, particularly if combustion is incomplete due to low flame temperature or restricted O_2 supply. The growing appreciation of these problems has led growers to favor pure CO_2 injection over the more traditional combustion systems.

3. Pure CO_2. The CO_2 enrichment method with the fewest side effects and problems is injection of pure CO_2 from pressurized tanks of liquid CO_2 directly into the greenhouse circulation system. Provided that the CO_2 is clean, many of the problems associated with fuel combustion are avoided. Commercial growers generally choose between liquid CO_2 and combustion on the basis of the relative cost of local fuel and compressed CO_2, as well as their proximity to a supplier of liquid CO_2. For research purposes, particularly controlled-environment studies, injection of pure CO_2 tends to be used almost exclusively. A few studies, notably an examination of field CO_2 enrichment by Allen (1979), have been conducted with dry ice (solid CO_2, b.p. −78.5°C). This form of CO_2 is not widely used because of the difficulties associated with its storage, handling, and dispensing. More detailed discussions of CO_2-enrichment technology may be found in current greenhouse management textbooks (Hanan *et al.* 1978; Nelson 1981).

4. Nontraditional Sources of CO_2. Other inexpensive forms of CO_2 have been tried experimentally as alternatives to the conventional sources. In some locations in the United States (Gulf states, Texas), CO_2 is a natural resource in deep geologic wells, although the gas from these wells may not be clean enough for plant growth without some processing. Buxton *et al.* (1979) noted that ventilation air from deep coal mines may contain up to 1300 μl/liter CO_2, and their analysis of air from an abandoned mine shaft in Kentucky indicated little contamination with ethylene, SO_2, NO_2, or CO. Numerous industries (e.g., steel, electrical power, cement, and petroleum product producers) exhaust CO_2 as a by-product of their manufacturing processes. Ironically, industrially generated CO_2 is labelled an environmental pollutant (Baes *et al.* 1977; Keeling *et al.* 1976; Lemon 1983; Revelle 1982; Strain and Armentano 1980). It has been frequently suggested that industrial CO_2 represents a massive source of "free" CO_2 that might be used for crop production. The concept is logical since many projects are already underway to scavenge industrial waste heat (White 1979), and large industrial areas are ready markets for fresh greenhouse produce. However, the presence of other gases that are toxic to plants may limit direct use of industrial CO_2 in greenhouses until adequate scrubbing techniques are developed.

There are still many untapped natural sources of CO_2. The oceans and lakes are one reservoir, as are carbonate sediments (e.g., limestone), which are available worldwide (Lemon 1983). It may be possible to develop ways of releasing this CO_2 at controlled rates into a greenhouse, perhaps by acidification, without the introduction of pollutants from the process.

B. Monitoring CO_2

Maximum benefits are obtained from CO_2 enrichment when the CO_2 is applied efficiently. The CO_2 must be free of contaminants and in a convenient form, and the rate of release and dispersal of the CO_2 into the crop canopy must be controlled. For optimal crop response, CO_2-enrichment systems should be linked both to environmental conditions (light, temperature, water and nutrient application, and ventilation) and to the growth and development of the crop (vegetative, flowering, and fruit development). Presently, commercial and experimental systems tend to link control of supplementary CO_2 with environmental variables such as light or temperature in the greenhouse rather than with the status of the plants themselves. However, efficient use of CO_2 may be more effectively achieved by monitoring the CO_2 concentration in the greenhouse, which reflects the photosynthetic activity and CO_2 demand of the crop.

Several methods can be used to measure CO_2 (Bauer *et al.* 1979). Of these, only electrical detection and infrared gas analysis have been incorporated into greenhouse and growth-chamber monitoring systems (Bailey *et al.* 1970; Pallas 1979; Tibbitts and Krizek 1978). Electrochemical detection is based on the linear increase in the electrical conductivity of water with increasing dissolved CO_2 and HCO_3^- content. A stream of sample air is bubbled through water of known temperature and pH, and variation in the CO_2 content is recorded as a change in solution conductivity. The error in measuring air levels of CO_2 by this method can be as great as 5–10%, even when the temperature dependence of the solubility of CO_2 in water is taken into account (Kimball and Mitchell 1979; Pallas 1979).

A more direct method of analyzing CO_2 concentrations is to directly measure the gaseous CO_2 content of the air on a continuous or intermittent basis. Several systems are now sold commercially which allow this approach in greenhouses. At the heart of these systems is an infrared gas analyzer (IRGA), which measures specific absorption by CO_2 in the infrared region of the spectrum. Modern IRGAs are relatively insensitive to water or other contaminating gases and have been designed to operate in either differential or absolute modes

(Kleuter 1979). In most commercial CO_2-monitoring systems, the IRGA operates in the absolute mode and is sensitive over a rather broad range of CO_2 concentrations (e.g., 0–2000 μl/liter). In contrast, the differential mode (the reference gas may be any known CO_2 mixture) is preferred for photosynthesis and gas exchange research purposes because greater accuracy can be achieved over a smaller range of CO_2 concentrations (Bravdo and Canvin 1979; Kleuter 1979).

The CO_2 content of an air sample may also be monitored colorimetrically. A discrete sample or a continuous stream may be bubbled through an indicator solution that changes color in response to CO_2-generated pH changes (Sharp 1964). Coupled with a photoelectric detection system, the CO_2 concentration may be monitored. Like the conductivity system, this method is relatively slow and temperature sensitive because equilibrium must be attained between the CO_2/HCO_3^- in the gas stream and the solution.

A technique that holds great promise as a monitoring system is gas chromatography. At present, miniature solid-state gas chromatographs are being developed as an extension of the silicon dioxide engineering industry (Angell *et al.* 1983). As in a standard gas chromatograph, the sample gas is injected into a stream of inert carrier gas flowing through a capillary column packed with adsorbent. As the gases pass through the minaturized capillary, they are repeatedly adsorded and desorbed in the column packing so that the component gases are separated. The separated gases are quantified by a thermal conductivity wafer linked to a microchip processor, which transfers the signal to the output device. The small volumes of sample gas required, the convenient size of these new instruments, and, in time, their low cost may soon make gas chromatography a standard method of assaying greenhouse air not only for CO_2 levels but also for levels of other components such as CO and ethylene. These devices are presently targeted at the industrial gas-monitoring market.

C. By-Products of CO_2 Enrichment

Despite the superiority of pure CO_2 over combustion as a source of CO_2 for plant enrichment, the cost and proximity of a pure CO_2 supply often make fuel combustion the most economical means of commercially enriching a greenhouse. Therefore, it is essential to consider the various by-products of combustion and the impact they may have on greenhouse management. As discussed earlier, the heat and water vapor produced by CO_2-enrichment burners must be considered in the overall temperature and humidity control in the greenhouse. Toxic by-products associated with combustion (e.g., SO_2, NO_x, CO, C_2H_4)

can significantly affect plant productivity and seriously reduce the quality (appearance) and value of fresh produce marketed directly to the consumer. If ventilation is reduced to maximize the response to CO_2 enrichment, such gaseous pollutants are not dispersed to the external atmosphere and remain in direct contact with the crop (Hanan *et al.* 1978; Hand 1982; Nelson 1981).

The effect of SO_2 on leaf photosynthesis has been studied extensively (Hallgren and Gezelius 1982; Majernik and Mansfield 1972; Loats *et al.* 1981; Takemoto and Noble 1982). In plants not previously acclimated to SO_2, 500–750 nl/liter SO_2 reduced the photosynthetic rate, although pretreatment with 250 nl/liter SO_2 eliminated the photosynthetic inhibition (Takemoto and Noble 1982). High CO_2 (600 or 1000 μl/liter) limited the adverse effects of exposure to 250 nl/liter SO_2 in C_3 plants (Carlson and Bazzaz 1982), presumably by closing stomata (Majernik and Mansfield 1972; Tibbitts and Kobriger 1983). Sulfur dioxide production during combustion can be controlled most simply by using fuels low (less than 0.1%) in sulfur (Hanan *et al.* 1978; Hand 1982).

The problems caused by nitrous oxides (NO_x) are more complex. Nitrous oxides are a product of fuel combustion in an O_2 and N_2 atmosphere. The concentration of NO_2 varies diurnally because the conversion of the initial by-product NO to NO_2 is mediated by light and the presence of O_2 (or O_3). The peak daily concentrations of NO_2 can reach 120–150 nl/liter, and total NO_x concentrations may exceed 400 nl/liter in a greenhouse as the result of propane or kerosene combustion (Capron and Mansfield 1975). Concentrations of 500 nl/liter of either NO or NO_2 reduced tomato leaf photosynthesis by approximately 30%, and although the two pollutants did not interact, their effects in mixtures were nearly additive (Capron and Mansfield 1976). Although 100 nl/liter NO_2 actually stimulated tomato seedling dry matter accumulation, higher concentrations of NO or NO_2 resulted in a reduction of 25–60% (Capron and Mansfield 1977). In the same study it was observed that the accumulation of NO_x to 400 nl/liter eliminated the growth advantage provided by 1000 μl/liter CO_2. Since the production of NO_x is proportional to the amount of CO_2 generated, the dangers of actual growth reduction with enrichment cannot be overlooked.

Although most studies emphasize the acute effects of high dosages of toxic substances on plant growth, chronic low-level exposure to a pollutant also may be injurious (MacLean 1983). Further, low-level combinations of pollutants (e.g., $SO_2 + NO_x$) may cause larger effects than either pollutant alone at higher levels (Bull and Mansfield 1974). This is of practical importance as many greenhouse structures are located in urban or industrial areas where airborne pollutants are a constant component of the background. The combined effects of such pollutants may be exaggerated by the in-house production of combustion by-products.

Carbon monoxide release from CO_2 burners does not present a serious problem in current crop production because the dangers are well recognized (Hanan *et al.* 1978; Hand 1982). Humans are much more sensitive to CO than plants, and the safety and health standards imposed for the protection of those working in the greenhouse preclude damage to crops by CO. Most CO production is due to poorly maintained burners or a restricted air supply to the burner. There is also some question as to whether all of the CO around dense vegetation comes from combustion. Luttge and Fischer (1980) have suggested that small amounts of CO are released from leaves of both C_3 and C_4 plants under photorespiratory conditions (i.e., high light, increased temperatures, and low CO_2 levels), a condition only partially alleviated by high CO_2 (Bravdo and Canvin 1979).

A number of alkenes are powerful growth regulators. Ethylene, an endogenous regulator (Abeles 1973), is present in the flue gases of burners (Hanan 1973; Hand 1982), and propylene, a C_2H_4 analog, is present in most commercial propane (Hand 1982). Pyrolytic production of C_2H_4 can be partially controlled by altering the air–gas mixture, but C_2H_4 seems to be an inevitable contaminant of greenhouse air enriched by combustion of fossil fuels. Venting with unpolluted air facilitates dispersal of the C_2H_4, but as pointed out already, venting reduces the benefits of CO_2 enrichment itself. Interestingly, CO_2 levels frequently encountered in closed greenhouses (0–2000 μl/liter) have been found to modify endogenous C_2H_4 metabolism in photosynthetic tissue (Grodzinski 1984; Grodzinski *et al.* 1982a,b, 1983; Woodrow 1982). Therefore, despite precautions designed to minimize the addition of C_2H_4 into a greenhouse, CO_2 enrichment itself may modify plant development by altering C_2H_4 metabolism and action.

A growing number of reports indicate that treatment with high CO_2 levels, even pure CO_2, can induce earlier senescence. St. Omer and Horvath (1983) observed that the life cycle of two winter desert annuals was compressed by growth in 1400 or 2100 μl/liter CO_2. Tobacco plants grown in 1000 μl/liter CO_2 underwent accelerated leaf senescence, expressed as an accelerated loss of chlorophyll and more rapid degradation of cellular organelles (J. F. Thomas, personal communication). Peet *et al.* (1983) observed earlier leaf yellowing of cucumber plants grown in high CO_2. In contrast to these reports, CO_2 appears to retard the senescence of illuminated excised leaf disks, apparently through its interaction with C_2H_4 (Aharoni and Lieberman 1979;

Horton *et al.* 1982; Satler and Thimann 1983). These experiments underscore the view (Grodzinski 1984; Grodzinski *et al.* 1982a,b, 1983) that the action of plant growth regulators such as C_2H_4 can be controlled by the availability of CO_2, which is itself mediated by major anabolic and catabolic processes (e.g., photosynthesis, photorespiration, dark respiration). At the whole-plant level, the effects of CO_2 enrichment must be interpreted in terms of the enhanced growth and development of the whole plant when exposed to the CO_2. More rapid "senescence," particularly of older leaf tissue, is perhaps not surprising in view of the life cycle acceleration and the associated development of new sinks, and presumably the faster rates of turnover and mobilization of metabolites from the older tissue, that occur with CO_2 enrichment.

Gaseous pollutants arising from combustion must be evaluated in relation to the benefits of enrichment; as noted, CO_2 may actually overcome some of the effects of toxic pollutants. It is important to determine the tolerance levels of the crops being grown for such compounds as SO_2, NO_x, and C_2H_4 and to take precautions to prevent accumulations above damaging levels. The use of sulfur-free fuel and possibly scrubbers, proper burner maintenance and an adequate O_2 supply to prevent incomplete combustion, and limited ventilation can limit or prevent crop damage from gaseous pollutants.

V. INTEGRATING CO_2 ENRICHMENT WITH ENVIRONMENTAL VARIABLES AND CROP DEVELOPMENT

The benefits of CO_2 enrichment may be in the form of larger yields, higher-quality products, and/or a shorter time to harvest, all factors that can improve the marketability of a crop. However, the key to successful horticultural production is the ability of growers to know their crop and market and then to integrate their knowledge of plant growth and development with efficient growing practices.

Many CO_2-enrichment injection systems link the injection of CO_2 with environmental inputs rather than crop development. Integrated greenhouse systems have been evolving for many years (Jensen 1977; White 1979). At present, most commercial and experimental systems inject CO_2 in response to signals from environmental monitors without a measurement of the CO_2 concentration within the crop canopy (see Section IV. B).

A simple system may be operated by timers set to turn a CO_2 burner on or to begin injecting pure CO_2 at dawn, and to switch the CO_2

source off at dusk. Photosynthetic carbon assimilation and efficiency are enhanced by supplementary CO_2 even on cloudy days, and such a system ensures that depletion of CO_2 below ambient levels (Fig. 8.3) does not occur. A further level of complexity links the supply of CO_2 to the "brightness" of the day through a light-measuring device. The light level should be measured within the greenhouse at a point where shading by the canopy or greenhouse structure does not interfere. A design of this type takes into account that, within certain limits, the carbon assimilation rate for any specific crop is dependent on the incident light intensity as well as the CO_2 concentration (Blackman 1905). The light response of the specific crop must be estimated first, and then CO_2 is added in response to a predetermined minimum light level.

Although increases in yield from CO_2 enrichment can be expected in structures with open vents and even in open-topped field chambers (see Rogers *et al.* 1984), it is relatively difficult to control the CO_2 concentration in a greenhouse canopy when the structure is "leaky" or well ventilated (Hand 1982; Kimball and Mitchell 1981). Greenhouse systems that prevent addition of CO_2 when vents are open fail to take advantage of the bright, warm conditions when CO_2 enrichment would be most useful. Many computerized systems have been designed to reduce venting when light intensity is high, thus maximizing the value of CO_2 enrichment. Although estimating the crop need for CO_2 by monitoring light intensity is a logical approach, it is an indirect method of control. With the introduction in recent years of CO_2-monitoring systems, CO_2 can be added when it is actually needed. Positioning of the intake lines for air sampling must be done with care. According to our understanding of canopy boundary layers, CO_2 gradient profiles can exist in dense vegetation. Therefore, intake lines for CO_2 analysis should be positioned inside the crop where the CO_2 concentration profile would be steepest (Fig. 8.1). The advent of inexpensive, minaturized sensing devices will make sampling at many points in the canopy easy and practical.

Efficient dispersal of CO_2 within a greenhouse is very important since some management practices (e.g., overhead vs. floor heating, misting, thermal blankets, and shade cloths) may create conditions that restrict air mixing analogous to those found in the field. Although solar heating may generate some convective air movement, this does not ensure adequate disruption of the canopy and leaf boundary layers (Monteith 1973; Oke 1978; Rosenberg 1974). As shown in Fig. 8.5, altered wind speed over the leaf can affect photosynthetic rates by reducing the boundary layer (Gaastra 1963) without a change in the CO_2 concentration. In Japan, circulation within dense greenhouse

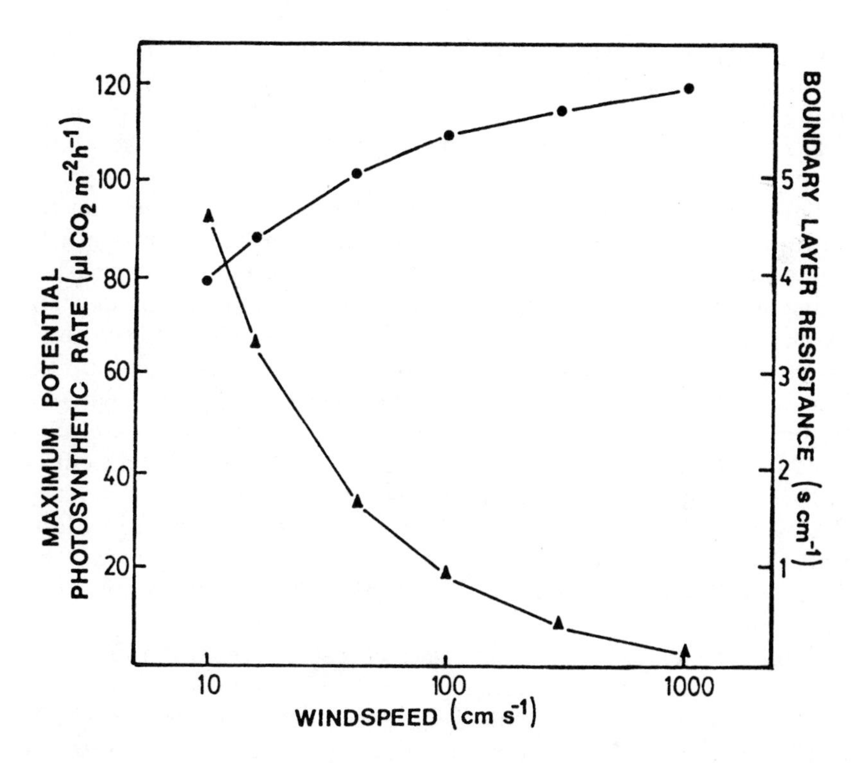

FIG. 8.5 The effect of wind speed on boundary layer resistance (—▲—) and maximum potential photosynthetic rate (i.e., maximum CO_2 diffusion rate into the leaf) of leaves at ambient CO_2 concentration (—●—).
Adapted from Gaastra (1966).

vegetation is increased by forcing air through flexible plastic tubes that are laid on the floor underneath the canopy (White 1982). Standard ventilation fans, which force air into the canopy, facilitate the dispersal of added CO_2, but their placement is critical (Walker and Duncan 1972; Hanan *et al.* 1978).

Good air circulation not only improves CO_2 distribution in the canopy, but also reduces variations in temperature and humidity. Recently, effective closed systems have been developed for arid climates (Ben-Jaacov *et al.* 1980; Kimball and Mitchell 1979; Wittwer 1981). The air within the greenhouse is circulated through water-saturated cooling pads or heat exchangers and enriched with CO_2, thus maintaining moderate temperatures, high humidities, and high CO_2. Misting with carbonated water has been investigated as alternative means of increasing CO_2 supply directly to the foliage (Carpenter 1966; Lin and Molnar 1980). Although the greenhouse CO_2 levels are very difficult to control with this method, it does provide both CO_2 enrichment and humidity control for the propagation of cuttings and may be of particular value in the future as more becomes known about the responsiveness of both photosynthetic and nonphotosynthetic tissue to higher than ambient levels of CO_2.

The degree of crop response to added CO_2 is related to the time and duration of its application. On a daily basis, application is clearly related to photosynthetic activity. However, as pointed out earlier, efficient application is also dependent on the stage of crop development. Calvert and Slack (1976) showed that added CO_2 had the greatest effect on tomatoes when supplied early in the day. Delaying the injection of CO_2 until after dawn reduced yield enhancement, whereas terminating enrichment a few hours before sunset did not diminish yield to the same extent. Eng *et al.* (1983, 1985) working with chrysanthemums growing under supplementary high intensity discharge (HID) lighting at commercial levels, obtained little benefit from nighttime CO_2 enrichment; however, growth of crassulacean acid metabolism (CAM) plants, which can fix CO_2 at night, may be improved by nighttime CO_2 enrichment (Allen 1979).

Table 8.1 clearly illustrates that crop responsiveness to high CO_2 levels varies with species. The harvest yield ratio expresses the relative advantage of growth with CO_2 enrichment over growth at ambient CO_2 for a particular crop and provides an indication as to whether the practice might be economically beneficial. Ultimately, the returns on an investment in CO_2 enrichment may depend on how precisely the supplemental CO_2 can be matched with the periods of greatest potential benefit during all stages of crop growth.

VI. FUTURE CONSIDERATIONS

In recent years there has been a growing awareness of the changes in our environment brought about by our own activities. We all enjoy in one way or another the benefits of increased industrialization, but modern technologies and intensive modernization have produced potentially detrimental side effects. One concern is "global CO_2 enrichment," which is attributed to increased industrialization, loss of natural vegetation (e.g., tropical forests), and alteration in the buffering capacity of the oceans (Bach 1980; Dahlman *et al.* 1983; Gifford 1979b; Keeling *et al.* 1976; Lemon 1983; Revelle 1982; Rogers *et al.* 1981, 1982; Rosenberg 1981, 1982; Rycroft 1982; Wittwer 1983). Data from several locations around the world, principally Mauna Loa in Hawaii, indicate that by the year 2050 AD the CO_2 level in the atmosphere may be over 600 μl/liter. Those who have taken notice of these data are also recognizing that plant scientists and particularly horticulturalists have been studying and documenting the role of atmospheric CO_2 in the growth and development of plants for many years.

Today, plant scientists can predict, with varying degrees of accuracy, the short-term effects of changes in CO_2 levels on the growth and yield of some crops in certain prescribed environments in which many variables (temperature, light, humidity) are controlled or can be predicted. However, as pointed out in this review, no one really knows what the long-term effect of elevated CO_2 levels will be on plant growth and global biomass production. Competition between C_3 and C_4 plants is unpredictable; the efficiency of water and nutrient use, seed production, and seed dispersal cannot be predicted on the basis of photosynthetic behavior. Plant and crop physiologists are still attempting to incorporate such processes as photoassimilate partitioning, water relations, and respiration (Elmore 1980; Gifford and Evans 1979; Kramer 1981) into accurate models of leaf photosynthesis (i.e., CO_2 fixation) and crop yield (Hesketh and Jones 1980). Thus, it seems somewhat premature to attempt to predict how complex ecosystems will respond to changes in ambient CO_2 levels. Although high CO_2 may facilitate carbon fixation and growth, the warming trends and changing precipitation patterns expected to accompany rising atmospheric CO_2 may override the direct effects of increased CO_2 on agricultural production.

Although the global effects of potentially rising CO_2 levels are unknown, it should not be forgotten that the practice of CO_2 enrichment in greenhouses is an old horticultural procedure that has provided valuable insight into the responsiveness of plants to CO_2. Substantial benefits can be expected from the controlled use of CO_2, although practical questions remain. The modification of existing computer-linked monitoring systems will undoubtedly provide better methods of measuring CO_2 fluxes in both the field and the greenhouse. Control of the gaseous environment of greenhouse structures is already cost effective for many high-value horticultural crops. Wider use of CO_2 enrichment in the field is likely, but in the short term some sort of shelter to prevent too-rapid wind dispersal of CO_2 is required. With an increased understanding of the responsiveness of crops to CO_2, more innovative applications of CO_2 in controlled environments will become possible. An interesting possibility is the breeding and selection of cultivars specifically adapted for optimal performance at high CO_2 concentrations. The effectiveness of high CO_2 concentrations as a selection tool has already been demonstrated in photorespiratory studies (Ogren 1984), and the concept could be extended to breeding for commercial cultivars. Cultivars in which high CO_2 levels would not result in stomatal closure might not show the CO_2 saturation effects currently seen in many crops. Characteristics such as fruit:leaf:root ratio, photoassimilate partitioning, and water use efficiency might all respond to selection at elevated CO_2.

There is still controversy about the long-term effect of high CO_2 exposures. The use of very high CO_2 levels in closed environments where humans must work continuously is not always practical (Tibbitts and Krizek 1978). However, we should not preclude the application of very high CO_2 concentrations (5000–10,000 μl/liter) for crop growth (Enoch *et al.* 1973). Several observations in the physiological and horticultural literature suggest that such high concentrations may prove useful in years to come. For example, in order to store many horticultural commodities, growers often use high CO_2 (greater than 5%) to reduce respiration and to inhibit premature ripening (Isenberg 1979). In addition, high CO_2 levels are known to block photosynthesis by closing stomata (Farquhar and Sharkey 1982; Goudriaan and van Laar 1978; Jones and Mansfield 1970; Raschke 1975). However, when the CO_2 concentration around leaf tissue is kept high, photosynthesis can still proceed at maximal rates because the gradient of CO_2 between the environment external to the leaf and the internal (leaf) CO_2 concentration is great enough to overcome stomatal resistance. At present, this knowledge is applied in laboratory experiments with leaf disks or leaves in closed systems (Delieu and Walker 1981; Grodzinski 1984) and has not really been tested on either a short-term or long-term basis with a crop. Furthermore, as Arteca *et al.* (1979) and Krizek *et al.* (1971) have demonstrated, the effect of very high CO_2 can persist for several months after enrichment ceases if the CO_2 is applied at an appropriate time and concentration.

The possibility that CO_2 is capable of major growth regulation beyond simple photosynthetic manipulation has not been exploited. As mentioned previously, soil CO_2 levels are very high, 1000–30,000 μl/liter (Lamborg 1983). Since root respiration seems to proceed under these conditions, CO_2 concentrations greater than those recommended for greenhouses (e.g., 1000–1500 μl/liter) may not be toxic to plants. The very fact that the CO_2 level at night in greenhouses and cloches rises considerably (Enoch *et al.* 1970, 1973) with no apparent phytotoxic effects underscores this point. In fact, we have no evidence to suppose that these high levels are not beneficial. We do have growing evidence that CO_2 alters ethylene metabolism and action (Abeles 1973; Lieberman 1979; Grodzinski 1984). Clearly, more research is required to determine the ways in which CO_2 may affect growth and development indirectly in addition to its *major* role as the substrate of photosynthesis and growth (Gifford and Evans 1979). The current reluctance to use high CO_2 levels because of their toxicity to humans, doubts about potential economic return, and reports of phytotoxic effects may be modified as greenhouse structures evolve, cultivation and harvesting become more automated, and understanding of the regulatory effects of CO_2 becomes more complete.

LITERATURE CITED

ABELES, F.B. 1973. Ethylene in plant biology. Academic Press, New York.

AHARONI, N.M. and M. LIEBERMAN. 1979. Ethylene as a regulator of senescence in tobacco leaf discs. Plant Physiol. 64:801-804.

AKITA, S. and D.N. MOSS. 1973. Photosynthetic responses to CO_2 and light by maize and wheat leaves adjusted for constant stomatal apertures. Crop Sci. 13:234-237.

ALLEN, L.H., JR. 1975. Shade-cloth microclimate of soybeans. Agron. J. 67:175-181.

ALLEN, L.H., JR. 1979. Potentials for carbon dioxide enrichment. p. 500-519. In: B.J. Barfield and J.F. Gerber (eds.), Modification of the aerial environment of plants. Monograph 2. Amer. Soc. Agr. Eng., St. Joseph, Michigan.

ALLEN, L.H., JR., S.E. JENSEN, and E.R. LEMON. 1971. Plant response to carbon dioxide enrichment under field conditions: a simulation. Science 173:256-258.

ANGELL, J.B., S.C. TERRY, and P.W. BARTH. 1983. Silicon micromechanical devices. Sci. Amer. 248:44-55.

AOKI, M. and K. YABUKI. 1977. Studies on the carbon dioxide enrichment for plant growth, VII. Changes in the dry matter production and photosynthetic rate of cucumber during carbon dioxide enrichment. Agr. Meteorol. 18:475-485.

AOKI, M., K. YABUKI, and H. KOYAMA. 1975. Micrometeorology and assessment of primary production of a tropical rain forest in West Malaysia. J. Agr. Meteorol. 31:115-124.

ARTECA, R.N., B.W. POOVAIAH, and O.E. SMITH. 1979. Changes in carbon fixation, tuberization, and growth induced by CO_2 applications to the root zone of potato plants. Science 205:1279-1280.

AUGUSTINE, J.J., M.A. STEVENS, and R.W. BREIDENBACH. 1979. Physiological, morphological, and anatomical studies of tomato genotypes varying in carboxylation efficiency. J. Amer. Soc. Hort. Sci. 104:338-341.

BACH, W. (ed.). 1980. The carbon dioxide problem. An interdisciplinary survey. Experientia 36:767-812.

BAES, C.F., JR., H.E. GOELLER, J.S. OLSON, and R.M. ROTTY. 1977. Carbon dioxide and climate: the uncontrolled experiment. Amer. Scientist 65:310-320.

BAILEY, W.A., H.H. KLUETER, D.T. KRIZEK, and N.W. STUART. 1970. CO_2 systems for growing plants. Trans. Amer. Soc. Agr. Eng. 13:263-268.

BAUER, C., G. GROS, and H. BARTELS (eds.). 1980. Biophysics and physiology of carbon dioxide. Springer-Verlag, New York.

BEN-JAACOV, J., A. HAGILADI, N. LEVAV, and N. ZAMIR. 1980. The hydrosolaric greenhouse—a new growing and propagating environment. Combined Proc. Intern. Plant Propagators Soc. 30:366-378.

BIGGIN, P. 1983. Tunnel cloches—development of a nursery technique for growing conifers. Forestry 56:45-60.

BIRAN, I., H.Z. ENOCH, N. ZIESLIN, and A.H. HALEVY. 1973. The influence of light intensity, temperature and carbon dioxide concentration on anthocyanin content and blueing of 'Baccara' roses. Scientia Hort. 1:157-164.

BISCOE, P.V., R.K. SCOTT, and J.L. MONTEITH. 1975. Barley and its environment. III. Carbon budget of the stand. J. Appl. Ecol. 12:269-293.

BJORKMAN, O. 1971. Comparative photosynthetic CO_2 exchange in higher plants. p. 18-32. In: M.D. Hatch, C.B. Osmond, and R.O. Slatyer (eds.), Photosynthesis and photorespiration. Wiley Interscience, New York.

BJORKMAN, O., M.M. LUDLOW, and P.A. MORROW. 1972. Photosynthetic performance of two rainforest species in their native habitat and analysis of their gas exchange. Carnegie Inst. Washington Yearb. 71:94-102.

BLACKMAN, F.F. 1905. Optima and limiting factors. Ann. Bot. 19:281-295.

BRAVDO, B. and D. CANVIN. 1979. Effect of carbon dioxide on photorespiration. Plant Physiol. 63:399-401.

BROWN, H.T. and F. ESCOMBE. 1902. The influence of varying amounts of carbon dioxide in air on the photosynthetic process of leaves and on the mode of growth of plants. Proc. Royal Soc. Lond. 70:397-413.

BROWN, K.W. and N.J. ROSENBERG. 1971. Turbulent transport and energy balance as affected by a windbreak in an irrigated sugar beet (*Beta vulgaris*) field. Agron. J. 63:351-355.

BROWN, K.W. and N.J. ROSENBERG. 1972. Shelter-effects on microclimate, growth and water use by irrigated sugar beets in the Great Plains. Agr. Meteorol. 9:241-263.

BRUN, W.A. and R.L. COOPER. 1967. Effects of light intensity and carbon dioxide concentration on photosynthetic rate of soybean. Crop Sci. 7:451-454.

BUGBEE, B. 1984. Effects of CO_2 concentration on plant productivity in controlled environments: effects of CO_2 depletion in growth chambers, growing rooms, and greenhouses. HortScience 19:505. (Abstr.)

BULL, J.M. and T.A. MANSFIELD. 1974. Photosynthesis in leaves exposed to SO_2 and NO_2. Nature 250:443-444.

BUXTON, J.W., J.N. WALKER, L. COLLINS, D. KNAVEL, and J.R. HARTMAN. 1979. Energy conservation by ventilating a greenhouse with deep-mine air. Scientia Hort. 11:19-30.

CABORN, J.M. 1973. Microclimates. Endeavour 34:30-33.

CALVERT, A. and G. SLACK. 1975. Effects of carbon dioxide enrichment on growth, development and yield of glasshouse tomatoes. I. Responses to controlled concentrations. J. Hort. Sci. 50:61-71.

CALVERT, A. and G. SLACK. 1976. Effect of carbon dioxide enrichment on growth, development and yield of glasshouse tomatoes. II. The duration of daily periods of enrichment. J. Hort. Sci. 51:401-409.

CANHAM, A.E. and W.J. McCAVISH. 1981. Some effects of CO_2, daylength and nutrition on the growth of young forest tree plants. I. In the seedling stage. Forestry 54:169-182.

CAPRON, T.M. and T.A. MANSFIELD. 1975. Generation of nitrogen oxide pollutants during CO_2 enrichment of glasshouse atmospheres. J. Hort. Sci. 50:233-238.

CAPRON, T.M. and T.A. MANSFIELD. 1976. Inhibition of net photosynthesis in tomato in air polluted with NO and NO_2. J. Expt. Bot. 27:1181-1186.

CAPRON, T.M. and T.A. MANSFIELD. 1977. Inhibition of growth in tomato by air polluted with nitrogen oxides. J. Expt. Bot. 28:112-116.

CARLSON, R.W. and F.A. BAZZAZ. 1982. Photosynthetic and growth response to fumigation with SO_2 at elevated CO_2 for C_3 and C_4 plants. Oecologia 54:50-54.

CARPENTER, W.J. 1966. Effect of intermittent misting with carbonated water on the growth of greenhouse chrysanthemums and lettuce. Abstr. 349. In: Proc. 17th Intern. Hort. Cong., Univ. of Maryland, College Park.

CHAPMAN, H.W., L.S. GLEASON, and W.E. LOOMIS. 1954. The carbon dioxide content of field air. Plant Physiol. 29:500-503.

COLLINS, W.B. 1976. Effect of carbon dioxide enrichment on growth of the potato plant. HortScience 11:467-469.

COOPER, R.L. and W.A. BRUN. 1967. Response of soybeans to a carbon dioxide-enriched atmosphere. Crop Sci. 7:455-457.

CUMMINGS, M.B. and C.H. JONES. 1918. The aerial fertilization of plants with carbon dioxide. Vermont Sta. Bull. 211.

DAHLMAN, R.C., B.R. STRAIN, and H.H. ROGERS. 1983. Research on the response of vegetation to elevated atmospheric carbon dioxide. p. 1-20. In: Rpt. Annu. Meeting Air Pollution Control Assoc., Atlanta, Georgia.

DELIEU, T. and D.A. WALKER. 1981. Polarographic measurement of photosynthetic O_2-evolution by leaf discs. New Phytol. 89:168-178.

DESJARDINS, R.L., L.H. ALLEN, JR., and E.R. LEMON. 1978. Variations of carbon dioxide, air temperature, and horizontal wind within and above a maize crop. Boundary Layer Meteorol. 14:369-380.

DULLFORCE, W.M. 1966. Analysis of the growth of lettuce in controlled environments with additional carbon dioxide. Abstr. 345. In: Proc. 17th Intern. Hort. Cong., Univ. of Maryland, College Park.

EHLERINGER, J. and O. BJORKMAN. 1977. Quantum yields for CO_2 uptake in C_3 and C_4 plants. Dependence on temperature, CO_2, and O_2 concentration. Plant Physiol. 59:86-90.

ELMORE, C.D. 1980. The paradox of no correlation between leaf photosynthetic rates and crop yields. p. 155-168. In: J.D. Hesketh and J.W. Jones (eds.), Predicting photosynthesis for ecosystem models, Vol. II. CRC Press, Boca Raton, Florida.

ENG, R.Y.N. 1982. Diurnal carbon dioxide enrichment of chrysanthemums under supplementary high pressure sodium lighting. M.S. Thesis, Univ. of Guelph, Guelph, Ontario.

ENG, R.Y.N., M.J. TSUJITA, B. GRODZINSKI, and R.G. DUTTON. 1983. Production of chrysanthemum cuttings under supplementary lighting and CO_2 enrichment. HortScience 18:878-879.

ENG, R.Y.N., M.J. TSUJITA, and B. GRODZINSKI. 1985. The effects of supplementary HPS lighting and carbon dioxide enrichment on the vegetative growth, nutritional status and flowering characteristics of *Chrysanthemum morifolium* Ramat. J. Hort. Sci. (in press).

ENOCH, H., I. RYLSKI, and Y. SAMISH. 1970. CO_2 enrichment to cucumbers, lettuce and sweet pepper plants grown in low plastic tunnels in a subtropical climate. Israel J. Agr. Res. 20:63-69.

ENOCH, H.Z., N. ZIESLIN, Y. BIRAN, A.H. HALEVY, M. SCHWARZ, B. KESLER, and D. SHIMSI. 1973. Principles of CO_2 nutrition research. Acta Hort. 32:97-118.

ENOCH, H., I. RYLSKI, and M. SPIGELMAN. 1976. CO_2 enrichment of strawberry and cucumber plants grown in unheated greenhouses in Israel. Scientia Hort. 5:33-41.

ENOCH, H.Z., I. CARMI, J.S. ROUNICK, and M. MAGARITZ. 1984. Use of carbon isotopes to estimate incorporation of added CO_2 by greenhouse-grown tomato plants. Plant Physiol. 76:1083-1085.

FARQUHAR, G.D. and T.D. SHARKEY. 1982. Stomatal conductance and photosynthesis. Annu. Rev. Plant Physiol. 33:317-345.

FINN, G.A. and W.A. BRUN. 1982. Effect of atmospheric CO_2 enrichment on growth, non-structural carbohydrate content, and root nodule activity in soybean. Plant Physiol. 69:327-331.

FISCHER, R.A. and M.I. AGUILAR. 1976. Yield potential in a dwarf spring wheat and the effect of carbon dioxide fertilization. Agron. J. 68:749-752.

FRENCH, C.J. and W.C. LIU. 1984. Seasonal variations in the effects of CO_2 mist and supplementary lighting from high pressure sodium lamps on rooting of English holly cuttings. HortScience 19:519-521.

FRYDRYCH, J. 1976. Photosynthetic characteristics of cucumber seedlings grown under two levels of carbon dioxide. Photosynthetica 10:335-338.

GAASTRA, P. 1963. Climatic control of photosynthesis and respiration. p. 113–140. In: L.T. Evans (ed.), Environmental control of plant growth. Academic Press, New York.

GATHERCOLE, R.W.E., K.J. MANSFIELD, and H.E. STREET. 1976. Carbon dioxide as an essential requirement for cultured sycamore cells. Physiol. Plant. 37:213–217.

GEISLER, G. 1963. Morphogenetic influence of (CO_2 + HCO_3^-) on roots. Plant Physiol. 38:77–80.

GIFFORD, R.M. 1979a. Growth and yield of CO_2-enriched wheat under water-limited conditions. Austral. J. Plant Physiol. 6:367–378.

GIFFORD, R.M. 1979b. Carbon dioxide and plant growth under water and light stress: implications for balancing the global carbon budget. Search 10:316–318.

GIFFORD, R.M. and L.T. EVANS. 1979. Photosynthesis, carbon partitioning, and yield. Annu. Rev. Plant Physiol. 32:485–509.

GOLDSBERRY, K.L. 1966. Effects of CO_2 on carnation stock plants. Colorado Flower Growers Assoc. Bull. 192:1–2.

GOUDRIAAN, J. and H.H. VAN LAAR. 1978. Relations between leaf resistance, CO_2-concentration and CO_2-assimilation in maize, beans, lalang grass and sunflower. Photosynthetica 12:241–249.

GRODZINSKI, B. 1984. Enhancement of ethylene release from leaf tissue during glycolate decarboxylation; a possible role for photorespiration. Plant Physiol. 74:871–876.

GRODZINSKI, B., I. BOESEL, and R.F. HORTON. 1982a. Ethylene release from leaves of *Xanthium strumarium* and *Zea mays*L. J. Expt. Bot. 33:344–354.

GRODZINSKI, B., I. BOESEL, and R.F. HORTON. 1982b. Effect of light intensity on the release of ethylene from leaves. J. Expt. Bot. 33:1187–1193.

GRODZINSKI, B., I. BOESEL, and R.F. HORTON. 1983. Light stimulation of ethylene release from leaves of *Gomphrena globosa*. Plant Physiol. 71:588–593.

GUTTORMSEN, G. 1972a. The effect of plastic tunnels on air and soil temperatures in relation to observations of cloud cover. J. Agr. Eng. Res. 17:99–106.

GUTTORMSEN, G. 1972b. The effect of perforation on temperature conditions in plastic tunnels. J. Agr. Eng. Res. 17:172–177.

HALLGREN, J.E. and K. GEZELIUS. 1982. Effects of SO_2 on photosynthesis and ribulose bisphosphate carboxylase in pine tree seedlings. Physiol. Plant. 54:153–161.

HANAN, J.J. 1973. Ethylene pollution from combustion in greenhouses. HortScience 8:23–24.

HANAN, J.J., W.D. HOLLEY, and K.L. GOLDSBERRY. 1978. Greenhouse management. Springer-Verlag, New York.

HAND, D.W. 1982. CO_2 enrichment, the benefits and problems. Scientia Hort. 33:14–43.

HAND, D.W. and K.E. COCKSHULL. 1975. Roses I: The effects of CO_2 enrichment on winter bloom production. J. Hort. Sci. 50:183–192.

HAND, D.W. and J.D. POSTLETHWAITE. 1971. The response of CO_2 enrichment of capillary-watered single-truss tomatoes at different plant densities and seasons. J. Hort. Sci. 46:461–470.

HAND, D.W. and R.W. SOFFE. 1971. Light-modulated temperature control and the response of greenhouse tomatoes to different CO_2 regimes. J. Hort. Sci. 46:381–396.

HARDMAN, L.L. and W.A. BRUN. 1971. Effect of atmospheric carbon dioxide enrichment at different developmental stages on growth and yield of components of soybeans. Crop Sci. 11:886–888.

HARPER, L.A., D.N. BAKER, J.E. BOX, JR., and J.D. HESKETH. 1973. Carbon dioxide and the photosynthesis of field crops: a metered carbon dioxide release in cotton under field conditions. Agron. J. 65:7–11.

HESKETH, J.D. and H. HELLMERS. 1973. Floral initiation in four plant species growing in CO_2 enriched air. Environ. Control in Biol. 11:51–53.

HESKETH, J.D. and J.W. JONES. 1980. Predicting photosynthesis for ecosystem models. CRC Press, Boca Raton, Florida.

HICKLENTON, P.R. and P.A. JOLLIFFE. 1978. Effects of greenhouse CO_2 enrichment on the yield and photosynthetic physiology of tomato plants. Can. J. Plant Sci. 58:801–817.

HICKLENTON, P.R. and P.A. JOLLIFFE. 1980a. Alterations in the physiology of CO_2 exchange in tomato plants grown in CO_2-enriched atmospheres. Can. J. Bot. 58:2181–2189.

HICKLENTON, P.R. and P.A. JOLLIFFE. 1980b. Carbon dioxide and flowering in *Pharbitis nil* Choisy. Plant Physiol. 66:13–17.

HO, L.C. 1977. Effects of CO_2 enrichment on the rates of photosynthesis and translocation of tomato leaves. Ann. Appl. Biol. 87:191–200.

HOFSTRA, G. and J.D. HESKETH. 1975. The effects of temperature and CO_2 enrichment on photosynthesis in soybean. p. 71–80. In: R. Marcelle (ed.), Environmental and biological control of photosynthesis. Dr. W. Junk Publishers, The Hague.

HOLLEY, W.D. and R.A. ALTSTADT. 1966. Adding CO_2 to carnation stock plants improves performance of the cuttings. Colorado Flower Growers Assoc. Bull. 192:2-3.

HOPEN, H.J. and S.K. RIES. 1962a. The mutually compensating effect of carbon dioxide concentrations and light intensities on the growth of *Cucumis sativus* L. Proc. Amer. Soc. Hort. Sci. 81:358-364.

HOPEN, H.J. and S.K. RIES. 1962b. Atmospheric carbon dioxide levels over mineral and muck soils. Proc. Amer. Soc. Hort. Sci. 81:365-368.

HORTON, R.F., L. WOODROW, I. BOESEL, and B. GRODZINSKI. 1982. Light, carbon dioxide and ethylene metabolism in photosynthetic tissue. p. 93-101. In: M.B. Jackson, B. Grout, and I.A. MacKenzie (eds.), Growth regulators in plant senescence. British Plant Growth Regulator Group, Monograph 8.

HUBER, S.C., H.H. ROGERS, and F.L. MOWRY. 1984. Effects of water stress on photosynthesis and carbon partitioning in soybean (*Glycine max* [L.] Mesr.) plants grown in the field at different CO_2 levels. Plant Physiol. 76:244–249.

HUGHES, A.P. and K.E. COCKSHULL. 1971. The variation in response to light intensity and carbon dioxide concentration shown by two cultivars of *Chrysanthemum morifolium* grown in controlled environments at two times of year. Ann. Bot. 35:933–945.

HUGHES, A.P. and K.E. COCKSHULL. 1972. Further effects of light intensity, carbon dioxide concentration and day temperature on the growth of *Chrysanthemum morifolium* cv. 'Bright Golden Anne' in controlled environments. Ann. Bot. 36:533–550.

HURD, R.G. 1968. Effects of CO_2 enrichment on the growth of young tomato plants in low light. Ann. Bot. 32:531–542.

IMAI, K. and D.F. COLEMAN. 1983. Elevated atmospheric partial pressure of carbon dioxide and dry matter production of konjak (*Amorphophallus konjac* K. Koch). Photosyn. Res. 4:331–336.

INOUE, E. 1965. On the CO_2 concentration profiles within crop canopies. J. Agr. Meteorol. 21:137-140.

IRELAND, C.R. and W.W. SCHWABE. 1982. Studies on the role of photosynthesis in the photoperiodic induction of flowering in the short-day plants *Kalenchoe blossfeldiana* Poellniz and *Xanthium pennsylvanicum* Wallr. I. The requirement for CO_2 during photoperiodic induction. J. Expt. Bot. 33:738-747.

ISENBERG, F.M.R. 1979. Controlled atmosphere storage of vegetables. Hort. Rev. 1:337-394.

JENSEN, M.H. 1977. Energy alternatives and conservation for greenhouses. HortScience 12:14-24.

JOHNSTON, R.E. 1972. A trial of glasshouse winter lettuce in Scotland. Hort. Res. 12:149-152.

JONES, R.J. and T.A. MANSFIELD. 1970. Increases in the diffusion resistances of leaves in a carbon dioxide-enriched atmosphere. J. Expt. Bot. 21:951-958.

KEELING, C.D., R.B. BACASTOW, A.E. BAINBRIDGE, C.A. EKDAHL, JR., P.R. GUENTHER, L.S. WATERMAN, and J.F.S. CHIN. 1976. Atmospheric carbon dioxide variations at Mauna Loa observatory, Hawaii. Tellus 28:538-551.

KIMBALL, B.A. 1982. Carbon dioxide and agricultural yield: an assemblage and analysis of 430 prior observations. U.S. Water Conservation Lab. Rpt. 11., USDA.

KIMBALL, B.A. 1983a. Carbon dioxide and agricultural yields: an assemblage of 770 prior observations. U.S. Water Conservation Lab. Rpt. 14., USDA.

KIMBALL, B.A. 1983b. Carbon dioxide and agricultural yield: an assemblage and analysis of 430 prior observations. Agron. J. 75:770-788.

KIMBALL, B.A. and S.B. IDSO. 1983. Increasing atmospheric CO_2: effects on crop yield, water use and climate. Agr. Water Mgt. 7:55-72.

KIMBALL, B.A. and S.T. MITCHELL. 1979. Tomato yields from CO_2-enrichment in unventilated and conventionally ventilated greenhouses. J. Amer. Soc. Hort. Sci. 104:515-520.

KIMBALL, B.A. and S.T. MITCHELL. 1981. Effects of CO_2 enrichment, ventilation, and nutrient concentration on the flavor and vitamin content of tomato fruit. HortScience 16:665-666.

KLUETER, H.H. 1979. Carbon dioxide: critique II. p. 235-240. In: T.W. Tibbitts and T.T. Kozlowski (eds.), Controlled environment guidelines for plant growth. Academic Press, New York.

KNECHT, G.N. 1975. Response of radish to high CO_2. HortScience 10:274-275.

KNECHT, G.N. and J.W. O'LEARY. 1974. Increased tomato fruit development by CO_2 enrichment. J. Amer. Soc. Hort. Sci. 99:214-216.

KNECHT, G.N. and J.W. O'LEARY. 1983. The influence of carbon dioxide on the growth, pigment, protein, chlorophyll, and mineral status of lettuce. J. Plant Nutr. 6:301-312.

KOTHS, J.S. and R. ADZIMA. 1966. Carnation quality as influenced by carbon dioxide enriched atmospheres. Abstr. 344. In: Proc. 17th Intern. Hort. Cong., Univ. of Maryland, College Park.

KRAMER, P.J. 1981. Carbon dioxide concentration, photosynthesis, and dry matter production. BioScience 31:29-33.

KRAMER, P.J., H. HELLMERS, and H.J. DOWNS. 1970. SEPEL: new phytotrons for environmental research. BioScience 20:1201-1208.

KRETCHMAN, D.W. and F.S. HOWLETT. 1966. The interrelationships of temperature and carbon dioxide enrichment on several quality attributes of greenhouse tomatoes. Abstr. 348. In: Proc. 17th Intern. Hort. Cong., Univ. of Maryland, College Park.

KRETCHMAN, D.W. and F.S. HOWLETT. 1970. CO_2 enrichment for vegetable production. Trans. Amer. Soc. Agr. Eng. 13:252-256.

KRIZEK, D.T. 1979. Carbon dioxide enrichment. p. 283-291. In: Proc. Beltwide Cotton Production Research Conferences, Phoenix, Arizona.

KRIZEK, D.T. 1984. Photosynthesis, dry matter production and growth in CO_2 enriched atmospheres. In: J. McD. Stewart and J.R. Mauney (eds.), Cotton physiology—a treatise. (In press). USDA.

KRIZEK, D.T., R.H. ZIMMERMAN, H.H. KLUETER, and W.A. BAILEY. 1971. Growth of crabapple seedlings in controlled environments: effect of CO_2 level and time and duration of CO_2 treatment. J. Amer. Soc. Hort. Sci. 96:285-288.

KRIZEK, D.T., W.A. BAILEY, H. KLUETER, and R.C. LIU. 1974. Maximizing growth of vegetable seedlings in controlled environments at elevated temperature, light, and CO_2. Acta Hort. 39:89-102.

LAMBORG, M.R., R.W.F. HARDY, and E.A. PAUL. 1983. Microbial effects. p. 131-176. In: E.R. Lemon (ed.), CO_2 and plants: the response of plants to rising levels of atmospheric carbon dioxide. AAAS Selected Symp. 84. AAAS, Washington, DC.

LEE, R.B. and C.P. WHITTINGHAM. 1974. The influence of partial pressure of carbon dioxide upon the carbon metabolism in the tomato leaf. J. Expt. Bot. 25:277-287.

LEMON, E.R. (ed.). 1983. CO_2 and plants: the response of plants to rising levels of atmospheric carbon dioxide. AAAS Selected Symp. 84. AAAS, Washington, DC.

LEMON, E.R. and J.L. WRIGHT. 1969. Photosynthesis under field conditions. XA. Assessing sources and sinks of carbon dioxide in a corn (*Zea mays* L.) crop using a momentum balance approach. Agron. J. 61:405-411.

LEMON, E.R., J.L. WRIGHT, and G.M. DRAKE. 1969. Photosynthesis under field conditions. XB. Origins of short-time CO_2 fluctuations in a cornfield. Agron. J. 61:411-413.

LIEBERMAN, M. 1979. Biosynthesis and action of ethylene. Annu. Rev. Plant Physiol. 30:533-591.

LIN, W.C. and J.M. MOLNAR. 1980. Carbonated mist and high intensity supplementary lighting for propagation of selected woody ornamentals. Combined Proc. Intern. Plant Prop. Soc. 30:104-112.

LIN, W.C. and J.M. MOLNAR. 1982. Supplementary lighting and CO_2 enrichment for accelerated growth of selected woody ornamental seedlings and rooted cuttings. Can. J. Plant Sci. 62:703-707.

LINDSTROM, R.S. 1965. Carbon dioxide and its effect on the growth of roses. Proc. Amer. Soc. Hort. Sci. 87:521-524.

LINDSTROM, R.S. 1968. Supplemental carbon dioxide and growth of *Chrysanthemum Morifolium* Ram. Proc. Amer. Soc. Hort. Sci. 92:627-632.

LIU, H.T. 1983. The effects of CO_2 enrichment on growth and cut flower production of *Gerbera jamesonii* H. Bolus ex Hook. F. M.S. Thesis, Univ. of Guelph, Guelph, Ontario.

LOATS, K.V., R. NOBLE, and B. TAKEMOTO. 1981. Photosynthesis under low level SO_2 and CO_2 enhancement conditions in three duckweed species. Bot. Gaz. 142:305-310.

LUTTGE, U. and K. FISCHER. 1980. Light dependent CO-evolution by C_3 and C_4 plants. Planta 149:59-63.

MacDONALD, K.B., E. de JONG, and H.J.V. SCHAPPERT. 1973. Soil physics I.

Soil respiration. Tech. Rpt. 14. Matador Project, Canadian Comm. for the Intern. Biological Programme, Univ. of Saskatchewan, Saskatoon, Canada.

MacLEAN, D.C. 1983. Air pollution and horticulture: an overview. HortScience 18:674-675.

MADORE, M. and B. GRODZINSKI. 1984a. Effect of oxygen concentration on ^{14}C-photoassimilate transport from leaves of *Salvia splendens* L. Plant Physiol. 76:782-786.

MADORE, M. and B. GRODZINSKI. 1984b. Transport of ^{14}C-labelled photoassimilates in a dwarf cucumber cultivar under CO_2 enrichment. HortScience 19:586 (Abstr.).

MADSEN, E. 1973. Effect of CO_2-concentration on the morphological, histological and cytological changes in tomato plants. Acta Agr. Scand. 23:241-245.

MAJERNIK, O. and T.A. MANSFIELD. 1972. Stomatal responses to atmospheric CO_2 concentrations during exposure of plants to SO_2 pollution. Environ. Pollut. 3:1- 7.

MARC, J. and R.M. GIFFORD. 1983. Floral initiation in wheat, sunflower, and sorghum under carbon dioxide enrichment. Can. J. Bot. 62:9-14.

MASTERSON, C.L. and M.T. SHERWOOD. 1978. Some effects of increased atmospheric carbon dioxide on white clover (*Trifolium repens*) and pea (*Pisum sativum*). Plants & Soil 49:421-426.

MATTSON, R.H. and R.E. WIDMER. 1971a. Effects of carbon dioxide during growth on vase life of greenhouse roses (*Rosa hybrida*). J. Amer. Soc. Hort. Sci. 96:284.

MATTSON, R.H. and R.E. WIDMER. 1971b. Year-round effects of carbon dioxide supplemented atmosphere on greenhouse rose (*Rosa hybrida*) production. J. Amer. Soc. Hort. Sci. 96:487-488.

MAUNEY, J.R., K.E. FRY, and G. GUINN. 1978. Relationship of photosynthetic rate to growth and fruiting of cotton, soybean, sorghum, and sunflower. Crop Sci. 18:259-263.

MILLER, D.R., N.J. ROSENBERG, and W.T. BAGLEY. 1973. Soybean water use in the shelter of a slat-fence windbreak. Agr. Meteor. 11:405-411.

MONTEITH, J.L. 1973. Principles of environmental physics. Edward Arnold Press, London.

MOORBY, J. and F.L. MILTHORPE. 1975. Potato. p. 225-258. In: L.T. Evans (ed.), Crop physiology. Cambridge Univ. Press, London.

MORTENSEN, L.M. and R. MOE. 1983a. Growth responses of some greenhouse plants to environment. V. Effect of CO_2, O_2 and light on net photosynthetic rate in *Chrysanthemum morifolium* Ramat. Scientia Hort. 19:133-140.

MORTENSEN, L.M. and R. MOE. 1983b. Growth responses of some greenhouse plants to environment. VI. Effect of CO_2 and artificial light on growth of *Chrysanthemum morifolium* Ramat. Scientia Hort. 19:141-147.

NELSON, P.V. 1981. Greenhouse operation and management. Reston Publ., Reston, West Virginia.

NEVINS, D.J. and R.S. LOOMIS. 1970. Nitrogen nutrition and photosynthesis in sugar beet (*Beta vulgaris* L.). Crop Sci. 10:21-25.

NEWTON, P. 1965. Growth of *Cucumis sativus*, variety Butcher's Disease Resistor, with two concentrations of carbon dioxide. Ann. Appl. Biol. 56:55-64.

NILWIK, H.J.M., W. GOSIEWSKI, and J.F. BIERHUIZEN. 1982. The influence of irradiance and external CO_2-concentration on photosynthesis of different tomato genotypes. Scientia Hort. 16:117-123.

OGREN, W.L. 1984. Photorespiration: pathways, regulation, and modification. Annu. Rev. Plant Physiol. 35:415–442.

OHTAKI, E. 1980. Turbulent transport of carbon dioxide over a paddy field. Boundary Layer Meteorol. 19:315–336.

OKE, T.R. 1978. Boundary layer climates. John Wiley & Sons, New York.

O'LEARY, J.W. and G.N. KNECHT. 1981. Elevated CO_2 concentration increases stomate numbers in *Phaseolus vulgaris* leaves. Bot. Gaz. 142:438–441.

PALLAS, J.E., JR. 1979. Carbon dioxide. p. 207–228. In: T.W. Tibbitts and T.T. Kozlowski (eds.), Controlled environment guidelines for plant research. Academic Press, New York.

PATTERSON, D.T. and E.P. FLINT. 1980. Potential effects of global atmospheric CO_2 enrichment on the growth and competitiveness of C_3 and C_4 weed and crop plants. Weed Sci. 28:71–75.

PATTERSON, D.T. and E.P. FLINT. 1982. Interacting effects of CO_2 and nutrient concentration. Weed Sci. 30:389–394.

PATTERSON, D.T., E.P. FLINT, and J.A. BEYERS. 1984. Effects of CO_2 enrichment on competition between a C_4 weed and a C_3 crop. Weed Sci. 32:101–105.

PEARCY, R.W. and J. EHLERINGER. 1984. Comparative ecophysiology of C_3 and C_4 plants. Plant, Cell, & Environ. 7:1–13.

PEET, M., S. HUBER, and D. PATTERSON. 1983. CO_2 enrichment effects on cucumber growth, fruiting, photosynthesis and enzyme activity. Plant Physiol. Suppl. 72:38.

PHILLIPS, D.A., K.D. NEWELL, S.A. HASSEL, and C.E. FELLING. 1976. The effect of CO_2 enrichment on root nodule development and symbiotic N_2 reduction in *Pisum sativum* L. Amer. J. Bot. 63:356–363.

PORTER, M.A. and B. GRODZINSKI. 1984. Acclimation to high CO_2 in bean. Carbonic anhydrase and RuBP carboxylase. Plant Physiol. 74:413–417.

PROCTOR, J.T.A. 1980. Some aspects of the Canadian culture of ginseng (*Panax quinquefolius* L.), particularly the growth environment. p. 39–47. In: Proc. 3rd Intern. Ginseng Symp. Korea Ginseng Res. Inst., Seoul, Korea.

PROULX, E.A. 1984. A case for the cloche. Org. Gardening 31:78–87.

PUROHIT, A.N. and E.B. TREGUNNA. 1974. Effects of carbon dioxide enrichment on *Pharbitis, Xanthium,* and *Silene* in short days. Can. J. Bot. 52:1283–1291.

RASCHKE, K. 1975. Stomatal action. Annu. Rev. Plant Physiol. 26:309–340.

REDMANN, R.E. 1974a. Photosynthesis, plant respiration and soil respiration measured with controlled environment chambers in the field: I. Methods and results. Tech. Rpt. 18. Matador Project, Canadian Comm. for the Internat. Biological Programme, Univ. of Saskatchewan, Saskatoon, Canada.

REDMANN, R.E. 1974b. Photosynthesis, plant respiration and soil respiration measured with a controlled environment chamber in the field: II. Plant CO_2 exchange in relation to environment and productivity. Tech. Rpt. 49. Matador Project, Canadian Comm. for the Intern. Biological Programme, Univ. of Saskatchewan, Saskatoon, Canada.

REDMANN, R.E. 1974c. Photosynthesis, plant respiration and soil respiration measured with controlled environment chambers in the field: III. Soil respiration. Tech. Rpt. 60. Matador Project, Canadian Comm. for the Intern. Biological Programme, Univ. of Saskatchewan, Saskatoon, Canada.

REVELLE, R. 1982. Carbon dioxide and world climate. Scientific Amer. 247:35–43.

RIPLEY, E.A. 1974. Micrometeorology: VII. Canopy CO_2 concentrations and fluxes. Tech. Rpt. 38. Matador Project, Canadian Comm. for the Intern. Biological Programme, Univ. of Saskatchewan, Saskatoon, Canada.

RIPLEY, E.A., and B. SAUGIER. 1975. Energy and mass exchange of a native grassland in Saskatchewan. In: D.A. deVries and N.H. Afgen (eds.), Heat and mass transfer in the biosphere. Part I. Transfer processes in plant environment. John Wiley & Sons, Toronto.

ROGERS, H.H., R.D. BECK, G.E. BINGHAM, J.D. CURE, J.D. DAVIS, W.W. HECK, J.O. RAWLINGS, A.J. RIORDAN, N. SIONIT, J.M. SMITH, and J.F. THOMAS. 1981. Response of vegetation to elevated carbon dioxide: field studies of plant responses to elevated carbon dioxide levels. Rpt. 005. U.S. Dept. of Energy, Carbon Dioxide Res. Div., and USDA–ARS, Washington, DC.

ROGERS, H.H., G.E. BINGHAM, J.D. CURE, B.G. DRAKE, W.W. HECK, S.C. HUBER, D.W. ISRAEL, F.L. MOURY, R.P. PATTERSON, J.O. RAWLINGS, J.F. REYNOLDS, N. SIONIT, J.F. THOMAS, J.M. ELLIS, R.B. KENNEDY, J.H. LEITH, S.A. PRIOR, W.A. PURSLEY, J.W. SMITH, and K.A. SURANO. 1982. Response of vegetation to elevated carbon dioxide levels. U.S. Dept. of Energy, Carbon Dioxide Res. Div., and USDA–ARS, Washington, DC.

ROGERS, H.H., N. SIONIT, J.D. CURE, J.M. SMITH, and G.E. BINGHAM. 1984. Influence of elevated CO_2 on water relations of soybeans. Plant Physiol. 74:233–238.

ROSENBERG, N.J. 1974. Microclimate: the biological environment. Wiley Intersciences, New York.

ROSENBERG, N.J. 1981. The increasing CO_2 concentration in the atmosphere and its implication on agricultural productivity. I. Effects on photosynthesis, transpiration and water use efficiency. Climatic Change 3:265–279.

ROSENBERG, N.J. 1982. The increasing CO_2 concentration in the atmosphere and its implication on agricultural productivity. II. Effects through CO_2-induced climatic change. Climatic Change 4:239–254.

ROWLEY, J.A. and A.O. TAYLOR. 1972. Plants under climatic stress. IV. Effects of CO_2 and O_2 on photosynthesis under high-light, low-temperature stress. New Phytol. 71:477–481.

RYCROFT, M.J. 1982. Analysing atmospheric carbon dioxide levels. Nature 295:190–191.

SATLER, S.O. and K.V. THIMANN. 1983. Metabolism of oat leaves during senescence. VII. The interaction of carbon dioxide and other atmospheric gases with light in controlling chlorophyll loss and senescence. Plant Physiol. 71:67–70.

SHADBOLT, C.A., O.D. McCOY, and R.L. WHITING. 1962. The microclimate of plastic shelters used for vegetable production. Hilgardia 32:251–266.

SHARP, R.B. 1964. A simple colorimetric method for the *in situ* measurement of carbon dioxide. J. Agr. Eng. Res. 9:87–94.

SIONIT, N., H. HELLMERS, and B.R. STRAIN. 1980. Growth and yield of wheat under CO_2 enrichment and water stress. Crop Sci. 20:687–690.

SIONIT, N., D.A. MORTENSON, B.R. STRAIN and H. HELLMERS. 1981a. Growth response of wheat to CO_2 enrichment and different levels of mineral nutrition. Agron. J. 73:1023–1027.

SIONIT, N., B.R. STRAIN, H. HELLMERS, and P.J. KRAMER. 1981b. Effects of atmospheric CO_2 concentration and water stress on water relations of wheat. Bot. Gaz. 142:191–196.

SKOK, J., W. CHORNEY, and W.S. BROECKER. 1962. Uptake of CO_2 by roots of *Xanthium* plants. Bot. Gaz. 124:118–120.

SKOYE, D.A. and E.W. TOOP. 1973. Relationship between temperature and mineral nutrition to carbon dioxide enrichment in the forcing of pot chrysanthemums. Can. J. Plant Sci. 53:609–614.

SKYTT-ANDERSEN, A. 1976. Regulation of apical dominance by ethephon, irradiance and CO_2. Physiol. Plant. 37:303–308.

SPARLING, J.H. and M. ALT. 1966. The establishment of carbon dioxide concentration gradients in Ontario woodlands. Can. J. Bot. 44:321–329.

SPITTLEHOUSE, D.L. and E.A. RIPLEY. 1974. Micrometeorology: X. Relationships between the canopy microclimate and structure of the Matador grassland. Tech. Rpt. 55. Matador Project, Canadian Comm. for the Intern. Biological Programme, Univ. of Saskatchewan, Saskatoon, Canada.

STOLWIJK, J.A.J. and K.V. THIMANN. 1957. On the uptake of carbon dioxide and bicarbonate by roots and its influence on growth. Plant Physiol. 32:513–520.

ST. OMER, L. and S.M. HORVATH. 1983. Elevated carbon dioxide concentration and whole plant senescence. Ecology 64:1311–1314.

STRAIN, B.R. and T.V. ARMENTANO. 1980. Environmental and societal consequences of CO_2 induced climate changes: response of "unmanaged" ecosystems. Natl. Tech. Inform. Ser., U.S. Dept. of Commerce, Springfield, Virginia.

SWABEY, J. 1977. Choice of coverage. p. 8–11. In: P.H. Brown (ed.), Protected strawberry cultivation. Grower Books, London.

TAKEMOTO, B.K. and R.D. NOBLE. 1982. The effects of short-term SO_2 fumigation on photosynthesis and respiration in soybean *Glycine max*. Environ. Pollut. (Ser. A) 28:67–74.

TAN, C.S., A.P. PAPADOPOULOS, and A. LIPTAY. 1984. Effect of various types of plastic films on the soil and air temperatures in 80 cm high tunnels. Scientia Hort. 23:105–112.

THOMAS, J.F. and C. HARVEY. 1983. Leaf anatomy of four species grown under continuous CO_2 enrichment. Bot. Gaz. 144:303–309.

THOMAS, J.F., C.D. RAPER, C.E. ANDERSON, and R.J. DOWNS. 1975. Growth of young tobacco plants as affected by carbon dioxide and nutrient variables. Agron. J. 67:685–689.

THOMPSON, S. and P. BIGGIN. 1980. The use of clear polythene cloches to improve the growth of one-year-old lodgepole pine seedlings. Forestry 53:51–63.

TIBBITTS, T.W. and J.M. KOBRIGER. 1983. Mode of action of air pollutants in injuring horticultural plants. HortScience 18:675–680.

TIBBITTS, T.W. and D.T. KRIZEK. 1978. Carbon dioxide. p. 80–100. In: R.W. Langhans (ed.), A growth chamber manual. Comstock, New York.

VERMA, S.B. and N.J. ROSENBERG. 1976. Carbon dioxide concentration and flux in a large agricultural region of the Great Plains of North America. J. Geophys. Res. 81:399–405.

VERMA, S.B. and N.J. ROSENBERG. 1981. Further measurements of carbon dioxide concentration and flux in a large agricultural region of the Great Plains of North America. J. Geophys. Res. 86:3258–3261.

VU, C.V., L.H. ALLEN, and G. BOWES. 1983. Effects of light and elevated atmospheric CO_2 on the ribulose bisphosphate carboxylase activity and ribulose bisphosphate level of soybean leaves. Plant Physiol. 73:729–734.

WADA, G. and T. KRISTOFFERSON. 1974. The effect of CO_2 application under various light and temperature regimes on growth and development of some florist crops. Rpt. 170. Agr. Univ. of Norway, Dept. of Floriculture and Greenhouse Crops.

WHITE, J.W. 1979. Energy efficient growing structures for controlled environment agriculture. Hort. Rev. 1:141–171.

WHITE, J.W. 1982. The greenhouses of Japan. Grower Talks (Oct.), p. 36–42.

WHITE, R.A.J. 1978. Response of tomatoes to low night–high day temperatures and carbon dioxide enrichment. Acta Hort. 76:141–145.

WIEST, S.C., G.L. GOOD, and P.L. STEPONKUS. 1976. Analysis of thermal environments in polyethylene overwintering structures. J. Amer. Soc. Hort. Sci. 101:687–692.

WILLEY, R.W. 1975. The use of shade in coffee, cocoa and tea. Hort. Abstr. 45:791–798.

WILLIAMS, L.E., T.M. DEJONG, and D.A. PHILLIPS. 1981. Carbon and nitrogen limitations on soybean seedling development. Plant Physiol. 68:1206–1209.

WILSON, C.C. 1948. Fog and atmospheric carbon dioxide as related to apparent photosynthetic rate of some broadleaf evergreens. Ecology 29:507–508.

WITTWER, S.H. 1970. Aspects of CO_2 enrichment for crop production. Trans. Amer. Soc. Agr. Eng. 13:249–251.

WITTWER, S.H. 1981. Advances in protected environments for plant growth. p. 679–715. In: J.T. Manassah and E.J. Briskey (eds.), Advances in food producing systems for arid and semi-arid lands. Academic Press, Toronto.

WITTWER, S.H. 1983. Rising atmospheric CO_2 and crop productivity. HortScience 18:667–673.

WITTWER, S.H. and W. ROBB. 1964. Carbon dioxide enrichment of greenhouse atmospheres for food crop production. Econ. Bot. 18:34–56.

WONG, S.C. 1979. Elevated atmospheric partial pressure of CO_2 and plant growth. I. Interactions of nitrogen nutrition and photosynthetic capacity in C_3 and C_4 plants. Oecologia 44:68–74.

WOO, K.C. and S.C. WONG. 1983. Inhibition of CO_2 assimilation by supraoptimal CO_2: effect of light and temperature. Austral. J. Plant Physiol. 10:75–85.

WOODROW, L. 1982. Ethylene release from leaves of *Ranunculus sceleratus* L. M.S. Thesis, Univ. of Guelph, Guelph, Ontario, Canada.

WOODWELL, G.M. and W.R. DYKEMAN. 1966. Respiration of a forest measured by carbon dioxide accumulation during temperature inversions. Science 154:1031–1034.

WYSE, R. 1980. Growth of sugarbeet seedlings in various atmospheres of oxygen and carbon dioxide. Crop Sci. 20:456–458.

YEATMAN, C.W. 1970. CO_2 enriched air increased growth of conifer seedlings. Forest Chronicle 46:229–230.

ZELITCH, I. 1971. Photosynthesis, photorespiration, and plant productivity. Academic Press, New York.

ZIESLIN, N., A.H. HALEVY, and Z. ENOCH. 1972. The role of CO_2 in increasing the yield of 'Baccara' roses. Hort. Res. 12:97–100.

ZIMMERMAN, R.H., D.T. KRIZEK, W.A. BAILEY, and H.H. KLUETER. 1970. Growth of crabapple seedlings in controlled environments: influence of seedling age and CO_2 content of the atmosphere. J. Amer. Soc. Hort. Sci. 95:323– 325.

9

Growth Regulators in Floriculture

Roy A. Larson
Department of Horticultural Science, North Carolina State University, Raleigh, North Carolina 27695

Horticultural Reviews, Volume 7

ISBN 0-87055-492-1

I. INTRODUCTION

Perhaps no segment of agriculture has made more extensive use of growth regulators in commercial crop production than floriculture. Growth regulators often have first been studied and used on floricultural crops rather than on food crops because of the less stringent governmental requirements for pesticide applications to nonfood crops. The high economic value of floricultural crops has also made them tempting targets for growth regulator applications, but the low volume of a growth regulator needed to accomplish its purpose in floriculture has not always provided a favorable economic return to chemical companies. Formulators of growth regulators have had to put high prices on their products, find wider uses for them on larger crop populations, or discontinue marketing products that are not sufficiently profitable. It is not unusual for growth regulators used primarily in floriculture to cost $25 or more per liter. Daminozide, for example, was very expensive when first introduced for use on potted flowering crops and only became widely used in floriculture after its widespread application on apple trees lowered the price. During the last 15 years, other promising compounds have been screened by chemical companies and tested by floriculturists at research institutions but not released commercially because of limited potential financial returns. The high costs of research, development, and label clearance discourage companies from marketing growth regulators unless high prices or large sales are assured.

Reviews describing the development of growth regulators, with special emphasis on compounds that control plant height, have been published by Adriansen (1983), Cathey (1964, 1975a), Sanderson (1969), Sachs and Hackett (1972), and Shanks and Link (1964). Wittwer (1968) discussed the use of growth regulators in horticulture. Weaver's (1972) book on the use of plant growth substances in agriculture, a comprehensive treatise, is an excellent introduction to the subject. The use of growth regulators in vegetable crop production has been reviewed by Thomas (1974, 1976). Luckwill (1981) discussed the various types of growth regulators, their morphogenetic effects, and physiological processes. The two volumes by Nickell (1983a,b) do not include floricultural crops, but many of the growth regulators he mentions have been tried or are used by floriculturists. The use of growth regulators on floricultural crops was reviewed by Dicks (1976), Wilkins *et al.* (1979), Seeley (1979), and Shanks (1982). The status of growth regulators used in commercial floriculture was reported by Ball (1975); Wikesjö (1975) discussed the use of growth regulators in potted plant and cut flower production in Sweden. Heins *et al.* (1979) and Wikesjö (1978) covered growth regulators in floriculture and listed chemicals, uses, and rates in tabular form; not all chemicals listed have label clearance for the uses described. In 1978, the Federal Insecticide, Fungicide, and Rodenticide Act was revised (Anon. 1979b) to be more lenient than the 1972 regulations. Heins *et al.* (1979) primarily discussed major floricultural crop applications, while Wikesjö (1978) included minor crops, such as *Beloperone* and *Pachystachys*. Guides to formulating growth regulator solutions (Hammer 1976) have been of practical value to growers.

Major improvements in application techniques in the near future should enhance the impact of growth regulators on the commercial production of agricultural commodities. Wareing (1976) has summarized the use of growth regulators to modify plant growth. Hudson (1976) forecast the future roles for growth regulators with emphasis on their use in tissue culture; he concluded that growth regulators would provide interim solutions to growth control problems, but that long-range solutions would result from genetic modification and new cultural techniques. The potential for growth regulators has been covered by Jeffcoat (1980) and Morgan (1980).

This review focuses on the application of growth regulators to floricultural crops before their removal from the greenhouse or field. Use of exogenous growth regulators in the postharvest period has been reviewed by Halevy and Mayak (1979, 1981) and Baker (1983).

There are several excellent British monographs on growth regulators, including the publication *Outlook on Agriculture*, which is very informative (Thomas 1976). The Plant Growth Regulator Society of America has excellent publications containing up-to-date information. Numerous innovative researchers have published the techniques and results of their work on growth regulators in weed control journals and monographs. There is also a vast body of European literature, which is largely unread by English-speaking scientists. Researchers in Denmark, Sweden, Norway, The Netherlands, and West Germany have been active in publishing reports on growth regulator research in their periodicals. Trade journals often give the first introduction to a new idea in the use of growth regulators in agriculture, usually before label clearance.

Although lack of label clearance prohibits use of some growth regulators in commercial production, it should not deter the inquisitive researcher. Floriculturists need to be more aggressive in obtaining minor use labels for growth retardants.

In this review, growth regulators are referred to by their common names. The corresponding trade names and chemical names for some compounds frequently used in floriculture are listed in Table 9.1.

II. BRIEF HISTORY OF GROWTH REGULATOR TECHNOLOGY

The discussion of the mode of action of many growth regulators throughout this review is limited because information is lacking on their mechanism. In this field, technology has preceded science in many instances. Some growth regulators effective in controlling an aspect of growth or flowering have become popular, even though their mechanism of action could not be explained satisfactorily. For example, poinsettia height was being controlled, azaleas were being chemically pinched, and bromeliads were being induced to flower before the action was understood. Recommendations on application techniques, concentrations, and volumes, and predictions of crop response, have been publicly proclaimed far in advance of an explanation as to why they are valid. There are exceptions. Bocion *et al.* (1975) and Bocion and de Silva (1976) explained the mechanism of action of dikegulac before the chemical was available commercially. Sachs *et al.* (1960) studied the cellular effects of AMO-1618 in 1960; unfortunately, the high cost and instability of this material precluded its wide use.

TABLE 9.1. Common, Trade, and Chemical Names of Growth Regulators Often Used or Studied in Floriculture

Common name	Trade name	Chemical name
Abscisic acid (ABA)		5-(1-hydroxy-2,6,6-trimethyl 4-oxo- 2-cyclohexenyl)-3-methyl-2,4-pentadienoic acid
AMO-1618	Carvadan	5-isopropyl-2-methyl-4-(piperidino-carbonyloxy) phenyl ammonium chloride
Ancymidol	A-Rest, Reducymol	α-cyclopropyl-α-(4 methoxyphenyl)-5-pyrimid methanol
BA		6-benzylamino purine (or benzyladenine)
Chlormequat	Cycocel	(2-chloroethyl) trimethylammonium chloride
Chlorphonium chloride	Phosfon D	2,4-dichlorobenzyl tributyl phosphonium chloride
Daminozide	Alar, B-Nine SP	butanedioic acid mono-(2,2-dimethyl hydrazide
Dikegulac	Atrinal	sodium salt of 2,3:4,6-bis-O-1-methyl-ethylidene)-*x*-L-xylo-2-hexulofur-anosonic acid
Ethephon	Ethrel, Florel	(2-chloroethyl) phosphonic acid
Gibberellins GA_3	Gibrel Gibsol	2,4*a*, 7-trihydroxyl-1-methyl-8-methylenegibb-3-ene-1,10-carbox-ylic acid 1-4 lactone
GA_{4+7}	Gibtabs in Promalin	
Methyl esters of fatty acids	Off-Shoot-O	methyl decanoate (*n*-decanol)
Paclobutrazol	Bonzi	(2*R*,3*R*+2*S*,3*S*)-1-(4-chlorophenyl-4, 4-dimethyl-2-(1,2,4-triazol-1-yl)
PBA	Accel	6-benzylamino-9-(2-tetrahydro-pyranyl)-9*H*-purine
Piproctanyl-bromide	Alden	1-allyl-1-(3,7-dimethyloctyl) piper-idinium bromide
Mepiquat chloride	BAS-106	5-(4-chlorophenyl-3,4,5,9,10-pentaaza-tetra-cyclo-5,4,1,O^2,O^6, $O^{8,11}$-dodeca-3,9-diene
Oxathiin	UBI-P293	2,3-dihydro-5,6-diphenyl 1,4-oxathiin
Triacontanol		1-hydroxytriacontane

Sachs *et al.* (1960) reported the effects of gibberellin (GA) and the growth retardant AMO-1618 on stem elongation, and the site in which the major activity occurred. Sachs and Kofranek (1963) included chlormequat and chlorphonium chloride, along with AMO-1618, in their experiments with chrysanthemum plants and demonstrated inhibition of longitudinal cell expansion and division in the subapical meristematic region. Sachs and Kofranek explained the increase in stem thickness of treated plants, first observed by researchers and growers, as a stimulation of transverse cell expansion and division, although the opposite effect was observed on longitudinal growth. They also showed that both effects could be nullified with exogenous applications of gibberellic acid.

Riddell *et al.* (1962) first reported research results on daminozide—a versatile compound that has become prominent in the floriculture industry. Its predecessor, a compound referred to as CO 11 (*N*-dimethylamino maleanic acid), had been highly regarded as a promising growth regulator because it seemed to increase heat and drought resistance and to retard cell elongation and division. However, CO 11 was not stable in water, and its formulator replaced it with B-995 [butanedioic acid mono-(2,2-dimethylhydrazide)], or daminozide, which is stable in water. Daminozide eventually was designated as B-Nine in the floriculture industry and Alar for pomological uses. The positive effects of daminozide on heat and drought tolerance have never been as well substantiated as have its impact on stem elongation of numerous floricultural crops and its promotion of flower bud initiation on azaleas. No satisfactory explanations for its effects have been published, although it has been reported to inhibit both GA and auxin biosynthesis.

Some growth regulators such as chlormequat are effective when applied either as a drench to the growing medium or as foliar sprays. Daminozide has been effective only when used as a foliar spray. Moore (1968) reported that it was apparently translocated with ease in both xylem and phloem of both monocotyledonous and dicotyledonous plants. It does elicit responses in a large number of floricultural crops but has not been effective on monocotyledonous plants such as Easter lilies (*Lilium longiflorum*). Moore does not consider this lack of sensitivity or response to be related to transport of the compound.

A promising new growth regulator, first called El 531, was identified in 1970 and has since been investigated on a broad spectrum of plants. The retardant was first named Quel (Anon. 1971), and early literature on the retardant refers to it by that name. The name was not approved, however, and the compound was named A-Rest (Anon. 1972), with the common name ancymidol. The formulator stressed the need for uniform distribution of spray applications, as the "spray to runoff" criterion is affected by leaf area, plant spacing, and similar factors. Spray solutions also can run down the stems into the potting medium, increasing the amount of ancymidol available to plants. Drench applications were recommended by the manufacturer because they are more accurate than sprays.

Ancymidol effectively controls the height of poinsettias, potted chrysanthemums, Easter lilies, and numerous other floricultural crops, when applied either as a drench or spray. It has not been widely accepted in commercial floriculture, perhaps because of growers' loyalty to earlier growth regulators, and the extreme sensitivity of some species to relatively low concentrations of ancymidol. Though ancymidol is not as popular among commercial flower growers as chlormequat or daminozide, it has been widely studied by researchers. Leopold (1971), in studies with lettuce hypocotyl bioassays, classified growth regulators based on their influence on gibberellin-induced growth, auxin-induced growth, or cytokinin-induced growth; ancymidol was found to affect growth induced by gibberellin. Coolbaugh and Hamilton (1976) confirmed Leopold's conclusions and demonstrated that the retarding effects caused by ancymidol application could be reversed with gibberellic acid. They concluded that ancymidol possibly inhibited the oxidation and conversion of ent-kaurene to ent-kaurenol, a preliminary step in the biosynthesis of gibberellic acid. This mode of action was confirmed by later studies of Coolbaugh *et al.* (1978). Researchers have revealed many uses of ancymidol that have not been adopted by commercial growers; these are listed in Table 9.2, along with selected references.

Ethephon has attracted the attention of many researchers, for a multitude of purposes. Shanks (1969) summarized some effects of ethephon including growth retardation, induction of abscission, stimulation of lateral or basal branching, and emasculation of flowers to facilitate pollination. Absorption of ethephon by roots, stems, or leaves makes it possible to apply ethephon as a spray or drench. The effectiveness of ethephon to induce flowering of bromeliads, to reduce incidence of toppling in hyacinths, and to enhance basal branching is considered in later sections on specific processes affected by growth regulators.

Application of methyl esters of fatty acids to overcome apical dominance and promote lateral shoot development was a major advance in the use of growth regulators to control growth of floricultural crops. Methyl esters of fatty acids, such as Off-Shoot-O and Emgard 2077, stimulate development of lateral shoots, but when applied to herbaceous plants, frequent instances of stem girdling have limited their use on crops such as chrysanthemums. They have been used successfully in commercial production of azaleas. Perhaps no growth regulators used in floriculture have been so variable in their effectiveness as the

TABLE 9.2. Uses of Growth Regulators on Floriculture Crops

Crops	Purpose of application	Growth regulator	References
Azalea	Flower development	GA_3, GA_{4+7}	Boodley and Mastalerz 1959; Joiner *et al.* 1982; Larson and Sydnor 1971; Nell and Larson 1974; Sydnor 1972
	Flower initiation	Chlormequat	Stuart 1961, 1964; McDowell and Larson 1966; Criley 1969
		Chlorphonium chloride	Stuart 1961
		Daminozide	Stuart 1964, McDowell and Larson 1966
	Lateral shoot development	Dikegulac	Anon. 1976; Arzee *et al.* 1977; Bocion *et al.* 1976; de Silva *et al.* 1976; Heursel 1975; Larson 1978
		Methyl esters of fatty acids	Anon. 1969; Carr and Lindstrom 1972; Cathey 1970; Furuta 1968; Joiner and Conover 1974; Nelson and Poteet 1969; Shu and Sanderson 1983; Sill and Nelson 1970; Stuart 1967
Bedding plants	Growth control	Ancymidol	Armitage *et al.* 1981; Carlson 1980; Cathey 1976; Wikesjo 1978
		Chlormequat	Read *et al.* 1974
		Daminozide	Armitage *et al.* 1981; Carlson 1980; Cathey 1976; McConnell and Struckmeyer 1970; Tjia and Buxton 1977; Wikesjo 1978
	Growth promotion	Aminoethoxyvinyl glycine (AVG)	Shanks 1980
	Protection from pollutants	Ethylene diurea	Cathey and Heggestad 1982a,b
Begonia	Growth control	Ancymidol	Heide 1969; Holcomb 1979; Krauskopf and Nelson 1979
		Chlormequat	Heide 1969; Holcomb 1979; Krauskopf and Nelson 1979; Wikesjo and Schusster 1982
		Daminozide	Heide 1969; Krauskopf and Nelson 1979
	Lateral shoot development	Dikegulac	Wikesjo 1978
Beloperone	Flower induction	Ethephon	Wikesjo 1978
	Growth control	Ancymidol	Adriansen 1977
		Chlormequat	Wikesjo 1978
		Daminozide	Wikesjo 1978
		Ethephon	Adriansen 1977
Bougainvillea	Delay of abscission	Silver thiosulfate (STS)	Cameron *et al.* 1981
	Flower induction	Chlormequat	Hackett and Sachs 1967

Bromeliads	Flower induction	Beta–hydroxyethyl hydrazine (BOH)	Cathey and Downs 1965; Systema 1969
		Ethephon	Anon. 1975a; Cathey and Taylor 1970; Poole and Conover 1975
		Naphthaleneacetic acid (NAA)	Zimmer 1969
Calceolaria	Delay of abscission	Silver thiosulfate (STS)	Cameron *et al*. 1981
	Growth control	Chlormequat	Hammer 1980; Johansson 1976
Carnation	Flowering	Abscisic acid (ABA)	Cathey 1969b
	Rooting	Daminozide	Read and Hoysler 1971
Chrysanthemum	Disbudding	Alkylnaphthalene	Cathey *et al*. 1966b
		Emulsifiable oils	Kofranek and Criley 1967
		Naphthalene	Kofranek and Markiewicz 1967
		Oxathiin	Anon. 1975b; Cathey 1976b; Menhenett 1980; Parups 1976; Zacharioudakis and Larson 1976
	Growth control	Ancymidol	Bonaminio and Larson 1978; Johnson 1974; Larson and Bonaminio 1973; Larson *et al*. 1974; McDaniel and Fuhr 1977; Tschabold *et al*. 1975
		Chlormequat	Lindstrom and Tolbert 1960
		Chlorphonium chloride	Anon. 1979; Jorgensen 1962
		Daminozide	Dicks 1972; Dicks 1972/73; McDaniel and Fuhr 1977; von Hentig and Hass 1982
		Paclobutrazol	Anon. 1982; Barrett 1982; Barrett and Bartuska 1982; Menhenett 1982; Shanks 1980
		Piproctanyl bromide	Bocion *et al*. 1978; Huppi *et al*. 1976; von Hentig and Hass 1982
	Lateral shoot development	Methyl esters of fatty acids	Cathey 1968, 1970; Cathey *et al*. 1966; Uhring 1971
		Oxathiin	Anon. 1975b; Menhenett 1978
	Rooting	Chlormequat	Read and Hoysler 1969
		Daminozide	Read and Hoysler 1969
	Protection from pollutants	Ancymidol	Johnson *et al*. 1978
Clerodendrum	Flower initiation	Ancymidol	Adriansen 1980
Cyclamen	Flowering	Chlormequat	Hildrum 1973; Koranski *et al*. 1978
		GA_3	Widmer 1978; Widmer *et al*. 1979

TABLE 9.2 (*cont.*)

Crops	Purpose of application	Growth regulator	References
Dahlia	Growth control	Ancymidol	De Hertogh and Blakely 1976; De Hertogh *et al*. 1981
		Chlormequat	De Hertogh and Blakely 1976
		Daminozide	De Hertogh and Blakely 1976
		Daminozide	Read *et al*. 1972
	Propagation	Chlormequat	Read *et al*. 1972
Easter lily	Growth control	Ancymidol	Larson *et al*. 1974; Lewis and Lewis 1980, 1981, 1982; Sanderson *et al*. 1975; Seeley 1975; Tjia *et al*. 1976a,b
		Chlorphonium chloride	Sanderson *et al*. 1975
Euphorbia fulgens	Flowering	Chlormequat	Runger and Albert 1975
Foliage plants	Flowering	GA_3	Henny 1983
	Growth control	Ancymidol	Blessington and Link 1980; Joiner *et al*. 1978; McConnell and Poole 1972, 1981; Vienravee and Rogers 1974
		Ethephon	Johnson *et al*. 1982
	Lateral shoot development	Benzylamino purine (BA)	Maene and DeBergh 1982
		Ethephon	Criley and Ika 1976
Fuchsia	Flowering	GA_3	Sachs and Bretz 1962
	Growth control	Ancymidol	Adriansen 1978
		Ethephon	Adriansen 1978
Geranium	Growth and flowering	Ancymidol	Miranda and Carlson 1980
		Chlormequat	Brisson and Larson 1983; Miranda and Carlson 1980; Semeniuk and Taylor 1970; Shanks 1980
		Chlorphonium chloride	Anon. 1979d
		Ethephon	Semeniuk and Taylor 1970
	Prevention of petal abscission	Silver thiosulfate (STS)	Anon. 1983b; Cameron and Reid 1983; Hausbeck *et al*. 1984
	Propagation	Chlormequat	Read and Hoysler 1969
		Daminozide	Read and Hoysler 1969
Gladiolus	Propagation	Ethephon	Magie 1971
	Protection from disease	Ethephon	Halevy *et al*. 1970
Gloxinia	Growth control	Ancymidol	Sydnor *et al*. 1972
		Daminozide	Sydnor *et al*. 1972
Hibiscus	Growth control	Chlormequat	Shanks 1969
		Ethephon	Shanks 1969
Hyacinth	Prevention of topple	Ethephon	De Hertogh 1982; Shoub and De Hertogh 1975

Hybrid lilies	Growth control	Ancymidol	Dicks *et al.* 1974; Hanks and Menhenett 1980; Menhenett and Hanks 1982b; Seeley 1982; Simmonds and Cumming 1977, 1978; Stuart 1971; White 1976
		Ethephon	Simmonds and Cumming 1977, 1978
		GA_3	Dicks *et al.* 1974
		Paclobutrazol	Menhenett and Hanks 1982a
Hydrangea	Growth control	Ancymidol	Tjia *et al.* 1976b
		Daminozide	Weiler 1980
		Ethephon	Shanks 1969; Weiler 1980
		Paclobutrazol	Scott 1982; Shanks 1980
	Induction of flower abscission	Ethephon	Tjia and Buxton 1976
Iris	Reduction of flower abortion	Ethephon	Kamerbeek *et al.* 1980
Kalanchoe	Growth control	Ancymidol	Carlson *et al.* 1977
		Daminozide	Nightingale 1970
Narcissus	Growth control	Ethephon	Briggs 1975; De Hertogh 1982; Moe 1980
New Guinea impatiens	Growth control	Ancymidol	Pasutti and Weigle 1980
		Chlormequat	Pasutti and Weigle 1980
		Daminozide	Pasutti and Weigle 1980
Opuntia	Lateral shoot development	Benzylamino purine (BA)	White *et al.* 1978
Orchids	Growth control	Ethephon	Adriansen 1980

TABLE 9.2 *(cont.)*

Crops	Purpose of application	Growth regulator	References
Poinsettia	Growth control	Ancymidol	Einert 1976; Holcomb *et al.* 1983; Larson *et al.* 1978; Tjia *et al.* 1976c; Wilfret *et al.* 1978
		Chlormequat	Anon. 1965; Cathey 1959; Holcomb *et al.* 1983; Holcomb and White 1970; Larson 1967; Larson *et al.* 1978; Lindstrom and Tolbert 1960; Shanks 1981a; Tjia *et al.* 1976c; White and Holcomb 1974
		Daminozide	Shanks 1981a
		Ethephon	Tjia *et al.* 1976c
		Paclobutrazol	Anon. 1982; Shanks 1980
	Lateral shoot development	Benzylamino purine (BA)	Carpenter *et al.* 1971; Carpenter and Beck 1972
	Propagation	Ancymidol	Beck and Sink 1974
		Chlormequat	Beck and Sink 1974
		Daminozide	Beck and Sink 1974; Read and Hoysler 1971
	Protection from pests	Chlormequat	Fischer and Shanks 1979
	Protection from pollutants	Ancymidol	Cathey and Heggestad 1973
		Chlormequat	Cathey and Heggestad 1973
	Reduction of epinasty	Aminoethoxyvinyl glycine (AVG)	Saltveit and Larson 1981; Shanks and Solomos 1980
		Cycloheximide	Saltveit and Larson 1983
		Silver thiosulfate (STS)	Reid *et al.* 1981; Saltveit *et al.* 1979
Primula	Seed germination	GA_3	Miller and Holcomb 1982
Rose	Lateral shoot development	Benzylamino purine (BA)	Carpenter and Rodriguez 1971; Faber and White 1977; Parups 1971
		Ethephon	Zieslin *et al.* 1972
Schlumbergera	Growth and flowering	Benzylamino purine (BA)	Heins *et al.* 1981
	Retardation of floral abscission	Silver thiosulfate (STS)	Cameron and Reid 1981; Cameron *et al.* 1981
Tulip	Growth control	Ancymidol	Briggs 1975; De Hertogh 1976; Hanks and Menhenett 1979; Kincaid and McDaniel 1979; Shoub and De Hertogh 1974
		Ethephon	Moe 1980
		Paclobutrazol	Menhenett and Hanks 1982a

methyl esters of fatty acids. Cultivar, environment, stage of growth, and even the method of solution preparation affect success or failure. Effectiveness was related to azalea bud morphology by Sill and Nelson (1970) and to environmental factors by Furuta *et al.* (1968). However, methyl decanoate was commercially available and accepted prior to publication of this information.

Dikegulac (Atrinal) was quite thoroughly investigated as a chemical pruning agent by Bocion *et al.*(1975). The compound is an intermediate substance in the commercial synthesis of L-ascorbic acid, and early investigations showed that it could retard stem elongation, promote lateral branching, and increase flower numbers on some plant species. It controlled height of herbaceous plants, shrubs, and even grasses. In contrast with methyl decanoate, dikegulac was absorbed by the foliage and translocated, and did not require physical contact with shoot apexes to overcome apical dominance. DNA synthesis was negatively affected by dikegulac.

Recently a combination of silver nitrate and sodium thiosulfate has been shown to perform several useful functions in floriculture. Silver thiosulfate inhibited floral abscission in *Schlumbergera* (Cameron and Reid 1981, 1983). Beyer (1976) showed that silver nitrate persistently blocked ethylene effects in etiolated pea plants with no phytotoxic consequences at effective concentrations. Veen and Van de Geijn (1978) found silver nitrate to be toxic at a saturation concentration and incorporated sodium thiosulfate to increase the mobility of silver without saturation. Silver nitrate alone has been phytotoxic when applied to poinsettias to reduce epinasty during shipping. The attributes and uses of silver thiosulfate have been summarized by Veen (1983). Hausbeck *et al.* (1984) reported a correlation between silver thiosulfate application to seedling geraniums and pythium infection.

A major hurdle in the commercial use of silver-containing compounds is the lack of label clearance. The heavy metal in these compounds could complicate clearance, yet silver is a vital component in photographic processing and widely used in the film industry. The benefits of silver thiosulfate in inhibiting floral abscission and prolonging the vase life of cut flowers should warrant application for label clearance by researchers who have adequate and appropriate data.

Gibberellins are endogenous plant growth substances that also can be applied exogenously, generally as GA_3 or GA_{4+7}. Kohl and Kofranek (1957) were among the first to investigate the possible uses of GA on floricultural crops. Gibberellin increases stem length and alters spray formation of cut chrysanthemums, but its most publicized use in floriculture has been as a chemical substitute for the cold requirement for azalea flowering (Boodley and Mastalerz 1959). It also has been used on cyclamen (Widmer 1978). There is no label clearance for use of GA on flower crops, but it is cleared for use on other agricultural commodities. Gibberellins are natural compounds and it is difficult to see why they would not be approved. Although there has been relatively little research on the use of gibberellins in floriculture, they have been thoroughly studied. Many excellent reviews on gibberellins have been published since the mid-1950s, including a particularly helpful one by Lang (1970) and the book edited by Krishnamoorthy (1975).

Morphactins are another group of growth regulators that have been tried on floricultural crops, but their adverse side effects have generally discouraged further exploration. Schneider's (1970) discussion of morphactins is enlightening. Parups (1970), who has conducted research on the uses of numerous growth regulators on floricultural crops, also has discussed the mechanism of action of morphactins in plant tissue.

Abscisic acid (ABA) has received considerable attention during the last 15 years, but floriculture researchers have shown little interest in it, perhaps because only limited quantities of synthetic ABA are available for experimentation. Retention of foliage and flowers is a goal of floriculture researchers and commercial growers, and ABA has not been effective in inducing such responses. Research by Addicott and his colleagues have provided much information on this compound (Addicott and Lyon 1969).

III. SPECIFIC USES

The popular use of growth regulators to limit stem elongation and encourage stocky growth of floricultural crops causes other contributions of growth regulators to the modern culture of flower crops to be overlooked. Major emphasis has been placed on height control, perhaps because refined cultural techniques, disease-free plants, improved vigor of cultivars, and other factors have resulted in vigorous plant growth. However, other plant processes and traits also are influenced by application of exogenous growth regulators; these are discussed individually in the following sections. The crops affected, purposes of application, the growth regulators used, and appropriate literature references are summarized in Table 9.2 as a quick reference. Wikesjö's (1978) summary of growth regulator treatments on floricultural crops is presented in Table 9.3.

A. Abscission

1. Inhibition. Premature loss of flowers and/or foliage is anathema to flower growers. Improper cultural practices that cause leaf or flower abscission must be avoided, but the inherent nature of some crops to lose leaves or flowers earlier than desired is not easily corrected. Most seedling geranium cultivars display a tendency for premature petal abscission, referred to as "shattering." Snapdragon florets will quickly abscise, or shatter, if exposed to low levels of ethylene. Most begonias, whether *B. semperflorens, B. tuber-hybrida, B. hiemalis*, or other species, have marked tendencies toward leaf and flower abscission before plants are ready to be sold. Holiday cacti are also very subject to early senescence and abscission.

Cameron and Reid (1981) showed that environmental stress accelerates senescence by promoting ethylene production. High temperatures and low light intensity can induce abscission. They treated *Schlumbergera* in the tight bud stage with a foliar application of approximately 5-10 ml of silver thiosulfate (STS) solution. Some plants were exposed to ethylene in chambers; the controls were left outside the chambers. Untreated plants that were exposed to 0.5 μl ethylene suffered abscission of flower buds within 24 hr, and 7 days later 85% of the flower buds or flowers had fallen. Control plants outside the chambers still had flower buds or flowers after 7 days. Plants treated with 4 m*M* silver thiosulfate and subjected to ethylene retained 90% of their flowers; a later experiment showed that 2 mM silver thiosulfate was effective and was not phytotoxic.

Cameron and Reid (1983) found no shattering on four seedling geranium cultivars sprayed with 0.5 m*M* silver thiosulfate 2–3 weeks prior to anthesis. The same concentration inhibited flower abscission from calceolaria plants, even when plants were placed in the dark for 4 days under dry conditions. Only 22% of the flowers abscised when plants had been treated with silver thiosulfate, but 83% of the flowers abscised on untreated plants. Bougainvillea plants also responded positively when they were treated with silver thiosulfate. After 3 days without watering, 56% of the flowers fell off untreated plants, whereas only 17% of the flowers fell from treated plants.

Cameron *et al.* (1981) estimated that silver thiosulfate sprays cost 0.1¢ per plant, based on 1981 silver nitrate prices. Researchers at Michigan State University (Anon. 1983b) found no differences in activity if the pH of the STS solution was 4, 7, or 10; if the mixing time was 5 or 50 sec; or if the temperature was 5°, 25°, or 50°C. Browning of the concentrated solution can occur, but that was not of concern to the

TABLE 9.3. Uses of Growth Regulators on Various Crops Compiled by K. Wikesjö (1978)

Crop	Purpose	Growth regulator	Method of application
POT PLANTS			
Abutilon	Growth control	Chlormequat	Spray
		Daminozide	Spray
Acalypha	Growth control	Chlormequat	Spray
Achimenes		Chlormequat	Spray
		Daminozide	Spray
Aechmea	Flower induction	Ethephon	Spray
Aphelandra	Growth control	Chlormequat	Spray or drench
Azalea	Flower bud initiation	Daminozide	Spray
	Lateral branching	Dikegulac	Spray
Begonia × *cheimantha*	Growth control	Chlormequat	Spray
Begonia × *hiemalis*	Growth control	Chlormequat	Spray
	Lateral branching	Dikegulac	Spray
Beloperone	Flower induction	Ethephon	Spray
	Growth control	Chlormequat	Spray
		Daminozide	Spray
Bougainvillea	Growth control	Daminozide	Spray
		Chlormequat	Drench
Browallia		Daminozide	Spray
Calceolaria		Chlormequat	Spray
Campanula isophylla		Daminozide	Spray
Chrysanthemum		Daminozide	Spray
Clerodendrum		Chlormequat	Spray
		Ancymidol	Spray
Euphorbia pulcherrima	Growth control single-stem	Chlormequat	Spray
	Pinched	Chlormequat	Spray
		Ancymidol	Drench
Fuchsia	Growth control	Daminozide	Spray
		Chlormequat	Spray
	Lateral branching	Dikegulac	Spray

Hibiscus	Growth control	Chlormequat	Spray
Hydrangea		Daminozide	Spray
		Ancymidol	Spray
Kalanchoe		Daminozide	Spray
	Lateral shoots	Dikegulac	Spray
Lilium	Growth control	Chlormequat	Drench
		Ancymidol	Spray
Nerium		Ancymidol	Drench
Pachystachys		Chlormequat	Spray
Pelargonium hortorum		Chlormequat	Spray
		Chlormequat	Drench
P. peltatum		Chlormequat	Spray
P. domesticum		Chlormequat	Spray
Rosa		Chlormequat	Spray
Stephanotis		Daminozide	Spray
Senecio		Daminozide	Spray
Sinningia		Chlormequat	Spray
CUT FLOWERS			
Chrysanthemum	Growth control	Daminozide	Spray
Dianthus		Chlormequat	Spray
Euphorbia fulgens		Chlormequat	Drench
BEDDING PLANTS			
Ageratum	Growth control	Daminozide	Spray
Begonia × tuberhybrida		Chlormequat	Spray
Chrysanthemum frutescens		Chlormequat	Spray or drench
Dahlia		Daminozide	Spray
Pelargonium (seedlings)		Chlormequat	Spray
Petunia		Daminozide	Spray
Salvia		Daminozide	Spray
Tagetes		Daminozide	Spray
Verbena		Dikegulac	Spray
Viola		Daminozide	Spray

researchers. Hausbeck *et al.* (1984) have reported an interaction between STS treatment and pythium disease development on the seedling geranium 'Ringo Scarlet.' Disease severity was increased at the highest levels of STS application.

Consumer acceptance of many flowering pot plants might increase if flower abscission were significantly delayed. Fuchsia, impatiens, mimulus, and hibiscus are representative floral crops plagued with excessive and accelerated floral abscission. The newly introduced 'Happy End' begonias, ideal for hanging basket culture, are very subject to premature flower abscission; prevention of this disorder would enhance their acceptance in the floral industry.

2. Promotion. Occasionally, flower growers could profit by promoting abscission of plant parts. For example, hydrangeas must be placed in cold storage after the flower buds are well developed because normal growth and development will not occur until plants have been exposed to temperatures of 1°–4°C for at least 6 weeks. Because plants without foliage are more responsive to the cold-temperature treatment (Post 1950), ethylene derived from apples was used to promote foliar abscission, but more controlled techniques were desired. Kofranek and Leiser (1958) tried chemical fumigants as defoliants, but the results were too variable for commercial application. Tjia and Buxton (1976) found that defoliation occurred within 8–9 days on hydrangea plants treated with an ethephon spray 2 weeks before plants were placed in cold storage. There was complete defoliation of plants of cultivars 'Rose Supreme' and 'Merveille' when 1000–5000 ppm of ethephon was applied. Some height reduction during the forcing period also resulted.

B. Flowering

One goal in the production of most floricultural products is to achieve anthesis as quickly as possible. Manipulation of photoperiod and temperature have been used to regulate flowering of several crops. There is a strong desire, however, for the discovery of some growth regulator that directly initiates flowering. Some growth regulators, such as daminozide and chlormequat, indirectly influence flowering by retarding vegetative growth. Bromeliads can be induced to flower when compounds that promote ethylene synthesis are applied. Growth regulators that overcome flower bud dormancy (e.g., gibberellic acid on azaleas) affect flowering but do not directly cause flowering to occur. In the subsection on flower initiation, growth regulators that rather directly influence flower initiation are discussed. The dormancy-breaking growth regulators are considered in the subsection on flower development.

1. Flower Initiation. Although flower initiation is usually not an objective of foliage plant producers, there are occasions (e.g., in breeding) when flowering is desired. Henny (1983) accomplished flowering in *Aglaonema commutatum* plants with GA_3. Untreated plants remained vegetative, whereas plants treated with 100, 200, or 400 ppm GA_3 flowered in 21 weeks, with five to seven inflorescences per plant. Similar responses have been reported for dieffenbachia (Henny 1980), spathiphyllum (Henny and Rasmussen 1981), and cordyline (Fisher 1980).

Cessation of vegetative growth and initiation of flower buds occur when azalea plants are treated with chlormequat, chlorphonium chloride, daminozide, ancymidol, or paclobutrazol. Stuart (1961) showed that 'Coral Bell' azalea plants would initiate flower buds when treated with chlorphonium chloride (Phosfon) or chlormequat (Cycocel). Daminozide also was shown to be effective in promoting azalea flower bud initiation (Stuart 1962). Treatment with daminozide did not delay flowering, whereas treatment with chlormequat did delay flowering of some cultivars. Results were so promising that Stuart was encouraged to conduct research on azalea flower bud initiation on a national level by enlisting the cooperation of researchers and growers in 30 states (Stuart 1964). In this project, growth regulator, daylength, and light intensity effects on flower bud initiation were studied. Plants were grown under diverse conditions, but some conditions were common to all sites. Daminozide proved to be most effective, i.e., it induced formation of more multiple flower buds and very uniform flowering compared with other compounds.

McDowell and Larson (1966) and Criley (1969) continued investigations on azalea flower bud initiation. McDowell and Larson found that flower bud initiation occurred more readily on plants of 'Red Wing' than of 'Alaska' and 'Hershey Red' when plants were treated with chlormequat or daminozide applied as foliar sprays. Criley (1969), who applied chlormequat as a drench to the potting medium, found that a combination of chlormequat and short days was more effective in accelerating initiation than was the use of short days alone. Ancymidol at 400 ppm was very effective in promoting flower bud initiation on azaleas (R. A. Larson, unpublished data), but the commercial formulation of ancymidol, which is only 264 ppm, was less effective.

Application of chlormequat or daminozide to promote flower bud initiation is now practiced by many azalea growers. Chlormequat is applied as a foliar spray 5 weeks after the final pinch (Larson 1980), at a concentration of 60–90 ml chlormequat in 3.8 liters of water. Daminozide is applied once, 5 weeks after pinching, at 2500 ppm or twice at 1500 ppm, with applications 1 week apart.

Heide (1969) investigated the use of growth regulators to promote flowering of begonia. Daminozide and chlorphonium chloride were ineffective, but plants treated with chlormequat flowered before untreated plants. High concentrations of chlormequat did cause a decrease in flower size, but size increased as temperatures declined from 24° to 18°C. Chlormequat promoted production of male flowers at 21°C. Heide suggested that chlormequat might inactivate endogenous gibberellins or other flower-inhibiting substances.

Hackett and Sachs (1967, 1968) investigated the effects of growth regulators on flower initiation and development of 'San Diego Red' bougainvillea. They showed that initiation was not affected by growth regulators, but that short days and chlormequat increased flowering, whereas long days and GA inhibited it. Flower initiation in 'Carmencita' bougainvillea was not induced by application of ancymidol, chlormequat, or daminozide but was affected by day length (Criley 1977). Growth regulators such as chlormequat can increase the effectiveness of photoinductive short days but cannot nullify the effects of long days. Chlormequat is widely used in Denmark on bougainvilleas to keep the plants compact, but short days are used to promote flower initiation.

Promotion of flowering in pineapple is a classical example of the chemical manipulation of flowering. Similar responses in other genera in the bromeliad family, although not as dramatic and of less economic impact, are practical and effective. Cathey and Downs (1965) induced flowering of *Billbergia pyramidalis* in 4–5 weeks by filling the center of rosetted leaves ("vases") with 0.1–0.4% beta-hydroxyethyl hydrazine (BOH or Omaflora) and then growing plants under long-day conditions at 18°C. Plants had to be at least 30 cm wide and 80 cm tall with a well-established root system for induction of flowering to occur. Phytotoxicity occurred on *B. pyramidalis* plants if the concentration of BOH exceeded 0.4%. Vases of each offshoot had to be treated to induce their flowering. Cathey and Downs also listed species that did not respond positively.

Systema (1969) also induced flowering of *Vriesea splendens* with BOH. The chemical was effective whether applied as a spray to the foliage or inserted in the vase (Table 9.4). The ability of plants to respond to exogenous applications of BOH depended on plant size, not age. Leaf dimensions, number of initiated leaves, or plant weight served as useful indexes of flowering potential. Systema stated that a 1% solution of BOH can be stored in brown flasks for approximately 3 months at 20°C, but weaker concentrations could not be stored that long.

Zimmer (1969) combined BOH and naphthaleneacetic acid (NAA) in attempts to induce flowering in *Vriesea* and *Guzmania* species.

TABLE 9.4. Response of *Vriesea Splendens* to Beta-Hydroxyethyl Hydrazine (BOH)

Treatment	Concentration	Flowering plants (%)
Control		0
BOH spray	0.075%	95
	0.150%	85
	0.300%	100
BOH in vase	5 mg/15 ml/plant	100
	5 mg/1.5 ml/plant	100

Source: Systema (1979).

Although NAA alone did not produce flowering plants, 1 mg of BOH per plant did result in flowering of all treated plants. Combinations of BOH and NAA produced the results shown in Table 9.5.

Ethephon is available commercially as Florel (Anon. 1975a) for use on bromeliads and other floricultural crops. Formulators of ethephon do stress the importance of daylength in flowering of bromeliads even when they are chemically treated. Some genera, such as *Ananas* and *Billbergia*, will respond under long days, but *Aechmea* and *Vriesea* will flower most consistently if plants are under short days at time of treatment. Heins *et al.* (1979) reported that bromeliads could be induced to flower if ethephon was applied to plants at least 18–24 months old. Flowering occurred about 2 months after 10 ml of a dilute 1.2% solution of ethephon was poured into the vase of the plants.

Early research showed that abscisic acid (ABA) could affect or control numerous plant processes such as flowering, dormancy, growth, and abscission, but applied use of ABA in commercial floriculture has not been realized. Synthetic ABA has not been readily available in the quantities needed for extensive investigations, so much of the work has been on tissue explants, barley endosperms, or other plant parts rather than intact flower crops. Synthetic ABA delayed petunia flowering and inhibited stem elongation of poinsettias but caused no responses in snapdragon (Cathey 1968). Cathey (1969b) found that 'Cardinal Sim' and 'Apollo' carnations responded to 500 and 1000 ppm ABA if plants were under short days and flower-

TABLE 9.5. Response of *Vriesea Splendens* to Beta-Hydroxyethyl Hydrazine (BOH) and Naphthaleneacetic Acid (NAA)

Treatment (mg/plant)			
BOH	NAA	Flowering plants (%)	Length of flower stalk (cm)
0.00	0.00	0	—
1.00	0.00	100	28
0.75	0.25	88	17
0.50	0.50	62	17
0.25	0.75	25	7
0.00	1.00	0	—

Source: Zimmer (1969).

ing was delayed. Greater quantities of ABA were required as daylengths were increased, and ABA only partially simulated short-day effects. Plants resumed normal growth as soon as ABA applications were discontinued, so applications had to be continued to maintain the desired response. Pale green foliage characterized the treated plants.

Some attempts have been made to manipulate chrysanthemum flowering with chemicals, but no growth regulator has been able to replace the short-day requirement for flowering. Skogen *et al.* (1982) obtained more flowers and hastened flowering when chrysanthemum plants were treated with triacontanol, a saturated long chain alcohol. Menhenett (1977, 1981) applied growth retardants to control growth, then applied several types of gibberellins to promote stem elongation and hasten flowering. Harada and Nitsch (1959) were able to initiate flowering of three Japanese chrysanthemum cultivars when they applied 50 g of GA_3 under warm, long-day conditions. These cultivars respond not to photoperiod but to vernalization, so GA was not a substitute for short days but for the cold requirement. Krishnamoorthy (1975) cited several investigations on chrysanthemums and other plant species in which gibberellins were used to replace daylength responses, but most results were negative or inconsistent. Similarly, the number of *Impatiens* plants that flowered increased when chlorphonium chloride was applied, while almost all control plants remained vegetative Nanda *et al.* (1967, 1969).

As exemplified by research conducted on *Clerodendrum*, it is often difficult to separate flower initiation from development unless anatomical studies are included in the investigations. Hildrum (1973) was among the first to investigate the flowering response of *Clerodendrum* to growth regulators. He reported that chlormequat promoted flower "development," whereas long days or GA applications delayed development. Koranski *et al.* (1978) reported that a 0.3 mg drench of ancymidol, applied to plants maintained under a noninductive photoperiod, accelerated flower bud "initiation" by 8 days but that flower "development" was not hastened over that of untreated plants. Adriansen (1980) found that ancymidol retarded vegetative growth, prompting a high percentage of plants to flower. In his unique method, ancymidol was applied through an injector system to plants placed on capillary mats. Plants treated in this manner were superior in quality to those sprayed with ancymidol or those not treated with the growth regulator.

Flower initiation of *Euphorbia fulgens* was promoted with chlormequat, even when the potting medium was dry (Runger and Albert 1975).

Gibberellic acid promotes stem elongation on fuchsia plants, and low concentrations of this growth substance also prevent differentiation of flower parts. Sachs and Bretz (1962) evaluated the response of 19 fuchsia cultivars to daylength, temperature, and GA treatments. Ten ppm of GA_3 inhibited differentiation of flower primordia, even when plants were grown under photoinductive long days. Heins *et al.* (1979) suggested four weekly applications of GA_3 at 250 ppm to promote stem elongation. Such high dosages of GA_3 should keep plants vegetative, even under long days, based on the findings of Sachs and Bretz (1962) using a much lower rate. Adriansen (1978) increased flowering of fuchsia plants 25–75% when ancymidol was applied at concentrations of 5, 15, and 25 ppm.

Kamerbeek *et al.* (1980) observed complex interactions in the response of 'Ideal' iris to ethephon. Dutch iris plants were sprayed on June 25, while they were still in the field and before foliage had senesced. They applied 1000 liters of a dilute solution (4 ml ethephon per liter) per hectare. The main axis apices of treated plants became reproductive earlier than those of untreated plants, as evidenced by a lower leaf number, and bud abortion decreased in treated plants. The researchers could not explain the observed effects of a low concentration of ethephon, applied in late June in the field, on flowering in winter.

2. Flower Development. It is difficult to distinguish flower initiation from development; classification in this review is based on present knowledge. No difficulty is encountered, however, in discussing the effect of GA on breaking of azalea flower bud dormancy, as flower buds are generally macroscopically visible when GA is applied.

Boodley and Mastalerz (1959) showed the promising possibilities of using GA as a substitute for 4–6 weeks of exposure to cool temperatures for breaking flower bud dormancy in azalea; 1000 and 2000 ppm were effective concentrations. Fewer GA_3 applications were needed to force plants after Christmas than before, perhaps indicating that later, more advanced stages of development responded better to GA_3 or that the plants had received some exposure to marginal dormancy-breaking temperatures. Cultivar differences also were observed: 'Hexe' was more responsive than 'Sweetheart Supreme.'

As discussed in the previous subsection, various researchers have studied the effects of growth retardants on azalea flower bud initiation. The ability to manipulate growth and flower initiation of azalea plants in the field or greenhouse opened prospects for year-round flowering and marketing; this, in turn, stimulated interest in breaking dormancy with chemicals, as natural cooling is unavailable during large segments of the year, even in the coolest areas of the country.

Research on the use of GA to break flower bud dormancy has been conducted by Larson and Sydnor (1971), Sydnor (1972), and Nell and Larson (1974). Positive results were usually achieved, including complete elimination of the need for cool-temperature storage, accelerated flowering, increased flower size on some cultivars, and a reduction in labor required for moving plants from the field or greenhouse to cold storage and back to the greenhouse for forcing. Chemical substitution is not without its problems, however.

Larson and Sydnor (1971) and Sydnor (1972) encountered cultivar differences in response to GA, as had Boodley and Mastalerz (1959). The stage of bud development had a critical effect on the response: Plants treated prematurely developed only rudimentary floral parts surrounded by enlarged bud scales. Nell and Larson (1974) reported that the style should be elongated before GA is applied. Most azalea production scheduling is based on numbers of weeks following the final pinch, but because environmental conditions vary considerably from season to season or from one geographical area to another, stage of floral bud development is a more reliable guideline. Growers do desire to have some chronological order in their production systems, however, and stage 6 (elongated style) can be expected 14 weeks after the final pinch, if plants were grown at a night temperature of 16°–18°C. Either 8 weeks of short days or one or two applications of daminozide or chlormequat prior to GA treatment helps assure success of the bud-breaking treatment. Five applications of GA_3 or GA_{4+7} at 1000 ppm each time, applied at weekly intervals, should suffice, but applications should cease when petal color can be detected on most of the flower buds. Application of too much GA results in poor flower color, soft petals, and weak pedicels. Joiner *et al.* (1982) in their studies on azalea used higher concentrations of either GA_3 or GA_{4+7} (2000 and 3000 ppm) because they believed that dormancy would be more difficult to overcome in Florida. They applied gibberellins weekly for 3 or 4 weeks and never did require 5 applications, to achieve bud break.

In general, GA_{4+7} has been shown to be as effective as GA_3, and in some experiments superior. Cost often is the determinant in deciding which compound to use. Commercial application of GA in azalea culture has not been strongly advocated by researchers, primarily because of its cost and lack of label clearance.

The use of GA to hasten flowering in cyclamen has been reported by Widmer *et al.* (1974), Widmer (1978), and Widmer *et al.* (1979). A single spray of GA at 50 ppm, applied 60–75 days prior to the desired date of anthesis, accelerated flowering by 4–5 weeks (Widmer *et al.* 1974). Flowering was hastened even more, but floral scapes were elongated and weak, when 100 ppm were used. Later recommendations (Widmer *et al.* 1979) indicated that 25 ppm GA_3 was satisfactory for many named cultivars and 10 ppm was best for F_1 hybrids. Later research prompted Lyons and Widmer (1983) to suggest that 15 ppm GA_3, applied 150 days after seed of hybrid and nonhybrid cultivars were sown, produced optimum effects and less possibility of undesirable side effects. They recommended spraying 8 ml of dilute solution per plant, on the crown below the leaves. This treatment accelerated flowering and resulted in more flowers being open at one time than on untreated plants. Parups (1979), however, obtained some flower distortion and undesirable stem elongation when he used GA at 25 or 50 ppm on three cyclamen cultivars. A substituted thalimide, designated as AC 94377 by the American Cyanamid Co., did not cause distortion, but its effects on flowering time were inconsistent.

Research on the chemical control of flowering of geraniums was stimulated by the introduction of seedling geraniums. Gibberellic acid at 25–50 ppm resulted in elongation of internodes, enhanced flowering, and increased size of petals and inflorescences (Shanks 1982). Carpenter and Carlson (1970), in experiments with six growth regulators applied to seedlings of 'Carefree Scarlet' geraniums, found that a chlormequat drench and daminozide spray were most effective in controlling height and hastening flowering, while ethephon promoted lateral shoot production. Chlormequat-treated plants were 3.5 cm shorter than untreated plants and were in flower 10 days earlier. Semeniuk and Taylor (1975) observed that ABA did not promote flowering of seedling geraniums, chlormequat controlled height and promoted development of basal branching, and ethephon suppressed vegetative growth and delayed flowering. Miranda and Carlson (1980) found that ancymidol and chlormequat caused earlier flowering of 'Sprinter Scarlet' and 'Sprinter Salmon' geranium seedlings, though ancymidol was probably more effective at reduced light intensities. They recommended application of 200 ppm ancymidol or 1500 ppm chlormequat 35–42 days after seed sowing.

Scattered chlorotic leaf areas, which often develop when chlormequat is applied to geranium foliage, affect esthetic appearance more than physiological processes, but it would be desirable if the chlorosis could be avoided. Conover (1970) reported that phytotoxicity of chlormequat was greater following foliar application to poinsettia plants grown at 27° or 32°C, compared with those grown at 21°C. Light intensity also affected foliar injury. The least amount of damage occurred when plants were treated under a 20% light reduction. Brisson (1983) attempted to develop spray techniques that would enable growers to achieve the desired effects with chlormequat without inducing chlorosis. Only half as much volume of solution was required to control height of 'Sprinter Scarlet' seedlings with spray droplets ranging in size from 100 to 400 μm than with smaller aerosol particles, but chlorosis was avoided with both formulations.

Growers often have been advised to withhold water from *Schlumbergera truncata* to promote flower initiation and development, but this practice has been shown to be ineffective. Heins *et al.* (1981), for example, subjected plants to water stress by continuing drought conditions for 21 days and observed no flower promotion. However, phylloclades increased when BA was applied at 100 ppm and plants maintained under long days, conducive to continued vegetative growth. The same concentration of BA applied 2 weeks after the start of photoinductive short days increased flower number.

Temperature, light intensity, photoperiod, age and/or size of plant, and nutrition all play major roles in determining if plants will remain vegetative or undergo transition to the reproductive state. Growth regulators that have rather direct effects on some physiological processes might assist indirectly in the regulation of flowering. The paper by Sachs and Hackett (1977) will introduce the reader to additional references on this topic not cited in this review.

C. Growth Control

1. Promotion. Heins *et al.* (1979) mentioned only two floricultural crops on which growth regulators are applied for the purpose of increasing elongation: that is, the use of GA to increase stem elongation of fuchsias and geraniums. One prominent crop missing from their compilation is cut chrysanthemum. Standard mums can be treated with 1-6 ppm of GA to promote growth if growers are concerned about excessively short stems; GA_3 is applied 1-3 days after planting and can be repeated 3 weeks later for this purpose (Byrne and Pyeatt 1976). Gibberellic acid can also be used to promote elongation of pedicels of some spray chrysanthemum cultivars; GA_3 at a concentration of 20 ppm is applied 4 weeks after the start of short days (Kofranek 1980).

Sachs and Bretz (1962) observed that 10 ppm GA prevented differentiation of fuchsia floral parts. Heins *et al.* (1979) suggested a 250 ppm spray treatment, applied four times at weekly intervals, for production of fuchsia tree forms, in which rapid growth is desirable.

Gibberellic acid also is useful in the production of tree-type geraniums. Response of geraniums to GA was first reported in the mid-1950s, and further research was conducted in the 1960s. Research results from those early studies were cited by Carlson (1982), and current suggestions for GA_3 treatment are based on those results. According to standard recommendations, GA_3 is applied as a foliar spray to geranium plants, starting 2 weeks after potting in 10-cm pots, at weekly intervals for five treatments. The most satisfactory concentration is reportedly 250 ppm, but internode elongation will be increased at a concentration as low as 25 ppm. There is a direct correlation between concentration and stem elongation, but distorted growth and plants of poor quality will be produced if too much GA_3 is applied. Flowering can be delayed with the gibberellin treatments, but Carlson (1982) did not consider the delay to be excessive. Flower size can be doubled with the use of GA_3, an added feature of the chemical promotion of growth. Shanks (1982b) suggested foliar spray applications of 25–50 ppm of GA_3 to promote internode elongation.

Kohl and Kofranek (1957) tried GA on several floricultural crops in an effort to promote growth and/or flowering. Crops varied in their responses: Leaf petioles elongated 300% when bird-of-paradise (*Strelitzia regina*) plants were treated, and hydrangea shoots also were longer on treated plants; however, no elongation was noted on stock (*Matthiola incana*).

Shanks (1980) found that aminoethoxyvinylglycine (AVG) promoted seedling growth when applied as a 60- or 180-ppm foliar spray or 0.05-mg soil drench. *Begonia semperflorens*, impatiens, salvia, and marigold were included in his study. He also found that the formulated mixture of BA and GA_{4+7} (Promalin) increased growth of roses when applied as monthly foliar sprays at 50 ppm.

Research on growth promotion by growth regulators is quite limited, particularly in comparison with the extensive research on growth retardation. There are times when promotion would be helpful to overcome problems encountered when excessive amounts of growth retardants are applied, when environmental factors delay growth, or when other situations arise that would make stem elongation desirable.

2. Retardation. The use of growth regulators to retard stem elong ation has been a very popular subject of research and commercial application ever since Cathey (1959) and Lindstrom and Tolbert (1960) showed that growth regulators could control poinsettia height. Cathey and Stuart (1961) reported the response of 54 plant species to the retardants AMO-1618, chlorphonium chloride, and chlormequat; Cathey (1975) compared the effects of ancymidol with those of four other growth regulators in 28 species. Articles by Adriansen (1972), Corcoran (1975), Dicks (1976), Sanderson (1969), Seeley (1979), Shanks (1982a), Shanks and Link (1964), and Weaver (1972) describe much of the history of chemical induced growth retardation.

Cathey (1964) defined a growth-retarding chemical as one that would delay cell division and elongation in shoot tissue without exhibiting any formative effects; i.e., growth control would be the major consequence from treatment with such retardants. In a few instances researchers and growers were startled to witness growth stimulation at low concentrations, but in the vast majority of cases, shorter plants were produced when growth retardants were applied. Cathey (1964) doubted that any one chemical could be formulated that would retard growth of all plants, and his prediction has proved accurate. Some growth retardants have a quite limited species range; in contrast, ancymidol affects a broad range of species, including monocotyledonous plants such as lilies and dicotyledonous plants such as chrysanthemums and poinsettias. Shanks (1980) has shown a very broad species range for paclobutrazol, and it is likely that a broad range of sensitive crops will be found for EL-800, an analog of the pyrimidinemethanol base of ancymidol.

The effectiveness of a growth retardant is influenced by its method of application. With poinsettia, a chlormequat drench application has greater residual activity than a foliar application, but three to six times more chlormequat is required with the drench application to achieve the same extent of height control. Foliar applications of ancymidol, chlormequat, and daminozide are successful if foliage is sprayed to runoff, but paclobutrazol works best if spray applications are directed at the stems. Ancymidol can be applied to lilies as a foliar spray or drench, or even as a bulb dip treatment. Ancymidol is not very effective when applied as a drench to plants growing in a bark medium (Larson and Bonaminio 1973; Larson *et al.* 1974; Tschabold *et al.* 1975; Bonaminio 1977).

Plant age and size also can influence the effectiveness of growth retardants. Spray applications to poinsettia foliage quickly reveal the waxy surfaces so prevalent on young foliage; the cuticular waxes slough off as foliage ages. Foliar runoff, a grower's guide to the volume of spray that should be applied, occurs very quickly on young foliage; older foliage retains and thus absorbs the growth regulator better. Large, crowded plants make uniform spray coverage very difficult, causing irregularity in uptake and response. Soil moisture, air temperature, cultivar, watering and fertilizing schedules, freedom from pests, and use of surfactants are other factors that can influence the effectiveness of growth retardant applications.

People often write or say that growth retardants can make plants shorter or reduce their height. The fallacy of such statements is obvious, as growth retardants cannot shrink plants to reduce height. Most readers do understand what is really meant by such language, but in this review I will try to refrain from using terms that imply growth retardants make tall plants shorter. Some of the papers cited, however, have titles with such terms as "reduction" or "shortened" in them. Cumbersome titles or sentences often are the consequence of authors' trying to be correct in their use of terms.

a. Bedding Plants. The ability to retard the growth of bedding plants would be desirable, because inclement weather in the spring months, which delays consumer demand, often results in tall or poor-quality bedding plants. In the past, withholding water and fertilizer or lowering temperatures were the only effective ways growers had to keep plants at the proper size. Growth retardants have lessened the need for such unacceptable cultural procedures. The responses of many bedding plant species to growth retardants were listed by Cathey (1976). Carlson (1980) summarized the responses of 24 bedding plant species to ancymidol and daminozide; 18 of the 24 species responded to daminozide spray applications, while ancymidol affected 22 species; *Viola* and *Gomphrena* did not respond to either retardant.

Useful guidelines for retarding the growth of bedding plants are available for both ancymidol and daminozide (Cathey 1976). Water should not be placed on the foliage of daminozide-treated plants for 24 hr following treatment to assure growth-retarding benefits; water does not reduce the effectiveness of ancymidol. Several bedding plant species in which growth control can be achieved with chlormequat, daminozide, or dikegulac are listed in Table 9.3.

Heins *et al.* (1979) also presented general recommendations for treating bedding plants with growth regulators. They suggest applying daminozide at concentrations of 2500–5000 ppm, 3–4 weeks after transplanting, with repeat applications if needed. Ancymidol may be applied as soon as adequate root systems have been produced. Daminozide was ineffective on celosia, coleus, and snapdragon; pansies, petunias, and verbena did not respond to ancymidol.

McConnell and Struckmeyer (1970) studied the response of marigolds to daminozide and two different daylength treatments. They observed darker colored foliage 2 days after plants were treated with the growth retardant. Internodes were shorter, flowering was delayed, and even leaf shape was altered. Growth retardant application would be more expensive than proper nutritional programs to assure dark green foliage, but the combined benefit of darker foliage and shorter plants is recognized and appreciated by growers. Use of a growth retardant does not lessen the need for proper fertilization, however.

In petunia, among the most popular bedding plants in the United States, increased basal branching is as important as height control for high quality. Treatment with daminozide will limit height but not cause an increase in basal branching. Ethephon, applied as a 250- to

500-ppm soil drench, increased basal branching and limited height (Tjia and Buxton 1977). The responses of two cultivars are shown in Table 9.6. Read *et al.* (1974) reported that plant height was controlled and plant quality was improved by application of encapsulated chlormequat.

Zinnia seedlings grow very quickly and plants can soon become too tall and spindly, which decreases their marketability. Armitage *et al.* (1981) found that three to four applications of ancymidol or daminozide to zinnia seedlings delayed flowering and reduced fresh weight.

Some bedding plant species have become popular as flowering potted plants. One of the first examples of such dual-purpose bedding plants was 'First Lady' marigold. Height control was necessary to maintain proper balance between pot size and plant height. Introduction of the 'Peter Pan' zinnia series also prompted interest in the use of bedding plants as flowering pot plants. Bedding plant species are also being evaluated as pot plants or as hanging basket material. Dikegulac, usually applied because it stimulates lateral branching, effectively limits shoot elongation of verbena and improves plant appearance (Wikesjö 1978, Anon. 1983a). Flowering is delayed at concentrations that are most effective in retarding shoot elongation (Table 9.7).

b. Begonia. The growth habits of most begonia species make plants in this genus less subject to excessive height than many other floricultural crops that have one main axis, such as unpinched chrysanthemums or poinsettias. There are occasions, however, when growth retardants are advantageous in begonia production.

Heide (1969) reported that the best height control of *Begonia* × *cheimantha* occurred at 18°C. Krauskopf and Nelson (1976) used ancymidol, chlormequat, and daminozide to retard growth of 'Schwabenland Red.' Their results were inconsistent, as time of year was a major factor in determining effectiveness. A dose of 0.125 mg of ancymidol per plant, applied as a drench, was effective throughout the year, regardless of light intensity; 0.250 mg per plant was excessive; and 0.062 mg was only beneficial during low-light conditions. The best chlormequat treatments were spray applications of 0.3% throughout

TABLE 9.6. Response of Two Petunia Cultivars to Ethephon Drench Treatment

Cultivar	Treatment (g/plant)	No. basal branches	Plant height (cm)	Days to flower
'Pink Cascade'	0	2.2	7.0	45
	0.5	6.5	0.8	60
'Royal Cascade'	0	3.3	11.4	43
	0.5	4.6	1.5	58

Source: Tjia and Buxton (1977).

TABLE 9.7 Effect of Dikegulac on Growth and Flowering of Verbena

Concentration (ppm)	Days to flower	Shoot length (cm)	No. lateral shoots
0	43	61	25
500	49	43	41
1000	59	37	45
1500	61	30	41

Source: From a talk by R. Moe, Aas, Norway, presented at the N.C. Commercial Flower Growers' Holiday Plant Day, Charlotte, NC in February, 1983 (Anon. 1983b).

the year, and 0.3 and 0.6% drench applications at low- and high-light intensity periods, respectively. Daminozide spray applications were ineffective, even at concentrations as high as 0.5%.

Holcomb (1979) used ancymidol and chlormequat to produce stocky Rieger begonias from terminal cuttings. Growth control was similar with both retardants, and lateral shoot development was not adversely affected. Wikesjö and Schussler (1982) achieved growth control with chlormequat applied as a foliar spray on the strong-growing 'Schwabenland,' 'Aphrodite,' 'Ballerina,' and 'Baluga'. Plants were treated two to three times at 8- to 10-day intervals, during long days, at concentrations of 0.08–0.09%. One application was used for weaker-growing cultivars for the purpose of forming evenly branched plants, not for height control.

c. Beloperone. The shrimp plant, although a minor floricultural crop, would benefit if stem elongation was limited. Adriansen (1977) reported that plants treated with 190 and 380 ppm ethephon were 40% shorter than untreated plants. Ancymidol was less effective. Wikesjö (1978) suggested chlormequat and daminozide treatments for height control of Beloperone.

d. Calceolaria. The typical growth habit of calceolaria is prevalent lateral shoots. Height control would be desirable when one shoot is dominant or when flowers are borne on spindly stems. Light green foliage also is frequently seen on calceolaria plants, even if nutrient programs seem to be appropriate; a growth retardant that also intensified foliage color would be very useful. Johansson (1976) reported that the best growth control was obtained with 0.8% chlormequat applied when flower buds were 1.5 mm in size. Hammer (1980) reported that two spray applications of 0.04% chlormequat or a 0.3% chlormequat drench provided excellent growth control.

e. Chrysanthemum. The high economic value of chrysanthemum probably explains the abundance of research reports on chemical growth control of this crop. Growers are willing to invest in effective chemicals that would ease the task of producing chrysanthemums on a weekly basis throughout the year. It was no coincidence that some of the very earliest growth retardant research was conducted on this crop (Lindstrom and Tolbert 1960).

Photoperiodic control of growth and flowering is precise enough that growers can quite accurately "tailor" potted chrysanthemum plants to any desired size. Plants can be kept short by early exposure to photoinductive cycles (short days), but inadequate foliage production will occur if not enough long days are scheduled first to induce vegetative growth. By using growth retardants, growers can provide enough long days for good leaf production without having plants become too tall for use in the home or office.

Although growth retardants offer the greatest benefits in production of potted chrysanthemums, they also are beneficial in cut chrysanthemum production with some cultivars and under certain conditions. Excessive stem elongation from the uppermost leaf to the base of the flower creates a sparse appearance, referred to as "neckiness." Spray applications of 2500 ppm of daminozide just after removal of lateral flower buds (disbudding) will inhibit cell elongation and division in the upper portion of the stem, causing the uppermost leaf and the base of the flower to be in close proximity to each other (Kofranek 1980; Anon. 1982/83). Development of yellow pigmentation in ray florets of some white-flowered cultivars occurs as a result of daminozide treatment, but those cultivars, such as 'Indianapolis White', have lost their popularity among growers. Their successors either have not exhibited the tendency toward "neckiness" and therefore are not treated with daminozide, or petal color is not affected when daminozide is used.

Potted chrysanthemums frequently require stem elongation control under the following conditions: vigorous cultivars, generous fertilization programs, successful disease control, optimum night temperatures, ample water, and possible crowding of plants to increase production for financial reasons. Chlormequat is no longer used to help control chrysanthemum height, but it was one of the first growth regulators evaluated (Lindstrom and Tolbert 1960). Plants treated with chlormequat were 6–16 cm shorter than untreated plants, their leaves were darker, larger, and thicker, and they were less likely to wilt under water stress. Sachs and Kofranek (1963) demonstrated that retardants reduced the number of cell divisions in the apical meristem of chrysanthemum plants and thereby caused shorter stems.

Chlorphonium chloride (Phosfon) was widely used in the early 1960s to control the height of potted chrysanthemums. Jorgensen (1962) classified chrysanthemum cultivars into five groups based on their growth habit or sensitivity to chlorphonium chloride; his schedules for pot mum production of the different cultivar groups are shown in Table 9.8. Chlorphonium chloride is still popular among some growers in Europe and Great Britain. Quality of cuttings reportedly has been improved by treatment of stock plants with this growth retardant (Anon. 1979d); stems and leaves on cuttings from treated plants become thicker than those from untreated plants.

Daminozide, which has been widely used in research and commercial pot mum production, is effective as a spray and no phytotoxicity has been noted. Dicks (1972, 1972/73) studied the uptake and distribution of daminozide, using a 0.5% solution on the chrysanthemum cultivar 'Lemon Princess Anne.' He found that more solution was taken up by young shoots than by the main shoot, but only 11% of the applied material was absorbed within 24 hr. The most rapid uptake occurred within the first few hours; absorption by young, lateral shoots declined abruptly after 8 hr; and no absorption occurred after 72 hr. Little translocation occurred from young shoots, but daminozide was more readily translocated when applied to the main stem.

Soaking rooted chrysanthemum cuttings in solutions of 2500 or 5000 ppm of daminozide for 60 sec prior to planting was as effective in controlling elongation as a spray application. Ancymidol dips caused excessive control. Misting of the soaked cuttings just after planting nullified the growth-retarding effect (McDaniel and Fuhr 1977).

Menhenett (1977, 1979, 1981) followed growth retardant treatment of potted chrysanthemums with an application of gibberellins in order to overcome any delays in anthesis caused by the growth retardant. He used the growth retardants daminozide (10,000 ppm) and piproctanyl bromide (Alden) (200 ppm) and the gibberellins A_1, A_3, A_{4+7}, A_5, and A_{13}. Gibberellins were applied 6 days after plants were treated with one of the retardants. Daminozide worked very rapidly, and growth already was inhibited when gibberellins were applied. Gibberellins began promoting stem elongation 5 days after they were applied. Gibberellins A_1, A_3, and A_{4+7} were the most effective; GA_{13} had the least effect on stem elongation. Gibberellins overcame delays caused by piproctanyl bromide more than those caused by daminozide. Results indicated that growth retardants caused a deficiency in gibberellins, which inhibited stem elongation and flowering.

The most often used growth retardant in commercial production of potted chrysanthemums probably is daminozide. Directions on its application (Anon. 1982/83) are based on extensive research, and adjustments for several contingencies are suggested. Cultivar response and environmental conditions are considered in the recommendations. Heins *et al.* (1979) recommended applying 0.25–0.50% daminozide as a foliar spray when lateral shoots on pinched plants

TABLE 9.8. Pot Mum Production Schedule, Including Chlorphonium Chloride (Phosfon) Applications, for Year–Round Flowering[1]

	Days from planting to pinching			Long days after planting			Application of 125–150 ppm Phosfon solution (days after pinching)		
Cultivar type[2]	Winter	Spring/Fall	Summer	Winter	Spring/Fall	Summer	Winter	Spring/Fall	Summer
Tall	14	12	12	7	5	3	3	3	1[3]
Relatively tall	14	12	12	10	7	5	3	3	1[3]
Medium	14	14	14	14	14	14	10	10	8
Short	21	14	14	28	21	14	17	10	8
Sensitive	14	14	14	14	14	14	17	17	15

Source: Taken from Jorgensen (1962).

[1] Seasons based on those at Shelby, North Carolina: Winter, Nov. 1–Feb. 15; Spring, Feb. 15–May 1; Summer, May 1–Sept. 1; Fall, Sept. 1–Nov. 1.

[2] Classification based on growth habit or sensitivity to growth regulator.

[3] Second treatment given 1 week later.

are 2.5–4 cm long; second applications, if needed for some cultivars, are applied 2–3 weeks later.

Lack of acceptance of morphactin in the flower industry is due to its negative side effects. Tjia *et al.* (1969) reported height control and increased shoot number but delayed flowering when a morphactin was applied to chrysanthemums. Usually, shoot number is unaffected by growth retardants, as the chemicals are applied after lateral shoots are 2–5 cm long.

Larson and Kimmins (1972) were among the first to report on the use of ancymidol on potted chrysanthemums. They showed that both drench and spray applications limited stem elongation on three cultivars in the 'Princess Anne' series (Table 9.9). The excessive growth control that occurred after some ancymidol treatments prompted more experimentation and elimination of the higher concentrations. Height control and marketable plants were obtained after treatment with lower amounts of ancymidol (Table 9.10). Flowering was delayed by as much as 1 week with some ancymidol treatments, particularly at the highest concentrations, but only a 1- to 2-day delay in flowering was caused by daminozide treatment.

Preliminary drench experiments on ancymidol were conducted in potting media with soil as a major ingredient. An ancymidol drench was not effective when applied to chrysanthemum plants growing in a medium primarily composed of composted pine bark humus, a popular ingredient in the southeastern part of the United States (Larson and Bonaminio 1973). 'Nob Hill' chrysanthemum plants grown in 3 parts pine bark humus:1 part sand:1 part acid moss peat medium (by volume) were 22 and 14 cm taller than plants grown in a medium composed of 2 parts acid moss peat:1 part soil:1 part perlite, when ancy-

TABLE 9.9. Effects of Daminozide and Ancymidol on Plant Height and Size of Pith Cells of Three Chrysanthemum Cultivars[1]

Treatment	Height change from control plants (%)			Size of pith cells cv. Gay Anne (μm)	
	'Bright Golden Anne'	'Gay Anne'	'Regal Anne'	Length	Width
Ancymidol spray					
125 mg/liter	−43	−20	−30	82	38
250 mg/liter	−43	−52	−50	76	29
500 mg/liter	−54	−58	−60	59	44
Ancymidol drench					
0.25 mg/15-cm pot	−34	−38	−32	79	46
0.50 mg/15-cm pot	−45	−46	−47	71	15
1.00 mg/15-cm pot	−55	−51	−55	47	18
Daminozide spray					
2500 mg/liter	−22	−20	−20	66	12
Control				115	26
LSD 5%				15	6

Source: Data from Larson and Kimmins (1972).
[1] Study begun on Dec. 9, 1970.

TABLE 9.10. Effects of Ancymidol and Daminozide on Growth and Flowering of Three Chrysanthemum Cultivars[1]

Treatment	Height change from control plants (%)			Delay in flowering compared with control (days)		
	'Golden Yellow Princess Anne'	'Gay Anne'	'Regal Anne'	'Golden Yellow Princess Anne'	'Gay Anne'	'Regal Anne'
Ancymidol foliar spray						
62 mg/liter	+17	−21	−18	2	2	1
125 mg/liter	−37	−27	−24	3	3	4
250 mg/liter	−47	−39	−20	6	7	2
Ancymidol drench						
0.12 mg/pot	−27	−21	−21	6	1	3
0.25 mg/pot	−43	−27	−26	3	7	1
0.50 mg/pot	−43	−42	−39	7	7	7
Daminozide spray						
2500 mg/liter	−20	−12	−13	1	2	3
Control	34 cm	34 cm	34 cm	May 11	May 12	May 12

Source: Data from Larson and Kimmins (1972).
[1]Study begun on March 4, 1971.

midol drenches of 0.25 mg and 0.50 mg/15-cm pot were used, respectively. Chrysanthemum plants sprayed with 132 ppm ancymidol averaged 47 cm in height when grown in the bark mix and only 35 cm in height when grown in the soil mix. Excess spray material moving down into the potting medium may have acted as a drench application, demonstrating the ineffectiveness of ancymidol drench treatments in a bark medium.

The results of using a combination of a pine bark medium and ancymidol drench treatments were consistent (Larson *et al.* 1974) and stimulated research, both on the chemistry of the growth retardant and the medium. Tschabold *et al.* (1975) found a 50% reduction in effectiveness when 0.5 mg ancymidol/15-cm pot was applied as a drench to plants growing in a 2:1 ratio of bark and sand compared with a 1:3 ratio. They reported that ancymidol has a water solubility of 650 ppm and that water moves rapidly through mineral soils. Leaching studies revealed that ancymidol was retained in the upper third of a bark:sand medium. These experiments also indicated that 48% of the ancymidol applied to a soil mix in a leaching column could be detected in the leachate, but only 0.2% could be found in the leachate from a bark medium. The chemical and physical properties of pine bark humus have now been thoroughly investigated and an excellent bibliography has been compiled by Pokorny (1982). Airhart *et al.* (1976) examined bark particles with a scanning electron microscope to learn more about water absorption. They noted that moisture was repelled by the suberized, waxy particle surfaces. Many other bark characteristics have been examined to determine probable reasons for the ineffectiveness of ancymidol drenches in bark medium. Bonaminio (1977) decreased the acidity of a bark medium from pH 4.3 to pH 6.2 by addition of 10 g of calcium hydroxide to 1 kg of pine bark medium, and recorded increased growth retardant activity. Addition of Ca(OH) would be a less expensive way to increase activity than to increase drench concentrations (Bonaminio and Larson 1978).

Ancymidol now is a commercially accepted growth retardant used in chrysanthemum production (Anon. 1974, 1982/83). Helpful application instructions have been prepared by the formulator for both spray and drench applications (Anon. 1974). Heins *et al.* (1979) suggested foliar treatments with 25–100 ppm ancymidol, depending on the inherent vigor of the cultivar, applied after pinching and the start of short days. Suggested ancymidol drench application rates are 0.125–0.5 mg/15-cm pot, applied as 180 ml of dilute solution per 15-cm pot to assure uniform distribution.

Several researchers have found piproctanyl bromide (Alden) to be a very active growth retardant on chrysanthemums. Huppi *et al.* (1976) found that one application of 100–200 ppm piproctanyl bromide was equivalent to two or three applications of daminozide at 3400 ppm. Menhenett (1977) compared daminozide, piproctanyl bromide, and ancymidol; all three compounds controlled height and delayed flowering (Table 9.11). The delay in flowering with daminozide was more than is usually observed in studies in the United States; Menhenett has demonstrated more typical delays in other experiments (Table 9.12).

Bocion *et al.* (1978) considered piproctanyl bromide to be as effective as daminozide for growth retardation in chrysanthemum. They stressed the importance of good spray coverage and the need to keep the solution in contact with the foliage at least 24 hr following application. They believed that absorption and translocation were superior in young, soft leaves near the apex than in older leaves lower on the plant.

Menhenett (1980) evaluated oxathiin, an experimental growth retardant formulated by Uniroyal (code designation UBI-P293), as a growth retardant on chrysanthemum; others have noted its growth-retarding effect when they were studying other features of the compound. Menhenett had good success in retarding stem elongation with only one spray application. Timing is critical in the application of oxathiin because it inhibits development of lateral shoots and lateral flower buds if applied at the same stage of shoot development as recommended for daminozide (Section III. D). Application of oxathiin was delayed by Menhenett until at least 23 but no more than 31 days after pinching. Some of Menhenett's results with 'Bright Golden Anne' are shown in Table 9.13.

ICI Americas has developed a promising growth retardant, paclobutrazol, with a broad spectrum of activity. The compound is an inhibitor of the biosynthesis of gibberellin (Anon. 1982). Internodes

TABLE 9.11. Effects of Daminozide, Piproctanyl Bromide, and Ancymidol on Shoot Growth and Flowering of 'Bright Golden Anne' Chrysanthemum

Treatment[1]	Amount used (mg ai)	Shoot length (cm)	Delay in flowering relative to control (days)
Control	0	19	—
Daminozide			
2500 ppm	37.2	11	6
5000 ppm	74.5	9	6
Piproctanyl bromide			
100 ppm	1.5	14	5
250 ppm	3.7	11	9
Ancymidol			
50 ppm	0.7	12	6
100 ppm	1.5	10	7

Source: Data from Menhenett (1977).
[1]Plants were pinched Nov. 14, 1974, and treated 15 days later with 15 ml solution/plant.

TABLE 9.12. Stem Length and Delay in Flowering of 'Bright Golden Anne' Chrysanthemums Treated with Growth Retardants

Treatment	Lateral shoot length (cm)	Delay in flowering compared with control (days)
Control—spray with water	19	0
Piproctanyl bromide		
100 ppm spray	9	6
250 ppm spray	8	7
Daminozide		
2500 ppm spray	12	1
5000 ppm spray	10	3
Control—Drench with water	19	0
Chlorphonium chloride		
77 ppm drench	10	4
154 ppm drench	8	7

Source: Data from Menhenett (1977, 1978).

are compressed on treated plants. The retardant is not translocated in the phloem. Spray applications are directed toward stems where penetration of the retardant occurs on dicotyledonous plants; the compound also can be absorbed by roots. Chrysanthemum height control can be achieved with one drench application of 0.125–0.50 mg active ingredient (Anon. 1982). Menhenett (1982) showed that 0.2–0.3 mg paclobutrazol/pot was equivalent in effect to 7.7 and 15.4 mg chlorphonium chloride; 30 ppm paclobutrazol applied as a spray was almost equivalent to 5000–7500 ppm daminozide. More uniform growth might be expected from the daminozide treatment because stem penetration of paclobutrazol is not as assured as foliar absorption of daminozide.

Barrett (1982) compared the effects of ancymidol, paclobutrazol, and daminozide on growth of chrysanthemum plants grown in a commercially prepared potting medium (Metro Mix 300) and in a medium containing pine bark humus (Table 9.14). Paclobutrazol was

TABLE 9.13. Stem Length and Delay in Flowering of 'Bright Golden Anne' Chrysanthemums Treated wtih Oxathiin

Treatment	Stem length (cm)	Delay in flowering compared with control (days)
Control	23	0
Oxathiin, 23 days after pinching		
2500 ppm	12	0
5000 ppm	9	10
Oxathiin, 28 days after pinching		
2500 ppm	14	1
5000 ppm	13	7
Oxathiin, 31 days after pinching		
2500 ppm	16	1
5000 ppm	15	4

Source: Menhenett (1980).

TABLE 9.14. Stem Length of 'Bright Golden Anne' Chrysanthemums Treated with Three Growth Regulators and Grown in Two Media

Treatment	Stem length (cm)	
	Metro Mix 300	Pine bark humus
Control	35	35
Ancymidol, 0.25 mg/15-cm pot	22	30
Paclobutrazol, 0.25 mg/15-cm pot	24	30
Diaminozide, 5000-ppm foliar spray, twice	21	21

Source: Barrett (1982).

similar to ancymidol in that it was less effective as a drench in medium containing pine bark humus. Barrett believed that hydrophobic surfaces on bark particles could be responsible for the reduced activity of drench applications of ancymidol and paclobutrazol. Barrett and Bartuska (1982) confirmed the advice that paclobutrazol sprays should be directed to stems. Plants treated with 0.125 mg/pot paclobutrazol as a drench had 11-cm-long shoots. Stems were 5 cm long when whole shoots were treated, 18 cm long when only leaves were treated, and 7 cm long when only stems were sprayed. Untreated plants had shoots that averaged 24 cm in length. The authors concluded that paclobutrazol, when applied as a drench, is readily translocated from roots to the apex in the xylem. Movement does not occur in the phloem when the retardant is applied as a foliar spray, and therefore little height control occurs.

Shanks (1980, 1982) found mepiquat chloride (BAS 106) to be quite effective in retarding growth of a number of ornamental crops but only moderately effective on chrysanthemums. Holcomb *et al.* (1983) compared BAS 106, in granular or wettable powder form, with daminozide, chlormequat, and ancymidol. Growth control was achieved with BAS 106, but some phytotoxicity did occur when high rates were used.

f. Dahlia. In the past, tuberous-rooted dahlia plants have not been feasible as flowering pot plants because of their excessive height. Cultivar selection for pot plant production has been improved, but some of the most attractive flowers still are borne on plants that are too tall. De Hertogh and Blakely (1976) tested ancymidol, chlormequat, and daminozide on dahlias and found that only ancymidol was an effective growth retardant. Date of anthesis, flower size, and number of shoots were not adversely affected by ancymidol and height was controlled. The authors recommended a drench application of ancymidol 2 weeks after planting, at 0.75–2.0 mg/pot, to a medium devoid of pine bark. While not much top growth has occurred at that time, further delay in application would lessen the growth control response. De Hertogh *et al.* (1981) now recommend a soil drench application of 1.0 mg ancymidol/15-cm pot; they include informative tables describing rates and volume for application of ancymidol in their report.

g. Easter Lily. Growth control of Easter lily plants challenges researchers and commercial growers because the annual variation in the date of Easter Sunday, ranging from late March to April 22, increases the impact of temperature, daylength, and light intensity on potential growth. Cold-storage treatments prior to forcing also affect growth; commercial growers often are unaware of the storage temperatures or duration and cannot predict if conditions are conducive to tall or short lilies. Most growth regulators have been successfully used primarily on dicotyledonous plants; the choice of suitable compounds for Easter lilies and other monocotyledonous plants is more restricted.

Wilkins (1980) advocated regulating growth of Easter lilies by environmental, not chemical, means. Such an approach, however, cannot always be achieved. One of the first successful growth regulators used for Easter lily growth control was chlorphonium chloride. Treated plants usually were much shorter than untreated plants, but undesirable side effects such as weak stems were often encountered (Wiggans *et al.* 1960).

A major advance in Easter lily growth control was realized when ancymidol was made available to researchers and growers (Anon. 1974). It was effective when used either as a drench or as a foliar spray. Sanderson *et al.* (1975) found that both ancymidol and chlorphonium chloride gave similar growth control and produced weakened stems. A soil drench of ancymidol usually was preferred, as the amount of material applied per plant was more accurately dispensed and applications were less dependent on environmental conditions (Tjia *et al.* 1976a). Ancymidol drenches were not very effective when used on Easter lily plants growing in media primarily consisting of pine bark humus, as is the case with chrysanthemums. This handicap may be partially overcome by incorporating moss peat in the medium to retain the growth retardant. Tjia *et al.* (1976a) applied ancymidol as a drench or spray on the 'Georgia' strain of Easter lilies; they considered the drench application to be superior in a soil medium. Untreated plants averaged 64 cm in height, plants treated with 100 or 200 ppm ancymidol applied as a spray averaged 56 and 49 cm, respectively, while plants treated with a drench at 0.5 mg/pot were only 36 cm tall. Tjia *et al.* (1976b) tried to improve spray applications by incorporating different surfactants, but drench treatments still were most satisfactory; additional surfactants were not beneficial to drench applications.

Seeley (1975) stressed that the quantity of drench or spray solutions

TABLE 9.15. Effect of Ancymidol Drench and Bulb Dip Treatments on Height of Easter Lilies

Treatment	Height (cm)
Control	29
Drench, 1.0 mg/pot	22
Bulb dip	
100 ppm	34
264 ppm	14

Source: Lewis and Lewis (1981)

applied per pot or plant could be as critical as the concentration of the ancymidol solution. In the past, concentration has been emphasized more than volume, particularly in spray applications. Recent increased concern for the environment and increased restrictions on polluted water leaving greenhouse premises have emphasized the need for application of minimum volumes of agricultural chemicals.

A new approach to ancymidol application is the lily bulb dip, first used on hybrid lilies. Lewis and Lewis (1980, 1981) achieved growth control with a quick dip (about 2 sec) in solutions of 100 and 264 mg/liter ancymidol (Table 9.15). Their later work prompted them to advocate bulb dipping prior to cold-storage treatment. (Lewis and Lewis 1982).

My colleagues and I have treated Easter lily bulbs, which had cold-storage treatment prior to submersion of bulbs, with more dilute solutions of ancymidol, but for longer durations, than those used by Lewis and Lewis. Excellent growth control occurred when bulbs were submerged for 60 min in 24 ppm ancymidol, but other treatments have been almost equally effective (Table 9.16). The potting medium was 3 parts pine bark humus:1 part sand:1 part acid moss peat (by volume). Bulbs were dipped and planted December 8, 1982. Cost of the 33-ppm bulb dip treatments in 1982 was estimated to be less than $0.03 per bulb, compared with $0.08 per bulb for a drench treatment of 0.5 mg active ingredient. Media composition was not a variable in these studies.

Plant form is not affected and basal leaves are not killed by the ancymidol bulb dip procedure, a frequent occurrence with other techniques of growth control. A major disadvantage of the bulb dip procedure is that growers have to commit themselves very early to a growth control program; also, at the present time there are no known remedial measures to promote stem elongation if excessive growth control occurs. Growers who consistently produce tall Easter lilies, perhaps because of crowding, low light intensity, or high temperatures near heating pipes or units, could dip bulbs prior to planting with little

TABLE 9.16. Effect of Different Ancymidol Treatments on Growth and Flowering of Two Easter Lily Cultivars

Cultivar	Treatment	Final plant height (cm)	Days to Visible bud	Days to Anthesis	No. flowers
'Nellie White'	Drench, 0.500 mg/15-cm pot	41	87	119	4
	Spray, 0.625 mg	40	87	119	5
	Bulb dip				
	16 ppm, 2 sec	41	86	118	5
	16 ppm, 60 min	36	88	120	4
	24 ppm, 2 sec	36	88	120	4
	24 ppm, 60 min	27	87	120	4
	33 ppm, 2 sec	28	90	124	4
	33 ppm, 60 min	30	95	125	4
	Untreated control	45	88	123	5
	LSD 5%	5	5	5	1
'Ace'	Drench, 0.500 mg/15-cm pot	46	72	102	4
	Spray, 0.625 mg	49	75	105	4
	Bulb dip				
	16 ppm, 2 sec	45	78	107	5
	16 ppm, 60 min	38	84	115	6
	24 ppm, 2 sec	50	85	116	7
	24 ppm, 60 min	23	78	112	5
	33 ppm, 2 sec	37	75	105	5
	33 ppm, 60 min	24	81	114	5
	Untreated control	53	72	101	5
	LSD 5%	8	6	6	1

Source: Unpublished data from N.C. State University (1983).

concern about excessive control. Lighting to promote flower initiation also can cause increased height, and such plants often require growth retardant treatments. Growers whose plants are inconsistent in height from one year to the next would want to try the bulb dip method on a limited basis.

Virus-free Easter lily plants have been evaluated at several universities and vigorous growth has always accompanied the virus-free trait. Growth control of Easter lily plants will become a greater challenge if bulbs from virus-free plants become readily available commercially. Virus-free and regular bulbs from Oregon State University have been compared at North Carolina State University for the past two years (Table 9.17).

TABLE 9.17. Growth of Virus-Free and Regular Easter Lily Plants[1]

Year	Type of bulb	Plant height (cm)	No. flowers
1982	Nellie White, regular	40	5.8
	Nellie White, virus-free	63	9.5
1983	Ace, regular	61	9.3
	Ace, virus-free	73	9.6
	Nellie White, regular	58	5.6
	Nellie White, virus-free	63	6.7

Source: Unpublished data from N.C. State University.
[1]Plants were not treated with growth regulators.

h. Foliage Plants. Research on growth control of foliage plants has lagged behind investigations on flowering potted plants. In many instances foliage plant growers would prefer rapid growth until plants have been purchased by customers. Blessington and Link (1980) believed that growth retardants might enhance acclimatization. Johnson *et al.* (1982) used ethephon and light intensity to control height and promote acclimatization of *Ficus benjamina* plants. They observed smaller and fewer leaves and drooping growth and poor quality in plants treated with 500 or 1000 ppm ethephon. Root growth was inhibited by ethephon when plants were grown in full sun, but root growth was unaffected when plants were grown in shade. Leaf abscission was most pronounced when plants were grown in full sun.

McConnell and Poole (1972) tried ancymidol very soon after it became available for research studies and reported positive control of vegetative growth of *Epipremnum aureum* plants. Their objective was to limit the vining habit so that excessive production space was not required. They found 100 ppm acymidol to be as satisfactory as 200 or 300 ppm.

Joiner *et al.* (1978) used four rates of ancymidol to control stem elongation of *Dieffenbachia maculata* plants; they observed increased chlorophyll content per unit area and improved acclimatization in treated plants. By testing ancymidol drench applications with different fertilizer application rates, they found an ancymidol drench of 0.66 mg ai/29-cm pot in combination with 1680 kg nitrogen/ha/year to be most effective.

Earlier research by Poole (1970) had shown that daminozide and UNI-F529 were effective on *Philodendron oxycardium, Epipremnum aureum,* and *Syngonium podophyllum*, whereas chlormequat and chlorphonium chloride had no effects. A concentration of 10,000 ppm daminozide was most effective.

McConnell and Poole (1981) listed the responses of about 70 species of foliage plants to chlorphonium chloride, daminozide, chlormequat, and ancymidol. Ancymidol was considered the most versatile growth regulator, as it was effective on more species than any other retardant.

Terrarium culture is not as popular now as it was a few years ago. A major criticism aimed at terrariums and the people who planted them was that improper plant material often was selected for terrarium use. Vigorously growing plants would quickly dominate the assortment of plants in the container, or all the plant material would grow too quickly and make the terrarium crowded and unattractive. The constantly humid atmosphere contributed to the problem of vigorous growth. Vienravee and Rogers (1974) tried to control growth of terrarium plants by ancymidol, daminozide, chlormequat, and chlorphonium chloride. Ancymidol was most effective on the six plant species tested. The growth regulator was incorporated in the medium in the terrarium, and then young plants were inserted. Proper plant selection still is the preferred approach for controlling growth in terrariums.

i. Geranium. Use of growth regulators to control flowering of geraniums has already been discussed (Section III. B. 2), but the effect on flowering often is secondary to that of growth. Articles by Carpenter and Carlson (1970), Semeniuk and Taylor (1975), and Miranda and Carlson (1980) are cited.

A more recent publication (Brisson and Larson 1983) reported on the use of chlormequat as an aerosol application to seedling geraniums. The advantages of this new approach are the ease and speed of application, even distribution in the greenhouse, and uniform coverage on plants. Chlormequat (2500 and 5000 ppm) was first applied to 'Sooner Red' and 'Smash Hit' geraniums on February 19 at rates of 1, 4.5, and 9.5 ml/m^3; a second application was made March 3. Treated plants flowered more quickly than untreated plants, and the 9.5 ml/m^3 treatment resulted in the shortest plants. One negative result was that the number of lateral shoots on 'Smash Hit' decreased as the chlormequat concentration increased. For example, in one experiment untreated plants averaged 14 axillary shoots, but plants receiving 9.5 ml/m^3 of chlormequat averaged only 9 axillary shoots. Axillary shoot development was unaffected on 'Sooner Red.' Treatment costs were $17.92 when clormequat was applied as a conventional foliar spray and $1.09 with the aerosol method.

Brisson (1983) studied the feasibility of using electrostatically charged particles of chlormequat. Uniform distribution of chlormequat was achieved with the Electrodyn-instrument. Both upper and lower leaf surfaces were covered with the charged particles unless plant density was excessive. The presence of plastic materials interfered with the coverage of the material, apparently because of the effect of plastic on grounding of the plants. Plants grown in plastic trays or placed on plastic sheets on the ground did not get covered with the spray as thoroughly as plants not surrounded or not in contact with plastic.

The original intent of the studies by Brisson and Larson (1983) was to control growth and flowering of seedling geraniums with chlormequat, without the blotchy yellowing of foliage that often is symptomatic of chlormequat spray applications. In these experiments, the least amount of yellowing occurred with the aerosol spray application, but phytotoxicity was also not a major problem with the more conventional methods of application.

The development of new and better cultivars has lessened the need for growth regulators to control vegetative development in geranium, but chlormequat is still used regularly by many growers for acceleration of flowering.

j. Gloxinia. The growth habit of gloxinias would seem to make use of growth regulators superfluous, but application of daminozide is recommended, particularly when the light intensity is less than 194 klx (Anon. 1980) and weak, elongated flower stems and leaf petioles are produced. A concentration of 0.1% daminozide applied 10–12 days after potting is recommended. Increased lateral shoot development, darker foliage, and shortened leaf petioles result from daminozide treatment. Sydnor *et al.* (1972) found that daminozide had the greatest effect on the chlorophyll content of leaves; it also affected plant height and diameter (Table 9.18) and even flower color. The anthocyanin content of flowers on 'Dwarf Delight' was much greater in plants treated with daminozide than in ancymidol-treated or control plants.

k. Hibiscus. Attractive flowers and foliage are among the attributes of hibiscus that make it promising as a pot plant. Growth must be regulated, to keep plants in balance with pot size and their expected use as a flowering houseplant. Ethephon spray applications controlled growth but stimulated lateral shoot growth (Shanks 1969); chlormequat controlled growth of both the main axis and lateral shoots and increased the number of flowers. Hibiscus is a popular flowering pot plant in Europe, and Wikesjö (1978) has suggested foliar

TABLE 9.18. Effects of Ancymidol and Daminozide on Growth and Chlorophyll Content of 'Dwarf Delight' and 'Royal Frosted Red' Gloxinias[1]

Treatment[2]	Plant height (cm)	Plant diameter (cm)	Leaf chlorophyll content[3] (OD at 665 nm)
Ancymidol spray			
62 mg/liter	19	41	0.162
125 mg/liter	19	40	0.164
250 mg/liter	18	38	0.171
Ancymidol pot drench			
0.12 mg/liter	19	41	0.165
0.25 mg/liter	19	42	0.177
0.50 mg/liter	18	40	0.177
Daminozide spray, 2500 mg/liter	16	37	0.198
Control	19	41	0.161
LSD 5%	1	2	0.011

Source: Sydnor *et al.* (1972).
[1]Data averaged for 2 cultivars because of similarity of results.
[2]Plants were grown in 15-cm plastic pots.
[3]Leaf chlorophyll content measured as optical density with a Bausch-Lomb Spectronic 30 Colorimeter at 665 nm.

application of chlormequat, at 700–1200 ppm, when lateral shoots are 3–4 cm long, with a second application 2–3 weeks later, depending on season of the year.

l. ***Hybrid Lilies.*** Interest in hybrid lilies has intensified in recent years, perhaps because of the wide assortment available and the versatility of hybrid lilies as cut flowers, potted, and garden plants. Growth control often is required, however, as the natural growth habits of many cultivars make them too tall for potted plants, and even excessively tall for cut flower production. Potential heights of some hybrid lily cultivars listed in a 1983 catalog are as follows:

For potted plants	For cut flower production
'Sunkissed' 45–55 cm	'Day Spring' 60–75 cm
'Connecticut Lemonglow' 45–60 cm	'Juliana' 60–75 cm
'Lovesong' 45–60 cm	'Peachblush' 60–75 cm
'Sunray' 45–60 cm	'Prelude' 60–75 cm
'Enchantment' 75–100 cm	'Harvest' 75 cm
	'Picasso' 75 cm
	'Firecracker' 75–90 cm
	'Goldrush' 75–90 cm
	'Matchless' 75–90 cm
	'Enchantment' 75–100 cm

These heights are the ranges expected without use of growth regulators. Plants exceeding 45 cm in height would require growth control to be acceptable as potted plants. There also is a maximum acceptable height for cut flower production, as plants that are too tall will not always remain erect and can be difficult to manage. Bulb suppliers have recommended drench applications of ancymidol at rates of 0.125–0.5 mg/15-cm pot in a 120 ml solution, applied at shoot emergence, for cut flower production. Potted lilies can be treated with ancymidol drench applications of 0.25–0.375 mg/pot when shoots emerge; a popular recommendation is to split application of ancymidol, applying the first treatment of 0.125 mg/pot at shoot emergence and the second one of 0.250 mg/pot 1 week later.

Stuart (1971) was one of the first to conduct growth regulator experiments on garden lilies as pot plants. Ancymidol drenches were effective at rates ranging from 0.25 to 1.0 mg/15-cm 3/4 pot; Five spray applications of 100-ppm solutions at weekly intervals were required for growth control. Dicks and Rees (1973) compared the effectiveness of ancymidol, chlormequat, chlorphonium chloride, and ethephon for growth control of 'Enchantment' and 'Joan Evans' hybrid lily treated in February when shoots were 6–7 cm long. Ancymidol-treated plants were 43–67% as tall as untreated plants, but several dead leaves were apparent at the bases of the stems. Dicks *et al.* (1974) applied ancymidol when inflorescences were initiated (plants were 2.6 cm tall), 10 days later (9.0 cm), and 20 days later (19.8 cm). The best time of treatment was 10 days after floral initiation; flower number was unaffected but anthesis was delayed by 4–6 days. Delayed application of ancymidol resulted in an increased number of dead leaves at the bases of stems.

Dicks *et al.* (1974) also evaluated drench applications, using vigorous root growth as a criterion. Gibberellic acid was tried in an attempt to counter ancymidol-induced growth retardation, if excessive. Pedicel length was increased over control plants when both chemicals were used, reflecting a synergistic reaction. Seeley (1975a,b) found that single ancymidol drench treatments, at concentrations of 0, 0.25, and 0.50 mg/pot, applied when lily plants were 5–7 cm tall, resulted in plants that were 35–50% shorter than untreated plants. Spray applications did not work well because leaf area was too limiting at the time of application. White (1976) tried to make Mid-Century hybrid lilies acceptable as potted plants by regulating their height. He found 0.5 mg ancymidol as a soil drench, applied at shoot emergence, to be satisfactory. He also used a 100-ppm ancymidol spray, applied when shoots were 7.5 cm long, and studied the relationship of cold storage and height.

The bulb dip approach to height control has been mentioned in the section on Easter lily growth control. Simmonds and Cumming (1977) applied chlormequat, ethephon, and ancymidol as a 12-hr dip on 'Enchantment' and 'Harmony' hybrid lilies after cold-storage treatment of the bulbs. The authors suggested that bulb suppliers could package chemically treated bulbs, rather than have forcers immerse bulbs in ancymidol solutions. Simmonds and Cumming (1978) later changed their approach and recommended treatment before cold-storage treatment (Table 9.19).

TABLE 9.19. Height of Hybrid Lily Cultivars Treated witn Ancymidol or Ethephon

		Height as % of controls	
Treatment	Cultivar	Pre-cold (0 weeks)	Post-cold (6 weeks)
Ancymidol, 10 ppm	Nutmegger	75	76
	Jamboree	48	48
	Black Beauty	48	33
Ethephon, 500 ppm	Nutmegger	41	44
	Jamboree	19	24
	Black Beauty	34	6

Source: Simmonds and Cumming (1978).

Hanks and Menhenett (1980) compared the effectiveness of several growth regulators in controlling height of Mid-Century hybrid lilies. Single drench applications were made 2–3 weeks after forcing was started. Ancymidol was the best growth retardant, but piproctanyl bromide and chlormequat also were active. Dikegulac controlled height but prevented flowering. Menhenett and Hanks (1982) found 5 mg paclobutrazol to be equivalent to 0.5 mg ancymidol per 15-cm pot when applied to 'Enchantment' plants.

Menhenett and Hanks (1982b) reported that virus-free hybrid lily plants were 62% taller than regular plants of the same cultivars, and leaf spread was 33% greater; in addition, floret number was doubled and there were fewer dried leaves at the base of virus-free plants. Two to three times more growth retardant was required to achieve height control on virus-free plants than on regular plants.

Seeley (1982) attempted to control garden lily height with drench treatments of ancymidol to bulbs planted in 12-cm pots in a medium composed of equal amounts of soil, perlite, and coarse moss peat. Two cultivars did not respond, but 'Enchantment' and 'Connecticut Lemonglow' plants were responsive to the ancymidol treatments (Table 9.20).

m. Hydrangea. The hydrangea has been a popular spring-flowering pot plant for many years, but excessive height has always been a potential problem. Plants marketed for Mother's Day usually present a greater height problem than the Easter crop, particularly if a tall-growing cultivar such as 'Rose Supreme' is grown. There is a movement toward the use of naturally low-growing plants, but some of the most popular cultivars now available do require some growth regulation. Major research studies on hydrangea growth control have been summarized by Weiler (1980).

Daminozide was the first growth regulator to successfully control stem elongation of hydrangeas, and it still is the most popular retard-

TABLE 9.20. Height of Four Hybrid Lily Cultivars Treated with Ancymidol Drench Treatments

	Plant height (cm)		
	Ancymidol drench treatment		
	0 mg/ 12-cm pot	0.25 mg/ 12-cm pot	0.50 mg/ 12-cm pot
Firecracker	46	44	39
Enchantment	44	19	11
Connecticut Lemonglow	45	37	20
Sans Souci	35	30	30

Source: Seeley (1982).

ant (Shanks 1975). Plants can be treated in the field by the propagator, but generally they are treated during the greenhouse forcing phase. Foliar applications in the summer months are usually in concentration ranges of 0.50–0.75% active ingredient, while concentrations of 0.25–0.50% are used during forcing. The recommended time of application during forcing is 2–4 weeks after initiation of forcing. However, many growers prefer to use plant growth as a guideline for application, making the first application when four to five pairs of leaves are evident and a second application, if needed, 1 week later.

Ancymidol also is effective in controlling hydrangea growth (Shanks 1975; Tjia *et al.* 1976). Concentrations of 50– 100 ppm have been effective, applied as sprays at the same time or stage of development deemed best for daminozide application.

The growth retardant paclobutrazol successfully controls stem elongation in hydrangea (Shanks 1980a). Scott (1982) applied either one or two applications of paclobutrazol at various concentrations to 'Rose Supreme' plants; the first application was made when four or five pairs of leaves had unfolded and the second application was made 2 weeks later. She found that the best paclobutrazol treatment on 'Rose Supreme' plants was two foliar applications at 50 ppm. Plants given this treatment averaged 37 cm in height, whereas untreated plants were 50 cm tall and daminozide-treated plants (2500 ppm twice) were 38 cm tall. Two paclobutrazol applications at 100 ppm or four applications at 50 ppm produced very short plants, small cymes, and delayed flowering, all evidence of overtreatment.

n. Kalanchoe. Tall-growing kalanchoe cultivars, such as 'Vulcan', were common before growth regulators became available, and 'Tom Thumb' became popular because of its compact growth habit. The use of growth regulators on more recently introduced cultivars has been reviewed by Love (1980).

Daminozide is probably the most frequently used growth regulator on kalanchoe. Nightingale (1970) reported that daminozide produced superior plant form on 'Mace'. Daminozide can be applied as a foliar spray 3–5 weeks after the beginning of photoinductive short days. Heins *et al.* (1979) recommended spray treatments of 5000 ppm daminozide, applied every 2 weeks after pinching tall-growing cultivars; they also reported a 1-week delay in flowering. Carlson *et al.* (1977) found the best daminozide treatment on 'Mace' to be 0.25–0.50 mg ai/10-cm pot, applied 2 weeks after the start of short days.

o. Narcissus. Briggs (1975) used ethephon as a soil drench to control growth of narcissus. A concentration of 960 ppm resulted in plants that were 34% shorter and had 30% shorter leaf length than untreated plants. Moe (1980) found the optimum ethephon drench concentration to be 100 mg/13-cm pot. The preferred time for application was when shoots were 8–10 cm long, as flowers aborted if ethephon was applied too early; late applications were ineffective. Guidelines have been published by De Hertogh (1982) for narcissus growth control with ethephon, including cultivar requirements and recommended time of application. He recommended applying ethephon when floral stalks are 10–12 cm long, with a possible second application 2–3 days later.

p. New Guinea Impatiens. Attractive flowers and foliage, versatility as a flowering potted plant or bedding plant, the ability to withstand full sun, and minimum labor requirements make New Guinea impatiens appealing to growers and gardeners. Inadequate spacing can cause plants to form undesirable shapes and appearance, and once plant shape has been adversely affected it is difficult to improve plant appearance. Pasutti and Weigle (1980) found that high rates of chlormequat, ancymidol, and daminozide were required to achieve the desired growth control. Ancymidol was the most effective growth regulator, but cultivar differences were so great that it would be difficult to make general recommendations. Chemical regulation of growth in this crop is not considered satisfactory.

q. Orchid. Very little research has been reported on the use of growth regulators to control vegetative growth of orchids. Adriansen (1980) did use growth retardants to control scape length of *Paphiopedilum*, but treatments were not too successful, and the duration of floral display was shortened. Untreated plants retained attractive flowers for 11 weeks, whereas flowers on ethephon-treated plants lasted only 7–9 weeks.

r. Poinsettia. Perhaps no other floral crop has been as thoroughly investigated for chemical growth regulation as poinsettia. Its high economic value may have some bearing on the popularity of poinsettia as a test species for evaluating growth regulators. Earlier cultivars were also very tall growing, and sweeping changes in poinsettia culture have also contributed to the need for growth control. Precise temperature control, generous watering and fertilizing practices, well-aerated growth media, and effective disease control stimulate poinsettias to grow rapidly and vigorously.

It was fortunate that some of the earliest research on the use of growth regulators was conducted on poinsettia, as positive results often were obtained, although isomerization of carvadan did result in inconsistent results (Cathey 1959). Pioneer investigators on the use of chlormequat to control poinsettia growth were Lindstrom and Tolbert (1960) and Lindstrom (1961). These workers found that 'Barbara Ecke Supreme' responded favorably to chlormequat drench applications and that short plants with dark green foliage were characteristic of plants treated with chlormequat. Since the early work of Cathey and of Lindstrom and Tolbert, almost every horticultural department in the United States has had some members doing research on poinsettia height control with chemicals. The development of the poinsettia as a commercial crop in Europe in the 1960s prompted growth control research at several European experimental stations and universities.

Chlormequat was the first consistently effective growth regulator used on poinsettias, and it still is the most widely used retardant almost 25 years after the initial demonstration of its effectiveness. The material originally only had label clearance for poinsettia cultivars with red bracts and was to be applied as a soil drench when root systems were well established. It now is permissible to use chlormequat on all poinsettia cultivars, as a foliar spray or as a drench. Only one-third as much chlormequat is needed with foliar applications as with drenches, and sprays are also easier to apply. Thus, foliar sprays now are used more frequently than chlormequat drenches. Synergistic effects of combining chlormequat with daminozide were reported by Shanks (1981a), and relaxed regulations on "tank mixes" could prompt use of this successful combination.

A comprehensive review of the first years of research on poinsettia growth control with chlormequat and daminozide was prepared by Larson (1967). Since then ancymidol has become commercially available, and there are several research reports pertaining to that compound. Other compounds (e.g., paclobutrazol) have also been formulated and tested. In addition, various application methods have been examined, and some innovative approaches have been reported.

Larson and McIntyre (1968) treated stock poinsettia plants in 15-cm clay pots with chlormequat, either as soil drenches or foliar applications at 2950 ppm, and then removed cuttings from the stock plants at various intervals. Plants from cuttings removed from 'Indianapolis Red' stock plants that had been treated with a chlormequat drench on August 17 elongated 17 cm before flowering, contrasted with 42 cm for cuttings from control plants. Even greater height control was achieved when cuttings removed from treated plants were treated again after potting. Chlormequat was effective longer when applied as a drench rather than as a spray. Drench applications to stock plants in large containers would be expensive. A modification in technique was to treat plants in 15-cm pots, remove the terminal growth as a pinch 1 week later, and realize growth control on both generations of plants.

Holcomb and White (1970) applied low concentrations of chlormequat at each irrigation, in a manner similar to fertilizer injection, which is commonly practiced in commercial floriculture. They noted that poinsettia plants could outgrow the effects of one or two applications at a higher concentration applied early in the season and a late application at high concentrations could have undesirable side effects. White and Holcomb (1974) later expanded this approach to applications of ancymidol as well as chlormequat. Multiple injections of growth regulators at low concentrations did not work as well as single applications at higher concentrations.

Chlormequat recommendations have not changed much since they were first developed. The expression of concentration on a ratio basis has made preparation of dilute solutions a relatively simple task, and probably has contributed to the popularity of chlormequat. Most drench applications consist of 1 part chlormequat (11.8% ai) to 40 parts of water, to yield a 2950-ppm solution. Yellowing of portions of the leaves would be quite pronounced if foliar applications were made at that concentration, so a 1:60 or even 1:80 dilution often is used, with more than one application, if necessary. Conover (1970) observed greater phytotoxicity when chlormequat was applied as a foliar spray to poinsettia plants grown at 27° or 32°C, compared with plants grown at 21°C. Light intensity also affected foliar injury, and the least amount of damage occurred when plants were treated under a 20% light reduction. Conover and Vines (1972) found that high fertilization rates after chlormequat foliar spray applications partially overcome the chlorosis.

The emergence of ancymidol as a major growth regulator in floriculture has been discussed in the sections on other crops. Ancymidol has not replaced chlormequat in poinsettia production, but it has stimulated interesting research on application techniques. Einert (1976) soaked new clay pots in ancymidol solutions of 5–100 ppm. A solution of 40 ppm was best for 'Eckespoint C-1 Red' and 'Eckespoint C-1 New Pink', whereas 20 ppm was phytotoxic to 'Eckespoint C-1 White' plants. Einert calculated that 10 ppm in the impregnated pot was equivalent to 1.0 mg ai/pot. Wilfret *et al.* (1978) used granular ancymidol at a concentration of 0.125 mg ai/10-cm pot and found it equivalent in effectiveness to a 0.5-mg ancymidol drench. The granular ancymidol was formulated by spraying the growth regulator on finely ground clay.

Tjia *et al.* (1976c) compared ancymidol, chlormequat, and ethephon soil drenches on poinsettias and concluded that the optimum concentrations for the different retardants were chlormequat, 2950 ppm (1:40 dilution); ancymidol, 0.5 mg ai/pot; granulated ethephon, 2.0, 3.5, and 7.0 g; and ethephon, 1250 ppm. They preferred drenches over foliar sprays because they considered drench treatments to be more precise and predictable, to cause less phytotoxicity, and to be less affected by such factors as relative humidity, temperature, and light. The larger volume required and higher cost of drench treatments were the major disadvantages. Treatment recommendations for growth control of poinsettia with ancymidol, chlormequat, or daminozide were summarized by Larson *et al.* (1978).

Paclobutrazol successfully controlled height of poinsettias when drench applications of 0.125–0.50 mg ai/15-cm pot were used (Anon. 1982). Shanks (1980) found paclobutrazol effective, but its cost of application on a commercial basis has not been established. Holcomb *et al.* (1983) found that mepiquat chloride (BAS 106), which is available as a granular material or a wettable powder, controlled height in poinsettia, but some early injury occurred at the highest rates. Node number on 'Paul Mikkelsen' was less on plants treated with BAS 106 than on plants treated with ancymidol or chlormequat, an unusual occurrence as node number is seldom affected by growth regulators. Shanks (1981) also showed the effectiveness of mepiquat chloride in controlling poinsettia height.

s. *Tulip.* Tulips are being grown in increasingly larger numbers as production scheduling is improved and cultivars are made available in an assortment of colors and flower forms. Plant height of tulips is usually not excessive while they are still in the forcing stage in the greenhouse, but plants often become too tall when placed indoors. Shoub and De Hertogh (1974) studied the interaction of ancymidol and gibberellic acid in growth regulation. Scape length was less when ancymidol was applied within the first 6 days in the greenhouse; GA_{4+7} reversed the growth-retarding effects, while GA_3 had no effect. Briggs (1975) found that ancymidol limited the stem length more than the leaf length of tulip plants, but the height of the topmost internode was scarcely affected. He suggested that ancymidol should be applied when stems are 5–10 cm tall.

According to De Hertogh (1976), ancymidol controlled total plant height, resulting in tulip plants 25–30 cm in height, and it reduced stem topple. Height was controlled because the lowest internode was shorter than on untreated plants. Ancymidol was more effective when applied in the greenhouse than when used in dark storage. Plants that were properly programmed or properly treated with ancymidol were of high quality. A potting mixture of peat:soil:perlite or sand was suggested.

Hanks and Menhenett (1979) found that time of tulip bulb harvest, bulb treatment, and cultivar affect plant height; ancymidol was the most effective compound of several tested for controlling height. Undesirable side effects of some growth regulator treatments include early petal drop and flower bud abortion.

Kincaid and McDaniel (1979) suggest three ways for tulip forcers to produce plants of the desired height: selection of low-growing cultivars, a shorter duration of low-temperature treatment, and use of ancymidol. They preferred the bulb dip method to soil drenches because with this method the retardant was present in the early stages of development, there was no persistence in the medium, and it was effective. Bulb dip concentrations of 0, 2, 4, 8, and 10 ppm were used.

Moe (1980), in studies with 10 tulip cultivars, found that the optimum time for ethephon application was when shoots were 8–10 cm long; flower bud abortion occurred if plants were treated earlier. Growth retardation was similar with ethephon solutions of 500, 1000, and 2000 ppm, applied at 15°, 18°, or 21°C. Paclobutrazol also has been shown to control tulip growth in the greenhouse and later, when plants were placed under simulated home conditions (Menhenett and Hanks 1982a).

D. Lateral Shoot Development

1. Disbudding. The practice of disbudding is primarily of concern to chrysanthemum and carnation growers. Because manual removal of lateral flower buds is costly, tedious, and can be damaging if done improperly, growers have searched for easier and less expensive methods of disbudding. Cultivar selection is one method, as cultivars vary greatly in lateral flower bud production. Chrysanthemum plants in the 'Princess Anne' series are quite easy to disbud, as there are relatively few leaves and buds on the upper plant parts. Disbudding also has been hastened by use of the center bud removal method, in which only the terminal flower buds are removed, but this method is not adequate on all cultivars. The long search for a "wonder" chemical to disbud chrysanthemums has not been successful.

Some of the earliest reports of chemical disbudding described incidences of lateral flower bud death caused by improper pesticide usage. However, attempts to duplicate this effect under controlled conditions usually gave inconsistent results. Cathey *et al.* (1966a) tried alkylnaphthalenes, solvents used in chlorinated insecticides, as chemical disbudding agents; Kofranek and Markiewicz (1967) tried naphthalenes. Kofranek and Criley (1967), using emulsifiable oils, found that English cultivars 'Fred Shoesmith' and 'Princess Anne' responded more favorably than the American cultivars 'Albatross', 'Good News', and 'Indianapolis White'. Disbudding agent effectiveness was dependent on temperature, time, and rate of application.

In 1975, Uniroyal introduced oxathiin (UBI-P293), which showed more promise as a chemical disbudding agent than any previous growth regulator (Anon. 1975b). Results from a coordinated effort to determine its effectiveness were reported by Cathey (1976), Parups (1976), and Zacharioudakis and Larson (1976). Cathey (1976), working with seven prominent chrysanthemum cultivars, observed the best results with 0.25–1.0% oxathiin, applied after 15–24 short days, but there was considerable variability in response. Parups (1976) treated six standard chrysanthemum cultivars and noted phytotoxicity in some treatments and smaller flower size. Oxathiin treatment followed by some manual disbudding 2 weeks later produced the best results. Zacharioudakis and Larson (1976) considered 0.5% oxathiin to be the optimum concentration when chrysanthemum plants were treated on the eighteenth short day, while 1.0% was best if plants were treated on the twenty-first short day.

These studies show that effective disbudding with oxathiin is influenced by cultivar, number of short days, temperature, relative humidity and light intensity. The use of calendar days as guidelines for treatment application is hazardous because lateral and terminal flower buds will not always be at the same stages of development after a given number of short days in different seasons. There was a very fine line in the studies between effective lateral flower bud removal and phytotoxicity or other adverse effects (e.g., damage to the terminal flower bud sometimes occurred). Menhenett (1980) later reported that a foliar application of 0.2% oxathiin eliminated half of the axillary flower buds on 'Bright Golden Anne'.

In recent years, there have been no reported studies on chemical disbudding. This is unfortunate, but the variability in results seems to have discouraged investigators and made them reluctant to make specific recommendations for use of oxathiin. Although oxathiin seemed to possess most of the attributes necessary for a chemical disbudding agent, it has been withdrawn.

2. Pinching. Apical dominance is encountered in the production of most floricultural crops, and cultural procedures often are geared to this phenomenon. The removal of shoot apexes to overcome apical dominance and to promote lateral shoot development is referred to as *pinching* in the United States and *stopping* in England. Pinched cut flower crops usually are more productive than single-stem crops, and pinched potted plants are more floriferous and shorter than single-stem plants. Pinching of chrysanthemums, poinsettias, carnations, and a few other crops is not a major task, as each plant has only one major axis, and removal of shoot apexes can be done rather quickly. Pinching requires more labor with azaleas, which have multiple shoots and a relatively long production time; roses also require considerable labor to pinch.

Some of the earliest research on chemical removal of apical dominance involved crops such as chrysanthemums, but the greatest commercial acceptance of the technique has occurred in azalea production. In some instances, chemical pinching improved lateral branching, justifying its use, even though manual pinching was not necessarily laborious or tedious.

Cathey *et al.* (1966a) were the first to report studies on chemical pinching or pruning of floricultural crops, but similar studies had previously been conducted in tobacco. The chemicals were methyl esters of fatty acids and the most prominent one was methyl decanoate. Procter and Gamble Distributing Company formulated one of the first commercially available chemical pruning agents, called Off-Shoot-O (Anon. 1969), and it has remained a prominent growth regulator.

Cathey (1970) discussed the plant characteristics that help determine the effectiveness of chemical pinching agents. Plants in a vegetative state, with elongated stems, and with foliage that readily absorbs chemicals respond quite readily to chemical pinching agents; on the other hand, plants in the reproductive state, with waxy foliage, and in which growth occurs in flushes do not respond much. Environmental factors also can affect response. For example, chemicals may be more effective at 30°C than at 13°–29°C because penetration is delayed at temperatures less than 13°C and greater amounts of pinching agents might be required to produce the same effects obtained at higher temperatures. Cathey listed the optimum dosage for chemical pinching of a large number of ornamental plants, and readers should refer to the list for further information.

a. Azalea. The use of chemical pinching in azalea production has been reviewed by Shu and Sanderson (1983) and Larson (1983). Stuart (1967, 1975) led the way in chemical pinching of azaleas and his research reports were excellent guides for later studies. It was soon apparent that chemical pinching with methyl decanoate did not produce as consistent effects as researchers had encountered with other growth regulators on other plant responses. It was already known that apex damage only occurred if the pinching agent physically contacted shoot tips, so translocation and other physiological processes were not involved. Chemical preparation, environmental factors, and cultivar characteristics were thought to influence the effectiveness of methyl decanoate.

Methyl esters of fatty acids require careful mixing to achieve a solution of uniform concentration. Some researchers and commercial growers have experienced excessive phytotoxicity on some plants, without apex damage. These problems can often be traced to poor preparation of the dilute solution. Cultivar differences also have frequently been observed when several cultivars were treated with the same concentration of methyl decanoate under the same environmental conditions. Furuta *et al.* (1968), for example, reported stimulated lateral branching on 'Coral Bell', but 'White Gish' plants did not show much response at the same concentration of methyl decanoate. They also found that the effectiveness of the chemical pinching agent was influenced by age of shoot and stage of apex development: Apexes that had undergone transition to the reproductive stage were not as responsive as vegetative shoot tips.

Nelson and Poteet (1969) and Sill and Nelson (1970) found that the cuticle thickness of leaves affected phytotoxicity of methyl decanoate. Concentrations that were regarded as optimum for apex damage and lateral shoot initiation resulted in leaf burn if cuticular layers were lacking (e.g. on newly unfurled foliage). Leaf length also influenced effectiveness and phytotoxicity. Thus, methyl decanoate was effective on 'Coral Bell', which has short leaves surrounding shoot apexes, but was ineffective on 'White Gish', which has longer leaves that protected shoot tips. Pubescence (trichomes) on the foliage also decreased apex damage, reducing the effectiveness of methyl decanoate.

Environment also has a major impact on the effectiveness of methyl decanoate. Carr and Lindstrom (1972) found that plants preconditioned at 10°C night and 15°C day temperatures responded better than plants preconditioned at 26°/32°C (night/day). Shoots 4–6 weeks old when treated were more readily affected than younger or older shoots. In areas where the relative humidity is low, one can expect rapid drying of the solution, and the chemical may evaporate before apexes are damaged. However, phytotoxicity may be excessive if the relative humidity is so high that the methyl decanoate solution requires a long period to dry. Type of spray pattern is not as crucial as type of nozzles and the spray droplet size.

Rather complete application instructions for Off-Shoot-O have been published by the Procter and Gamble Distributing Company (Anon.

1969). Researchers and growers are urged to treat a few plants with several different concentrations and to look for apex damage or leaf injury 30–45 min after application. The best treatment then can be repeated on a larger scale the following day if weather conditions remain the same. Suggested trial concentrations are listed in Table 9.21. The use of volume ratios of methyl decanoate to water makes preparation of dilute solutions quite simple.

Proper observance of label directions should result in successful chemical pinching of azaleas with methyl decanoate in a relatively short time. Application of chemical pinching agents requires only a few hours, whereas manual pinching of the same number of azalea plants could require several weeks of diligent work. When undesirable growth, such as very long shoots, is present, it usually is removed manually before methyl decanoate is applied to promote lateral shoot growth on the vast majority of shoots. The interval between manual pinching of plants and application of methyl decanoate should be short enough so that lateral shoots have not begun to emerge before chemical application; such young shoots could be severely damaged by methyl decanoate, limiting new shoot development. Chemical pinching can also be done first, closely followed by manual pinching and shaping of plants. Manually pinched plants usually display new lateral shoots before plants that have been treated with methyl decanoate, but shoot numbers should be increased with the chemical treatment. More leaf axils remain intact with chemical pinching.

Dikegulac, another chemical pinching agent, has a completely different mode of action than methyl decanoate. It is readily translocated to shoot apexes when applied as a foliar spray, so direct contact with shoot tips is not necessary. There are other differences between methyl decanoate and dikegulac. Shoot apex injury is apparent within an hour or so after methyl decanoate is applied, whereas the effects of dikegulac may not be apparent for 2 weeks or longer, when some chlorosis or red pigmentation occurs. Lateral shoots will initiate and develop faster on manually pinched azalea plants than on plants treated with dikegulac of methyl decanoate, but lateral growth is

TABLE 9.21. Suggested Methyl Decanoate Concentrations

Methyl decanote to use (ml)		Active ingredient (%)	Volume ratio—methyl decanoate to water
To 945 ml of emulsion	To 3 liters of emulsion		
60	240	2.8	1:15
75	300	3.5	1:12
90	360	4.2	1:10
105	420	5.0	1:8

Source: Anon. (1969).

delayed more by dikegulac than by methyl decanoate. Apexes tend not to abort on plants treated with dikegulac, but lateral shoots arise at the leaf axils, often extending down to the seventh or eighth node. New foliage is often abnormally thin and develops symptoms similar to those of copper deficiency; foliage eventually widens, but leaves on dikegulac-treated plants never reach the size of leaves on plants that have been pinched manually or with methyl decanoate.

Dikegulac was developed in the mid-1970s (Anon. 1976) and immediately showed potential for pinching several ornamental plant species. It has successfully promoted lateral branching on herbaceous plants such as Rieger begonia, calceolaria, kalanchoe, peperomia, salvia, pachystachys, and verbena (Anon. 1979a). Spray concentrations of 0.05–0.16% dikegulac (ai) have been adequate for herbaceous plants. More woody species, such as bougainvillea, fuchsia, clerodendrum, *Hedera helix*, and *Ficus*, require concentrations of 0.1–0.3%. The formulator recommends using 0.23–0.66% for greenhouse azalea plants.

Heursel (1975), working in Belgium where azaleas are very important economically, was among the first to investigate the possible use of dikegulac as a chemical pinching agent. Other early reports on its mechanism of activity and characteristics were published by Bocion *et al.* (1975), Bocion and de Silva (1976), de Silva *et al.* (1976), Arzee *et al.* (1977), and Adriansen (1979). Their research indicated that environmental conditions did not influence the activity of dikegulac as much as they influenced methyl decanoate. The best results could be expected when plants were grown at 16°C day and 12°C night temperatures, at relative humidities of 70 and 90% for day and night, respectively (Bocion and de Silva 1976). Several suggestions for the use of dikegulac as a chemical pinching agent can be summarized as follows:

- Only vigorous, healthy plants should be treated.
- Temperature of the dikegulac solution should not exceed 30°C.
- Plants should not be watered for 24 hr after treatment.
- Repeat applications should be avoided.
- Dikegulac should be applied alone.
- Dikegulac should be applied on the day it is prepared.
- Actively growing shoots, 3–5 cm long, should be treated.

Concentrations of 0.23–0.39% are now suggested for Belgian and Southern indica types, and 0.39 and 0.66% for Kurume, Gish, and Stewartsonian azaleas. The same plants reportedly can be treated two or three times per year, at 8-week intervals.

Larson (1978) was one of the first researchers in the United States to study dikegulac as a pinching agent on azaleas. Combinations of manual pinching and dikegulac application, or methyl decanoate followed by dikegulac applications, were tried with several cultivars. Best results were obtained when the most prominent shoots were manually pinched and dikegulac was applied within a week. With this treatment, lateral shoots were produced in increased numbers, and plants could be tailored to a desirable form. The use of chemical pinching reduced but did not eliminate hand labor. Several disadvantages of dikegulac were noted, including delayed appearance of lateral shoots and narrow, young foliage. Dikegulac acts as a growth retardant as well as a pinching agent, so shorter shoots can be expected on chemically treated plants. Efforts to decrease the delay in lateral shoot development and to overcome the narrowness of new foliage have not yet been successful.

b. Chrysanthemum. The economic importance of chrysanthemums and the large number of pinched plants that are produced commercially made them a logical crop to include in the earliest studies on chemical pinching (Cathey *et al.* 1966; Cathey 1968, 1970). Stem girdling discouraged commercial use of methyl decanoate for removal of apical dominance. Uhring (1971) examined plants histologically and detected damage 1 min after application of methyl decanoate. A frequent cause of injury was the accumulation of methyl decanoate in leaf axils; axillary buds were killed following chemical pinching, nullifying the potential benefits desired.

Menhenett (1978) used oxathiin at concentrations of 0.4–0.8% ai to chemically pinch chrysanthemums. The chemical was applied to shoot apexes a few days after planting. Its effectiveness as a pruning agent was influenced by time of application, as it is for disbudding (Section III. D. 1).

c. Foliage Plants. The reasons for desiring lateral shoot development on foliage plants vary, depending on the plant species, its use, and the objectives of researchers and growers. Maximum numbers of lateral shoots often are required for stock plants, and additional shoots sometimes improve plant shape of finished plants.

Criley and Oka (1976) applied cytokinins and ethephon on the *Dieffenbachia picta* cultivar 'Rudolph Roehrs' and found that rooting was depressed when 250 and 500 ppm of a cytokinin were used, while ethephon slightly improved rooting. Maene and Debergh (1982) tried to promote shoot development of *Cordyline terminalis* 'Celestine Queen' plants with 6-benzylaminopurine (BA) dissolved in 1% dimethyl sulfoxide (Table 9.22); they concluded that 250 ppm BA was

TABLE 9.22. Lateral Shoot Development of *Cordyline Terminalis* Treated with 6-Benzylamino Purine (BA)

BA treatment[1] (ppm)	No. shoots	Shoot length (cm)	Rooting (%)	True to type (%)
100	7	4.0	100	100
250	25	3.4	86	78
500	35	2.8	48	45

Source: Maene and Debergh (1982).
[1]BA dissolved in 1% dimethyl sulfoxide.

the best treatment. Higaki and Rasmussen (1979) increased the growth and flowering of anthurium plants when they applied 1000 ppm BA to promote adventitious shoot production. Chlorosis and leaf abscission were caused by applications of 1000 and 1500 ppm ethephon. Chemical pinching agents undoubtedly will be used more extensively in the future on foliage plants, perhaps primarily to increase propagative units of plant species in which terminal cuttings are used.

d. Opuntia. Although opuntia is of minor importance in commercial floriculture, successful chemical stimulation of new growth of this succulent-type plant might serve as a model system for similar species. White *et al.* (1978) increased secondary shoot growth with applications of GA. Cell expansion was stimulated and even spines elongated; new growth followed. The cytokinin PBA[6-(benzylamino)-9-(2-tetrahydropyranyl)-9*H*-purine] caused effects almost completely opposite to those prompted by GA.

e. Poinsettia. Multistemmed poinsettia plants are becoming increasingly popular with commercial growers because of their lower production costs. A fuller plant shape also is achieved with lateral branching. Poinsettias can be manually pinched quite rapidly, but chemicals might stimulate increased branching of some cultivars, such as the 'Eckespoint C-1' series, that do not break readily when shoot apexes are removed. A chemical pinching agent might increase either the number of lateral shoots or improve the quality of shoots that have initiated. Cultivars in the Hegg or Gutbier series do not require additional stimulation of lateral shoots, as almost every leaf axil below the pinch seems to give rise to an acceptable shoot in these cultivars.

Carpenter *et al.* (1971) applied PBA (200 and 300 ppm) and BA (500 and 1000 ppm) to 'Eckespoint C-1 Red', 'Annette Hegg', and 'Dark Red Annette Hegg' plants. Lateral branching was improved on all three cultivars. Ethephon also increased lateral shoots, but not as effectively as cytokinins did. Some chlorosis occurred when BA was used, but the disorder disappeared in 2–3 weeks. Carpenter and Beck (1972) used similar chemicals on stock plants. Cuttings removed from stock plants that had been treated with BA 30 days before removal were delayed in rooting. Shanks and Purohit (1970) tried numerous chemicals to promote lateral branching; 3000 ppm dikegulac was found to be most effective on 'Eckespoint C-1 Red' plants. Shanks (1981) later reported that dikegulac stunted plant growth and could only be used under good growing conditions.

f. **Rose.** Greenhouse rose plants usually remain in production for at least 5 years, and rose growers become concerned about the development of basal breaks as plants mature. Pruning practices vary, but the objectives are to maximize yield of salable flowers per plant and to keep plant height within bounds.

Application of cytokinins did not increase basal branching in roses, but axillary shoot development was promoted (Carpenter and Rodriguez 1971). Parups (1971) used BA and adenine to rejuvenate 'Forever Yours' and 'Regal Gold' plants. Chemicals were applied as lanolin pastes at each bud. He found that rose stems must be wounded above and below the buds for penetration of the chemicals to occur, and suggested that the chemicals should be applied to vigorously growing plants. The numbers of breaks induced by BA and adenine increased under high light intensity, high fertility, and high CO_2 conditions.

Zieslin *et al.* (1972) found that similar environmental conditions promoted basal branching on 3-year-old plants of 'Baccara' treated with ethephon; one application at 1000 ppm was optimum. Their technique for applying ethephon was to wound shoots with a saw blade that had been dipped in ethephon. The number of renewed shoots per plant after various treatments was as follows: control, 0.3; ethephon alone (7500 ppm), 0.3; scoring alone, 0.9; and scoring plus ethephon, 1.6. Spraying 'Baccara' rose plants with ethephon to promote basal branching has become a commercial practice in Israel.

Faber and White (1977) used PBA on 'Red American Beauty' and were dissatisfied with the results. They believed that their plants were too mature for the experiment as only a limited number of basal shoots were produced.

E. Propagation

The application of growth regulators to promote rooting has been a commercially accepted practice for many years. I will not discuss the numerous reports on the use of hormone-like rooting agents (e.g., IBA), but will focus on studies conducted with nonauxin growth regulators.

Some fungicides have been reported as stimulatory or inhibitory for rooting.

1. Vegetative Propagation. Magie (1971) increased corm production by 50% and cormel production by 90% in gladiolus by applications of ethephon with a fungicide. The purpose of the fungicide was to control *Fusarium*; the ethephon apparently acted synergistically to decrease losses. Corm size was reduced by ethephon, but that disadvantage was outweighed by the benefits of increased corm and cormel yields. *Fusarium* promotes ethylene synthesis in bulb crops, causing unexpected morphological and physiological changes.

Read *et al.* (1972) reported increased tuber size and number in *Dahlia pinnata* after treatment with 2500 ppm daminozide applied 23 days after potting and repeated on the thirty-ninth day, and with 2360 ppm chlormequat applied 23 days after potting. Simmonds and Cumming (1976) used 5μM BA and 5μM NAA as 12-hr bulb dips, along with fungicide treatment, to increase production of bulblets of hybrid lilies. The bulb dip was only effective on clones of cultivars that were naturally poor producers of bulblets; the heavy bulblet producer 'Empress of India' produced fewer bulblets after application of growth regulators.

Read and Hoysler (1969) reported increased root length and branching of geranium, chrysanthemum, and dahlia from cuttings dipped in 2500 ppm daminozide; chlormequat was ineffective. Stem base treatment was more consistent than foliage dips. Read and Hoysler (1971) later expanded their investigation to include carnation and poinsettia cuttings. When they dipped the basal 25 cm of the cuttings in 2500 or 5000 ppm daminozide, rooting in both species was promoted by a 15-sec dip. Beck and Sink (1974) were unable to stimulate rooting of poinsettia cuttings by applying ancymidol, chlormequat, or daminozide.

2. Seed Propagation. Miller and Holcomb (1982) soaked primula seed for 21 hr in GA_3 solutions at concentrations as high as 1000 ppm; they found 250 ppm to be most effective for increasing germination. Temperatures above 21°C inhibited germination and only 61% of untreated seed germinated; germination increased to 76% after GA_3 treatment at the same temperature. Anderson and Widmer (1975) reported that cyclamen seed soaked in 100 ppm GA_3 showed accelerated germination, but this benefit was negated by increased embryo expulsion and death.

F. Protection

Increased stem and leaf thickness and other anatomical changes that occur in stressed plants have prompted researchers to determine whether growth regulators could act as protectants for floricultural crops under stress. The decreased water loss observed after application of some growth regulators suggests that increased drought and/or heat tolerance could be achieved; indeed, some early technical data sheets emphasized the increased heat and drought tolerance of chemically treated plants. It is not surprising, therefore, that researchers have investigated growth regulators to improve plant resistance to adversity caused by environmental factors or by chemical pollutants.

1. From Pollutants. The protection from pollutants provided by growth regulators is an added bonus to their growth-controlling effects. Plants that had been treated with growth regulators have been observed to withstand the effects of pollutants at their final destination. Since greenhouses often are located very close to metropolitan areas, with the accompanying threat of damage from ozone and other pollutants, pollutant protection during crop production might be very advantageous.

Cathey and Heggestad (1973) treated plants of eight poinsettia cultivars with ancymidol and chlormequat to determine if they provided protection from ozone and sulfur dioxide. Pronounced differences were noted in cultivar sensitivity to SO_2: 'Annette Hegg' was affected by exposure for 3 hr to 3 ppm of SO_2, whereas 'Eckespoint C-1 Red' plants were unaffected. Both retardants reduced visible injury by the pollutants. Leaf blades were smaller on treated plants, with smaller cells and smaller intercellular spaces, which apparently inhibited the penetration of the pollutants.

Cathey and Heggestad (1982a) determined that ethylene diurea protected petunia plants exposed to ozone more effectively than did daminozide. Protection from ozone damage was evident within 24 hr after application of ethylene diurea and lasted for 14 days. Ethylene diurea could be applied as a foliar spray at 500 ppm or as a soil drench at 100 ml/10-cm pot. Visible changes in plant form or function were not evident, but the authors speculated that ethylene diurea increased production of certain enzymes and nullified the tendency of ozone to accelerate senescence. Ethylene diurea did not provide protection against SO_2. Cathey and Heggestad (1982b) have summarized the response of many herbaceous plants to ethylene diurea and its modification of sensitivity to ozone.

Johnson *et al.* (1978) studied the ability of ancymidol to mitigate SO_2 damage to 'Hurricane' and 'Shasta' chrysanthemums. Plants exposed to SO_2 for 2 and 8 hr had less leaf damage if ancymidol had been applied, but at exposures as long as 24 hr, the SO_2 concentration had to be 0.2 ppm or less for ancymidol to reduce injury.

2. From Drought. Technical data sheets for some promising growth regulating chemicals often state that increased drought resistance could be expected. This trait is not too important for greenhouse-produced floricultural crops but is of interest to growers producing crops outdoors and probably would be beneficial for plants used in the landscape after purchase.

Halevy (1967) reported that plants treated with growth retardants lost as much moisture through transpiration as untreated plants. Martin and Lopushinsky (1966) did not observe much increase in the drought resistance of plants treated with daminozide compared with untreated. They found that water deficits and wilting continued to be troublesome even for treated plants, although plants treated with daminozide did seem to recover more quickly when watered after severe drought.

3. From Pests. Anatomical changes that provide protection against penetration by pollutants might also be expected to provide protection against insects and disease organisms. Halevy *et al.* (1970) found that ethephon, used as a 30-min dip, partially protected gladiolus corms from *Fusarium*. Sprouting of corms was enhanced when 1000 ppm ethephon was applied to corms that had been stored 0–4 weeks at 5°C, but the advantages were lost with 6–8 weeks of storage. Corm splitting was also increased, however, and flowering was delayed. Cormel production was increased, as was mentioned earlier. Sammons *et al.* (1981) inoculated begonia, chrysanthemum, poinsettia, and rose plants with appropriate disease organisms, applied growth regulators, and failed to achieve significant protection.

Tauber *et al.* (1971) reported that daminozide prevented an increase in the population of white fly (*Trialeurodes vaporarium*) on petunias. Eggs were present on the foliage of treated plants, but population increases did not occur. The researchers did not know if the results were caused directly by daminozide toxicity to the pest or because of induced changes in the host plants. Fischer and Shanks (1979) noted reduced white fly populations on 'Annette Hegg Lady' poinsettia when plants were treated with 0.72 mg ancymidol or 1500 mg chlormequat per 13-cm pot, but white fly infestation was higher on 'Eckespoint C-1 Red' plants that received the same treatment.

4. From Physiological Disorders. Growth-regulating chemicals manifest their activities morphologically, anatomically, and physiologically, so it is not surprising that growth regulators reportedly have decreased the severity of some physiological disorders.

a. Epinasty in Poinsettia. Several poinsettia cultivars exhibit pronounced leaf and bract epinasty after plants have been sleeved and sent to retail outlets. Ethylene has been implicated in this abiotic disorder (Sacalis 1978; Staby *et al.* 1978; Saltveit *et al.* 1979), and ways have been sought to eliminate or at least lessen petiole bending. Saltveit and Larson (1981, 1983) reduced epinasty by using chemicals known to inhibit ethylene evolution. For example, application of silver ions, which inhibit ethylene action, 24 hr before sleeving reduced epinasty, as did aminoethoxyvinylglycine (AVG), which acts as an inhibitor of ethylene synthesis. Shanks and Solomos (1980) controlled epinasty by applying 400 ppm AVG 4 hr or more before sleeving. They found that application of 200 ppm AVG completely inhibited ethylene synthesis for 26 hr and observed no phytotoxic effects from AVG applications.

Reid *et al.* (1981) found that epinasty could be prevented by applying silver thiosulfate 18 hr before plants were exposed to ethylene. Saltveit and Larson (1983) reported that cycloheximide, applied 24 hr before sleeving, prevented epinasty but caused phytotoxic effects.

Unfortunately, satisfactory control compounds do not have label clearance for control of epinasty and cannot be recommended for this purpose in commercial production (Hammer *et al.* 1981).

b. Stem Topple of Hyacinth. Stem topple renders hyacinth plants unsalable and has hindered the acceptance of potted hyacinths, either by commercial growers or by customers. Shoub and De Hertogh (1975) found 'Pink Pearl' a nontopple type, whereas 'Blue Giant' was topple-prone; pith cells are narrower and shorter in 'Pink Pearl' than in 'Blue Giant.' These workers applied ancymidol drenches of 2 mg and 4 mg ai/15-cm pot in 100 ml of water on the first day of forcing. Ancymidol altered cell shape and size and reduced floral stalk length and the scape:inflorescence ratio.

De Hertogh (1982) later reported that ethephon was the best growth regulator to use to prevent stem topple in potted hyacinths. Spray applications at 1000 or 2000 ppm, applied before the appearance of flower color, were effective; premature senescence occurred if ethephon was applied after flower color was evident. Leaves on floral stalks were 7.5–10 cm long at the time of application. De Hertogh cautioned growers to be certain that foliage was dry when ethephon was applied and not to moisten leaves for at least 12 hr after treatment. He observed cultivar differences: 'Eros' did not require ethephon for topple control, whereas 'Delft Blue' had to be treated twice at 2000 ppm to prevent topple.

IV. NEW APPLICATION TECHNIQUES

Many researchers have tried innovative ways to apply growth regulators—as granulates, coated with permeable coverings, impregnated in clay pots, and as dips—but foliar spray and drench applications remain the most popular methods. A few of the newer application techniques are discussed here. Some of them were tried several years ago but not accepted in commercial floriculture because of cost or lack of publicity or knowledge about the techniques.

Read *et al.* (1974) used encapsulated chlormequat as a growth regulator on tomatoes, petunias, snapdragons, and marigolds. He found that 12–25 g of encapsulated chlormequat per liter of medium controlled height as effectively as spray or drench applications. Dark, green foliage and thick stems were other consequences of the treatments. Plants treated with encapsulated chlormequat were better shaped than those receiving more conventional treatments, as the effect was gradual over an extended period and would last longer in the home.

Spray applications are preferred by many growers because they do not have to wait for the potting medium to dry out before treatment, spraying can be accomplished quickly, and less material is generally required with spray applications than with drenches. The effectiveness of spray applications can vary greatly, depending on the volume applied, droplet size, use of surfactants, and other variables. Although droplet size and volume applied are affected by the spraying equipment used, most research reports do not indicate nozzle orifice size, the pressure with which the material was applied, or droplet size. Cathey (1969) studied application techniques to determine the best way to apply growth regulators. He found that fine droplets achieved by applying the growth regulator as a fog or mist with a paint sprayer were better than larger droplets. He also used an ultra-low volume, high-concentration sprayer, which caused daminozide to be retained on the foliage as a "glistening film." However, chlormequat worked best if applied as a high-volume dilute solution with large droplets until runoff occurred. Brisson (1983) also compared the effectiveness of daminozide and chlormequat with low-volume, high-concentration or high-volume, low-concentration applications. Chrysanthemums were treated with daminozide, and poinsettias and seedling geraniums were treated with chlormequat. Similar results were obtained with both methods of application.

Use of electrostatically charged particles of growth regulators or other pesticides is not a new concept, but improved equipment has rekindled interest in it (Anon. 1979c; Lindquist 1983; Brisson and Larson 1983). The negatively charged spray droplets are attracted to the positively charged plant surfaces. There should be little wasted material with this method, and excellent coverage can be expected unless plants are too crowded or are poorly grounded. Plastic film on the ground, or even the plastic trays and containers used in bedding plant production, will impede grounding of the plant material and cause poor spray coverage. The most widely used equipment now available for electrostatic spraying is the Electrodyn, which is best suited for field use. The long wand of the device is difficult to handle in greenhouses with bench supports.

Impregnating clay pots with ancymidol (Einert 1976) has already been mentioned (Section III.D.2.r). Von Hentig and Hass (1982) soaked rock wool blocks and peat capsules in growth regulators, as a new approach to application. Cutting losses often were high. The combination of rock wool and daminozide did not work at low concentrations and caused phytotoxicity at high concentrations.

Another approach, tried many years ago but abandoned because of governmental regulations, is the use of combinations of growth regulators, such as daminozide and chlormequat tank mixes to control poinsettia height. More relaxed regulations might now prompt interest in such an approach. Shanks (1981a) achieved as much height control of poinsettias with one spray application of daminozide (5000 ppm) and chlormequat (2000 ppm) in combination as with separate spray applications of 2000 ppm chlormequat. No phytotoxicity was noted with the combination spray, whereas chlorosis occurred with chlormequat alone.

V. FUTURE PROSPECTS

The search for new chemicals to manipulate growth and flowering of floricultural crops will continue. Ever-increasing labor costs should stimulate research to discover chemical means to perform tasks now done manually; high fuel costs may stimulate attempts to use chemicals to replace warm-temperature requirements. Several new compounds are being evaluated at present by companies well known for their leadership in agricultural chemicals.

Increased knowledge of endogenous substances should stimulate investigations on their enhancement or substitution by exogenous growth regulators. There are still exciting areas open to new ideas. For example, no exogenous or synthetic chemical has been found that can take the place of long days to keep short-day plants vegetative, or act as a substitute for natural short days or pulling black cloth to promote flowering. Steroid hormones were once considered as substitutes for long days, but research to evaluate this concept was discouraging. One might hope for the eventual discovery of synthetic chemicals to control flowering—the least understood physiological process in horticulture, yet the basis of the floricultural industry.

The international scene in floriculture has changed drastically in the last 10 years. Countries previously ignored in floricultural surveys suddenly are in positions of leadership in the production of cut flowers. Changes in protective quarantine regulation could lead to shifts in the world production of potted plants, as well. The advantages that some countries have in terms of climate and labor costs might be nullified by use of growth regulators that modify the influence of environment or reduce tedious work. We might see more use of new plant materials in floriculture if plant growth regulators permitting easier methods of culture or management were available. Floriculturists can look to the future with anticipation and excitement.

LITERATURE CITED

ADDICOTT, F.T. and J.L. LYON. 1969. Physiology of abscisic acid and related substances. Annu. Rev. Plant Physiol. 20:139–164.

ADRIANSEN, E. 1972. Kemisk vaekstregulaering af potteplanter. Tidsskrift for Planteavl 76:725–841.

ADRIANSEN, E. 1977. Retardation of growth in *Beloperone guttata* with ethephon and ancymidol (in Danish). Tidsskrift for Planteavl 81:381–384.

ADRIANSEN, E. 1978. The effect of ethephon and ancymidol in *Fuchsia* × *hybrida* (in Danish). Tidsskrift for Planteavl 82:429–432.

ADRIANSEN, E. 1979. Kemisk knibning af azalea med Off-Shoot-O, Atrinal, og UBI-P293. Tidsskrift for Planteavl 83:205–212.

ADRIANSEN, E. 1980a. Virkning of daglaengde og tilfoselsmetode for ancymidol pa blomstring og vaekst hos *Clerodendrum thomsoniae* Balf. Tidsskrift for Planteavl 84:399–413.

ADRIANSEN, E. 1980b. The effect of growth retardants on the scape of *Paphiopedilum* Pfitz (in Danish). Tidsskrift for Planteavl 84:531–535.

ADRIANSEN, E. 1983. Kemisk vaekstregulaering. In: Potteplanter I. Produktion, metoder, midle. Gartner Info, Copenhagen.

AIRHART, D.L., N.J. NATARELLA, and F.A. POKORNY. 1978. The structure of processed pine bark. J. Amer. Soc. Hort. Sci. 103:404–408.

ANDERSON, R.G. and R.E. WIDMER. 1975. Improving vigor expression of cyclamen seed germination with surface disinfestation and gibberellin treatments. J. Amer. Soc. Hort. Sci. 100:597–601.

ANON. 1965. Cycocel: plant growth regulant. Tech. Dept. Cyanamid International, Wayne, New Jersey.

ANON. 1969. Off-Shoot-O. Chemical pinching agent for ornamentals. For use on azaleas. Procter and Gamble Distributing Co., Cincinnati, Ohio.

ANON. 1971. Technical report on Quel. A plant growth regulator. Elanco Products Co., Indianapolis, Indiana.

ANON. 1972. Technical report on A-Rest (ancymidol). A plant growth regulator. Elanco Products Co., Indianapolis, Indiana.

ANON. 1974. A-Rest: linear response plant growth regulator for broad spectrum use. Elanco Bull. Specialty Products. Elanco Products Co., Indianapolis, Indiana.

ANON. 1975a. Florel: plant growth regulator. Amchem Products, Inc., Ambler, Pennsylvania.

ANON. 1975b. UBI-P293: plant growth regulator. Uniroyal Chemical Div., Uniroyal, Inc., Bethany, Connecticut.

ANON. 1976. Atrinal: plant growth regulator. Dr. R. Maag Ltd., CH-8157, Dielsdorf, Switzerland.

ANON. 1979a. Atrinal: plant growth regulator. Maag Agrochemicals Tech. Data Sheet. HLR Sciences, Inc., Vero Beach, Florida.

ANON. 1979b. Federal Register Document 79–17970. Washington, DC.

ANON. 1979c. New spray technique promises better plant coverage. Agrichemicals Age 23(8):20–21.

ANON. 1979d. Phosfon developments. Perifleur Times. Perifleur Ltd., West Sussex, England.

ANON. 1980. Small's hybrid gloxinia seedlings. Earl J. Small Growers, Inc., Pinellas Park, Florida.

ANON. 1982. Technical information, PP333. Agr. Chemicals Div., ICI Americas, Inc., Goldsboro, North Carolina.

ANON. 1982/83. Gloeckner chrysanthemum manual. Fred C. Gloeckner Co., Inc., New York.

ANON. 1983a. Energy and new crops are discussed in Charlotte. N. C. Flower Growers' Bull. 27(2):6–8.

ANON. 1983b. STS-mixing methods tested. Flor. Rev. 172(4462):37.

ARMITAGE, A.M., R.E. BASS, W.H. CARLSON, and L.C. EWART. 1981. Control of plant height and flowering of zinnia by photoperiod and growth retardants. HortScience 16:218–220.

ARZEE, T., H. LANGENAVER, and J. GRESSEL. 1977. Effects of dikegulac, a new growth regulator, on apical growth and development of three compositae. Bot. Gaz. 138:18–28.

BAKER, J.E. 1983. Preservation of cut flowers. In: L.G. Nickell (ed.), Plant growth regulating chemicals, Vol. II. CRC Press, Boca Raton, Florida. p. 77–191.

BALL, V. 1975. Round up on retardants. Grower Talks 39(3):1–14.

BARRETT, J.E. 1982. Chrysanthemum height control by ancymidol, PP333, and EL-500 dependent on medium composition. HortScience 17:896–897.

BARRETT, J.E. and C.A. BARTUSKA. 1982. PP333 effects on stem elongation dependent on site of application. HortScience 17:737–738.

BOCION, P.F. and W.H. DE SILVA. 1976. Some effects of dikegulac on the physiology of whole plants and tissues; interactions with plant hormones. In: Proc. 9th Intern. Conf. on Plant Growth Substances, Lausanne, Switzerland.

BOCION, P.F., W.H. DE SILVA, G.A. HUPPI, and W. SZKRYBALO. 1975. Group of new chemicals with plant growth regulatory activity. Nature 258:142–143.

BOCION, P.F., W.H. DE SILVA, and H.R. WALTHER. 1978. Growth retardation activity of piproctanyl bromide on *Chrysanthemum morifolium* Ramat. HortScience 13:184–185.

BONAMINIO, V.P. 1977. Ancymidol effects on selected floricultural crops grown in controlled environment chambers. Ph.D. Thesis, North Carolina State Univ., Raleigh.

BONAMINIO, V.P. and R.A. LARSON. 1978. Influences of potting media, temperature, and concentration of ancymidol on growth of *Chrysanthemum morifolium* Ramat. J. Amer. Soc. Hort. Sci. 103:752–756.

BOODLEY, J.W. and J.W. MASTALERZ. 1959. The use of gibberellic acid to force azaleas without a cold temperature treatment. Proc. Amer. Soc. Hort. Sci. 74:681–685.

BRIGGS, J.B. 1975. The effects on growth and flowering of the chemical growth regulator ethephon on narcissus and ancymidol on tulip. Acta Hort. 47:287–296.

BRISSON, L.D. 1983. The effects of spray droplet size on application of growth retardants to floricultural crops and an investigation of spray coverage from electrostatically charged droplets. M.S. Thesis, North Carolina State Univ., Raleigh.

BRISSON, L.D. and R.A. LARSON. 1983a. Application of cycocel as an aerosol on seedling geraniums. N. C. Flower Growers' Bull. 27(4):5–9.

BRISSON, L.D. and R.A. LARSON. 1983b. Charge! Electrostatic sprayers being put to test. Greenhouse Manager 2(11):122–124.

BYRNE, T.G. and L.E. PYEATT. 1976. Gibberellin sprays to increase stem length of 'May Shoesmith' chrysanthemums. Flor. Rev. 157(4077):31–32,75.

CAMERON, A.C. and M.S. REID. 1981. The use of silver thiosulfate anionic complex as a foliar spray to prevent flower abscission of Zygocactus. HortScience 16:761–762.

CAMERON, A.C. and M.S. REID. 1983. Use of silver thiosulfate to prevent flower abscission from potted plants. Scientia Hort. 19:373–378.

CAMERON, A.C., M.S. REID, and G.W. HICKMAN. 1981. Using STS to prevent flower shattering in potted flowering plants—progress report. Univ. of Calif. Flower & Nursery Rpt. (Fall):1–3.

CARLSON, W.H. 1980. Bedding plants. p. 477–522. In: R.A. Larson (ed.), Introduction to floriculture. Academic Press, New York.

CARLSON, W.H. 1982. Tree geraniums. p. 158–160. In: J.W. Mastalerz and E.J. Holcomb (eds.), Geraniums III. Pa. Flower Growers, University Park.

CARLSON, W.H., S. SCHNABEL, J. SCHNABEL, and C. TURNER. 1977. Concentration and application time of ancymidol for growth retardation of *Kalanchoe blossfeldiana* Poellniz. cv. Mace. HortScience 12:568.

CARPENTER, W.J. and G.R. BECK. 1972. Growth regulator induced branching of poinsettia stock plants. HortScience 7:405–406.

CARPENTER, W.J. and W.H. CARLSON. 1970. The influence of growth regulators and temperature on flowering of seed propagated geraniums. HortScience 5:183–184.

CARPENTER, W.J. and R.C. RODRIGUEZ. 1971. The effect of plant growth regulating chemicals on rose shoot development from basal and axillary buds. J. Amer. Soc. Hort. Sci. 96:389–391.

CARPENTER, W.J., R.C. RODRIGUEZ, and W.H. CARLSON. 1971. Growth regulator induced branching of non-pinched poinsettias. HortScience 6:457–458.

CARR, L.H. and R.S. LINDSTROM. 1972. The influence of environmental and morphological factors on chemical pinching of greenhouse azaleas. J. Amer. Soc. Hort. Sci. 97:407–410.

CATHEY, H.M. 1959. Poinsettia study. Flor. Rev. 132(3426):19–20, 83–84.

CATHEY, H.M. 1964. Physiology of growth retarding chemicals. Annu. Rev. Plant Physiol. 15:271–302.

CATHEY, H.M. 1968. Report of cooperative trial on chemical pruning of chrysanthemums with fatty acid esters. Flor. Rev. 141(3663):19–21, 87–90.

CATHEY, H.M. 1969a. Enhancing the activity of chemical growth retardants. II. A method of applying growth retardants. Flor. Rev. 144(3720):28–29, 61–63.

CATHEY, H.M. 1969b. Response of *Dianthus caryophyllus* L. (carnation) to synthetic abscisic acid. Proc. Amer. Soc. Hort. Sci. 93:560–568.

CATHEY, H.M. 1969c. Response of some ornamental plants to synthetic abscisic acid. Proc. Amer. Soc. Hort. Sci. 93:693–698.

CATHEY, H.M. 1970. Modern chemical pruning of plants. Florist & Nursery Exchange 153(15):6–12.

CATHEY, H.M. 1975a. Comparative plant growth-retarding activities of ancymidol with ACPC, Phosfon, chlormequat, and SADH on ornamental plant species. HortScience 10:204–216.

CATHEY, H.M. 1975b. Growth retardants. Comparing A-Rest plus 4 others. Grower Talks 39(2):1–29.

CATHEY, H.M. 1976a. Growth regulators. p. 177–189. In: J.W. Mastalerz (ed.), Bedding plants. Pa. Flower Growers, University Park.

CATHEY, H.M. 1976b. Influence of a substituted oxathiin, a localized growth inhibitor, on the stem elongation, branching and flowering of *Chrysanthemum morifolium* Ramat. J. Amer. Soc. Hort. Sci. 101:599–604.

CATHEY, H.M. and R.J. DOWNS. 1965. Guidelines for regulating flowering of bromeliads. The Exchange 143(10):27–29.

CATHEY, H.M. and H.E. HEGGESTAD. 1973. Effect of growth retardants and fumigations with ozone and sulfur dioxide on growth and flowering of *Euphorbia pulcherrima* Willd. J. Amer. Soc. Hort. Sci. 98:3–7.

CATHEY, H.M. and H.E. HEGGESTAD. 1982a. Ozone and sulfur dioxide sensitivity of petunia: modification by ethylene diurea. J. Amer. Soc. Hort. Sci. 107:1028–1035.

CATHEY, H.M. and H.E. HEGGESTAD. 1982b. Ozone sensitivity of herbaceous plants: modification by ethylene diurea. J. Amer. Soc. Hort. Sci. 107:1035–1042.

CATHEY, H.M. and N.W. STUART. 1961. Comparative plant growth-retarding activity of Amo-1618, Phosfon and CCC. Bot. Gaz. 123:51–57.

CATHEY, H.M. and R.L. TAYLOR. 1970. Flowering of bromeliads with spray applications of 2-chloroethane phosphonic acid (Ethrel). Flor. Rev. 146(3790):38–39, 82–86.

CATHEY, H.M., G.L. STEFFENS, N.W. STUART, and R.H. ZIMMERMAN. 1966a. Chemical pruning of plants. Science 153:1382–1383.

CATHEY, H.M., A.H. YOEMAN, and F.F. SMITH. 1966b. Abortion of flower buds in chrysanthemum after application of a selected petroleum fraction of high aromatic content. HortScience 1:61–62.

CONOVER, C.A. 1970. Phytotoxic responses of *Euphorbia pulcherrima* to light, temperature, and Cycocel levels. Proc. Florida State Hort. Soc. 83:459–461.

CONOVER, C.A. and H.M. VINES. 1972. Chlormequat drench and spray applications to poinsettias. J. Amer. Soc. Hort. Sci. 97:316–320.

COOLBAUGH, R.C. and R. HAMILTON. 1976. Inhibition of ent-kaurene oxidation and growth by alpha-cyclopropyl-x-(p-methoxyphenyl)-5-pyrimidine methyl alcohol. Plant Physiol. 57:245–248.

COOLBAUGH, R.C., S.S. HIRANO, and C.A. WEST. 1978. Studies on the specificity and site of action of alpha-cyclopropyl-x-[p-methoxyphenyl]-5-pyrimidine methyl alcohol (ancymidol), a plant growth regulator. Plant Physiol. 62:571–576.

CORCORAN, M.R. 1975. Gibberellin antagonists and antigibberellins. p. 289–332. In: H.N. Krishnamoorthy (ed.), Gibberellins and plant growth. Wiley Eastern Ltd., New Delhi, India.

CRILEY, R.A. 1969. Effect of short photoperiods, Cycocel, and gibberellic acid upon flower bud initiation and development in azalea 'Hexe.' J. Amer. Soc. Hort. Sci. 94:392–395.

CRILEY, R.A. 1977. Year around flowering of double bougainvillea: effect of day-length and growth retardants. J. Amer. Soc. Hort. Sci. 102:775–778.

CRILEY, R.A. and S. OKA. 1976. Stimulating bud break on dieffenbachia cane pieces. Hort. Dig. (Univ. of Hawaii) 29:1.

DE HERTOGH, A.A. 1976. Guidelines for utilization of ancymidol (A-Rest) for height control of tulips forced as potted plants. Holland Flower Bulb Tech. Services Bull. 1. Netherlands Flower-Bulb Institute, Hillegom, Holland.

DE HERTOGH, A.A. 1982. Guidelines for utilization of Florel (ethephon) for reduction of stem topple of potted hyacinths and reduction of total plant height of potted daffodils. Holland Flower Bulb Tech. Services Bull. 10. Netherlands Flower-Bulb Institute, Hillegom, Holland.

DE HERTOGH, A.A. and N. BLAKELY. 1976. The influence of ancymidol, chlormequat and daminozide on the growth and development of forced *Dahlia variabilis* Willd. Scientia Hort. 4:123–130.

DE HERTOGH, A.A., J.W. LOVE, and J. BARRETT. 1981. Guidelines for greenhouse forcing of Dutch-grown tuberous-rooted dahlias as pot plant. Holland Flower Bulb Tech. Services Bull. 7. Netherlands Flower-Bulb Institute, Hillegom, Holland.

DE SILVA, W.H., P.F. BOCION, and H.R. WALTHER. 1976. Chemical pinching of azalea with dikegulac. HortScience 11:569–570.

DICKS, J.W. 1972. Uptake and distribution of the growth retardant, daminozide, in relation to control of lateral shoot elongation in *Chrysanthemum morifolium*. Ann. Appl. Biol. 72:313–326.

DICKS, J.W. 1972/73. Growth retardants and pot plants. Scientific Hort. 24:164–174.

DICKS, J.W. 1976. Chemical restriction of stem growth in ornamentals, cereals and tobacco. Outlook on Agr. 9:69–75.

DICKS, J.W. and A.R. REES. 1973. Effects of growth-regulating chemicals on two cultivars of Mid-Century hybrid lily. Scientia Hort. 1:133–142.

DICKS, J.W., J. GILFORD, and A.R. REES. 1974. The influence of timing of application and gibberellic acid on the effects of ancymidol on growth and flowering of Mid-Century hybrid lily cv. Enchantment. Scientia Hort. 2:153–163.

EINERT, A.E. 1976. Slow-release ancymidol for poinsettia by impregnation of clay pots. HortScience 11:374–375.

FABER, W.R. and J.W. WHITE. 1977. The effect of pruning and growth regulator treatments on rose plant renewal. J. Amer. Soc. Hort. Sci. 102:223–225.

FISCHER, S.J. and J.B. SHANKS. 1979. Whitefly infestation on chrysanthemum and poinsettia treated with plant and insect growth regulators. J. Amer. Soc. Hort. Sci. 104:829–830.

FISHER, J.B. 1980. Gibberellin-induced flowering in Cordyline (Agaraceae). J. Expt. Bot. 31:731–735.

FURUTA, T., L. PYEATT, E. CONKLIN, and J. YOSHIHASHI. 1968. Environmental conditions and effectiveness of chemical pinch agents on azalea. Flor. Rev. 142(3690):23,61–62.

HACKETT, W.P. and R.M. SACHS. 1967. Chemical control of flowering in bougainvillea 'San Diego Red.' Proc. Amer. Soc. Hort. Sci. 90:361–364.

HACKETT, W.P. and R.M. SACHS. 1968. Experimental separation of inflorescence development from initiation in bougainvillea. Proc. Amer. Soc. Hort. Sci. 92:615–621.

HALEVY, A.H. 1967. Effect of growth retardants on drought resistance and longevity of various plants. p. 277–283. In: Proc. 17th XVII Intern. Hort. Cong., Vol. III, College Park, Maryland.

HALEVY, A.H. and S. MAYAK. 1979. Senescence and postharvest physiology of cut flowers. Part I. Hort. Rev. 1:204–236.

HALEVY, A.H. and S. MAYAK. 1981. Senescence and postharvest physiology of cut flowers. Part 2. Hort. Rev. 3:59–143.

HALEVY, A.H., R. SHILO, and S. SIMCHON. 1970. Effect of 2-chloroethane phosphonic acid (Ethrel) on health, dormancy, and flower and corm yields of gladioli. J. Hort. Sci. 45:427–434.

HAMMER, P.A. 1976. Guide to making growth retarding solutions. Ext. Serv. Publ. HO-130. Purdue Univ., West Lafayette, Indiana.

HAMMER, P.A. 1980. Other flowering pot plants. p. 435–475. In: R.A. Larson (ed.), Introduction to floriculture. Academic Press, New York.

HAMMER, A., R. LARSON, M. REID, J. SACALIS, M. SALTVEIT, and G. STABY. 1981. How to reduce petiole bending of poinsettia plants. Ohio Florists' Assoc. Bull. 626:1–4.

HANKS, G.R. and R. MENHENETT. 1979. Response of potted tulips to new and established growth retarding chemicals. Scientia Hort. 10:237–254.

HANKS, G.R. and R. MENHENETT. 1980. An evaluation of new growth retardants on Mid-Century hybrid lilies. Scientia Hort. 13:349–359.

HARADA, H. and J.P. NITSCH. 1959. Flower induction in Japanese chrysanthemums with gibberellic acid. Science 129:777–778.

HAUSBECK, M.K., C.T. STEPHENS, and R.D. HEINS. 1984. STS/pythium interaction—does it exist? BP News 15:1–2.

HEIDE, O.M. 1969. Interaction of growth retardants and temperature in growth, flowering, regeneration, and auxin activity of *Begonia* × *cheimantha* Everett. Physiol. Plant. 22:1001–1012.

HEINS, R.D., R.E. WIDMER, and H.F. WILKINS. 1979. Growth regulators effective on floricultural crops. Dept. of Horticulture and Landscape Architecture, Univ. of Minnesota, St. Paul.

HEINS, R.D., A.M. ARMITAGE, and W.H. CARLSON. 1981. Influence of temperature, water stress and BA on vegetative and reproductive growth of *Schlumbergera truncata*. HortScience 16:679–680.

HENNY, R.J. 1980. Gibberellic acid (GA_3) induces flowering in *Dieffenbachia maculata* 'Perfection.' HortScience 15:613.

HENNY, R.J. 1983. Flowering of *Aglaonema commutatum* 'Treubii' following treatment with gibberellic acid. HortScience 18:374.

HENNY, R.J. and E.M. RASMUSSEN. 1981. Promotion of flowering in Spathiphyllum 'Mauna Loa' with gibberellic acid. HortScience 16:554–555.

HEURSEL, J. 1975. Results of experiments with dikegulac used on azaleas (*Rhododendron simsii* Planch). Mededelingen van de Faculteit Landbouwwetenschappen Rijkuniversiseit Gent 40:849–857.

HIGAKI, T. and H.P. RASMUSSEN. 1979. Chemical induction of adventitious shoots in anthuriums. HortScience 14:64–65.

HILDRUM, H. 1973. The effect of daylength, source of light, and growth regulators on growth and flowering of *Clerodendrum thomsonae* Ball. Scientia Hort. 1:1–11.

HOLCOMB, E.J. 1979. Growing Reiger begonia from terminal cuttings. Flor. Rev. 165(4256):26–27, 72–73.

HOLCOMB, E.J. and J.W. WHITE. 1970. A technique for soil application of a growth retardant. HortScience 5:16–17.
HOLCOMB, E.J., S. REAM, and J. REED. 1983. The effect of BAS 106, ancymidol, and chlormequat on chrysanthemum and poinsettia. HortScience 18:364–365.
HUDSON, J.P. 1976. Future roles for growth regulators. Outlook on Agr. 9(2):95–98.
HUPPI, G.A., P.F. BOCION, and W.H. DE SILVA. 1976. A new quaternary ammonium compound with plant growth regulatory activity. Experientia 32:37–38.
JEFFCOAT, B. 1980. Aspects and prospects of plant growth regulators. Monograph 6. British Plant Growth Regulator Group.
JOHANSSON, J. 1976. The regulation of growth and flowering in *Calceolaria* × *speciosa* Lilja. Acta Hort. 64:239–244.
JOHNSON, C.R., J.N. JOINER, and D.N. GOWER. 1978. Ancymidol decreases chrysanthemum sensitivity to SO_2 fumigation. HortScience 13:35.
JOHNSON, C.R., D.B. McCONNELL, and J.N. JOINER. 1982. Influence of ethephon and light intensity on growth and acclimatization of *Ficus benjamina*. HortScience 17:614–615.
JOINER, J.N., R.T. POOLE, C.R. JOHNSON, and C. RAMCHARAM. 1978. Effects of ancymidol and N, P, K on growth and appearance of *Dieffenbachia maculata* 'Baraquiniana.' HortScience 13:182–184.
JOINER, J.N., O. WASHINGTON, C.R. JOHNSON, and T.A. NELL. 1982. Effect of exogenous growth regulator on flowering and cytokinin levels in azaleas. Scientia Hort. 18:143–151.
JORGENSEN, S. 1962. The practical use of Phosfon-D as a drench in pot mum culture. N.C. Flower Growers' Bull. (April):3–5.
KAMERBEEK, G.A., A.J.B. DURIEUX, and J.A. SCHIPPER. 1980. An analysis of the influence of Ethrel on flowering of iris 'Ideal': an associated morphogenetic physiological approach. Acta Hort. 109:235–241.
KINCAID, S. and G.L. McDANIEL. 1979. Effects of ancymidol soil drench and time of application of bulb dips on 'Paul Richter' tulips. HortScience 14:276–277.
KOFRANEK, A.M. 1980. Cut chrysanthemums. p. 3–45. In: R.A. Larson (ed.), Introduction to floriculture. Academic Press, New York.
KOFRANEK, A.M. and R.A. CRILEY. 1967. Emulsifiable oils as disbudding agents for chrysanthemums. Flor. Rev. 139(3611):24.
KOFRANEK, A.M. and A.T. LEISER. 1958. Chemical defoliation of *Hydrangea macrophylla*. Proc. Amer. Soc. Hort. Sci. 71:555–562.
KOFRANEK, A.M. and L. MARKIEWICZ. 1967. Selected naphthalenes as disbudding agents for chrysanthemums. Flor. Rev. 140(3617):20–21, 54–57.
KOHL, H.C., JR. and A.M. KOFRANEK. 1957. Gibberellin on flower crops. Calif. Agr. 11(5):9.
KORANSKI, D.S., B.E. STRUCKMEYER, and G.E. BECK. 1978. The role of ancymidol in *Clerodendrum* flower initiation and development. J. Amer. Soc. Hort. Sci. 103:813–815.
KRAUSKOPF, D.M. and P.V. NELSON. 1976. Chemical height control of Reiger Elatior begonia. J. Amer. Soc. Hort. Sci. 101:618–619.
KRISHNAMOORTHY, H.N. 1975. Role of gibberellins in juvenility, flowering and sex expression. p. 114–143. In: H.N. Krishnamoorthy (ed.), Gibberellins and plant growth. Wiley Eastern Ltd., New Delhi, India.
LANG, A. 1970. Gibberellins: structures and metabolism. Annu. Rev. Plant Physiol. 21:537–570.
LARSON, R.A. 1967. Chemical growth regulators and their effects on poinsettia height control. N.C. Agr. Expt. Sta. Tech. Bull. 180.
LARSON, R.A. 1978. Stimulation of lateral branching of azaleas with dikegulac-sodium (Atrinal). J. Hort. Sci. 53:57–62.

LARSON, R.A. 1980. Azaleas. p. 237–260. In: R.A. Larson (ed.), Introduction to floriculture. Academic Press, New York.

LARSON, R.A. 1983. Enhancement of azalea production with plant growth regulators. Scientific Hort. 34:102–109.

LARSON, R.A. and V.P. BONAMINIO. 1973. Application of A-Rest to chrysanthemums growing in a bark medium. N.C. Flower Growers' Bull. 17(5):5–7.

LARSON, R.A. and R.K. KIMMINS. 1972. Response of *Chrysanthemum morifolium* Ramat to foliar and soil applications of ancymidol. HortScience 7:192–193.

LARSON, R.A. and M.L. McINTYRE. 1968. Residual effect of Cycocel in poinsettia height control. Proc. Amer. Soc. Hort. Sci. 93:667–672.

LARSON, R.A. and T.D. SYDNOR. 1971. Azalea flower bud development and dormancy as influenced by temperature and gibberellic acid. J. Amer. Soc. Hort. Sci. 96:786–788.

LARSON, R.A., J.W. LOVE, and V.P. BONAMINIO. 1974. Relationship of potting mediums and growth regulators in height control. Flor. Rev. 155(4017):21,59–62.

LARSON, R.A., J.W. LOVE, D.L. STRIDER, R.K. JONES, J.R. BAKER, and K.F. HORN. 1978. Commercial poinsettia production. N.C. Agr. Ext. Serv. Manual AG-108.

LEOPOLD, A.C. 1971. Antagonism of some gibberellin actions by a substituted pyrimidine. Plant Physiol. 48:537–540.

LEWIS, A.J. and J.S. LEWIS. 1980. Response of *Lilium longiflorum* to ancymidol bulb dips. Scientia Hort. 13:93–97.

LEWIS, A.J. and J.S. LEWIS. 1981. Improving ancymidol efficiency for height control of Easter lily. HortScience 16:89–90.

LEWIS, A.J. and J.S. LEWIS. 1982. Height control of *Lilium longiflorum* Thunb. 'Ace' using ancymidol bulb dips. HortScience 17:336–337.

LINDQUIST, R.K. 1983. Preliminary experiment with an electrostatic applicator for insect and mite control on greenhouse-grown plants. Ohio Florists' Assoc. Bull. 639:4–6.

LINDSTROM, R.S. 1961. The response of floricultural crops to various growth regulators. I. Effect of (2-chloroethyl)-trimethylammonium chloride on poinsettia (*Euphorbia pulcherrima* Willd.). Quart. Bull. Mich. Agr. Expt. Sta. 43:839–847.

LINDSTROM, R.S. and N.E. TOLBERT. 1960. (2-chloroethyl)-trimethylammonium chloride and related compounds as plant growth substances. IV. Effect on chrysanthemums and poinsettias. Quart. Bull. Mich. Agri. Expt. Sta. 42:917–928.

LOVE, J.W. 1980. Kalanchoes. p. 411–434. In: R.A. Larson (ed.), Introduction to floriculture. Academic Press, New York.

LUCKWILL, L.C. 1981. Growth regulators in crop production. Studies in Biology 129. Edward Arnold Pub. Ltd., London.

LYONS, R.E. and R.E. WIDMER. 1983. Effects of GA_3 and NAA on leaf lamina unfolding and flowering of *Cyclamen persicum*. J. Amer. Soc. Hort. Sci. 108:759–763.

MAENE, L.J. and P.C. DEBERGH. 1982. Stimulation of axillary shoot development of *Cordyline terminalis* 'Celestine Queen' by foliar sprays of 6-benzylamino purine. HortScience 17:344–345.

MAGIE, R.O. 1971. Effects of ethephon and benzinidazoles on corm and cormel production by gladiolus cormels. HortScience 6:351–352.

MARTIN, G.C. and W. LOPUSHINSKY. 1966. Effect of N-dimethyl amino succinamic acid (B-995), a growth retardant, on drought tolerance. Nature (London) 209:216–217.

McCONNELL, D.B. and R.T. POOLE. 1972. Vegetative growth modification of *Scindapsus aureus* by ancymidol and PBA. Proc. Florida State Hort. Soc. 85:387–389.

McCONNELL, D.B. and R.T. POOLE. 1981. Growth regulators. p. 307–325. In: J.N.

Joiner (ed.), Foliage plant production. Prentice-Hall, Englewood Cliffs, New Jersey.
McCONNELL, D.B. and B.E. STRUCKMEYER. 1970. Effect of succinic acid 2,2-dimethyl hydrazide on the growth of marigold in long and short photoperiods. HortScience 5:391–393.
McDANIEL, G.L. and S.B. FUHR. 1977. Height control of chrysanthemum by soaking rooted cuttings in growth retardants. HortScience 12:461–462.
McDOWELL, T.C. and R.A. LARSON. 1966. Effects of (2 chloroethyl)trimethylammonium chloride (Cycocel), N-dimethyl sucinamic acid (B-Nine), and photoperiod on flower bud initiation and development in azaleas. Proc. Amer. Soc. Hort. Sci. 88:600–605.
MENHENETT, R. 1977. A comparison of the effects of a new quaternary ammonium growth retardant with those of other growth retarding chemicals on the pot chrysanthemum (*Chrysanthemum morifolium*). Ann. Appl. Biol. 87:451–463.
MENHENETT, R. 1978. Chemical "pinching" of pot chrysanthemums with 2,3-dihydro-5,6-diphenyl-1,4-oxathiin. Scientia Hort. 8:81–89.
MENHENETT, R. 1979. Effects of growth retardants, gibberellic acid and indol-3-yl-acetic acid on stem extension and flower development in the pot chrysanthemum (*Chrysanthemum morifolium* Ramat). Ann. Bot. 43:305–318.
MENHENETT, R. 1980. Chemical control of shoot growth and axillary inflorescence development in *Chrysanthemum morifolium* Ramat with 2,3-dihydro-5,6-diphenyl-1,4-oxathiin. J. Hort. Sci. 55:239–246.
MENHENETT, R. 1981. Interactions of the growth retardants daminozide and piproctanyl bromide, and gibberellins A_1, A_3, A_{4+7}, A_5, and A_{13} in stem extension and inflorescence development in *Chrysanthemum morifolium* Ramat. Ann. Bot. 47:359–369.
MENHENETT, R. 1982. Chrysanth trials with PP 333 successful. Grower 98(24):30–31,33,35.
MENHENETT, R. and G.R. HANKS. 1982a. New retardant shows promise for pot-grown lilies and tulips. Grower 97(16):17–18,20.
MENHENETT, R. and G.R. HANKS. 1982b. The response of virus-free and virus-infected lily 'Enchantment' to the retardants ancymidol, chlormequat, mepiquat chloride and BTS 44584, a ternary sulphonium carbamate. Scientia Hort. 17:61–70.
MILLER, E.A. and E.J. HOLCOMB. 1982. Effect of GA_3 on germination of *Primula vulgaris* Huds. and *Primula* × *polyantha* Hort. HortScience 17:814–815.
MIRANDA, R.M. and W.H. CARLSON. 1980. Effect of timing and number of applications of chlormequat and ancymidol on the growth and flowering of seed geraniums. J. Amer. Soc. Hort. Sci. 105:273–277.
MOE, R. 1980. The use of ethephon for control of plant height in daffodils and tulips. Acta Hort. 109:197–204.
MOORE, T.C. 1968. Translocation of the growth retardant N,N-dimethylaminosuccinamic acid - ^{14}C (B-995-^{14}C). Bot. Gaz. 129:280–285.
MORGAN, P. 1980. Synthetic growth regulators: potential for development. Bot. Gaz. 141:337–346.
NANDA, K.K., H.N. KRISHNAMOORTHY, T.A. ANURADHA, and K. LAL. 1967. Floral induction by gibberellic acid in *Impatiens balsamina*, a qualitative short day plant. Planta 76:367–370.
NANDA, K.K., H.N. KRISHNAMOORTHY, and K. LATA. 1969. Effect of decapitation, Phosfon D and Cycocel on the flowering of *Impatiens balsamina* exposed to varying numbers of short days. Plant & Cell Physiol. 10:357–362.
NELL, T.A. and R.A. LARSON. 1974. The influence of foliar applications of GA_3,

GA_{4+7}, and PBA on breaking flower bud dormancy on azalea cvs. Redwing and Dogwood. J. Hort. Sci. 49:323–328.

NELSON, P.V. and L.Z. POTEET. 1969. A preliminary investigation of chemical pruning of plants—some common questions and answers. Flor. Rev. 145(3752):19,50–53.

NICKELL, L.G. 1983a. Plant growth regulating chemicals. I. CRC Press, Boca Raton, Florida.

NICKELL, L.G. 1983b. Plant growth regulating chemicals. II. CRC Press, Boca Raton, Florida.

NIGHTINGALE, A.E. 1970. The influence of succinamic acid 2-2 dimethylhydrazide on the growth and flowering of pinched vs. unpinched plants of the *Kalanchoe* hybrid 'Mace.' J. Amer. Soc. Hort. Sci. 95:273–276.

ORTON, P.J. 1979. The influence of water stress and abscisic acid on the root development of *Chrysanthemum morifolium* cuttings during propagation. J. Hort. Sci. 54:171–180.

PARUPS, E.V. 1970. Effect of morphactin on the gravimorphism and the uptake, translocation and spatial distribution of indol-3-yl-acetic acid in plant tissues in relation to light and gravity. Physiol. Plant. 23:1176–1186.

PARUPS, E.V. 1971. Use of 6-benzylamino purine and adenine to induce bottom breaks in greenhouse roses. HortScience 6:456–457.

PARUPS, E.V. 1976. Use of 2,3-dihydro-5,6-diphenyl-1,4-oxathiin for disbudding of standard chrysanthemums. Can. J. Plant Sci. 56:531–537.

PARUPS, E.V. 1979. Flowering of cyclamen as affected by gibberellic acid and a substituted thalimide. HortScience 142:279–280.

PASUTTI, D.W. and J.L. WEIGLE. 1980. Growth regulator effect on New Guinea impatiens hybrids. Scientia Hort. 12:293–298.

POKORNY, F.A. 1982. Horticultural uses of bark. Softwood and hardwood. A bibliography. Univ. Ga. Expt. Sta. Res. Rpt. 402.

POOLE, R.T. 1970. Influence of growth regulators on stem elongation and rooting response of foliage plants. Proc. Florida State Hort. Soc. 83:497–502.

POST, K. 1950. Florist crop production and marketing. Orange Judd Publ. Co., New York.

READ, P.E. and V. HOYSLER. 1969. Stimulation and retardation of adventitious root formation by application of B-Nine and Cycocel. J. Amer. Soc. Hort. Sci. 94:314–316.

READ, P.E. and V. HOYSLER. 1971. Improving rooting of carnation and poinsettia cuttings with succinic acid-2,2-dimethylhydrazide. HortScience 6:350–351.

READ, P.E., C.W. DUNHAM, and D.J. FIELDHOUSE. 1972. Increasing tuberous root production in *Dahlia pinnata* Cav. with SADH and chlormequat. HortScience 7:62–63.

READ, P.E., V.L. HERMAN, and D.A. HENG. 1974. Slow release chlormequat: a new concept in plant growth regulators. HortScience 9:55–57.

REID, M.S., Y. MOR, and A.M. KOFRANEK. 1981. Epinasty of poinsettias—the role of auxin and ethylene. Plant Physiol. 67:950–952.

RIDDELL, J.A., H.A. HAGEMAN, C.M. J'ANTHONY, and W.L. HUBBARD. 1962. Retardation of plant growth by a new group of chemicals. Science 136:39.

RÜNGER, W. and G. ALBERT. 1975. Influence of temperature, soil moisture and CCC on the flowering of *Euphorbia fulgens*. Scientia Hort. 3:393–403.

SACALIS, J.N. 1978. Ethylene evolution of petioles of sleeved poinsettia plants. HortScience 13:594–596.

SACHS, R.M. and C.F. BRETZ. 1962. The effect of daylength, temperature and gibberellic acid upon flowering in *Fuchsia hybrida*. Proc. Amer. Soc. Hort. Sci. 80:581–588.

SACHS, R.M. and W.P. HACKETT. 1972. Chemical inhibition of plant height. HortScience 7:440–447.

SACHS, R.M. and W.P. HACKETT. 1977. Chemical control of flowering. Acta Hort. 68:29–49.

SACHS, R.M. and A.M. KOFRANEK. 1963. Comparative cytohistological studies on inhibition and promotion of stem growth in *Chrysanthemum morifolium*. Amer. J. Bot. 50:772–779.

SACHS, R.M., A. LANG, C.F. BRETZ, and J. ROACH. 1960. Shoot histogenesis: subapical meristematic activity in a caulescent plant and the action of gibberellic acid and AMO-1618. Amer. J. Bot. 47:260–266.

SALTVEIT, M.E., JR. and R.A. LARSON. 1981. Reducing leaf epinasty in mechanically stressed poinsettia plants. J. Amer. Soc. Hort. Sci. 106:156–159.

SALTVEIT, M.E., JR. and R.A. LARSON. 1983. Effect of mechanical stress and inhibitors of protein synthesis on leaf epinasty in mechanically stressed poinsettia plants. J. Amer. Soc. Hort. Sci. 108:253–257.

SALTVEIT, M.E., JR., D.M. PHARR, and R.A. LARSON. 1979. Mechanical stress induces ethylene production and epinasty in poinsettia cultivars. J. Amer. Soc. Hort. Sci. 104:452–455.

SAMMONS, B., J.F. RISSLER, and J.B. SHANKS. 1981. Effects of growth regulators on diseases of begonia, chrysanthemum, poinsettia and rose. HortScience 16:673–674.

SANDERSON, K.C. 1969. Growth retardants for florist crops—A to Z. Flor. Rev. 145(3757):29–30,52–54.

SANDERSON, K.C., W.C. MARTIN, JR., K.A. MARCUS, and W.E. GOSLIN. 1975. Effects of plant growth regulators on *Lilium longiflorum* Thunb. cv. Georgia. HortScience 10:611–613.

SCHNEIDER, G. 1970. Morphactins: physiology and performance. Annu. Rev. Plant Physiol. 21:499–536.

SCOTT, B. 1982. Hydrangeas respond to new growth regulator. N.C. Flower Growers' Bull. 26(4):10–12.

SEELEY, J.G. 1975a. Both concentration and quantity of retardant affect height of lilies. Flor. Rev. 157(4073):19,67–70.

SEELEY, J.G. 1975b. Height control for garden lilies as potted plants. Flor. Rev. 157(4071):45,86–89.

SEELEY, J.G. 1979. Interpretation of growth regulator research with floriculture crops. Acta Hort. 91:83–92.

SEELEY, J.G. 1982. Tailoring garden lilies as potted plants. Flor. Rev. 170:21–23,75,77–78.

SEMENIUK, P. and R. TAYLOR. 1970. Effects of growth retardants on growth of geranium seedlings and flowering. HortScience 5:393–394.

SHANKS, J.B. 1969. Some effects and potential uses of Ethrel on ornamental crops. HortScience 4:56–58.

SHANKS, J.B. 1975. Hydrangeas. p. 352–368. In: V. Ball (ed.), The Ball red book. 13th ed. Geo. J. Ball, Inc., West Chicago.

SHANKS, J.B. 1980a. Chemical dwarfing of several ornamental greenhouse crops with PP 333. p. 46–51. In: Proc. 7th Annu. Mtg. Plant Growth Regulator Working Group, Dallas, Texas.

SHANKS, J.B. 1980b. Promotion of seedling growth with AVG. Plant Growth Regulator Working Group Bull. 6(4):5–6.

SHANKS, J.B. 1981a. Poinsettias—the Christmas flower. Md. Flor. 231:17–20.

SHANKS, J.B. 1981b. Promotion of greenhouse cut roses with GA_{4+7} and benzyladenine. p. 224–230. In: Proc. 8th Annu. Mtg. Plant Growth Regulator Soc. of Amer., St. Petersburg, Florida.

SHANKS, J.B. 1982a. Fundamentals of plant growth regulation. N.C. Flower Growers' Bull. 26(5):4–11.

SHANKS, J.B. 1982b. Growth regulating chemicals. p. 106–113. In: J.W. Mastalerz and E.J. Holcomb (eds.), Geraniums III. Pa. Flower Growers, University Park.

SHANKS, J.B. and C.B. LINK. 1964. The chemical regulation of plant growth for florists. Md. Flor. 108:1–16.

SHANKS, J.B. and A. PUROHIT. 1978. Chemical promotion of branching of poinsettia (*Euphorbia pulcherrima* Willd. ex Klotzsch) and chrysanthemum (*C.* × *morifolium* Ramat.) in the greenhouse. p. 252–259. In: Proc. 5th Annu. Mtg. Plant Growth Regulator Working Group, Blacksburg, Virginia.

SHANKS, J.B. and T. SOLOMOS. 1980. Observations on the effect of AVG on stress ethylene, epinasty, and lasting of poinsettia. p. 114–120. In: Proc. 7th Annu. Mtg. Plant Growth Regulator Working Group, Blacksburg, Virginia.

SHOUB, J. and A.A. DE HERTOGH. 1974. Effects of ancymidol and gibberellins A_3 and A_{4+7} on *Tulipa gesneriana* L. cv. Paul Richter during development in the greenhouse. Scientia Hort. 2:55–67.

SHOUB, J. and A.A. DE HERTOGH. 1975. Floral stalk topple: a disorder of *Hyacinthus orientalis* L. and its control. HortScience 10:26–28.

SHU, L.J. and K.C. SANDERSON. 1983. Greenhouse azaleas: a continually improving greenhouse crop. Greenhouse Manager 1(11):48,52–54.

SILL, L.Z. and P.V. NELSON. 1970. Relationship between azalea bud morphology and effectiveness of methyl decanoate, a chemical pruning agent. J. Amer. Soc. Hort. Sci. 95:270–273.

SIMMONDS, J.A. and B.G. CUMMING. 1976. Propagation of Lilium hybrids. I. Dependence of bulblet production on time of scale removal and growth substances. Scientia Hort. 5:77–83.

SIMMONDS, J.A. and B.G. CUMMING. 1977. Bulb-dip application of growth regulating chemicals for inhibiting stem elongation of 'Enchantment' and 'Harmony' lilies. Scientia Hort. 6:71–81.

SIMMONDS, J.A. and B.G. CUMMING. 1978. The interaction of a dormancy-breaking cold treatment, ancymidol, and ethephon in relation to stem elongation and flower production of Lilium cultivars. Scientia Hort. 8:57–64.

SKOGEN, D., A.B. ERIKSEN, and S. NILSEN. 1982. Effect of triacontanol on production and quality of flowers of *Chrysanthemum morifolium* Ramat. Scientia Hort. 18:87–92.

STABY, G.L., J.F. THOMPSON, and A.M. KOFRANEK. 1978. Postharvest characteristics of poinsettias as influenced by handling and storage procedures. J. Amer. Soc. Hort. Sci. 103:712–715.

STUART, N.W. 1961. Initiation of flower buds in Rhododendron after application of growth retardants. Science 134:50–52.

STUART, N.W. 1962. Azalea growth rate regulated by chemicals. Flor. Rev. 130:35–36.

STUART, N.W. 1964. Report on cooperative trial on controlling flowering of greenhouse azaleas with growth retardants. Flor. Rev. 133(3477):37–39,74–76.

STUART, N.W. 1967. Chemical pruning of greenhouse azaleas with fatty acid esters. Flor. Rev. 140(3631):26–27,68.

STUART, N.W. 1971. Reduction in height of eight garden lily hybrids following use of a chemical growth regulator, ancymidol. The Lily Yearb. 24:37–42.

SYDNOR, T.D. 1972. The effects of exogenous gibberellins and the location, identification and estimation of four endogenous growth-regulatory compounds in Rhododendron sp. cv. Gloria azalea. Ph.D. Thesis, North Carolina State Univ., Raleigh.
SYDNOR, T.D., R.K. KIMMINS, and R.A. LARSON. 1972. The effects of light intensity and growth regulators on gloxinia. HortScience 7:407–408.
SYSTEMA, W. 1969. Regulating the flowering of *Vriesea splendens* by BOH and acetylene. Acta Hort. 14:237–244.
TAUBER, M.J., B. SHALUCHA, and R.W. LANGHANS. 1971. Succinic acid-2,2-dimethylhydrazide (SADH) prevents whitefly population increase. HortScience 6:458.
THOMAS, T.H. 1974. Physiological and economic targets for plant growth regulators. p. 949–958. In: Proc. 12th British Weed Control Conference, Vol 3. Brighton, England.
THOMAS, T.H. 1976. Growth regulation in vegetable crops. Outlook on Agr. 9(2):62–68.
TJIA, B. and J. BUXTON. 1976. Influence of ethephon spray on defoliation and subsequent growth on *Hydrangea macrophylla* Thunb. HortScience 11:487–488.
TJIA, B. and J. BUXTON. 1977. Ethrel studies on *Petunia hybrida* Vilm. Scientia Hort. 7:269–275.
TJIA, B., P.C. KOZEL, and D.C. KIPLINGER. 1969. Morphactins useful for increasing lateral branching and reducing ultimate height of chrysanthemums. Flor. Rev. 144(3733):21–23,55–56.
TJIA, B., M.S. SANDHU, and J. BUXTON. 1976a. Height control of Easter lilies with ancymidol (A-Rest) applied as foliar spray, soil drench, and soil-incorporated slow release capsules. Flor. Rev. 157(4079):27–28.
TJIA, B., L. STOLTZ, M.S. SANDHU, and J. BUXTON. 1976b. Surface active agent to increase effectiveness of surface penetration of ancymidol on hydrangea and Easter lily. HortScience 11:371–372.
TJIA, B., S. OHPANAYIKOOL, and J. BUXTON. 1976c. Comparison of soil applied growth regulators on height control of poinsettia. HortScience 11:373–374.
TSCHABOLD, E.E., W.C. MEREDITH, L.R. GUSE, and E.V. KRUMKALNS. 1975. Ancymidol performance as altered by potting media composition. J. Amer. Soc. Hort. Sci. 100:142–144.
UHRING, J. 1971. Histological observations on chemical pruning of chrysanthemum with methyl decanoate. J. Amer. Soc. Hort. Sci. 96:58–64.
VEEN, H. 1983. Silver thiosulfate: an experimental tool in plant science. Scientia Hort. 20:211–224.
VEEN, H. and S.C. VAN DE GEIJN. 1978. Mobility and ionic form of silver as related to the longevity of cut carnations. Planta 140:93–96.
VIENRAVEE, K. and M.N. ROGERS. 1974. Growth control of terrarium plants with ancymidol. Flor. Rev. 155(4008):24–26,67–68.
VON HENTIG, W.W. and I. HASS. 1982. Early treatment of potted chrysanthemums with retarding substances. Acta Hort. 125:179–191.
WAREING, P.F. 1976. Introduction—modification of plant growth by hormones and other growth regulators. Outlook on Agr. 9(2):42–45.
WEAVER, R.J. 1972. Plant growth substances in agriculture. W.H. Freeman and Co., San Francisco.
WEILER, T.C. 1980. Hydrangeas. p. 353–372. In: R.A. Larson (ed.), Introduction to floriculture. Academic Press, New York.
WHITE, J.W. 1976. Lilium sp. 'Mid-Century Hybrids' adapted to pot use with ancymidol. J. Amer. Soc. Hort. Sci. 101:126–129.

WHITE, J.W. and E.J. HOLCOMB. 1974. Height control methods for poinsettia. HortScience 9:146–147.

WHITE, J.W., C. ALLEN, M. EASLEY, and P. STROCK. 1978. Effects of growth regulants on *Opuntia basilarus*. HortScience 13:181–182.

WIDMER, R.E. 1978. Cyclamen gibberellic acid treatment. Minn. Sta. Flor. Bull. (December 1):1.

WIDMER, R.E., L.C. STEPHEN, and M.V. ANGELL. 1974. Gibberellin accelerates flowering of *Cyclamen persicum* Mill. HortScience 9:476–477.

WIDMER, R.E., R.E. LYONS, and M.C. STUART. 1979. Fast-crop cyclamen '79. Flor. Rev. 164(4240):34–35,77.

WIGGANS, S.C., R.N. PAYNE, and R.P. EALY. 1960. The effect of Phosfon, a growth retardant, on lilies, caladiums and abelias. Okla. Agr. Expt. Sta. Processed Series 365:1–15.

WIKESJÖ, K. 1975. Production programs for pot plants and cut flowers in Sweden. Hort. Advisory Bull. 89. Agr. College, Alnarp, Sweden.

WIKESJÖ, K. 1978. Vaxtreglerande medel till krukvaxter, snittblommor och utplanteringsvaxter. Tradgard 130. Swedish Univ. Agr. Sci., Alnarp.

WIKESJÖ, K. and H. SCHUSSLER. 1982. Growing Reiger begonia year-round. Flor. Rev. 170(4397):30,72–74.

WILFRET, G.J., B.K. HARBAUGH, and T.A. NELL. 1978. Height control of pixie poinsettia with a granular formulation of ancymidol. HortScience 13:701–702.

WILKINS, H.F. 1980. Easter lilies. p. 329–352. In: R.A. Larson (ed.), Introduction to floriculture. Academic Press, New York.

WILKINS, H.F., W.E. HEALY, and R.D. HEINS. 1979. Past, present, future plant growth regulators. Acta Hort. 91:23–32.

WITTWER, S.H. 1968. Chemical regulators in horticulture. HortScience 3:163–167.

ZACHARIOUDAKIS, J.N. and R.A. LARSON. 1976. Chemical removal of lateral buds of *Chrysanthemum morifolium* Ramat. HortScience 11:36–37.

ZIESLIN, H., A.N. HALEVY, Y. MOR, A. BACHRACH, and I. SAPIR. 1972. Promotion of renewal canes in roses by ethephon. HortScience 7:75–76.

ZIMMER, K. 1969. Control of flowering in bromeliads. Acta Hort. 14:229–235.

10

Hydroponic Vegetable Production

Merle H. Jensen and W. L. Collins[1]
Environmental Research Laboratory, University of Arizona, Tucson, Arizona 85706

[1] The authors acknowledge grant support by the National Science Foundation and other sources.

Horticultural Reviews, Volume 7

ISBN 0-87055-492-1

I. INTRODUCTION

Controlled-environment agriculture (CEA) involves all aspects of modifying the natural environment to achieve optimum plant and animal growth. For plants, this means control at both the aerial and root levels. Since regulating the environment is a major objective of CEA systems, production takes place inside enclosures that allow for control of air temperature, light, water, and plant nutrition, and that provide climatic protection. In the early 1970s, the continuance and expansion of CEA in all areas of the world was threatened by the rising cost and limited availability of fuels to heat and cool greenhouse units. Since this period, significant advances have been made in greenhouse structures, environmental control systems, and cultural techniques. Though soil is still the predominant growing medium in the world, this review deals primarily with hydroponics because of the increasing interest in and development of many new types of hydroponic processes.

A. Definitions

Hydroponics is a technology for growing plants in nutrient solutions (water and fertilizers) with or without the use of an artificial medium (e.g. sand, gravel, vermiculite, rockwool, peat moss, sawdust) to provide mechanical support. *Liquid* hydroponic systems have no other supporting medium for the plant roots; *aggregate* systems have a solid medium of support. Hydroponic systems are further categorized as *open* (i.e., once the nutrient solution is delivered to the plant roots, it is not reused) or *closed* (i.e., surplus solution is recovered, replenished, and recycled).

Some regional growers, agencies, and publications persist in confining the definition of hydroponics to liquid systems only. This exclusion of aggregate hydroponics serves to blur statistical data and may lead to underestimation of the extent of the technology and its economic implications.

Virtually all hydroponic systems are enclosed in greenhouse-type structures in order to provide temperature control, to reduce evaporative water loss, to better control disease and pest infestations, and to protect hydroponic crops against the elements of weather, such as wind and rain. Thus, while hydroponics and CEA are not synonymous, CEA usually accompanies hydroponics, and their potentials and problems are inextricable.

Although hydroponics is widely used to grow flower, foliage, and bedding plants as well as certain high-value food crops, this review focuses on vegetable crops.

B. Attributes

The principal advantages of hydroponic CEA include high-density maximum crop yield, crop production where no suitable soil exists, a virtual indifference to ambient temperature and seasonality, more efficient use of water and fertilizers, minimal use of land area, and suitability for mechanization and disease control. A major advantage of hydroponics, as compared with growth of plants in soil, is the isolation of the crop from the underlying soil, which often has problems associated with disease, salinity, or poor structure and drainage. The costly and time-consuming tasks of soil sterilization and cultivation are unnecessary in hydroponic systems and a rapid turnaround of crops is readily achieved.

The principal disadvantages of hydroponics are the high costs of capital and energy inputs, relative to conventional open-field agriculture (OFA), and the high degree of competence in plant science and engineering skills required for successful operations. Because of its significantly higher costs, successful applications of hydroponic technology are limited to crops of high economic value, to specific regions, and often to specific times of the year when comparable OFA crops are not readily available.

With CEA capital costs several orders of magnitude higher than those for OFA, the types of food crops feasible for hydroponics are severely limited by potential economic return. Agronomic crops are totally inappropriate; a decade ago, it was calculated that the highest market prices ever paid would have to increase by a factor of five for hydroponic agronomy to break even (Anon. 1977). Since then, CEA costs have more than doubled, while crop commodity prices have remained constant. Indeed, in the United States, open-field agronomic crops are usually in surplus and a significant percentage of the available cropland is deliberately idled. Repeated pricing studies have shown that only high-quality, garden-type vegetables-tomatoes, cucumbers, and specialty lettuce—can provide breakeven or better revenues in hydroponic systems. These are, in fact, the principal (and virtually only) hydroponic CEA food crops grown today in the United States; in Europe and Japan these vegetables along with eggplant, peppers, melons, strawberries and herbs are grown commercially in hydroponic systems.

II. HISTORY

The development of vegetable production in controlled environments has not been rapid. Although the first use of CEA was the growing of off-season cucumbers under "transparent store" for the Roman Emperor Tiberius during the first century, the technology is believed to have been used little if at all for the following 1500 years.

Greenhouses (and experimental hydroponics) appeared in France and England during the seventeenth century; Woodward grew mint plants without soil in England in the year 1699. The basic laboratory techniques of nutrient solution culture were developed (independently) by Sacks and Knap in Germany about 1860. Further experiments in liquid and aggregate systems were reported in New Jersey and California in the mid-1920s and early 1930s (Dalrymple 1973), at a time when greenhouses *per se* (with conventional soil) were becoming common in Europe and the midwestern United States. Gericke (1940) published a description of a quasi-commercial use of the liquid technique and apparently coined the word *hydroponics* in passing. The technology was used in a few limited applications on Pacific islands during World War II. After the war, Purdue University popularized hydroponics (called *nutriculture*) in a classic series of extension service bulletins (Withrow and Withrow 1948) describing the precise delivery of nutrient solution to plant roots in either liquid or aggregate systems. (Purdue's appellation was not widely accepted; the definition of hydroponics was simply broadened to include all such systems.) In the mid-1950s, greenhouses covered with plastic film appeared in Kentucky (Courter and Carbonneau 1966) and evaporative cooling was used in Texas. Air-supported plastic greenhouses were first used in Washington in 1959 (Dalrymple 1973). Many of these systems used hydroponic growing techniques.

Greenhouse area began to expand significantly in Europe and Asia during the 1950s and 1960s, and large hydroponic CEA systems were developed in the deserts of California, Arizona, Abu Dhabi, and Iran about 1970 (Fontes 1973; Jensen and Teran 1971). In these desert locations, the advantages of the technology were augmented by the duration and intensity of the solar radiation, which maximized photosynthetic production.

Precise data are elusive, but extrapolating from Dalrymple's (1973) global survey and the Japanese (Shimizu 1979) and European developments reported herein, it is believed that there are presently more than 110,000 ha of CEA in the world, a significant but unknown percentage of which also represents the use of hydroponics. Japan, the Netherlands, the USSR and its satellites, and Italy have the largest areas of CEA; the majority of their greenhouses are used for food production. During the 1970s, when commercial interest in CEA waned somewhat, there was increasing scientific and technical fascination with hydroponics. A computer analysis of the literature in the data base of the National Agricultural Library shows that during the past 10 years, there have been 1500 hydroponic citations added to the literature, more than 300 of which appeared during 1981 and 1982. These publications originated in approximately 40 nations in which hydroponics and CEA are permanently practiced, but the largest number of reports came from the United States, Japan, Holland, Italy, and the USSR, in that order. The anomaly in this is that although U.S. scientists have published extensively, this country has a much smaller area in CEA than the major producers, and little of it is used for food crops.

A. Status of Controlled Environment Agriculture in Western Europe

The 1981 estimated greenhouse area in Western Europe and the area planted using soilless techniques are listed in Table 10.1. The

TABLE 10.1. Estimated Greenhouse Area and Area Planted with Soilless Techniques in Western Europe—1981

Country	Greenhouse area (ha)	Soilless area (ha)	Soilless as % of total greenhouse area
Netherlands	8,755	255	2.9
United Kingdom	2,139	65	3.0
Scandinavia (excluding Denmark)	769	55	7.2
Denmark	526	45	8.6
France	4,300	27	0.6
West Germany	2,829	21	0.7
Spain	34,900	17	—
Belgium	1,863	12	0.6
Greece/Italy	17,400	3	—
TOTAL	73,481	500	0.7

Source: Chris J. Graves, personal communication.

estimated area of the crops grown using soilless techniques are listed in Table 10.2. Escalating oil prices from 1973 to 1983 ultimately increased the costs of CEA heating and cooling by one to two orders of magnitude, causing the industry to contract. Although the absolute area "under glass" (or, more often, plastic film) probably peaked about 1979, it is projected that in Western Europe additional greenhouse hectarage will change from soil to soilless systems of cultivation during the rest of the 1980s (Table 10.3).

B. Status of Controlled Environment Agriculture in the United States

The earliest American greenhouse is believed to be a structure built on the Lyman estate in Waltham, Massachusetts, about 1800 (Compton 1960). This structure, which has recently been restored, has a low, sloping roof of sash windows, which can be raised for ventilation.

In 1867, a committee of the Massachusetts Horticultural Society noted the rapid growth of vegetables under glass and suggested that prizes be offered to encourage the practice (Anon. 1880). Yet by 1900, when there were approximately 880 ha of greenhouses in the United States, only about 40 ha were used for the production of winter vegetables (Galloway 1899), principally in the Boston area. (Even then, the crops were tomatoes, cucumbers, and lettuce—at high prices.) By 1912, American greenhouses for vegetables had increased to 104 ha (Walt 1917); Dalrymple (1973) reported that of 2000 ha of greenhouse structures, 89% were used to grow ornamental and bedding plants—only 233 ha were devoted to food crop production. There has been little change during the past decade. The total greenhouse area was believed to have enlarged slightly by 1976 (Jensen 1977), but by the end

TABLE 10.2. Estimated Area Planted to Various Salad Crops and Flowers Using Soilless Techniques in Western Europe—1981

Country/region	Area (ha)					
	Total	Cucumbers	Tomatoes	Lettuce	Other[1]	Flowers and plants
Netherlands	255	212	25	5	9	4
Scandinavia (Including Denmark)	100	66	30	2	1	1
United Kingdom	65	20	35	7	1	2
Southern Europe[2]	47	38	5	—	4	—
Northern Europe[3]	33	15	6	4	2	6
TOTAL	500	351	101	18	17	13

Source: Chris J. Graves, personal communication.
[1]Peppers, eggplant, summer squash, strawberries, etc.
[2]France, Spain, Italy, and Greece.
[3]West Germany and Belgium.

TABLE 10.3. Estimated Hectares of Soilless Cultivation Systems in Western Europe—1981 to 1990

Country/region	Year				
	1981	1982	1984	1986	1990
Netherlands	255	400	1025	1650	1980
Northern Europe[1] (excluding Netherlands)	198	232	300	368	504
Southern Europe[2]	47	56	70	248	566
TOTAL	500	688	1395	2266	3050

Source: Chris J. Graves, personal communication.
[1]United Kingdom, Scandinavia, West Germany, and Belgium.
[2]France, Spain, Italy, and Greece.

of 1982, only 253 ha of American greenhouses were being used for food crops (Lieberth 1982), hardly changed from Dalrymple's earlier survey.

Vegetables are now grown in CEA systems (both hydroponic and with soil) in every state in the nation, ranging from Ohio's 65 ha to an insignificant 0.05 ha in North Dakota. Fourteen states, located primarily in middle and southern latitudes of the nation, account for more than 80% of the CEA vegetable acreage in the United States (Table 10.4). The number of liquid hydroponic systems, however, diminished sharply from 564 operators in 1980 to 268 in 1982. These figures, however, are misleading because some of the operators merely switched from liquid to aggregate systems (e.g., sand culture or bag culture).

At first glance, the paucity of hydroponics in the United States is puzzling; Americans are seldom voluntary technical laggards. The contemporary public interest in "health foods" has significantly increased the consumption of fresh vegetables, particularly those that are economically appropriate to hydroponics. Popular interest in the technology remains high. It is environmentally impeccable. In addition, hydroponics is adaptable to space vehicle life support systems

TABLE 10.4. States in the United States with More Than 5 ha of CEA Vegetable Systems—1982

State	Area (ha)	State	Area (ha)
Arizona	8.7	Kansas	7.2
Arkansas	14.9	Kentucky	6.1
California	34.5	Michigan	8.4
Florida	8.0	New York	7.4
Hawaii	13.4	No. Carolina	9.3
Illinois	6.1	Ohio	65.1
Indiana	10.2	Texas	12.0
		TOTAL[1]	211.2

Source: Lieberth (1982).
[1]Accounts for 82.5% of the 256 ha in CEA vegetable production (both soil and hydroponics).

and is an appropriatae technology for the United States to export to arid Third World regions. Thus, the key factors would seem to be in place for an extensive American development of hydroponics, although the promise so far has exceeded the reality.

III. LIQUID (NONAGGREGATE) HYDROPONIC SYSTEMS

Liquid systems are, by their nature, closed systems in which the plant roots are exposed to the nutrient solution, without any type of growing medium, and the solution is reused.

A. Nutrient Film Technique (NFT)

The nutrient film technique was developed during the late 1960s by Dr. Allan Cooper at the Glasshouse Crops Research Institute, Littlehampton, England (Winsor *et al.* 1979); a number of subsequent refinements have been developed at the same institution (Graves 1983). Together with the modified systems to which it has given rise (which will be described subsequently), NFT appears to be the most rapidly evolving type of liquid hydroponic system today.

In a nutrient film system, a thin film of nutrient solution flows through plastic lined channels, which contain the plant roots; the walls of the channels are flexible, to permit them being drawn together around the base of each plant to exclude light and prevent evaporation. The main features of a nutrient film system are shown in Fig. 10.1.

Nutrient solution is pumped to the higher end of each channel and flows by gravity past the plant roots to catchment pipes and a sump; the solution is monitored for replenishment of salts and water before it is recycled. Capillary material in the channel prevents young plants from drying out, and the roots soon grow into a tangled mat.

A principal advantage of this system in comparison with others is that a greatly reduced volume of nutrient solution is required, which may be more easily heated during winter months to obtain optima temperatures for root growth (Hanger 1982a) or cooled during hot summers in arid regions to avoid bolting and other undesirable plant responses. Although only 50 ha are in commercial NFT production in England, the technology is increasing in popularity elsewhere in Europe and is also in wide use experimentally (Graves 1983).

A convenient size for an NFT unit, or module, is said to be 0.4 ha (Spensley *et al.* 1978). An installation of this size for producing toma-

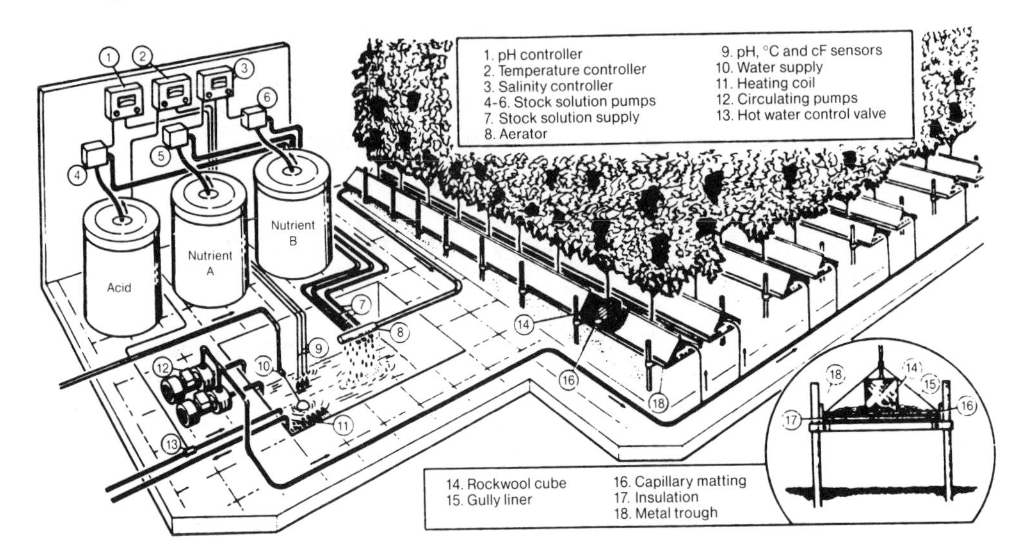

FIG. 10.1. Main features of a nutrient film hydroponic system.
From Graves (1983).

toes (Fig. 10.2), which typically contains 12,000 plants, will require 15,000 liters of nutrient solution and consume 15,000–16,000 liters of water daily during the summer; i.e., the solution is effectively replaced daily. A nutrient film system, like other closed systems, is vulnerable to water of poor quality, and toxic ions will accumulate in the water (Asher 1981; Howeler *et al.* 1982) if the solution is not totally changed periodically. Some growers are reported to deliberately introduce leakage rates as high as 30% in their systems to reduce the accumulation of harmful ions; however, this practice appears to negate the inherent water conservation feature of the system. A system for growing lettuce, another typical NFT crop, is shown in Fig. 10.3.

The slope of the channels in an NFT unit need not be severe: A drop of 1 in 50 to 1 in 75 appears suitable, although 1 in 100 is not sufficient (Spensley *et al.* 1978). Depressions in the channel floors must be avoided because ponding of immobile solution will lead to oxygen depletion and growth retardation. This peril has led to some "raised" NFT designs: rigid channels of galvanized steel lined with plastic film and using adjustable stands to obtain optimum slope levels. Although effective in eliminating ponding, and also in rendering NFT more susceptible to mechanization, this design increases capital costs.

For crops like tomatoes and cucumbers, the width of the channel is usually 25–30 cm (to avoid a thick root mat), although narrower channels (15 cm) do not affect yields of tomatoes (Spensley *et al.* 1978). The maximum length should not be greater than 20 m. In a lev green-

FIG. 10.2. Tomato production using the nutrient film technique. Insert shows root growth within channels.

FIG. 10.3. Lettuce production in NFT system with plastic-lined channels.

house, longer runs could restrict the height available for plant growth and may accentuate problems of poor solution aeration. To assure good aeration, the nutrient solution could be introduced into the channels at two or three points along their length.

The biological effects (and concomitant energy savings) of heating and cooling plant roots rather than manipulating the temperature of the much larger air volume in CEA structures are evident if not widely understood (Moorby and Graves 1980; Orchard 1980). In any event, NFT minimizes the amount of nutrient solution to be treated.

After the energy crunch in the early 1970s, hydroponic and CEA operators naturally experimented with reduced nighttime air temperatures. These lower air temperatures resulted in reduced yield of early tomatoes (first month harvesting), which normally represent only about 7% of total yield but as much as 15% of total crop value; however, total yield and total crop value often proved higher than normal with lower air temperatures (Hurd and Graves 1983; Pollock 1979), apparently because the delayed appearance of fruit permitted the development of a larger plant structure (Verkerk 1966).

As nighttime air temperature and temperature of the nutrient solution appear to influence fruit production independently (Orchard 1980), NFT permits growers to use lower and more economical night air temperatures, which may increase crop yield and total value, while more affordably heating the modest volume of nutrient solution to prevent reduced plant uptake of nutrients (Moorby and Graves 1980). Other crops, including winter lettuce (Morgan *et al.* 1980; Moustafa and Morgan 1981), also appear to benefit from root heating when the plants are small and close to the heated solution. However, in a recent large-scale lettuce trial at the Environmental Research Laboratory at the University of Arizona, there was no increase in growth due to increased solution temperature when 2.5-cm-thick expanded plastic sheeting was present between the plants and the solution, apparently because of the insulative effect of the plastic.

Conversely, in desert regions, NFT may permit an economical cooling of plant roots, avoiding the more expensive cooling of the entire greenhouse aerial temperature. Problems occur in hydroponics when the temperature of the nutrient solution exceeds 35°C (less with lettuce, which begins to bolt at root temperatures much above 20°C). Channel surfaces in NFT systems may be colored white to maintain lower nutrient temperatures. Cooling the liquid for lettuce production not only reduces bolting but lessens the incidence of the damping off fungus *Pythium aphanidermatum*, which also affects the establishment and yield of CEA tomato and cucumber crops. These and other phenomena clearly indicate the need for additional studies on the physiological responses of plants and diseases as a function of temperature variations within the rhizosphere.

The capital costs of an NFT growing system have recently been established at $81,000/ha (not including the cost of construction labor and of the greenhouse enclosing the system), with annual operating costs of $22,000/ha (Van Os 1983) for replacement of plastic troughs and other items.

B. Modified NFT

1. Cast Concrete. Covering a greenhouse floor with cast concrete shaped into NFT channels is one type of modified NFT system. Capital investment is higher, but maintenance is reduced compared with a standard system. In a typical installation for year-round lettuce production (Lauder 1977), the concrete was formed into parallel channels 10 cm wide, 2.5 cm deep, and 45 m long, on a slope of 1 to 50. The surface of the concrete was painted with epoxy resin to isolate it from the nutrient solution (Fig. 10.4). In such an installation in Cheshunt, England (Jensen 1982), 4 ha of concrete NFT channels have permitted a high degree of mechanization in planting and harvesting, with resulting lower production costs (14.5¢/head in summer; 27.0¢/head in winter, higher due to heating costs).

FIG. 10.4. Lettuce production in a modified NFT system with shallow troughs of cast concrete.

2. Movable Channels. In a modified NFT system first proposed by Prince *et al.* (1981) for lettuce production, the channels can be spread apart to increase area as a function of plant growth and size. This variable plant spacing technique is designed to maximize space utilization and leaf interception of radiation. There are limited uses of this system in Canada and the United States.

Movable benches covered with corrugated sheets have been used as lettuce planting troughs in some experiments in the United States and Denmark (Fig. 10.5). Plants are set in the corrugations, through which nutrient solution flows, at close intervals when young, and spread out as more growing space is needed. Benches are movable to allow access to growing areas.

Flat sheets of expanded polystyrene, approximately 2 m × 4 m × 2.5 cm, which can be transported on movable benches, have been used for lettuce production by Varley and Burrage (1981). The sheets, framed in wood, are covered with plastic film to form a wide water trough; additional polystyrene sheets drilled with holes for lettuce seedlings (20/m^2) are fitted into the troughs. This system has been further simplified in a Dutch application (Fig. 10.6).

Schippers (1982) has suggested that tomato and cucumber production could be maximized in a limited area by mounting channels on

FIG. 10.5. Modified NFT system with movable benches covered with corrugated material as channeling.

FIG. 10.6. Modified NFT system with polyethylene stretched over wood framing.

casters and arranging them without intermediate pathways. This would increase plant density by approximately 25%. When cultural operations are required, channels may be moved and separated to create a pathway.

Giacomelli developed a design at Rutgers University for planting tomatoes in flexible plastic tubes suspended by a movable cable system (Giacomelli *et al.* 1982), thus transferring the weight of the rows to greenhouse structural members. Such easily moved rows would permit close spacing of young plants and wider spacing as plants mature, with a potential 25% increase in annual yield per unit area of greenhouse space.

Van Os (1983) has described a completely mechanized NFT system to produce cut chrysanthemums in which mobile planting tables or "transporters" are moved from room to room for planting, growing, and machine harvesting, with a reduced number of interior pathways (Fig. 10.7). Potter and Sims (1975) have referred to a system of movable channels for such crops.

3. Multicropping. In Holland, research engineers at Wageningen are using NFT in a system to multicrop lettuce with CEA tomatoes. The system is normally used to grow tomatoes, but during winter months, additional 25-cm-wide troughs in each bay of the greenhouse

FIG. 10.7. Modified NFT system with movable channels on benches for mechanization of flower production.

are used to produce automated-harvest NFT lettuce. The heads are planted through holes in a flexible plastic material that covers each trough. At harvest, a winching machine pulls the covering material, lettuce and all, up an incline to be rolled up on a spool; as the plastic moves upslope toward the winch, a cutting mechanism severs lettuce heads from roots. The lettuce moves off on a conveyor belt toward the packing station, as the roots are removed on a different conveyor and the plastic cover is slowly wound up on the drum (Jensen 1982).

Similar, if less mechanized, multicropping systems have been developed in England (Starkey 1980). Tomatoes or cucumbers are grown in 35-cm-wide NFT channels during spring, summer, and fall; double rows of lettuce, celery, or Chinese cabbage are grown in the same troughs during winter months.

4. Pipe Systems. An A-frame system developed by Morgan and Tan (1982) provides for high-density lettuce production. Seedlings are planted in sloping (drop of 1 in 30) plastic tubes, 30 mm in diameter, arranged in horizontal tiers resembling A-frames in end-view. This system, developed for use with Dutch Venlo glasshouses, effectively doubles the usable growing surface and accommodates a plant density of 40/m^2. Another A-frame system of tiered NFT channels, developed for strawberry production, is reported to facilitate spraying and picking, with quick crop turnaround and reduced labor requirements (White 1980).

Schippers (1978) developed a vertical pipe system in which small-diameter plastic pipes, 1.3 m long, are suspended by overhead wires above a nutrient solution-collecting channel. Germinated lettuce plants are squeezed into holes (20–28/pipe) in the sides of the tubes, and nutrient solution is pumped into the top of each pipe, to drip down through the tubing and plant roots. A similar system with aluminum irrigation piping has been used in Israel to produce vegetables and flowers. Vertical NFT systems using plastic or aluminum piping (Fig. 10.8) are relatively costly, and they have not been widely adopted in commercial operations.

5. Moving Belts. To maximize lettuce production in a limited area, Whittaker Agri-Systems developed a highly mechanized NFT system (Rogers 1983) in a 0.8-ha CEA complex in California. Twenty-day-old seedlings are transplanted into movable belts in NFT troughs stacked two high (Fig. 10.9) and mechanically harvested after 19–35 days of growth; conveyor belts transport the lettuce to packing stations and refrigerated storage. The upper tiers or troughs obviously shade out plants on the lower tiers, causing some reduction in yield and quality of the latter; this effect could be reduced by eliminating some of the upper tiers to permit the entrance of more radiation. This capital-

FIG. 10.8. Modified NFT system with a vertical piping. Nutrient solution is periodically sprayed into the top of the hollow pipes and drips downward past plant roots.

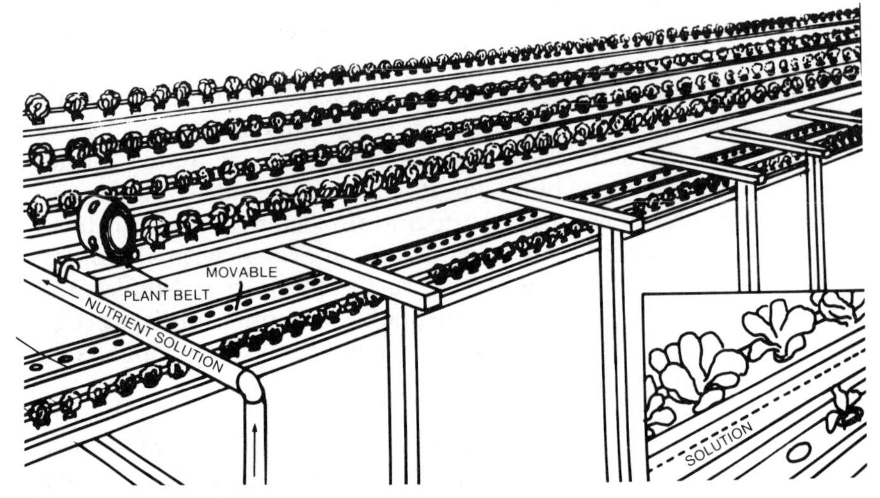

FIG. 10.9. Lettuce production in modified NFT system with movable belts in troughs stacked two high.

intensive system is not yet in widespread use; its economic viability remains to be demonstrated.

A self-contained CEA module called the Ruthner System is being marketed in Austria. Figure 10.10 shows a diagram of this system, in which two vertical conveyor belts move within a prefabricated, artificially lighted steel structure that is insulated and lined with reflective material. Plants are moved up and down within a three-dimensional light grid and supplied with nutrient solution at the bottom of the unit. As is the case with earlier CEA systems based totally on artificial light sources, it is difficult, despite marketing claims, to imagine a competitive advantage for such an energy-intensive technology in the everyday world.

C. Floating Hydroponics

There have been several reports on floating hydroponic systems during the past decade. These deal primarily with seed germination in floating materials for subsequent transplanting, or with the use of beds of floating particles to cover the entire surface of the nutrient solution, to provide some plant support.

Farnsworth (1975) in California obtained patents beginning in 1975 for the complete growout of individual plants in separate floating collars. These could be moved easily on the surface of the nutrient solution, and the collars periodically enlarged as plant size increased, to maximize the use of surface area and leaf interception of radiation.

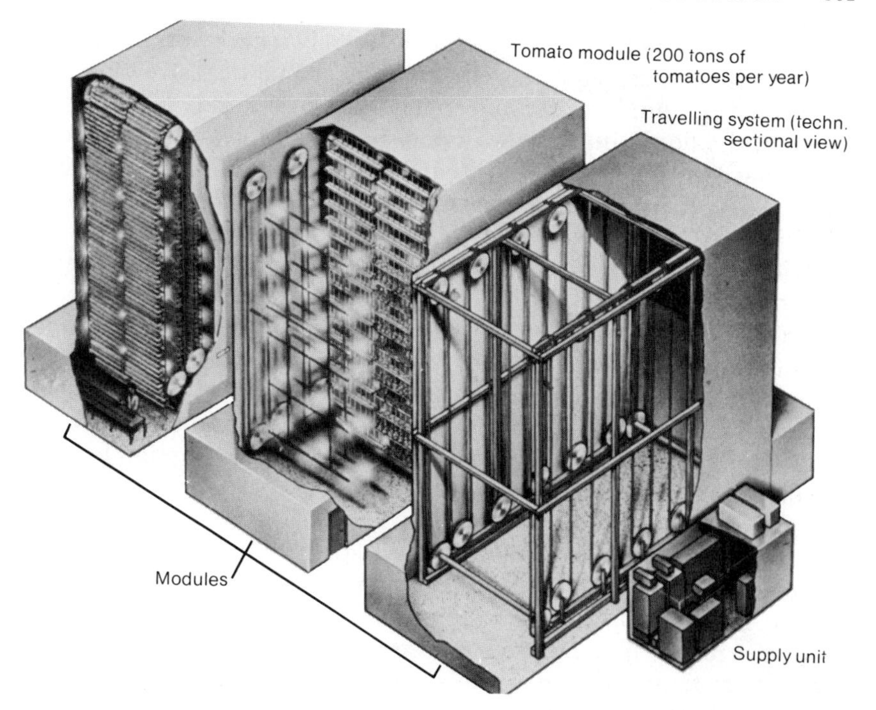

FIG. 10.10. Self-contained, artificially lighted hydroponic modules (Ruthner System) with vertical conveyor belts.

(Independently, research personnel at the Environmental Research Laboratory at the University of Arizona had developed a similar experimental system, but did not publish their results.) There have been no reported commercial applications of this technique.

In 1976, a method of growing a number of heads of lettuce or other leafy vegetables on a floating raft of expanded plastic was developed independently by Jensen (1980) in Arizona and Massantini (1976) in Italy. Jensen's experiments led to a large-scale, 2-year production and marketing trial in Arizona, most details of which are reported, in the remainder of this section, for the first time.

In cooperation with Kraft, Inc., researchers at the University of Arizona, in 1980, designed and developed a prototype hydroponic CEA *raceway* lettuce production system, which was operated during 1981–1982. A 0.5-ha greenhouse containing 0.26 ha of water surface in 10 production raceways, as well as nursery, packaging, and research areas, was operated year-round. The design production rate was 4.5 million heads/ha/year, and the design production cost for the prototype system was 21.4 cents/head.

Raceway is a perhaps unfortunate borrowing from aquaculture nomenclature; in this context it means merely a horizontal, rectangular tank (as contrasted with a circular tank or a vertical structure). The term does not imply water movement, although it may sound as if it should; the liquid in a raceway may be completely static. The raceways used in the prototype system were each 4 m × 70 m, and 30 cm deep. The nutrient solution was monitored, replenished, recirculated, and aerated. A raceway system has two distinct advantages: The nutrient pools are virtually frictionless conveyor belts for planting and harvesting movable floats, and the plants are spread in a single horizontal plane, so that interception of sunlight per plant is maximal.

Four commercial cultivars of lettuce were grown on the floats: a short-day leafy type ('Waldemann's Green'), and three cultivars of summer butterhead ('Ostinata', 'Salina', and 'Summer Bibb'). Adjusted for system changes, the average monthly production level exceeded design parameters by 120%, and the cost per head was 7% lower than the projected cost.

Two- to three-week-old seedlings were transplanted to holes in the 2.5-cm-thick plastic floats in staggered rows with approximately 300 cm^2/plant (Fig. 10.11). (The original concept was to plant more heads with narrower spacings and subsequently transfer them in mid-growout to floats with fewer holes and wider spacings. This concept was modeled by not executed.) Raceway growout time to harvest was 4–6 weeks. As a crop of several floats was harvested from one end of a raceway, new floats with transplants were introduced at the other end. Long lines of floats with growing lettuce were moved easily at the touch of a finger (Fig. 10.12 and 10.13).

FIG. 10.11. Two- to three-week-old lettuce seedlings transplanted into polystyrene floats in University of Arizona raceway hydroponic system.

FIG. 10.12. Long lines of floats with growing lettuce are moved easily on nutrient pools, which are virtually frictionless conveyor belts for planting and harvesting, in University of Arizona raceway system.

FIG. 10.13. Interception of sunlight is maximized in the University of Arizona's floating hydroponic (raceway) system.

Growth rates of all cultivars correlated positively with levels of available light and did so up to the highest levels measured, although radiation levels in the Arizona desert are two to three times that of more temperate climes (Glenn 1983). This finding was surprising in that OFA lettuce is saturated by relatively low levels of light, and growth is inhibited as radiation increases. Additionally, greenhouse lettuce in other regions is usually regarded as a cool-season crop. A further finding was that crops grown during autumn, when daylight hours are decreasing, used available light two to three times more efficiently than winter or spring crops.

The reason for the latter effect is that daytime air temperatures also correlated positively with growth; therefore, fall crops, grown under higher temperatures, were more efficient than spring crops. (However, during the summer monsoon season when evaporative cooling systems are ineffective, a combination of high-temperature and high-light levels caused lettuce to bolt. Several experiments were conducted to chill the nutrient solution, which reduced bolting). The best predictor of lettuce growth in the prototype raceway system was the product of daytime temperature and the log of radiation.

Concurrent with the operation and modification of the raceway production systems, packaging and marketing experiments were conducted in association with Kraft, Inc. Packaging individual heads in air-sealed plastic bags extended shelf-life up to 3 weeks and provided protection during transportation (Fig. 10.14); this procedure has also been effective in Norway (Lawson 1982). (Sealing the heads in a CO_2 atmosphere had no apparent beneficial effect.) The lettuce was also sealed with the roots intact, as researchers at General Mills (Mermelstein 1980) had reported this permitted plants to stay alive and unwilted for extended periods; this procedure has also been used by ICI in England (Shakesshaft 1981). In the Arizona experiments, however, the roots-on package did not appear to increase shelf-life, was three times more expensive to prepare and pack, increased product volume and weight for transportation, and was not particularly popular with wholesalers or retailers.

The Kraft–University of Arizona project was terminated as scheduled in December 1982, at the end of the second year. It was not announced if further uses of the technology were planned.

D. Aeroponics

1. Root Mist Technique. In an unusual application of closed-system hydroponics, plants are grown in holes in panels of expanded polystyrene or other material with the plant roots suspended in midair

FIG. 10.14. Packaging individual heads of lettuce in air-sealed plastic bags extends shelf-life up to 3 weeks.

beneath the panel and enclosed in a spraying box (Fig. 10.15). The box is sealed so that the roots are in darkness (to inhibit algal growth) and in saturation humidity. A misting system sprays the nutrient solution over the roots periodically; the system is normally turned on for only a few seconds every 2–3 min. This is sufficient to keep the roots moist and the nutrient solution aerated.

Aeroponic systems of this type, with inclined polystyrene surfaces (Fig. 10.16), were first described in detail in 1970 by Massantini (1973) in Italy. Similar systems were developed subsequently by M.H. Jensen in Arizona for lettuce (Fig. 10.17), spinach (Fig. 10.18), and even tomatoes (Fig. 10.19), although the latter was judged not to be economically viable. In fact, there are no known large-scale commercial aeroponic operations in the United States, although several small companies market systems for home use. For some types of advanced horticulture, such as rooted cuttings of imported foliage plant materials without soil, aeroponics may prove to be suitable.

The A-frame aeroponic system developed in Arizona (Fig. 10.17) for low, leafy crops may be feasible for commercial food production. Inside a CEA structure, these frames are oriented with the inclined slope facing east–west. The expanded plastic panels are standard-sized (1.2 m × 2.4 m), mounted lengthwise, and spread 1.2 m at the base to form an end-view equilateral triangle. The A-frame rests atop a panel-sized

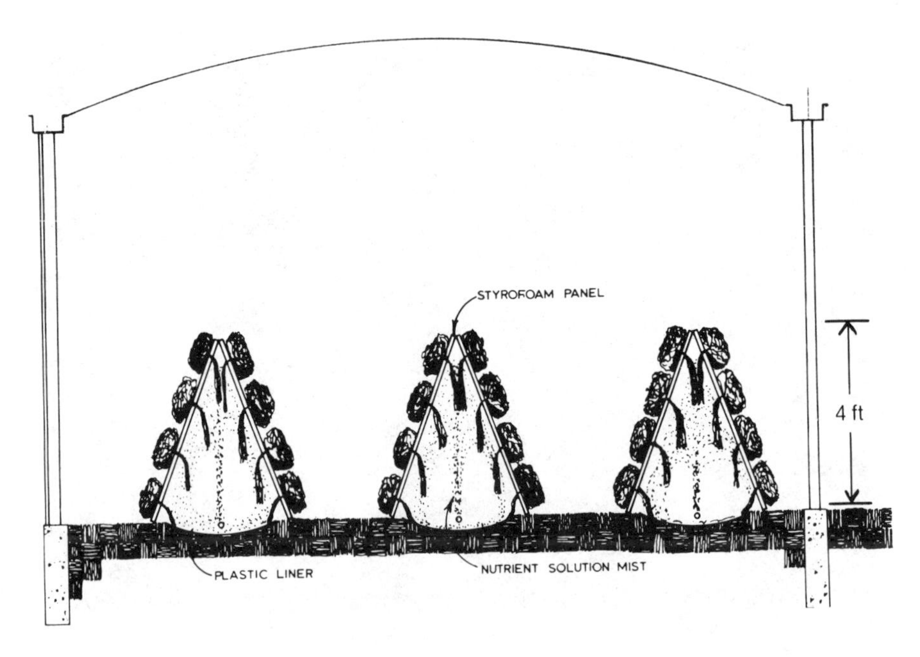

FIG. 10.15. In root mist aeroponic system, plant roots are suspended in midair and sprayed with a nutrient mist.

FIG. 10.16. Original aeroponic system for strawberry production was developed at the University of Pisa in Italy.

FIG. 10.17. Aeroponic A-frame unit, developed at the University of Arizona, makes better use of greenhouse space than original designs.

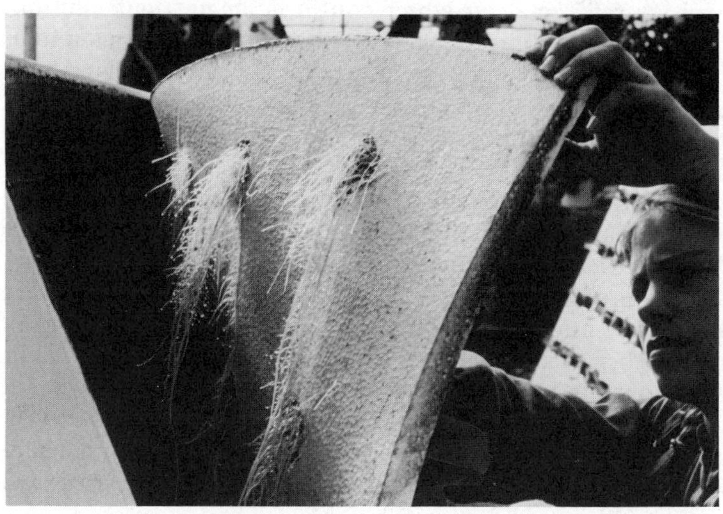

FIG. 10.18. Spinach roots hanging in midair inside closed aeroponic A-frame unit.

FIG. 10.19. Movement of nutrients was studied in aeroponic tomato experiments at the University of Arizona in 1970.

watertight box, 25 cm deep, which contains the nutrient solution and misting equipment (Jensen 1980). Young transplants in small cubes of growing medium are inserted into holes in the panels, which are spaced at intervals of 18 cm on center. The roots are suspended in the enclosed air space and misted with nutrient solution as described previously.

An apparent disadvantage of such a system is uneven growth resulting from variations in light intensity on the inclined crops (Giboney 1980); further studies on variance of slope are therefore indicated. An advantage of this technique for CEA lettuce or spinach production is that twice as many plants may be accommodated per unit of floor area as in other systems; i.e., as with vine crops, the cubic volume of the greenhouse is better utilized. Unlike the small test systems described here, larger plantings could utilize A-frames more than 30 m in length, sitting atop a simple, sloped trough that collects and drains the nutrient solution to a central sump; greenhouses, furthermore, could be designed to be much lower in height.

Another potential commercial application of aeroponics, in addition to the rooting of foliage plant cuttings, would be the production of leafy vegetables in locations with extreme restrictions in area and/or weight. This would specifically include very large, manned space vehicles for extended occupancy. Jensen's (1980) work has included the development of aeroponic systems in a revolving "space drum" that simulates extraterrestial gravities (Fig. 10.20). Later, the Boeing Co. further developed these concepts as shown in Fig. 10.21.

FIG. 10.20. Experimental aeroponic "space drum," developed by M.H. Jensen, permits growth of lettuce and other crops in extraterrestrial gravities.

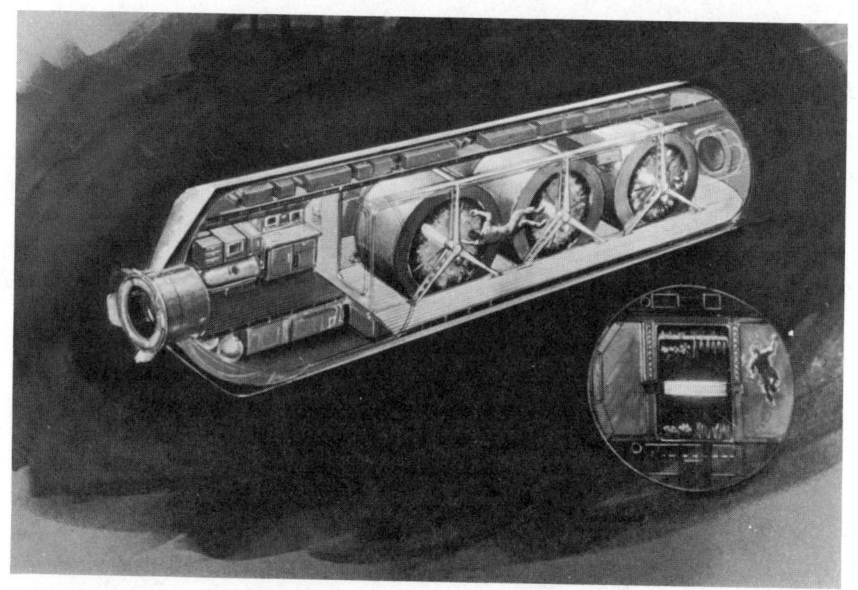

FIG. 10.21. Boeing Co. design concept for revolving "space drum" for crop production in a space vehicle.

2. Ein Gedi Technique. A technology developed at the Kibbutz Ein Gedi in Israel for the production of tomatoes, cucumbers, and ornamental plants (Soffer and Levinger 1980) is superficially similar to a floating hydroponic system but is really more akin to an aeroponic system. In this application, the narrow raceways are 20 cm wide and 14 m long, with a depth of 10 cm (Fig. 10.22). The planting boards are supported by the sides of the raceways, and plant roots are suspended into an air space between the boards and the nutrient solution. The latter is monitored, replenished, and recycled with aeration. More than 30 such units are reported to be in experimental operation in Israel, Holland, Norway, and Italy, to evaluate the system.

IV. AGGREGATE HYDROPONIC SYSTEMS

In aggregate hydroponic systems, a solid, inert medium provides support for the plants. As in liquid systems, the nutrient solution is delivered directly to the plant roots. Also, aggregate systems may be either open or closed, depending on whether, once delivered, surplus amounts of the solution are recovered and reused. Open systems do not recycle the nutrient solution; closed systems do.

A. Open Systems

In most open hydroponic systems, excess nutrient solution is recovered; however, the surplus is not recycled to the plants but is disposed of in evaporation ponds or used to irrigate adjacent landscape plant-

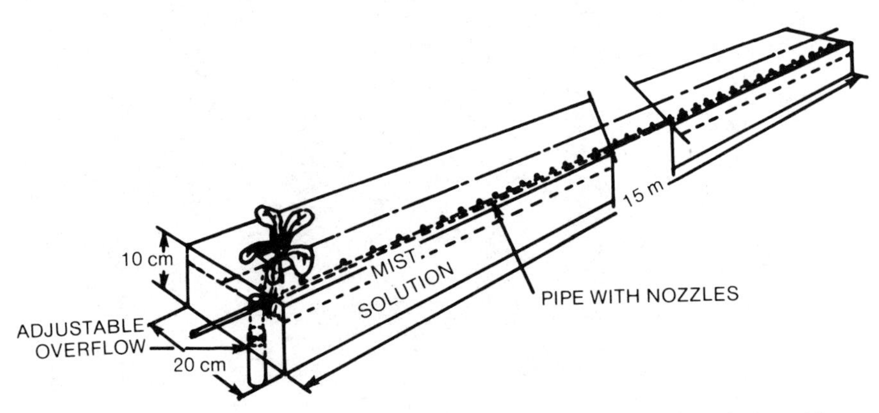

FIG. 10.22. Aeroponic system developed at the Kibbutz Ein Gedi in Israel has planting boards fixed atop narrow raceways. Plant roots are suspended in the air between the boards and nutrient solution.

ings or windbreaks. Because the nutrient solution is not recycled, such open systems are less sensitive to the composition of the medium used or to the salinity of the water. This in turn has given rise to experimentation with a wide range of growing media and development of lower-cost designs for containing them. In addition to wide growing beds in which a sand medium is spread across the entire greenhouse floor, troughs, trenches, and bags are also used, as well as slabs of porous horticultural grade rockwool.

Fertilizers may be fed into the irrigation water with proportioners (Fig. 2.23, Plan A), which are discussed in Section V. B. 1, or may be mixed with the irrigation water in a large tank or sump (Fig. 2.23, Plan B). Irrigation is usually programmed through a time clock, and in larger installations, solenoid valves are used to allow only one section of a greenhouse to be irrigated at a time (permitting use of smaller-sized mechanical systems).

1. Trough or Trench Culture. Some open aggregate systems involve relatively narrow growing beds, either as abovegrade troughs (Fig. 2.24) or subgrade trenches (Fig. 2.25), whichever are more economical to construct at a given site. In both cases the beds of growing media are separate from the rest of the greenhouse floor and confined

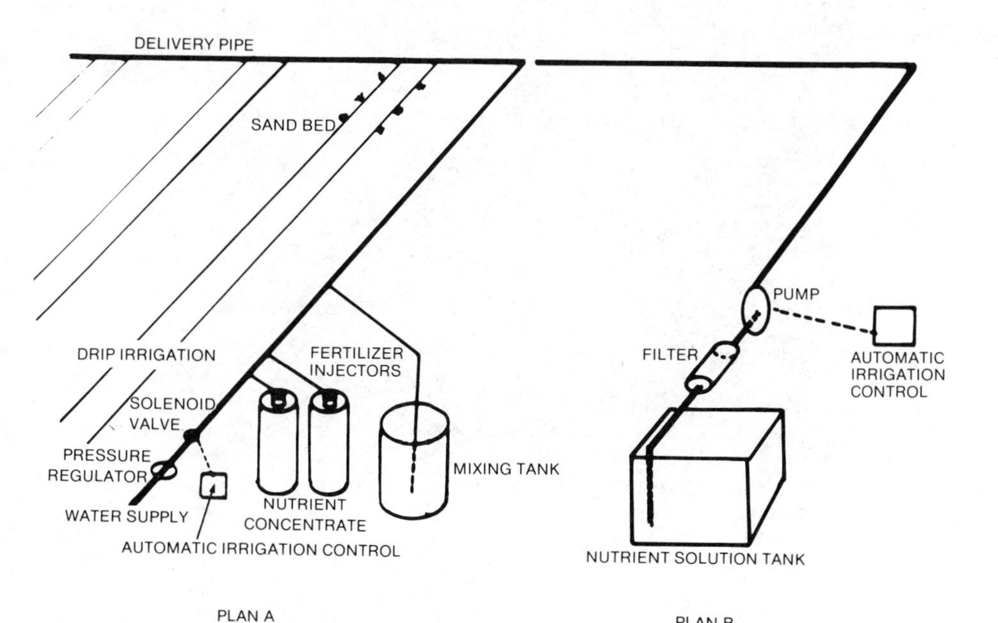

FIG. 10.23. Nutrient introduction in an open hydroponic systems. Plan A uses a fertilizer proportioner; plan B uses a simple mixing tank.

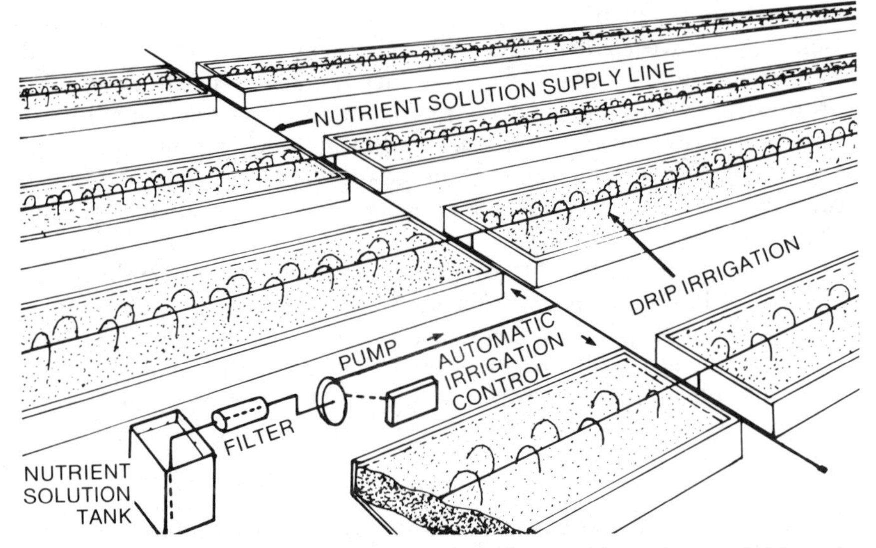

FIG. 10.24. Open aggregate system using abovegrade, waterproofed troughs and drip irrigation.

FIG. 10.25. Open aggregate system using belowgrade trenches with plastic-lined sand beds.

within waterproof materials. For ease of description, this will henceforth be called trough culture.

Concrete is a traditional construction material for permanent trough installations (it may be covered with an inert paint or epoxy resin); fiberglass or plyboard covered with fiberglass is also used. To reduce costs, polyethylene film, at least 0.01 cm in thickness, is now commonly used. The film, usually in double layers to avoid leakage (pinholes in either layer will seldom match up), is placed atop a sand base and supported by either planks, cables, or concrete blocks (Jensen 1968; Sheldrake and Dallyn 1969).

The size and shape of the growing bed are dictated by labor efficiencies rather than by engineering or biological constraints. Vine crops such as tomatoes usually are grown in troughs only wide enough for two rows of plants, for ease in pruning, training, and harvesting. Low flowering plants are grown in somewhat wider beds, with a midpoint a worker can conveniently reach at arm's length. Bed depth varies with the type of growing media, but about 25 cm is a typical minimum; shallower beds of 12–15 cm are not uncommon but require more attention to irrigation practices. Length of the bed is limited only by the capability of the irrigation system, which must deliver uniform amounts of nutrient solution to each plant, and by the need for lateral walkways for work access. A typical bed length is about 35 m. The slope should have a drop of at least 15 cm per 35 m for good drainage, and there should be a well-perforated drain pipe of agriculturally acceptable material inside the bottom of the trough beneath the growing medium.

As open systems are less sensitive than closed systems to the type of growing medium used, a great deal of regional ingenuity has been displayed in locating low-cost inert materials for trough culture. Typical media include sand (Jensen 1971), vermiculite (Harris 1976), sawdust (Adamson and Maas 1981; Maas and Adamson 1981), perlite, peat moss (Maher 1976), mixtures of peat and vermiculite (Jensen 1968), and sand with peat or vermiculite (Smith 1982). One of the most effective of these materials is a mixture of peat moss and vermiculite developed at Cornell University (Sheldrake and Boodley 1965) to provide trough growers with a sterile soil. The composition of this mixture is given in Table 10.5. The peat moss–vermiculite medium is customarily used with drip or trickle irrigation systems; a fertilizer regime for plants grown in this medium is discussed in Section V. B.

2. Bag Culture. Bag culture is similar to trough culture, except that the growing medium is placed in plastic bags, which are formed in lines on the greenhouse floor, thus avoiding the costs of troughs or

TABLE 10.5. Composition of Peat Moss–Vermiculite Medium for Trough Culture

Component	Quantity
Shredded sphagnum peat	388 liters
Horticultural vermiculite (#2 or #3)	388 liters
Limestone (not hydrated lime)	4540 g
Superphosphate (20% P_2O_5)	908 g
Potassium nitrate (14–0–44)[1]	454 g
Borax (11% B)	10 g
Iron	25 g

Source: Sheldrake and Boodley (1965).
[1]Instead of potassium nitrate, 1362 g of 5N–4.3P–8.3K

trenches and complex drainage systems. The bags also may be used for at least two years and are much easier and less costly to steam sterilize than bare soil.

The bags are typically of UV-resistant polyethylene, which will last in a CEA environment for 2 years (Sheldrake 1981); they have a black interior. The exterior of the bag should be white in deserts and other regions of high light levels in order to reflect radiation and inhibit heating the growing medium; conversely, a darker exterior color is preferable in northern, low-light latitudes to absorb winter heat. Bags used for horizontal applications (the most common) are usually 50–70 liter in capacity. Growing media for bag culture include peat, vermiculite, or a combination of both to which may be added polystyrene beads, small waste pieces of polystyrene, or perlite to reduce the total cost. When used horizontally, bags are sealed at both ends after being filled with the medium.

The bags are placed flat on the greenhouse floor at normal row spacing for tomatoes or other vegetables (Judd 1982), although it would be beneficial to first cover the entire floor with white polyethylene film. M. H. Jensen demonstrated with trough culture in New Jersey that 86% of the radiation falling on a white plastic floor is reflected back up to the plants (Fig. 10.26) compared with less than 20% of the light striking bare soil. Such a covering may also reduce relative humidity and the incidence of some fungus diseases.

Paired rows of bags are usually placed flat, 1.5 m apart (from center to center) and with some separation between bags; i.e., each row is not end-to-end. Holes are made in the upper surface of each bag for the introduction of transplants, and two small slits are made low on each side for drainage or leaching. Some moisture is introduced into each bag before planting. Less commonly, the bags are placed vertically with open tops for single-plant growing; these have the disadvantages of being less convenient to transport, requiring more water, and maintaining less even levels of moisture.

Drip irrigation of the nutrient mix is recommended, with a capillary

FIG. 10.26. Trough culture beds with white plastic film covering the aisles to maximize reflection of sunlight into the crop.

FIG. 10.27. Tomato plants at Ohio State University growing in bags filled with peat and vermiculite and supplied with nutrients by drip irrigation.

tube leading from the main supply line to each plant (Fig. 10.27). Plants growing in high-light, high-temperature conditions will require up to 2 liters of nutrient solution per day. Moisture near the bottom of the bagged medium should be examined often, and it is best to err on the wet side (Sheldrake 1981).

The most commonly grown crops in bag culture are tomatoes and cucumbers (also cut flowers). When tomatoes are grown, each bag is used for two crops per year for at least 2 years. It has not yet been established how many crops may be grown before the bags must be replaced or steam-sterilized, but the latter is performed by moving the bags together under a tarpaulin, at an estimated cost of less than $1000/ha (Sheldrake 1981).

In 1983, the typical cost of bags filled with growing medium was approximately $30,000/ha (W. R. Grace Co., personal communication). This relatively high cost probably is the primary factor inhibiting use of bag culture in the United States, where hydroponics struggles to remain cost competitive. In Europe, however, where hydroponic CEA is used more widely, bag culture is replacing trough culture because of its greater ease in operations and in moving of material into and out of greenhouses.

3. Rockwool Culture. The use of horticultural rockwool as the growing medium in open hydroponic systems has been increasing rapidly; such systems now receive more attention from a research standpoint than any other type in Europe (Hanger 1982). Cucumbers and tomatoes are the principal crops grown on rockwool; in Denmark, where rockwool culture originated in 1969 (Verwer 1976), virtually all cucumber crops are grown on rockwool. This technology is the primary cause of the rapid expansion of hydroponic systems in Holland, which increased from 25 ha in 1978 to 80 ha in 1980, and increased again to more than 500 ha by the end of 1982 (Van Os 1983).

In the Westland region of Holland, which has the highest concentration of CEA greenhouses in the world, soil was used and steam-sterilized as necessary until the late 1970s; until then, inexpensive natural gas was available for sterilization. When fuel costs increased, Westland growers turned to methyl bromide as a soil fumigant. But bromides began to build up in the groundwater (and salinity increased due to saltwater intrusion from expansion of regional ship canals), so the Dutch government began to frown on the use of methyl bromide (Van Os 1983). Lacking other inexpensive means of soil sterilization, increasing numbers of Dutch operators turned to hydroponics, experimenting first with such growing media as soil, peat, and even bales of straw, and with NFT and bag culture, until the rockwool culture was developed.

Rockwool, first developed as an acoustical and insulation material, is made from a mixture of diabase, limestone, and coke that is melted at high temperature, extruded in small thread (0.005 mm in diameter), and pressed into sheets weighing about 80 kg/m^3. Insulation rockwool (and fiberglass batting as well) is inappropriate for horticulture. For use as a growing medium, rockwool must be modified in a process that has remained confidential (Hanger 1982). By the early 1970s, horticultural rockwool was being manufactured in Denmark, primarily for cucumber production, and the technology has spread to West Germany, Holland, England, Canada, and (in small test facilities) Israel.

As a growing medium, rockwool is not only relatively inexpensive but is also inert and biologically nondegradable, takes up water easily, is approximately 96% "pores" or interstitial air spaces, has evenly sized pores (which has important consequences for water retention), lends itself to simplified and lower-cost drainage systems, and is easy to bottom-heat during winter. Its versatility is such that rockwool is used in plant propagation and potting mixes, as well as in hydroponics.

In open hydroponic systems, plants are usually propagated by direct seeding in small (40 mm^3) rockwool cubes that have a small hole punched in the top and are saturated with nutrient solution. These cubes are usually transplanted into larger rockwool cubes (75 mm^3), manufactured specifically to receive the germinating cubes, and side-wrapped with black plastic film. The larger cubes are subsequently placed atop rockwool slabs placed on the greenhouse floor. The slabs are usually 15–30 cm wide, 75–100 cm long, and 75 mm thick.

A typical CEA layout for open-system rockwool culture is shown in Fig. 10.28. Level ground is covered with white polyethylene film for good hygiene and light reflection. A bed normally consists of two rows of rockwool slabs, each individually wrapped in white film, with the rows spaced 30 cm apart. The slabs should have a slight inward tilt toward a central drainage swale. If bottom heat is required, the slabs are placed atop polystyrene sheets, grooved in the upper surface to accommodate hot water pipes, as shown in Fig. 10.28. Because of the porosity of the rockwool, almost all of the nutrient solution remains in the slab for plant use if an appropriately modest irrigation schedule is used; if there is a surplus, it will drain out of the slab and into the shallow channel.

Before seedlings are transplanted, the rockwool slabs are soaked with nutrient solution. The plants, remaining in the small rockwool cubes in which they were established, are simply set atop the slabs through holes cut in the plastic film (Fig. 10.29). If the root system is well developed in the cubes, the roots will move into the slab within 2–3 days. Each plant receives nutrient solution via individual drippers,

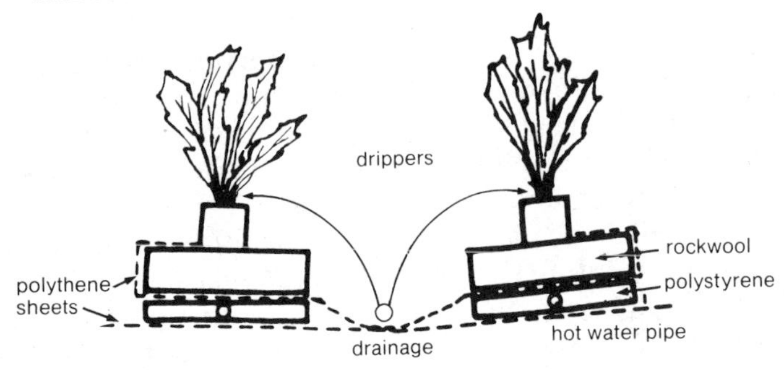

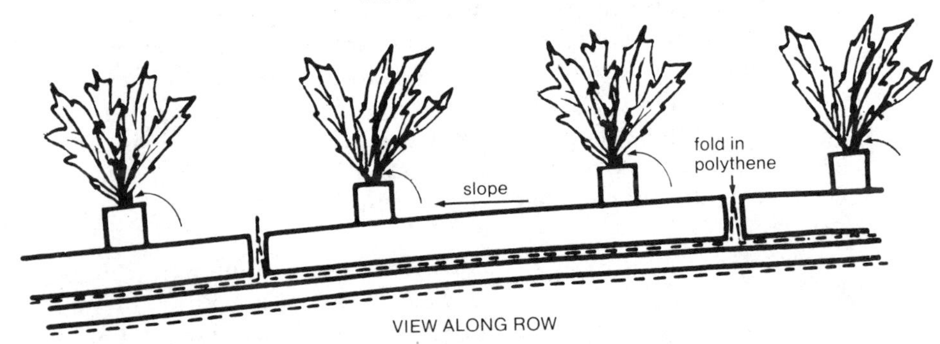

FIG. 10.28. Typical layout for rockwool culture. Rockwool slabs are positioned for ease of drainage and use of bottom heating.
From Hanger (1982b).

FIG. 10.29. Plants in small rockwool cubes set atop rockwool slabs in holes cut in the covering plastic. Each plant is drip irrigated with nutrient solution.

with irrigation rates varied as a function of plant demand and environmental conditions.

Rockwool has several inherent advantages as an aggregate: It is lightweight when dry, easily handled, simple to bottom heat, and easier to steam-sterilize than many other types of aggregate materials; or it may simply be discarded or incorporated as a soil amendment after crops have been grown in it for several years. In addition, an open system with rockwool permits accurate and uniform delivery of nutrient solution; requires less equipment, fabrication, and installation costs; and entails less risk of crop failure due to the breakdown of pumps and recycling equipment. The obvious disadvantage is that rockwool may be relatively costly unless manufactured within the region.

If only for the latter reason, there is no significant application of rockwool culture in the United States, where no horticultural rockwool or similar material is manufactured. Nor is there much industrial interest in such development, given the unattractive market represented by the small and dwindling number of American hydroponic operations, although one or two small research grants from private industry have been solicited.

4. Sand Culture. Concurrent with the beginning of rockwool culture in Denmark in 1969, a type of open-system aggregate hydroponics using pure sand as the growing medium was under development by researchers at the University of Arizona, initially for desert applications (Fontes *et al.* 1971; Jensen 1971, 1973). It was logical to investigate such a potential. Because other types of growing media must be imported to desert regions and may require frequent renewal, they are more expensive than sand, a commodity usually available in profuse abundance.

The Arizona researchers designed and tested several types of sand-based hydroponic systems. The growth of tomatoes and other greenhouse crops in pure sand was compared with their growth in nine other mixtures (e.g., sand mixed in varying ratios with vermiculite, rice hulls, redwood bark, pine bark, perlite, peat moss); there were no significant differences in yield (Jensen 1973; Jensen 1975). Unlike many other growing media, which undergo physical breakdown during use, sand is a permanent medium; it does not require replacement every 1 or 2 years.

Pure sand can be used in trough or trench culture. However, in desert locations, it is often more convenient (and less expensive) to cover the greenhouse floor with polyethylene film and a system of perforated drainage pipes, and then backfill the area with sand to a depth of approximately 30 cm (Fig. 10.30). If the depth of the sand bed is shallower, moisture conditions may not be uniform and plant roots may grow into the drain pipes. The areas to be used as planting beds may be level or slightly sloped; supply manifolds for nutrient solution must be sited accordingly.

Different types of desert and coastal sands with a variety of physical and chemical properties were used successfully by the University of Arizona workers. The size distribution of sand particles is not critical, except that exceptionally fine material, such as mortar sand, does not drain well and should be avoided. The particle-size distribution of sand typically used in this technology is given in Table 10.6. Standard drip or trickle irrigation systems are used; the nutrient solutions, as well as proportioner or mix-tank delivery systems, are reviewed in Sections V. B and C. In the larger Arizona-type systems, fertilizer proportioners usually are used and irrigation is programmed through a time clock system (Fig. 10.31). If the sand used is highly calcareous, increased amounts of chelated iron must be applied to the plants. Sand growing beds should be fumigated annually because of possible introduction of soilborne diseases and nematodes.

In considering any new location for a closed hydroponic CEA facility, a water agricultural suitability analysis must be done. It is important to know the sodium absorption ratio (SAR), salinity, and pH of the water, and whether any undesirable elements (e.g., boron or fluo-

FIG. 10.30. In sand culture, the entire greenhouse floor is backfilled with sand to a depth of about 30 cm over a plastic membrane and perforated drainage pipes.

FIG. 10.31. In large sand culture system developed at the University of Arizona, timed proportioning system injects nutrient solution into the irrigation water for drip irrigation.

ride) are present. Irrigation water containing more than 500 ppm (approx. 0.78 mhos/cm $\times 10^3$) of salts not used by plants in large amounts should be avoided.

Irrigation practices are particularly critical in the desert during the high-radiation summer months, when crops may have to be irrigated up to eight times per day. Proper irrigation is indicated by a small but continuous drainage (4–7% of the application) from the entire growing area. Evaporation of water around small summer tomato transplants is often high, which can lead to a slight buildup of fertilizer salts in the planting beds. Extra nitrogen causes excessive vegetative growth, decreasing the number of marketable fruits; this can be avoided by reducing the amount of nitrogen in the nutrient solution from the time of transplant until the appearance of the first blossoms. Drainage

TABLE 10.6. Particle-Size Distribution of Typical Sand Used in Hydroponic Systems

Particle Size (mm)	Size Distribution (%)
Over 4.760	1
2.380–4.760	10
1.190–2.380	26
0.590–1.190	20
0.277–0.590	25
0.149–0.277	15
0.074–0.149	2
Less than 0.074	1

Source: Jensen (1971).

from the beds should be tested frequently, and the beds leached when drainage salts exceed 3000 ppm.

In the United States, the principal crops grown in sand culture systems are tomatoes and seedless cucumbers (Fig. 10.32); yields of both crops have been high (Jensen 1980). Cucumber production has exceeded 700 MT/ha. In the Middle East, a wider variety of table vegetables has been grown in sand culture installations (Fontes 1973), as OFA fresh produce is relatively scarce in remote desert regions.

The 1983 capital cost of a sand culture hydroponic system in Tucson, exclusive of CEA (i.e., not including greenhouses, heating/cooling devices, controls, cold storage, etc.) is $50,000/ha. This compared favorably with the capital costs of systems employing rockwool or NFT (see Table 10.12).

B. Closed Systems

1. Gravel. Closed systems using gravel as the aggregate material were commonly employed for commercial and semicommercial (family) hydroponic CEA facilities 20 years ago. Though a number of these installations remain in operation, this technology has, for the most part, been superseded by NFT systems or by the newer, open aggregate systems described in the previous section. As in all closed

FIG. 10.32. Cucumbers grown in a sand culture system with drip irrigation. In Arizona, cucumber yields in this system have exceeded 700 MT/ha.

systems, great care must be taken to avoid the buildup of toxic salts and to keep the system free of nematodes and soilborne diseases. The latter can be a problem and is a big disadvantage of closed systems. Once certain diseases are introduced, the infested nutrient solution will contaminate the entire planting.

A typical subirrigated gravel closed system of the early type is illustrated in Fig. 10.33. The main components of such a system are (a) a growing bed filled with rock, extremely coarse sand, or any other similar inert nonphytotoxic medium, and with a subgrade piping system; (b) a supply tank or sump to contain the nutrient solution; (c) a pump to deliver the solution to the growing bed; and (d) a timing mechanism to actuate the pump. In these systems, the solution is customarily flooded into the bed to within about 2 cm of the surface of the medium and then drains by gravity back into the sump for reuse. Drip irrigation from the surface has also been used in these systems, particularly when sand is the growing medium; care must be taken to provide enough irrigation lines to wet the aggregate uniformly as there is very little lateral movement of the solution.

Such systems are capital intensive because they require leakproof growing beds, as well as subgrade mechanical systems and nutrient storage tanks. Waterproofed concrete was used initially, and later fiberglass (Fig. 10.34) or heavy-gauge plastic film, to form the growing beds.

2. NFT and Rockwool. A more recent development in Europe is the combination of NFT and rockwool culture, which is, by definition, a closed aggregate system. In this application, which is not widespread, plants are established on small rockwool slabs positioned in channels containing recycled nutrient solution (Hanger 1982). This procedure

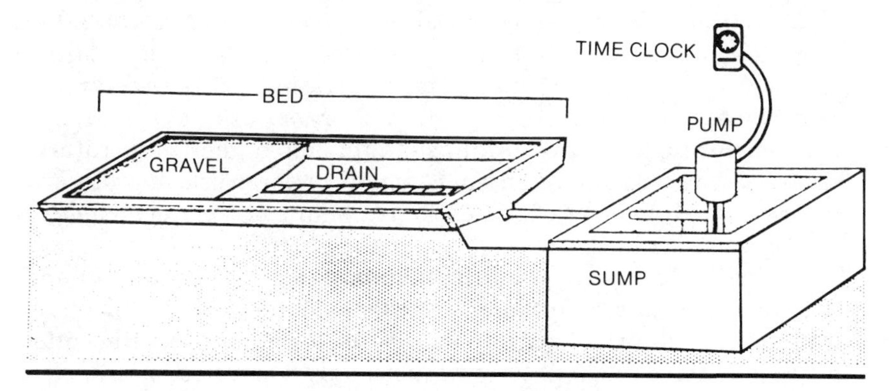

FIG. 10.33. Early type of subirrigated gravel closed system, which has generally been superseded by NFT or open aggregate systems.

FIG. 10.34. Although expensive, fiberglass trays and tanks have been used in closed gravel culture of tomatoes and cucumbers.

reduces the amount of rockwool needed in comparison with the open systems described in Section IV. A. 3. This combined technology is believed by its advocates to be conservative of nutrient solution because the rockwool acts as a nutrient reservoir in case of pump failure and helps anchor the plants in the nutrient channel.

V. NUTRIENT SOLUTIONS

Many formulae for hydroponic nutrient solutions have been published, but they are all quite similar, differing mostly in the ratio of nitrogen to potassium (Table 10.7). It is unlikely there are any "secret ingredients" that make one formula better than another; and the rationale for differing nitrogen/potassium ratios is no particular mystery: Plants need less nitrogen during short or dark days, and more nitrogen during long days, bright sunlight, and higher temperatures. Theoretically, each plant in each region of the world has its own nutritional requirements, but in practice, plants fortunately display a forebearing tolerance.

There can be a significant difference in the cost, purity, and solubility of the chemicals comprising a nutrient solution, depending on the grade (pure, technical, food, or fertilizer) used. Smaller operators often buy ready-mixed nutrient formulations; only water need be added to prepare the nutrient solution. Larger facilities prepare their own solutions to standard or slightly modified formulae. The commonly used

TABLE 10.7. Composition of Hydroponic Nutrient Solutions

Reference	Element (mg/kg)											
	N	P	K	Ca	Mg	S	Fe	Mn	Cu	Zn	B	Mo
Hoagland and Arnon (1950)	210	31	234	160	48	64	0.6	0.5	0.02	0.05	0.5	0.01
Schwartz (1978)	126	93	312	124	43	160	—	—	—	—	—	—
Resh (1981)	175	65	400	197	44	197	2	0.5	0.03	0.05	0.5	0.02
Verwer (1976)	173	39	280	170	25	103	1.7	1.1	0.017	0.25	0.35	0.058
Graves (1983)	175	50	400	225	50	—	3	1	0.1	0.1	0.4	0.05

Source: Resh (1981).

weight factors in grams required to make 1000 liters of a 1-ppm solution are given in Table 10.8; it is necessary only to multiply the factor for a chemical by the number of ppm desired in the formula to obtain the number of grams to be used per kiloliter.

The local availability and cost of fertilizers often determine what components are used. Using Table 10.8, one can easily prepare virtually any formula; if a readily available chemical is not listed, its weight factor can be calculated from the atomic weight.

The preparation of typical nutrient solutions for tomato and cucumber culture in open or closed hydroponic systems is outlined in Table 10.9. A micronutrient solution designed to supplement these basic solutions is described in Table 10.10 (Ellis *et al.* 1974). For larger hydroponic systems, chemicals are weighed out individually to an accuracy of ±5% (smaller deviations generally have no apparent effect on plant growth) and arranged near the mixing tanks in a manner that precludes workers from double-weighing any component. The chemicals are simply added to the tanks and stirred vigorously; the

TABLE 10.8. Weight Factors for Calculating the Amounts of Chemicals (in Grams) Needed to Prepare 1000 Liters of a 1-ppm Hydroponic Nutrient Solution

Chemical compound[1]	Principal essential element supplied	Grams in 1000 liters
Ammonium sulfate (21–0–0)	Nitrogen	4.76
Calcium nitrate (15.5–0–0)	Nitrogen	6.45
	Calcium	4.70
Potassium nitrate (13.75–0–36.9)	Nitrogen	7.30
	Potassium	2.70[2]
Sodium nitrate (15.5–0–0)	Nitrogen	6.45
Urea (46–0–0)	Nitrogen	2.17
Nitro phoska (15–6.6–12.5)	Nitrogen	6.60
	Phosphorus	15.00
	Potassium	8.30
Monopotassium phosphate (0–22.5–28)	Potassium	3.53
	Phosphorus	4.45
Potassium sulfate (0–0–43.3)	Potassium	2.50
Potassium chloride (0–0–49.8)	Potassium	2.05
Monocalcium phosphate (triple super) (0–20.8–0) 13 ca	Phosphorus	4.78
Monoammonium phosphate (11–20.8–0)	Phosphorus	4.78
Calcium sulfate (gypsum)	Calcium	4.80
Boric acid	Boron	5.64
Copper sulfate	Copper	3.91
Ferrous sulfate	Iron	5.54
Chelated iron, 9%	Iron	11.10
Manganese sulfate	Manganese	4.05
Magnesium sulfate (Epsom salts)	Magnesium	10.75
Molybdenum trioxide	Molybdenum	1.50
Sodium molybdate	Molybdenum	2.56
Zinc sulfate	Zinc	4.42

[1]Chemical compounds vary in percentage of ingredients. These figures are within the workable tolerance of most available fertilizers or chemicals listed. Figures in parentheses indicate percentages of N, P, and K, respectively.
[2]2.7 g KNO_3 in 1000 liters of water equals 1 ppm K and 0.36 ppm N.

TABLE 10.9. Preparation of Nutrient Solutions for Tomato and Cucumber Culture in Closed or Open Hydroponic Systems[1]

Chemical compound (fertilizer grade)	Principal element supplied	Tomato: Soln. A, Seedlings to first fruit set (g/1000 liters)	Tomato: Soln. B, Fruit set to harvesting (g/1000 liters)	Cucumber: Soln. C[2], Seedlings to first fruit set (g/1000 liters)	Cucumber: Soln. D, Fruit set to harvesting (g/1000 liters)
Magnesium sulfate $MgSO_4 \cdot 7H_2O$ (Epsom salt grade)	Mg	500	500	500	500
Monopotassium phosphate KH_2PO_4 (0–22.5–28.0)	K, P	270	270	270	270
Potassium nitrate KNO_3 (13.75–0–36.9)	K, N	200	200	200	200
Potassium sulfate[3] K_2SO_4 (0–0–43.3)	K	100	100	—	—
Calcium nitrate[4] $Ca(N)_3)_2$ (15.5–0–0)	N, Ca	500	680	680	1357
Chelated iron[5] FE 330	Fe	25	25	25	25
Micronutrients[6]		150 ml	150 ml	150 ml	150 ml

[1]Final nutrient concentrations in mg/kg: Soln. A—Mg (50), K (199), P (62), N (113), Ca (122), and Fe (2.5); Soln. B—Mg (50), K (199), P (62), N (144), Ca (165), and Fe (2.5); Soln. C—Mg (50), K (154), P (62), N (144), Ca (165), and Fe (2.5); Soln. D—Mg (50), K (154), P (62), N (260), Ca (330), and Fe (2.5). All solutions are supplemented with micronutrients.

[2]Solution C can be used for other vegetable crops; adjust N levels to 200 ppm for leafy vegetables such as lettuce.

[3]The use of potassium sulfate is optional.

[4]Adjust N levels to 200 ppm for leafy vegetables such as lettuce.

[5]Up to 50 g/1000 liters may be necessary if a calcareous growing medium is used.

[6]See Table 10.10 for preparation of micronutrient solution.

TABLE 10.10. Preparation of Micronutrient Solution for Tomato and Cucumber Culture in Closed or Open Hydroponic Systems[1]

Chemical compound	Element supplied	Grams to use[2]
Boric acid (H_3BO_3)	B	7.50
Manganous chloride ($MnCl_2 \cdot 4H_2O$)	Mn	6.75
Cupric chloride ($CuCl_2 \cdot 2H_2O$)	Cu	0.37
Molybdenum trioxide (MoO_3)	Mo	0.15
Zinc sulfate ($ZnSo_4 \cdot 7H_2O$)	Zn	1.18

[1]Final nutrient concentrations in mg/kg: B (0.44), Mn (0.62), Cu (0.05), Mo (0.03), and Zn (0.09).
[2]Add water to mixture of micronutrients to make 450 ml of stock solution (heat to dissolve). Use 150 ml of this micronutrient solution with each 1000 liters of nutrient solution (Table 10.9).

order of addition is not important, but it is easiest to dissolve the most insoluble salts first (monocalcium phosphate and calcium sulfate).

A. Closed Systems

Closed systems (such as nutrient film systems or technique, NFT) are economical in the use of nutrients but require frequent monitoring and adjustment of the nutrient solution. Electrical conductivity is a convenient measure of the total salt concentration, but it provides no indication of the concentration of major elements and is virtually unaffected by the quantity of trace elements present. Thus, periodic chemical analyses are required, usually every 2–3 weeks for major elements (N, P, K, Ca, Mg, S) and every 4–6 weeks for micronutrients (Cl, B, Cu, Fe, Mn, Mo, Zn) (Graves 1983). It is essential that the relative concentrations of nutrients in the nutrient solution approximate crop uptake ratios; otherwise, some nutrients accumulate while others are depleted. Chemical additions to the nutrient solution may be required weekly or even daily to maintain a proper balance of nutrients.

Many operators, particularly of smaller closed systems, find such a schedule of monitoring, analysis, and chemical adjustment undesirable. A not uncommon practice is to begin with a new solution, at the end of a week to add one-half of the original formula, and at the end of the second week, to dump the remaining mixture from the tanks or sumps, and start all over again. This procedure has a simplicity that may compensate for any lack of precision.

B. Open Systems

As the nutrient solution is not recovered and recycled in open systems, it does not require monitoring and adjustment; once mixed, it is generally used until depleted. In addition, the quality of the irrigation water used is less critical than in closed systems. Up to 500 mg/kg of extraneous salts is easily tolerated; for some crops (tomatoes, for example), even higher extraneous salinities are permissible, although not desirable.

Though the nutrient solution *per se* does not require monitoring in open systems, the growing medium may, particularly if the irrigation water is relatively saline or if the hydroponic facility is located in a warm, high-sunlight region. To avoid salt accumulation in the medium, enough irrigation water must be used to allow a small drainage from the planting beds. This drainage should be collected and tested periodically for total dissolved salts. If the salinity of the drainage is 3000 ppm or above, the planting beds must be leached free of salts (using the in-place irrigation system), or at least to a point equal to the salt content of the water used.

1. Fertilizer Proportioners. An automatically controlled open system utilizes fertilizer proportioners, manufactured devices that inject specific amounts of nutrient solution into the irrigation water. For such usage, the solution must be highly concentrated and is prepared in two separate mixtures: one containing calcium nitrate and iron, the other containing the balance of the dissolved chemicals. (This avoids combining calcium nitrate and magnesium sulfate in concentration, which will precipitate calcium sulfate.) A twin-head proportioner is thus required.

The rate of injection by the proportioner heads determines the necessary concentration of the nutrient solutions. For example, if each head injects 1 liter of stock solution into each 200 liters of water passing through the irrigation system, the stock solutions must be 200 times the concentration listed in Table 10.9.

Twice weekly, the total dissolved salts of the nutrient solution/water actually delivered to the plants must be checked; the proportioner pumps must be examined separately to verify that the pumps are operating correctly. Merely testing for total salts at the end of the system is not sufficient; if one head is ejecting too much solution, and the other too little, total salts may appear to be appropriate, but the ratio of elements may be badly skewed.

C. Nutritional Disorders

Generally, there are no nutritional disorders specific to hydroponics; i.e., they do not differ in cause and effect from such disorders in field agriculture. Nutritional disorders are more likely to occur in closed hydroponic systems than in open systems because impurities or unwanted ions in the recycled liquid, or from the chemicals used, may more easily destroy the balance of the formulation and accumulate to toxic levels.

The most common nutritional disorders in hydroponic systems are caused by too much ammonium and zinc and too little potassium and calcium. High levels of ammonium, which cause various physiological disorders in tomatoes, can be avoided by providing no more than 10% of the necessary nitrogen from ammonium, it is best to completely avoid ammonium in nutrient solutions. Low levels of potassium (less than 100 ppm in the nutrient solution) can affect tomato acidity and reduce the percentage of high-quality fruit (Winsor and Massey 1978). Low levels of calcium induce blossom-end rot in tomatoes and tipburn on lettuce. Zinc toxicity, caused by dissolution of the element from galvanized pipework in the irrigation system, can be avoided by using plastic or other materials suitable for agriculture.

Nutrient-related disorders of crop plants can be avoided by maintaining careful control of the composition of the nutrient solution, particularly in closed systems (Graves 1983).

VI. DISEASE AND INSECT CONTROL

There is a common—and incorrect—assumption that hydroponic CEA systems must be relatively free of diseases and insect pests because the technology is enclosed. Unfortunately, the introduction of pathogens and insect pests can occur when doors are opened and closed and people and materials moved in and out. Indeed, pest populations can increase with alarming speed in CEA installations because of the lack of natural environmental checks, and root diseases in closed liquid systems can spread quickly to all plants. Conversely, the enclosure of the growing area should make it easier than in OFA to practice control methods.

For the past 40 years, control of agricultural diseases and insects has depended on the use of chemicals. In Europe, and in most other regions where CEA is practiced on a large scale, this has been true for greenhouse crops as well as field crops. There, many apparently effective pesticides and chemicals (none produced specifically or exclusively for hydroponics) are available and legal for use on greenhouse crops.

However, in the United States, where so many of the world's agricultural chemicals have been invented, few have been cleared for greenhouse use. This is not only because the effects of such chemicals inside CEA structures may be different and more dangerous than they are in open-field crops, but primarily because chemical manufacturers are quite aware of the limited use of CEA in American food production. With such a small potential market, they see little justification to spend the large sums necessary to obtain documentation for federal and state certification for the use of such materials on human food crops grown in greenhouses.

A new factor has been introduced by the frightening propensity of some insects to develop resistance to chemical pesticides. This has revived worldwide interest in the concept of biological control—the deliberate introduction of natural enemies of insect pests—particularly when used in association with horticultural practices, plant genetics, and other control mechanisms. This combined approach, called integrated pest management (IPM), is of particular interest to Americans in CEA because of the paucity of pesticides with legal clearance for use in greenhouses.

A. Root Diseases

Closed hydroponic systems permit a rapid spread of root diseases. With tomato, the most common crop, some natural root death occurs when the extension growth of the roots ceases as the first fruits begin to swell. At that time of natural stress, the plant is vulnerable to root pathogens (Evans 1979), which may be introduced by contaminated seed, infected propagating mixtures, or even adjacent greenhouse soil.

Common root pathogens causing tomato wilts are *Fusarium* and *Verticillium* spp.; these diseases are most easily avoided by using resistant cultivars. Young plants may be destroyed, especially at higher root temperatures, by *Pythium* and *Phytophthora* spp. (Price 1980; Stanghellini *et al.* 1982); these fungi need water to reproduce and kill all but the largest roots of adult plants. These latter fungi have been controlled in England by using a wettable powder formulation of etridiazol at a concentration of 20 mg/kg in the nutrient solution (Staunton and Cormican 1980); even better results have been reported using 20 mg/kg etridiazol with 5 mg/kg copper as copper oxinate, added to the nutrient solution monthly (Price and Dickinson 1980). However, these procedures are not approved for greenhouse use in the United States.

The use of ultraviolet radiation to reduce bacterial plant pathogens in nutrient solutions appeared promising in some NFT systems in England (Buyanousky *et al.* 1981; Ewart and Chrimes 1980) but was not effective in preliminary trials at the University of Arizona. There, *in situ* ultraviolet irradiation lowered bacterial counts for only 48 hr, after which populations returned to original levels, even with continued irradiation. Furthermore, though irradiation appeared to destroy the zoospores of *Pythium* spp. in nutrient solutions, it rendered iron into a form unavailable to plants, making it necessary to add additional iron after every irradiation (Stanghellini *et al.* 1982).

Bacterial canker (*Corynebacterium michiganense*) reportedly infected several NFT crops in England (Davies 1980), but this is primarily a seedborne disease, and there was no evidence it was spread through the nutrient solution. Lettuce crops have been decimated by the root-affecting fungus *Olpidium brassicae*, called "big-vein virus"; it is apparently a problem during periods of low light levels and slow growth. The zoospores can be killed by the use of certain wetting agents applied at a concentration of 20 ppm in the nutrient solution every 4 days until symptoms disappear (Tomlinson and Faithfull 1980). This procedure is not approved for greenhouse use in the United States.

B. Foliage Diseases and Insects

There are no foliage-damaging plant diseases or insect pests associated exclusively with hydroponics, or with CEA; i.e., the same diseases and pests infect tomatoes, cucumbers, and lettuce whether grown in hydroponic systems or open fields. The problem is that although many chemicals have been approved in the United States for use on OFA crops, few are registered for greenhouse applications, for the reason given earlier. At the same time, the CEA environment is ideal for the rapid proliferation of unwanted pathogens and insects.

One result of this, in such a marginal industry as American CEA, is the possibly widespread but seldom-acknowledged illegal use of unregistered chemicals for disease and pest control in greenhouses. A grower in financial straits, facing loss of a critical crop, may not hesitate to use a compound approved for use on similar crops in OFA. This may be of substantial danger to both the applicator and the ultimate consumer, as these potent chemicals break down differently in a greenhouse environment than in the open air and when exposed to unaltered sunlight. There are no data available on the extent of such violations, but the supposition that violations are common underscores the need for some funding mechanism to support the costs of screening and registering pesticides for use in CEA.

C. Integrated Pest Management (IPM)

There is no standard definition or technique of IPM, but generally it consists of a carefully structured and monitored combination of biological controls, plant genetics, culture practices, and use of chemicals. The concept is that insects can be controlled largely by encouraging their natural predators and parasites, by using resistant cultivars, and by using proper plant spacings and other cultural practices—and that pesticides should be used only as a last resort. The rationale is that pesticides are expensive, often reduce maximum possible yield (Burgess 1974), and may lead to the evolution of pesticide-resistant insect species. Integrated pest management is also ideologically attractive for those who espouse the consumption of "natural" foods or who are concerned with the dispersion of toxic materials. The individual components of IPM are discussed in the remainder of this section.

1. Biological Control. The use of biological control with greenhouse crops was first reported in England in the 1920s. Whitefly infestation of tomatoes was countered by the introduction of the parasite *Encarsia formosa* Gahan; results were mixed, and the practice was discontinued after the development of synthetic organic pesticides after World War II (Gould *et al.* 1975; McClanahan 1973).

By the 1960s, large populations of two-spotted spider mites (a severe cucumber pest) became resistant to many pesticides. Whiteflies also began to show the same characteristic (Wardlow *et al.* 1972). The predatory mite *Phytoseiulus persimilis* Anthias-Henriot was used successfully to control spider mites (French *et al.* 1976; Gould *et al.* 1975), and *Encarsia* once again was used against whiteflies (Gould *et al.* 1975; Parr *et al.* 1976). A number of other natural predators and parasites are now used for controlling pests in CEA (Bravenboer 1974). Most current procedures were developed by researchers at the Glasshouse Crops Research Institute in England and the Research Station for Vegetables and Fruit Under Glass in The Netherlands. Scientists in this field meet every three years and publish a volume of conference proceedings.

An obvious difficulty with IPM is the integration of biological controls with even limited pesticide use; the chemicals usually kill the introduced enemies as well as the target pests, requiring reintroductions. In addition, it is improbable that growers could keep at hand, or have access to, predators and parasites for all major pests, or be trained to use them. In part, biological control resembles hydroponics itself—it is an attractive concept, easier to promote than to practice.

2. Cultural Control. If plants are spaced too closely together (a bad practice to begin with, affecting plant growth and crop yield), air circulation and light penetration are inhibited, permitting insect populations to increase and restricting access to them. Wider spacing facilitates monitoring and control of insects. So does a general program of greenhouse sanitation—removing trash, dead plant material, and other debris that may shelter pests.

3. Physical Control. A CEA lighting system should be of a type and placement to avoid attracting outside insects into the greenhouses. Certain kinds of interior lighting (UV or yellow mercury vapor) can be used as insect traps, and objects painted certain colors and sprayed with adhesive materials serve as traps for some pests. As an example of the latter, boards measuring approximately 30 cm^2 and painted yellow (Rustoleum No. 659) attracted whiteflies in Arizona.

4. Genetic Control. For hydroponic CEA crops, as well as OFA crops, cultivars have been developed for tolerance or resistance to certain insects, viruses, and fungi. In actual use, however, such cultivars seldom provide maximum yields. For the grower, a trade-off may be necessary between yield potential and resistance; selecting cultivars should be done case by case, taking into account region, local pest populations, etc.

5. Chemical Control. If all else fails, chemical controls are used as a measure of last resort in an IPM program. This is appropriate in OFA, with so many pesticides and chemical procedures approved for open-field crops, but difficult for hydroponic CEA operators because so few materials are certified for greenhouse use.

VII. STRUCTURES AND ENVIRONMENTAL CONTROL

Hydroponic growing systems are not, technically, the same as CEA; however, the structures and techniques used in the latter are vital to most hydroponic applications. Total costs of hydroponic vegetable production are determined more by greenhouse expenses than by any other direct or indirect expenses. It would be unrealistic, therefore, to assess hydroponics without a brief review of the structures associated with hydroponic systems.

As a structure for growing plants, a greenhouse is a contradiction of requirements. It must admit the visible light portion of solar radiation for plant photosynthesis and therefore must be transparent. At the same time, to protect the plants, a greenhouse must be ventilated or cooled during the day because of the heat load from the radiation, and yet the structure must be heated or insulated during cold nights. These are difficult criteria. (Technically, it would be possible to satisfy these contradictory criteria by growing plants in a windowless, insulated, heated or air-conditioned box using artificial illumination only. Several commercial hydroponic CEA systems of this type have been designed and tested. It is obvious, however, that this design increases energy and capital costs, and it is difficult to imagine that such systems could be competitive with simpler structures utilizing no-cost sunlight.)

The primary greenhouse requirement—the admission of visible light—must be accommodated. The ensuing problems of daytime heat gain and nighttime heat loss were handled traditionally by mechanical heating or cooling—as much as necessary. This presented no insurmountable difficulty when fuel was relatively inexpensive, (although heating/cooling costs often approximated 20% of CEA operating budgets), but the situation became far different beginning in the mid-1970s. As fuel prices escalated, heating greenhouses in Ohio, which has the largest area of CEA in the nation, became formidably expensive. Heating by natural gas cost more than $100,000 ha/winter, and heating by oil more than $128,000 ha/winter (Gerhart 1976). (Coal would have cost less, but was impossible to burn economically and still satisfy federal antipollution requirements.) Some CEA rose growers in Massachusetts paid nearly $250,000 ha/winter for heating oil at the zenith of oil prices, and in Florida, the cost of heating oil increased 600% in 6 years (Baird and Mears 1976).

Many older greenhouses (and some newer ones) are constructed of single glazings of glass panes, framed in wood or metal, and supported by internal structures of the same material. Newer greenhouses more typically use plastics in place of glass; they consist of rigid supporting structures (wood, galvanized steel, or aluminum), roofed with one or two layers of transparent plastic film, and are usually walled with rigid plastic panels.

A. Plastic Glazings

Polyethylene sheet film was developed in England in 1938 and used for the first time as a greenhouse covering by Emory at the University of Kentucky in the mid-1950s (Courter and Carbonneay 1966). Use of double-layer, air-inflated polyethylene film to reduce heat transfer (Roberts and Mear 1966; Sheldrake and Langhans 1961) subsequently became standard in the industry and was responsible for savings of up to 40% in fuel for winter heating.

Other plastics have been used for greenhouse glazings, but with indifferent technical or economic results. Corrugated fiberglass panels, polyvinylchloride (PVC) film, Mylar, Tedlar, etc., proved either more unsuitable, inconvenient or—in most cases—expensive than polyethylene, even though the latter may have to be replaced more frequently. Newer materials—polycarbonates, acrylics, double-skinned panels, light-selective films laminated to polyethylene or Tedlar, etc.- are now being studied, but their technical promise is offset at present by high costs.

The ideal greenhouse "selective film" should do the following (Anon. 1977):

1. Transmit the visible light portion of the solar radiation spectrum, the only portion utilized by plants for photosynthesis.
2. Absorb the small amount of ultraviolet in the spectrum (3%) and cause some of it to fluoresce into visible light, useful to plants.
3. Reflect or absorb infrared radiation (49% of the spectrum), which plants cannot use and which causes greenhouse interiors to overheat.
4. Minimize cost, and have a 10- to 20-year usable life.

Such a film would obviously improve CEA performance, increasing light levels and crop yields and reducing solar heat load. The first three criteria present no extraordinary technological challenge. However, in the context of the small American greenhouse industry, the inventiveness of plastics manufacturers has been inhibited by their understandable doubts about the market potential for such a product.

B. Thermal Conservation

Whatever the energy source for winter heating in greenhouses, the conservation and more efficient use of interior heat have received considerable attention. Some of the major techniques and improvements developed, and relative average fuel savings realized, are summarized in Table 10.11.

An exterior insulating curtain of double-layer, air-inflated polyethylene for placement over roof glass was developed at Ohio State University (Bauerle and Short 1976) for reducing heat loss and increasing daytime humidity. A number of different systems for interior insulating curtains, which can reduce fuel consumption by more than one-third, are also in use. For example, at night, Japanese CEA growers place a removable sheet of polyethylene film over the crop and an individual cover over each plant row (Fig. 10.35), pulling the plastic aside during the day to maximize incoming radiation. Some Dutch growers use an interior subroof curtain of inflated polyethylene tubes to reduce heat loss; the tubes are deflated and pushed aside during the day (Fig. 10.36) and extended and inflated at night to cover the crop (Fig. 10.37). White *et al.* (1976) tested 20 different materials for interior

TABLE 10.11. Potential Annual Fuel Savings Realized by Maintenance, Modification, and Miscellaneous Energy Conservation Methods in Greenhouses Relative to Standard Glass Greenhouses

Method	Annual fuel saving (%)	
	Range	Avg
MAJOR MODIFICATIONS:		
Double plastic film over glass	40–60	50
Glass lap sealants	5–40	15
Single plastic film over glass	5–40	30
Double-layer plastic film	30–40	35
Curtains	20–60	33
Polystyrene beads	60–90	
Liquid foam	40–75[1]	
OTHER MODIFICATIONS		
Sidewall insulation	5–10	
Foundation insulation	3–6	
Insulating ventilation fans	1–5	
Heating systems:		
Automatic firetube cleaners	6–20	
Turbulators	8–16	
MAINTENANCE:		
Structure	3–10	
Heating system	10–20	

Source: Badger and Poole (1979).
[1]More recent data from University of Arizona experiments indicated 70–80% fuel savings.

FIG. 10.35. To conserve heat at night, some Japanese CEA growers cover crops and individual plant rows with sheets of polyethylene film, which is pulled back during the day.

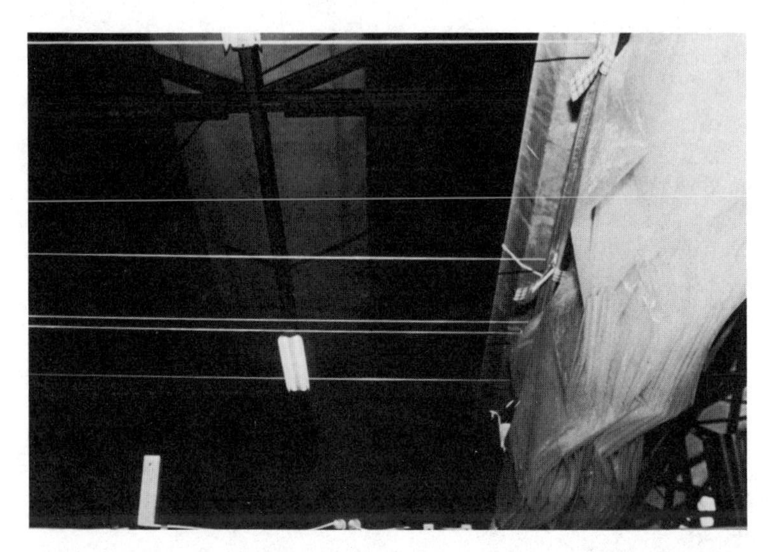

FIG. 10.36. During the day, heat-conserving polyethylene tubes, used in some Dutch greenhouses, are deflated and drawn aside to admit light and air.

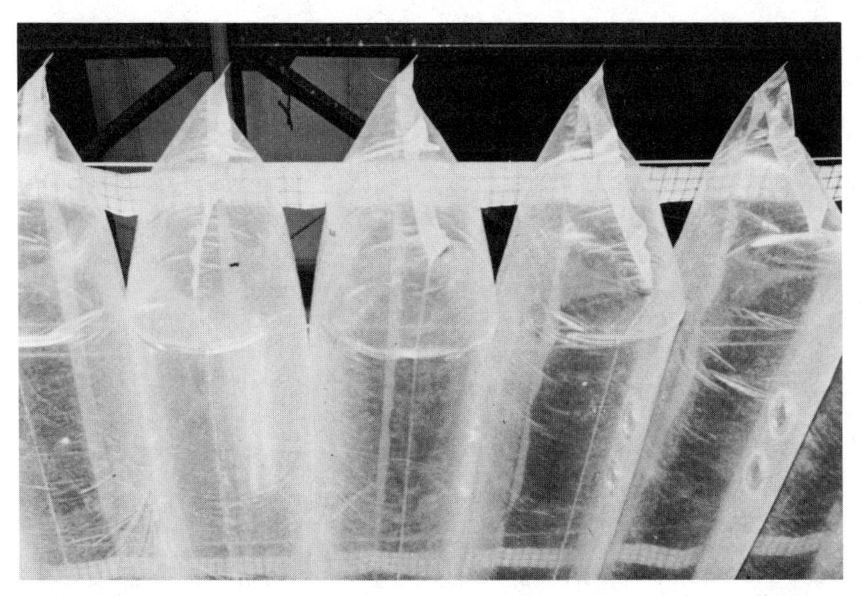

FIG. 10.37. At night, polyethylene tubes are extended and inflated to reduce heat loss.

curtain use and found that foylon (a polyester–aluminum foil fabric) could reduce glasshouse fuel requirements 57% (Fig. 10.38).

The disadvantages of curtain systems (in addition to their cost) include potentially awkward installation, problems of daytime storage, relocation of heating systems, and the possibility of increasing snow loads in certain regions. The big advantage is that they work and result in substantial fuel savings.

Several techniques have been developed for introducing recoverable insulation materials, which can be removed during daytime hours, into the interstitial space between double-layer polyethylene roofs. The use of polystyrene beads, a concept first developed by Zomeworks in New Mexico for residential glass-wall applications, has been tested at the Ohio Agricultural Research and Development Center; this technique reduces winter nighttime heat loss by up to 90% (Elwell *et al.* 1984). Horizontal roofs, however, are more complex to "bead or unbead" than vertical walls, and daytime storage of the beads presents problems in transport and storage.

A liquid foam system was developed in Arizona for recoverable interstitial insulation (Groh 1976). Material with a liquid-to-foam expansion ratio of 1 to 1000 is blown into the space between roof glazings (Fig. 10.39), reducing nighttime heat loss 70–80%. The foam collapses

FIG. 10.38. Foylon curtains have been shown to reduce greenhouse fuel requirements by more than 50% in research at Pennsylvania State University.

FIG. 10.39. Reusable liquid foam system developed at the University of Arizona reduces nighttime heat loss in greenhouses by 70–80%.

in about 1 hr, so a greenhouse must be refoamed during the night; the small amount of reconstituted liquid is collected in gutters to drain back into a central tank for reuse. The same system may also be used to provide partial daytime shading for summer crops sensitive to high light levels. This concept appears to have substantial promise and is still under development.

As a theoretical ideal, a hydroponic CEA structure would have a transparent roof, or none at all, during daytime hours, and a different, heavily insulated roof during cold nights. This is technically possible, particularly for small, special-use structures, but would at present be impractical for commercial food production operations. However, innovative design may lead to practical implementation of this or other, nearly ideal, concepts for obtaining maximum radiation during the day and maintaining adequate temperatures at night.

C. Alternative Heat Sources

1. Solar Energy. Solar technologies appear logical for CEA winter heating. Greenhouses are solar collectors during daytime; indeed, infrared radiation warms interiors so much, particularly in desert regions of high light levels, that fan ventilation is usually required to reject the excess heat. It has been estimated that up to 80% of the heating needs in U.S. greenhouses could be met by solar energy, depending on geographic location (Jensen 1977). However, using the

the chimney would not work. However, if the chimney is designed and oriented to serve as a solar heater itself and/or if there is a preheater section of the greenhouse interior between the end of the crop growing zone and the chimney (perhaps merely a sheet of black plastic film on the ground, beneath the transparent roof), the mean density of the chimney air would be reduced, and the proper amount of draft provided to induce airflow.

The reverse type of induced natural ventilation would be a downdraft chimney, in which ambient air is evaporatively cooled, increasing its density. Calculations indicate that a downdraft chimney need not be as high as an updraft chimney to achieve a given air velocity, and that the use of a fogging system, rather than conventional cooling pads, would greatly reduce frictional losses. A combination of updraft and downdraft chimneys at opposite ends of a greenhouse has been theorized to achieve a "push-pull" effect for ventilation.

A different type of alternative ventilation was studied at the University of Arizona for (but not published in) the 1977 technology assessment. In this system, conventional ventilation fans were replaced with a hinged, bellows-type greenhouse wall. It was calculated that a bellows-wall would theoretically permit lower cost "fan" ventilation because of reduced horsepower requirements to move a given amount of air.

E. Carbon Dioxide Enrichment

Elevating the amount of CO_2 in the aerial environment of unventilated CEA/hydroponic structures has given spectacular increases in crop yield in the northern latitudes of the United States and western Europe (Wittwer 1967). Bewley (1963) reported a 30% increase in tomato yield during 1926–1930 from CO_2 enrichment, but there were few commercial applications of the technique until 1961, when a Dutch grower stimulated the wide use of closed-vent kerosene heating stoves in greenhouses of the Westland region, which greatly improved lettuce quality (Anon. 1962). Research in the United States developed from the reports of Chapman and Loomis (1953) and Gaastra (1959). Kimball (1982) correlated the data on CO_2 enrichment from more than 360 observations of 24 different crops in 50 reports. This study showed (again) that doubling atmospheric CO_2 increased agricultural yields an average of 30%. The use of CO_2 enrichment in northern latitudes, however, diminished after the mid-1970s because of the high cost of the energy needed to generate the gas. The increased productivity was not sufficient to compensate for the higher costs of the enrichment.

In southern latitudes, higher radiation levels and ensuing higher temperatures require daytime ventilation of greenhouses even during

greenhouse itself (and its growing plants) to collect and store large amounts of thermal energy, though plausible, is impractical because the plants cannot tolerate the higher air temperatures necessary. Other types of interior solar energy collection and storage systems have been proposed (Liu and Carlson 1976) but would substantially reduce usable growing area in greenhouses.

A variety of external solar collectors (both flat-plate and concentrators) and associated transportation and storage systems has been proposed or developed, some more ingenious than others. Lockhead used hot-water solar collectors to provide 75–80% of greenhouse heating (McCormick 1976), and scientists at Rutgers University developed a low-cost polyethylene collector and a concept of thermal storage in a porous concrete greenhouse floor (Roberts *et al.* 1976). However, the typical poorly insulated greenhouse would require a collector area approximately half that of the greenhouse itself (Price *et al.* 1976), and such systems require either large capital investment or costly annual maintenance, or both. As is true for most solar heating and cooling applications, the payback based on energy savings is measured in years.

Daytime collection and storage of solar energy for same-night use means that only a fraction of the *annual* solar energy available is being used, thus severely reducing the cost-efficiency of such systems. A subsequent concept is Seasonal Thermal Energy Storage (STES), i.e., the collection and storage of solar energy year-round for seasonal use during winter months. This would greatly reduce the collector area required, but greatly increase the costs of storage capacity (probably in underground water tanks). An STES system has not yet been developed.

2. Geothermal Energy. Iceland is noted for a variety of applied uses of geothermal energy, including electric power generation and the heating of residential and commercial buildings. At least 5 ha of CEA glasshouses in Hveragerdi, Iceland, are heated geothermally, and there are plans to increase the area up to 30 ha, for the production of both vegetables and ornamental plants.

Several small applications of geothermal greenhouse heating are known in the western United States, particularly in rural areas (e.g., San Luis Valley, Colorado) where heated aquifers of potable water are close to the surface and are commonly used for residential heating and even domestic water. Such facilities have not been expanded because their locations are inappropriate for transporting and marketing food products.

The quality of geothermally heated water has been the principal deterrent to the use of such energy to heat American greenhouses.

Geological maps of western states are studded with known or probable locations of such warmed aquifers; though some are merely too deep for cost-efficient recovery, many are charged with highly corrosive compounds and heavy metals. These are difficult and risky fluids for many industrial applications, including electric power generation, and even more formidable for agriculturists.

Heat recovery from such aqueous solutions is well within existing technology. But if the use of geothermally heated water in greenhouses required well-head gas seals, complex heat-exchanger systems, protection against silica deposition in mechanical equipment, careful disposal or reinjection of used thermal fluids, etc., then the purpose of the application—cost savings for marginal CEA operations—would be rapidly defeated. For this reason, the application of this type of energy to greenhouse heating is more easily advocated than realistically anticipated.

3. Waste Heat. Probably the most feasible sources of low-cost thermal energy are industrial plants that produce large amounts of waste heat, principally steam-electric generating stations. Annual waste heat from generating stations in the United States is the equivalent of more than 2×10^9 barrels of oil, or 15–20% of all national energy consumption (Madewell *et al.* 1975). Such heat is released from condenser cooling water to the atmosphere via cooling towers, or the warmed water is returned to its source, usually a river, an impoundment, or the sea. An estimated 2 kWh of waste energy are produced for every kWh of electrical power produced (Skaggs *et al.* 1976).

There are several test installations in the United States utilizing waste heat for greenhouse heating, most notably the Tennessee Valley Authority's Brown's Ferry Nuclear Plant in Alabama. The technology itself exists. Because such generating plants are operated to provide electricity, the amounts and temperatures of cooling water may vary somewhat. However, large base-load plants cycle out so much heated water that only a fraction of the rejected heat could be utilized at best; and though condensers and mechanical systems must be cleaned and chemically treated periodically, the changes in water quality caused by such processes can be avoided by the use of heat exchangers. The cost of heat-exchanger systems are an inhibiting factor, but the most important barrier to waste heat utilization is probably land location and ownership. It is economically infeasible to pipe and pump large volumes of low-grade heat for any appreciable distance; i.e., only sites adjacent to generating stations (or other sources of waste heat) can be considered for greenhouse applications.

Such lands are customarily owned by the public or private power

companies that operate the generating stations, and use covenan may be complex. Most agriculturists are simply uninterested in dev oping food production facilities on such lands and under complicat restrictions; indeed, many power companies are reluctant even consider such diversions. The concept of governmental, munici "waste heat industrial parks" designed around thermal generat stations has been advocated by Widmer (1979), to provide for a var of productive uses for the rejected heat, including CEA structu Such applications have not become widespread.

D. Alternative Ventilation and Cooling Methods

Natural ventilation is common in European greenhouses inadequate in many lower-latitude locations in the United S where conventional fan ventilation, with or without associate porative cooling, is standard. In greenhouses employing exhaus conventional evaporative cooling is achieved by having the m air enter the structure through pads of wood shavings or che treated corrugated fiber products kept moistened with water. also an increasing use of fogging systems, which emplo pressure water and an overhead network of pin-nozzles to f interior atmosphere with pervasive clouds of minute water reducing dry-bulb temperature more efficiently than do cooli

Lower-cost techniques of ventilating and cooling hydropo structures in the United States have been studied, but the conceptual designs have yet to be implemented. A technolo ment of CEA in 1977 (Anon. 1977) proposed the use of a sol chimney as a lower-cost ventilation device. The design pr "natural draft" have been in the engineering literature for the concept appears to have been first proposed for greenh cations by P. Solari in 1975 (personal communication).

When interior greenhouse air becomes heated by solar expands in volume and decreases in density and may be a denser gas. If a greenhouse roof is designed with a wide one side of the structure, the warmed air has a tendency chimney (a tendency that increases with the height of th (makeup) air, being cooler and denser, flows into the displace it. This establishes a natural draft. A portion used to maintain the velocity of the moving air; the rem to overcome the resistance of the chimney.

If evaporative cooling is required, the draft must als resistance of the cooling pads, which is substantial, an reaching the chimney may be denser than ambient ai

winter months; in such open structures, maintaining elevated amounts of CO_2 is not feasible. The gas could be confined in a southern greenhouse if it was sealed and the interior temperature kept down by mechanical refrigeration, which is technically simple but economically absurd. Research at the University of Arizona also indicated that in a well-sealed greenhouse, trace contaminants in the CO_2 inhibited plant growth or proved lethal to plants, requiring the use of medical-quality bottled gas for experiments.

These phenomena suggest that although the salubrious effects of CO_2 enrichment on crop yield are well documented, there are several constraints on its economic application in CEA. First, CO_2 enrichment is only practical in northern latitudes, where greenhouses do not require daytime ventilation. Second, greenhouses that are poorly sealed or have structural leaks or gaps would not be suitable for CO_2 enrichment. Finally, inexpensive techniques of generating reasonably uncontaminated CO_2 are necessary.

VIII. CONTEMPORARY COSTS AND PRODUCTIVITY

A comparison of the estimated 1983 capital costs of several hydroponic systems, exclusive of the CEA structures necessary to contain them, is presented in Table 10.12. The range is relatively narrow for bag, rockwool, and sand culture, with an average cost of $48,000/ha; however, the cost figure for sand culture includes construction labor costs, whereas the figures for bag and rockwool culture do not. NFT is nearly twice as costly (again, labor costs are excluded), and the experimental raceway lettuce system used in Arizona (including construction labor) is costlier yet. Although differences in the underlying data make direct comparisons of these cost figures difficult, we believe the ranking of these systems is accurate, if sand culture is moved to the

TABLE 10.12. Estimated Capital Costs of Contemporary Hydroponic Systems Exclusive of Costs for Associated CEA Facilities—1983

Type of system	$/ha
Bag culture	38,000[1]
Sand culture	50,000[2]
Rockwool	55,000[1]
NFT	81,000[1]
Raceway (floating)	205,000[2]

[1]Excludes construction labor costs. Glasshouse Crops Research Institute, Littlehampton, England, 1983.
[2]Includes construction labor costs.

low end. In addition, it must be understood that the interior costs of hydroponic systems are but the tip of the iceberg if new CEA facilities must be constructed to enclose them.

Published greenhouse costs are often deceptive: They may consist of structural and equipment costs only, without adequate allowance for the costs of construction labor, utilities, roads, fences, lighting, tools, vehicles, working capital, etc. A 1983 estimate of such a total, turnkey cost in Arizona for a modern and sophisticated CEA system, exclusive of land and of the interior hydroponic system, is $85/m^2$. If costs of the interior hydroponic system are included, the turnkey expense is at least $100/m^2$.

Balanced against these costs is the significantly higher productivity of hydroponic CEA in comparison with OFA. Yield data have been reported in the literature for years; typical yields for crops grown hydroponically in desert greenhouses in the American Southwest and in the Middle East are compared with typical "good yields" for open-field crops in Table 10.13. As the data indicate, the yield per crop is usually higher in hydroponic CEA than in OFA because of the optimal growing conditions, balanced plant nutrients, etc., provided in controlled environments. Furthermore, from 2 to 10 crops per year, depending on the vegetable grown, are possible with CEA, whereas only 1 crop per year generally is possible with OFA.

In the United States, given the high cost of hydroponics and its awkward competitive stance, it is not surprising that only a few large CEA facilities for food production have been constructed in recent years and that the total area under cover has actually diminished. Many growers have survived only because their facilities are older and capital costs have been amortized. In Europe, Japan, and the USSR, where hydroponics can be more profitable for out-of-season

TABLE 10.13. Yields of Vegetable Crops Grown Hydroponically in Desert Greenhouses (CEA) and in Open Fields (OFA)

	Hydroponic CEA[1]			OFA[2]
Crop	Yield/crop (MT/ha)	No. crops/year	Total yield (MT/ha/year)	Total yield MT/ha/year)
Broccoli	32.5	3	97.5	10.5
Bush beans	11.5	4	46.0	6.0
Cabbage	57.5	3	172.5	30.0
Chinese cabbage	50.0	4	200.0	—
Cucumber	250.0	3	750.0	30.0
Eggplant	28.0	2	56.0	20.0
Lettuce	31.3	10	313.0	52.0
Pepper	32.0	3	96.0	16.0
Tomato	187.5	2	375.0	100.0

[1]Source: University of Arizona, unpublished results (1983).
[2]Source: Knott (1966).

produce, newer installations, or renovations of older facilities, are more common.

IX. DETERRENTS TO HYDROPONICS IN THE UNITED STATES

In many regions of the world, it is difficult for hydroponic vegetables to compete with field crops. In the United States it is almost impossible. There are five separate but interacting reasons for this:

1. The diverse climate in North America (unlike the fairly homogeneous climates of Japan and Europe) permits conventional field production of low-cost vegetables somewhere on the continent during any time of the year.
2. A rapid and effective nationwide transportation system exists, making local food production for individual regions or communities unnecessary.
3. The already high capital and energy costs for hydroponics have increased alarmingly since the mid-1970s, with no concomitant breakthroughs in CEA design or materials to mitigate these costs.
4. Few chemicals for disease and pest control have been cleared by the federal government for use in greenhouses, and the market is too small for manufacturers to undertake the expense of certification.
5. The technical and economic perils of hydroponics require an unusual management mixture of biologic and engineering sciences, as well as informed and aggressive marketing. Such combinations have been rare.

The impact of the first of these reasons is significant, and the least possible to manipulate. Were no other vegetables readily available in the United States at low prices during domestic "off seasons," the higher costs of hydroponics would not be such a competitive disadvantage. But North America's climate is unique, in good part due to the Rocky Mountains, the single most important terrestrial feature in global climate. Parts of the southern United States and northern Mexico can and do grow winter vegetables. During the past 10 years, Mexico greatly increased OFA off-season exports to the United States, particularly of tomatoes, one of the most profitable crops for hydroponics (Anon. 1978–1982). These Mexican imports have reduced American off-season peak prices, leading to even smaller profitability margins.

The second factor is not fully appreciated by the average citizen. No region of the nation has to grow its own food. Probably the most singular aspect of the American transportation system, in the context of fresh produce, is the interstate highways and related networks. Low-cost, rapid, frequent, precisely timed delivery of open-field vegetables from any growing area to any market region anywhere in the United States is now taken for granted. As there is no need for any given community to provide its own vegetables, the potential for off-season, local hydroponic production is greatly blunted.

The third reason for the static industry—the high cost of capital structures and energy inputs—has remained technically intractable because of minimal market demand. Few answers have been found because few have been sought. In the last full-scale technology assessment of CEA (Anon. 1977), a theoretical new design was postulated to reduce capital costs 60–80%, while reducing inputs of purchased energy to nearly zero by utilizing solar energy. No such structure has been built. The most significant technical improvement required, a longer-lived, lower-cost, selectively-transparent plastic film, is believed to be within the capability of manufacturers. No such film has been made available.

The fourth reason listed requires no further explanation. While European greenhouse operators use a variety of chemical controls for plant diseases and pests, and while a large number of pesticides and herbicides are cleared for OFA use in the United States, there are few chemicals approved for American greenhouses. Chemical manufacturers are unwilling to spend the large sums necessary to obtain clearance when the market is small.

The fifth reason for a dormant U.S. hydroponics industry—the lack of interdisciplinary support and management—is not merely a contemporary "hard times" phenomenon. This is evident by a review of hydroponic CEA commercial failures during the brief halcyon era (if it could be called that) of American hydroponics during the late 1960s. At that time, market demand was growing, purchased energy was cheap, structural materials were inexpensive, competitive field crops were not extensively imported, and high-risk investment funds were readily available. Even then, the failure rate was high.

Hydroponics is an inherently attractive, often oversimplified technology, which is far easier to promote than to sustain. Aside from outright unscrupulosity—which has not been unknown—a weakness in any of a number of technical or economic links snaps a complex chain. And weaknesses due to management inexperience or lack of scientific and engineering support have been common. The list of problems plagueing hydroponic operations is long: low yields, nutrient-deficient and unattractive crops, plant diseases, insect infestation, summer overheating, winter chilling, undercapitalization, odd promotion schemes, indifferent cost accounting, and, the most lethal, an unawareness of the subtleties of produce marketing. The energy crunches and embargoes of the early 1970s were less a mortal wound to many hydroponics operators than merely a coup de grace.

Given this background, one of the most surprising things about U.S. vegetable hydroponics is that it exists. There are still strong levels of popular and corporate support. During the past five years, several American corporations have developed new prototype commercial hydroponic systems. These do not appear to have been economic successes, yet at least three other large corporations, undaunted, are known to be planning projects of their own.

Symbolic of this continuing enchantment is a phenomenon at EPCOT Center, Walt Disney World, Florida—a theme park which opened in October 1982, after 16 years of planning and construction. Attendance immediately exceeded all expectations, and one of the most popular pavilions proved to be "The Land," featuring a 2.6-ha futuristic display of highly sophisticated hydroponic CEA vegetable production. This is a straightforward, adult presentation. Within 20 months, more than 15 million visitors had toured the greenhouse areas, often standing in line for hours to gain admission.

X. SUMMARY AND RESEARCH NEEDS

"Improved" versions of hydroponics proliferate, bewilderingly. Some new systems seem to be ingeniously simple, others marvelously complex, some improbably bizarre. The technology lends itself to tinkering. Of even greater wonderment, virtually all such systems "work." This is a testimony to the persistence of photosyntheses.

However, the major elements of hydroponics, considered separately from the structural and environmental parameters of greenhouses and CEA, remain unchanged. The precise composition of a nutrient solution—the mixture of water and fertilizers that is the essence of the technology—varies somewhat from author to author, or from one grower to another (Sections V. A,B). But there is an obvious similarity to all of these formulae, and most growing plants are congenially tolerant of the range represented, particularly in open systems. Further, developments in aggregate systems—new types or mixtures of growing media—may have significant impact on regional economic efficiencies, although they are not breakthroughs in the basic technology (i.e., they represent changes in degree, not in kind).

It is almost impossible to catalog all claimed hydroponic inventions, devices and experiments of the past decade; the significant work would be concealed in the tangled thicket of gadgetry and unsupported assertion. What is important is what has occurred in the crucible of commercial applications or in the documented research of university groups and industrial investigators. A brief summary of a few of the major developments, which have been discussed in this review, follows:

1. Two somewhat similar open aggregate systems—bag culture (Section IV. A. 2) and rockwool culture (Section IV. A. 3)—are rapidly gaining favor in Europe for cucumber and tomato production.
2. The nutrient film technique (Section III. A), a type of closed liquid hydroponic system, was devised in England and is being used in the production of tomatoes, cucumbers, and lettuce. There are now about 50 ha of NFT systems in England, and similar installations are appearing elsewhere in Europe and the United States.
3. Although tomato and cucumber have been the most common hydroponic crops in the United States, a supposed national trend toward increased salad consumption has stimulated corporate interest in off-season lettuce. At least four large commercial trials have recently used closed liquid systems for production of Boston, or leaf lettuce: variants of NFT in Illinois, California, and Washington (Section III. B) and a floating hydroponic system in Arizona (Section III. C). The Illinois system is bench-mounted; in the California trial, tiers of lettuce were planted in movable belts atop narrow NFT channels stacked two deep; in the Arizona experiment, lettuce was planted on plastic panels floating atop long, shallow nutrient pools (raceways). Neither the California nor the Arizona system appeared to sustain corporate interest; the others are still under evaluation.
4. Sand culture (Section IV. A. 4) is an open aggregate system developed in Arizona for large facilities in the deserts of the southwestern United States and the Middle East. Coarse sand (atop a drainage system) is a permanent growing medium covering the entire greenhouse floor, without trenches, troughs, or bags; plants are individually drip-irrigated. There has been no known expansion of this technology since the mid-1970s; indeed, there have been contractions. High energy costs and OFA imports closed some U.S. installations, and industrial development or war damage has affected those in the Middle East. However, the technology remains attractive for deserts or other locations with indigenous sand.

What is needed to advance hydroponic CEA is not so much any dramatic scientific breakthrough (although such an improbability would be more than welcome) but a series of relatively modest technological improvements that would decrease the capital and operating costs (in particular, improvements in the greenhouse envelopes protecting the crops) and development of new cultivars more tolerant of wider temperature swings, more disease resistant, and more appropriate to mechanization.

More specifically, the costs of heating and cooling must be greatly reduced, while providing for optimum daytime radiation levels for photosynthesis. This suggests an initial concentration on the design and testing of low-capital, low-energy "greenhouses of the future" to provide optimal environments more economically. It also suggests an accelerated program of plant engineering and plant breeding to develop cultivars specifically adapted to CEA production. The biological objectives might include the development of short determinant tomatoes for fully automated systems; tomato and cucumber cultivars to flourish under lower, more economical CEA temperatures at northern latitudes or higher elevations; other cultivars for hot CEA temperatures in arid regions; yet other cultivars more tolerant of saline water; and so on.

A particular area in which government intervention is suggested (in addition to biological programs) is the improvement of integrated pest management for greenhouse systems. Somehow, the negative feedback loop inhibiting the approval of chemicals for greenhouse control of diseases and pests (manufacturers are reluctant to undertake such expenses because the market is small; the market is small because, in part, there are few chemicals cleared for use) must be eliminated.

A summary of research needs includes (but is not limited to) the following:

1. Design engineering of a lowest-cost CEA structure to be ventilated, heated, and cooled as much as possible by solar and solar-effect phenomena, and other alternative energy sources.
2. Materials engineering, particularly in the development of a low-cost, long-lived, selective transparent plastic film for CEA roof structures.
3. Development of lower-cost nighttime insulation devices and techniques.
4. Design of a plant bioengineering program to develop new, temperature-tolerant, machine-harvestable, disease-resistant hydroponic CEA cultivars.
5. Root temperature studies to determine influences on growth rates and plant development.
6. Disease control of waterborne pathogens in closed hydroponic systems (filtration, UV radiation, etc.).

7. Integrated pest management systems for hydroponic CEA applications in order to minimize the need for pesticides.
8. The development and governmental approval of chemicals for disease and pest control in greenhouses.
9. New aggregate material(s) (e.g., a counterpart of European rockwool) for lower-cost installation and maintenance.
10. Wider utilization of industrial waste heat for greenhouse heating.

XI. THE FUTURE

There seems to be a kind of technological imperative driving development of hydroponic CEA. Agriculture, like manufacturing, generally moves toward higher-technology, more capital-intensive solutions to problems. Hydroponics epitomizes high technology and is capital-intensive. It is highly productive, suitable for automation, conservative of water and land, protective of the environment, and yet, for most of employees, requires only basic agricultural skills. It can be argued (and has been) that hydroponic CEA is "the next logical step" after traditional OFA.

Given present circumstances, however, there seems to be no rational basis for anticipating a much wider or faster diffusion of hydroponic technology than has already occurred. In affluent nations, the comparative economic data are stark and conclusive; hydroponic CEA, even with postulated improvements, cannot compete with field crops, if and when the latter are easily available. In less affluent nations, hunger problems are now acknowledged to be the result of a maldistribution of wealth, not of any world shortage of food. In addition, complex high technologies do not transplant well unless nurtured by an even more complex social infrastructure of modern education, industry, transportation, communications, information, etc.

Despite these deterrents, there are other forces that may reduce present capital and operating costs, making hydroponic CEA more competitive; invalidate current economic projections based exclusively on comparative costs and pricing; and render the spectrum of "present circumstances" obsolete.

Continuing research and development, for example, may lead to more cost-efficient structures and materials; to reduced requirements of purchased energy; to new cultivars more appropriate to controlled environments and mechanized systems; to better control (including improved plant resistance) of diseases and pests. To the extent that these improvements increase crop yield and reduce unit costs of production, hydroponic CEA will become more competitive.

The economic prospects for hydroponic CEA may change if governmental bodies determine that, in some circumstances, politically desirable effects of hydroponic CEA merit subsidy for the public good. Such beneficial effects may include the conservation of water in regions of scarcity or food production in hostile environments; governmental support for these reasons has occurred in the Middle East (Fontes 1973; Hicks 1973). Another desirable societal effect can be the provision of income-producing employment for chronically disadvantaged segments of the population entrapped in economically depressed regions; such employment produces tax revenues as well as personal incomes, reducing the impact on welfare rolls and improving the quality of life. Government subsidy for this reason has occurred at least once in the rural United States (Quechan Environmental Farms, Ft. Yuma Reservation, Winterhaven, CA) and is under consideration for additional, urban applications.

An example of the third factor that may substantially mitigate the current dismal prospects for hydroponic CEA is the 3°C increase in global temperatures by the year 2050 projected by scientists at the National Center of Atmospheric Research (NCAR) in Colorado (Kellogg and Schware 1981), due to carbon dioxide buildup in the atmosphere (i.e., the "greenhouse effect"). Such a temperature increase would lead to tremendous and historic changes in global patterns of wind and rainfall. The NCAR modelling is said to suggest (Roberts *et al.* 1982) that the vast croplands in the central United States could become a Dust Bowl once again, forcing the country to become a net importer of food. Should the central United States experience debilitating aridity after the turn of the century and the abundance of its field crops diminish, hydroponic CEA would become more cost competitive, important to the public good, and more rapidly deployed than can now be imagined.

Hydroponic CEA is now a technical reality. The technology is being increasingly used for commercial vegetable production in certain areas (notably western Europe) where field-grown fresh vegetables are unavailable for much of the year. Further technological improvements, public subsidy of new installations for desirable social and political reasons, and projected increases in the global atmospheric temperature, if they occur, are likely to encourage the adoption of hydroponic CEA for production of fresh vegetables in many countries of the world.

LITERATURE CITED

ADAMSON, R.M. and E.F. MAAS. 1981. Soilless culture of seedless greenhouse cucumbers and sequence cropping. Pub. 1725E. Agr. Canada, Ottawa.

ANON. 1880. p. 227–253. In: History of the Massachusetts Horticultural Society, 1829–1878. Massachusetts Horticultural Society, Boston.

ANON. 1962. Dutch now treat 4,000 acres of lettuce with CO_2. Grower (Dec. 8):904–906.

ANON. 1977. An assessment of controlled environment agriculture technology. NSF Contract C-1026. Intern. Research & Technology Corp., McLean, Virginia.

ANON. 1978–1982. Fresh fruit and vegetable shipments. Agr. Mktg. Serv., USDA.

ASHER, C.J. 1981. Limiting external concentrations of trace elements for plant growth: use of flowing solution culture techniques. J. Plant Nutr. 3:163–180.

BADGER, P.C. and H.A. POOLE. 1979. Conserving energy in Ohio greenhouses. Ohio Agr. Res. and Dev. Center, Wooster, Special Cir. 102, Ext. Bul. 651, Ohio State Univ., Columbus.

BAIRD, C.D. and D.R. MEARS. 1976. Performance of hydroponic solar greenhouse heating system in Florida. p. 110–128. In: Proc. Solar Energy Fuel–Food Workshop, Univ. of Arizona, Tucson.

BAUERLE, W. and T. SHORT. 1976. American experience suggests that plastic-over-glass is on for winter tomatoes. Grower Tech. (Suppl.) 8:22–23.

BEWLEY, W.F. 1963. Commercial glasshouse crops. Country Life Ltd., London W.C. 2. p. 63.

BRAVENBOER, L., Ed. 1974. Integrated control in greenhouses. Organisation Internationale de Lutte Biologique/S.R.O.P. Antibes, France.

BURGES, H.D. 1974. Modern pest control in glasshouses. Span 17:32–34.

BUYANOUSKY, G., J. GALE, and N. DEGANI. 1981. Ultra-violet radiation for the inactivation of micro organisms in hydroponics. Plant & Soil 60(1):131–136.

CHAPMAN, H.W. and W.E. LOOMIS. 1953. Photosynthesis in the potato under field conditions. Plant Physiol. 28:703–716.

COMPTON, A.E. 1960. The Lyman greenhouses. Old-Time New England (Winter) p. 83.

COURTER, J.W. and M.C. CARBONNEAU. 1966. Versatility key to today's greenhouse glazing. Flor. & Nursurey Exchange (July):5–8.

DALRYMPLE, D.G. 1973. A global review of greenhouse food production. USDA Rpt. 89.

DAVIES, J.M.L. 1980. Disease in NFT. Acta Hort. 98:299–305.

ELLIS, N.K., M.H. JENSEN, J. LARSEN and N.F. OEBKER. 1974. Nutriculture systems. Station Bull. 44, Purdue Univ., West Lafayette, IN.

ELWELL, D.L., T.H. SHORT and R.P. FYNN. 1984. A double-plastic greenhouse with a polystyrene-pellet energy screen and floor heating for winter tomato production. Acta Hort. 148:461–467.

EVANS, S.G. 1979. Susceptibility of plants to fungal pathogens when grown by the nutrient film technique (NFT). Plant Path. 28(1):45–48.

EWART, J.M. and J.R. CHRIMES. 1980. Effects of chlorine and ultra-violet light in disease control of NFT. Acta Hort. 98:317–323.

FARNSWORTH, R.S. 1975. Process and apparatus for growing plants. U.S. Pat. 3,927,491.

FONTES, M.R. 1973. Controlled-environment horticulture in the Arabian Desert at Abu Dhabi. HortScience 8:13–16.

FONTES, M.R., J. O'TOOLE, and M.H. JENSEN. 1971. Vegetable production under plastic on the desert seacoast of Abu Dhabi. p. 93–102. In: Proc. 10th Natl. Agr. Plastics Conf., Chicago.

FRENCH, N., W.J. PARR, H.J. GOULD, J.J. WILLIAMS, and S.P. SIMMONDS. 1976. Development of biological methods for the control of *Tetranychus urticae* on tomatoes using *Phytoseiulus persimilis*. Ann. Appl. Biol. 83:177–189.

GAASTRA, P. 1959. Photosynthesis of crop plant as influenced by light, carbon dioxide, temperature and stomatal diffusion resistance. Meded Landbouwhoogeschool, Wageningen 59:1–68.

GALLOWAY, B.T. 1899. Progress of commercial growing of plants under glass. p. 575. In: U.S. Dept. of Agr. Yearbook.

GERHART, A.W. 1976. Practical application of energy saving ideas. p. 11–14. In: Proc. Solar Energy Fuel–Food Workshop, Univ. of Arizona, Tucson.

GERICKE, W.F. 1940. The complete guide to soilless gardening Prentice-Hall, Englewood Cliffs, New Jersey.

GIACOMELLI, G.A., W.J. ROBERTS, D.R. MEARS, and H.W. JANES. 1984. Greenhouse tomato production in a movable cable supported growing system. Acta Hort. 148:89–95.

GIBONEY, P.M. 1980. Nutrient uptake by aeroponically grown Bibb lettuce as related to nutrient solution concentration. M.S. Thesis, Univ. of Arizona, Tucson.

GLENN, E.P. 1984. Seasonal effects of radiation and temperature on growth of greenhouse lettuce in a high isolation desert environment. Scientia Hort. 22:9–21.

GOULD, H.J., W.J. PARR, H.C. WOODVILLE and S.P. SIMMONDS. 1975. Biological control of glasshouse whitefly (*Trialeurodes vaporariorum*) on cucumbers. Entomophaga 20:255–292.

GRAVES, C.J. 1983. The nutrient film technique. Hort. Rev. 5:1–44.

GROH, J.E. 1976. Liquid foam insulation system for greenhouses. p. 213–222. In: Proc. Solar Energy Fuel–Food Workshop, Univ. of Arizona, Tucson.

HANGER, B. 1982a. Hydroponics—what, why and how. Austral. Hort. 80(12)59–71.

HANGER, B. 1982b. Rockwool in horticulture—a review. Austral. Hort. 80(5):7–16.

HARRIS, D.A. 1976. A modified drip culture method for the commercial production of tomatoes in vermiculite. p. 85–90. In: Proc. Intern. Working Group on Soilless Culture. 4th Intern. Congr. on Soilless Culture, Las Palmas, Canary Islands, Spain.

HICKS, N. 1973. A guide to Khark environmental farms (Iran). Technical Rpt. Environmental Res. Lab., Univ. of Arizona, Tucson.

HOAGLAND, D.R. and D.I. ARNON. 1950. The water-culture method for growing plants without soil. Cir. 347. Calif. Agr. Expt. Sta., Univ. of California, Berkeley.

HOWELER, R.H., D.G. EDWARDS, and C.J. ASHER. 1982. Micronutrient deficiencies and toxicities of cassava plants grown in nutrient solutions 1. Critical tissue concentrations. J. Plant Nutri. 5:1059–1076.

HURD, R.G. and C.J. GRAVES. 1983. Interactions between air and root temperature effects on tomatoes in nutrient film culture. In: Annu. Rpt. Glasshouse Crops Res. Inst., Littlehampton, England. (In Press)

JENSEN, M.H. 1968. Ring and trough culture for greenhouse tomato production. Tech. Rpt. 1. Environmental Res. Lab., Univ. of Arizona, Tucson.

JENSEN, M.H. 1971. The use of polyethylene barriers between soil and growing medium in greenhouse vegetable production. p. 144–149. In: Proc. 10th Natl. Agr. Plastics Conf., Chicago. p. 144–149.

JENSEN, M.H. 1973. Exciting future for sand culture. Amer. Veg. Grower 21(11):33–34, 72.

JENSEN, M.H. 1975. Arizona research in controlled environment agriculture. p. 13–82. In: Proc. Tenn. Valley Greenhouse Veg. Workshop, Chattanooga, Tennessee.

JENSEN, M.H. 1977. Energy alternatives and conservation for greenhouses. HortScience 12:14–24.

JENSEN, M.H. 1980. Tomorrow's agriculture today. Amer. Veg. Grower 28(3):16–19, 62, 63.

JENSEN, M.H. 1982. Review of greenhouse leafy vegetable industry in Holland, England, Norway and Denmark. Environmental Res. Lab., Univ. of Arizona, Tucson.

JENSEN, M.H. and M.A. TERAN. 1971. Use of controlled environment for vegetable production in desert regions of the world. HortScience 6:33–36.
JUDD, R. 1982. Bag culture. Amer. Veg. Grower 30(11):42, 44.
KELLOGG, W.W. and R. SCHWARE. 1981. Climate change and society. Westview Press, Boulder, Colorado.
KIMBALL, B.A. 1982. Carbon dioxide and agricultural yield: an analysis of 360 prior observations. Agr. Res. Service, USDA.
KNOTT, J.E. 1966. Handbook for vegetable growers. J. Wiley & Sons, New York.
LAUDER, K. 1977. Lettuce on concrete. Grower 87(8):40–41, 44–46.
LAWSON, G. 1982. New air filled pack extends shelf life to three weeks. Grower 97(26):47–48.
LIEBERTH, J.A. 1982. Greenhouse industry survey reveals. Amer. Veg. Grower 30(11):10,12.
LIU, R.C. and G.E. CARLSON. 1976. Proposed solar greenhouse design. p. 129–141. In: Proc. Solar Energy Fuel–Food Workshop, Univ. of Arizona, Tucson.
MAAS, E.F. and R.M. ADAMSON. 1981. Soilless culture of commercial greenhouse tomatoes. Pub. 1460. Agr. Canada, Ottawa.
MADEWELL, C.E., L.D. KING, J. CARTER, J.B. MARTIN and W.K. FURLONG. 1975. Using power plant discharge water in greenhouse vegetable production. Progress Rpt. Bull. Z-56. Tenn. Valley Authority, Muscle Shoals, Alabama.
MAHER, M.J. 1976. Growth and nutrient content of a glasshouse tomato crop grown in peat. Scientia Hort. 4:23–26.
MASSANTINI, F. 1973. l'Aeroponica: realizzazioni pratiche, studi e ricerche. p. 83–97. In: Proc. Intern. Working Group on Soilless Culture, 3rd Intern. Congress on Soilless Culture, Sassari, Italy.
MASSANTINI, F. 1976. Floating hydroponics; a new method of soilless culture. p. 91–98. In: Proc. Intern. Working Group on Soilless Culture, 4th Intern. Congress on Soilless Culture. Las Palmas, Canary Islands, Spain.
McCLANAHAN, R.J. 1973. Integrated control of greenhouse pests. Agr. Canada, Ottawa. p. 34–35.
McCORMICK, P.O. 1976. Performance of non-integral solar collector greenhouses. p. 51–60. In: Proc. Solar Energy Fuel–Food Workshop, Univ. of Arizona, Tucson.
MERMELSTEIN, N.H. 1980. Innovative packaging of produce earns 1980 IFT food technology industrial achievement award. Food Technol. 34(16):42–48.
MOORBY, J. and C.J. GRAVES. 1980. The effects of root and air temperatures on the growth of tomatoes. Acta Hort. 98:29–43.
MORGAN, J.V. and A. TAN. 1982. Production of greenhouse lettuce at high densities in hydroponics. p. 1620 (Abstr.). In: Proc. 21st Intern. Hort. Cong., Hamburg, Vol. 1.
MORGAN, J.V., A.T. MOUSTAFA, and A. TAN. 1980. Factors affecting the growing-on stage of chrysanthemum in nutrient solution culture. Acta Hort. 98:253–261.
MOUSTAFA, A.T. and J.V. MORGAN. 1981. Root zone warming of spray chrysanthemums in hydroponics. Acta Hort. 125:217–226.
ORCHARD, B. 1980. Solution heating for the tomato crops. Acta Hort. 98:19–28.
PARR, W.J., H.J. GOULD, N.H. JESSOP and F.A.B. LUDLAM. 1976. Progress towards a biological programme for glasshouse whitefly (*Trialeurodes vaporariorum*) on tomatoes. Ann. Appl. Biol. 83:349–363.
POLLOCK, R.D. 1979. Early tomatoes mean high fuel costs. Hort. Ind. (Nov.):17–19.
POTTER, R.F. and T. SIMS. 1975. New developments in chrysanthemum culture in nutrient film. Natl. Chrysanthemum Soc. Bull. 87:13–15.

PRICE, D. 1980. Fungal flora of tomato roots in nutrient film culture. Acta Hort. 98:269–275.

PRICE, D. and A. DICKINSON. 1980. Fungicides and the nutrient film technique. Acta Hort. 98:277–282.

PRICE, D.R., G.E. WILSON, D.P. FROEHLICH and R.W. CRUMP. 1976. Solar heating of greenhouses in the Northeast. p. 173–190. In: Proc. Solar Energy Fuel–Food Workshop. Univ. of Arizona, Tucson.

PRINCE, R.P., J.W. BARTOK, and D.W. PROTHEROE. 1981. Lettuce production in a controlled environment plant growth unit. Amer. Soc. Agr. Eng. Trans. 24:725–730.

PRINCE, R.P., W. GIGER, JR., J.W. BARTOK, JR., and T.L. LOGEE. 1976. Controlled environment plant growth. Dept. Agr. Eng., Univ. of Connecticut, Storrs.

RESH, H.M. 1981. Hydroponic food production. Woodbridge Press Publ. Co., Santa Barbara, California.

ROBERTS, W.J. and D.R. MEARS. 1969. Double covering a film greenhouse using air to separate film layers. Amer. Soc. Agr. Eng. Trans. 12:32–33, 38.

ROBERTS, W.J., J.C. SIMPKINS, and P. KENDALL. 1976. Using solar energy to heat plastic film greenhouses. p. 142–159. In: Proc. Solar Energy Fuel–Food Workshop, Univ. of Arizona, Tucson.

ROBERTS, W.O. and E.J. FRIEDMAN. 1982. Living with the changed world climate. Aspen Inst. Humanistic Studies, New York.

ROGERS, H.T. 1983. Greenhouse of the future. Greenhouse Grower 1(1):72, 74, 76.

SCHIPPERS, P.A. 1978. A vertical hydroponic system. Amer. Veg. Grower 26(5):20–21.

SCHIPPERS, P.A. 1982. Developments in hydroponic tomato growing. Amer. Veg. Grower 30(11):26–27.

SCHWARZ, M. 1978. Guide to commercial hydroponics. Israel Univ. Press, Jerusalem.

SHAKESSHAFT, R.G. 1981. A kitchen harvest of living lettuce. Amer. Veg. Grower 29(11):10,12.

SHELDRAKE, R., JR. 1981. Money bags? Amer. Veg. Grower. 29(11):42, 44.

SHELDRAKE, R., JR. and J.W. BOODLEY. 1965. Commercial production of vegetable and flower plants. Ext. Bull. 1056. Cornell Univ., Ithaca, New York.

SHELDRAKE, R., JR. and S. DALLYN. 1969. The ring culture method for production of greenhouse tomatoes. Veg. Crops Dept. Mimeo 149. Cornell Univ., Ithaca, New York.

SHELDRAKE, R., JR. and R.W. LANGHANS. 1961. Heating study with plastic greenhouses. p. 16–17. In: Proc. 2nd Natl. Hort. Plastics Conf., State College, Mississippi.

SHIMIZU, S. 1979. Greenhouse horticulture in Japan. Japan FAO Association, Tokyo.

SKAGGS, R.W., D.C. SANDERS, and C.R. WILLEY. 1976. Use of waste heat for soil warming in North Carolina. Amer. Soc. Agr. Eng. Trans. 19:159–167.

SMITH, I.E. 1982. Soil cultivation of cucumbers and tomatoes under protection in Natal. Ph.D. Thesis, Univ. of Natal, Pietermaritzburg, South Africa.

SOFFER, H. and D. LEVINGER. 1980. The Ein Gedi system—research and development of a hydroponic system. p. 241–252. In: Intern. Soc. for Soilless Culture, Proc. 5th Intern. Cong. on Soilless Culture, Wageningen, the Netherlands.

SPENSLEY, K., G.W. WINSOR, and A.J. COOPER. 1978. Nutrient film technique—crop culture in flowing nutrient solution. Outlook on Agr. 9:299–305.

STANGHELLINI, M.E., L.J. STOWELL, and L. BATES. 1982. Pythium root rot of spinach and lettuce. Plant Pathology Dept., Univ. of Arizona, Tucson.

STARKEY, N. 1980. Station aims for more flexible NFT cropping. Grower 94(4):8.

STAUNTON, W.P. and T.P. CORMICAN. 1980. The effect of pathogens and fungicides on tomatoes in a hydroponic system. Acta Hort. 98:293–297.

TOMLINSON, J.A. and E.M. FAITHFULL. 1980. Studies on the control of lettuce big-vein disease in recirculated nutrient solutions. Acta Hort. 98:325–332.

VAN OS, E.A. 1983. Dutch developments in soilless culture. Outlook in Agr. (in press).

VARLEY, M.J. and S.W. BURRAGE. 1981. New solution for lettuce. Grower 95(15):19–21, 23, 25.

VERKERK, K. 1966. Temperature response in early tomato production. Acta Hort. 4:26–31.

VERWER, F.L.J.A.W. 1976. Growing horticultural crops in rockwool and nutrient film. p. 107–119. In: Proc. Intern. Working Group on Soilless Culture, Las Palmas, Canary Islands, Spain.

WALT, R.L. 1917. Vegetable forcing. Orange Judd Co., New York.

WARDLOW, L.R., F.A.B. LUDLAM, and N. FRENCH. 1972. Insecticide resistance in glasshouse whitefly. Nature (London) 239:164–165.

WHITE, H. 1980. NFT in the air makes for easier management. Grower 94(2):28, 31–32.

WHITE, J.W., R.A. ALDRICH, K. VEDAM, J.L. DUDA, S.M. REBUCK, G.R. MARINER, and J.R. SMITH. 1976. Energy conservation systems for greenhouses. p. 191–212. In: Proc. Solar Energy Fuel–Food Workshop. Univ. of Arizona, Tucson.

WIDMER, R.E. 1979. Commercial greenhouse heating with reject heat from electric generating plants. HortScience 14:566,675.

WINSOR, G.W. and D.M. MASSEY. 1978. Some aspects of the nutrition of tomatoes grown in recirculating solution Acta Hort. 82:121–132.

WINSOR, G.W., R.G. HURD, and D. PRICE. 1979. Nutrient film technique. Growers Bull. 5. Glasshouse Crops Res. Inst., Littlehampton, England.

WITHROW, R.B. and A.P. WITHROW. 1948. Nutriculture. S.C. 328. Purdue Univ. Agr. Exp. Sta., Lafayette, Ind.

WITIWER, S.H. 1967. Carbon dioxide and its role in plant growth. p. 311–322. In: Proc. 17th Intern Hort. Congr., Vol. 3. Univ. of Maryland, College Park, Maryland.

Index (Volume 7)

Cumulative Index (Volumes 1 - 7)

R

Contributor Index (Volumes 1 – 7)